American Institute of Chemical Engineers

5th Topical Conference on Natural Gas Utilization 2005

Held at the 2005 AIChE Spring National Meeting

April 10-14, 2005
Atlanta, Georgia, USA

Printed from e-media with permission by:

Curran Associates, Inc.
57 Morehouse Lane
Red Hook, NY 12571
www.proceedings.com

ISBN: 978-1-60423-469-5

Some format issues inherent in the e-media version may also appear in this print version.

American Institute of Chemical Engineers

Topical Conference on Natural Gas Utilization
2005

Table of Contents

TI - 5th Topical Conference on Natural Gas Utilization

Mixed Fluid Power Rankine Cycle using LNG as Heat Sink with LPG Extraction

John Y. Mak
Technical Director, Process Engineering

Dick Nielsen
Executive Director, Process Engineering

Curt Graham
Principal Engineer, Process Engineering

Energy and Chemicals Group, Fluor Enterprises, Inc., Aliso Viejo, California

AIChE Spring Meeting, April 2005
5th Topical Conference on Natural Gas Utilization
Atlanta April 10-14, 2005

Abstract

There are numerous technical and environmental challenges in the design of LNG regasification terminals. Most LNG receiving terminals are limited to standalone conventional LNG regasification facility, and have not taken advantages of the inherent cold energy from LNG. The other challenge in LNG terminals is that the facility may be required to accept LNG with heating values higher than North American gas pipeline specifications. This paper describes LNG regasification technologies (patent pending) that allow the LNG terminal to accept non-compliant LNG, while reducing environmental impacts, and producing quality pipeline gas and liquid products for sales. This process also provides revenue opportunity from power generation utilizing LNG in the power cycle. In addition, some importers may desire to extract propane for the LPG liquid fuel market and butane plus for gasoline blending even if the import LNG meets the pipeline gas specifications.

Introduction

Approximately 8 to 10% of the energy in natural gas is consumed in the LNG liquefaction process and during LNG transport. A typical LNG liquefaction process requires about 230 kW to liquefy one MMscfd of high-pressure natural gas. About 280 MW of power is required in a 1,200 MMscfd liquefaction plant. Theoretically, some of the power consumed in LNG liquefaction is recoverable at the LNG receiving terminal if the LNG is used as a heat sink for power generation.

There are significant synergies between power generation and LNG regasification. The low temperature refrigeration available during LNG regasification can be used for as a heat sink for power generation. Waste heat from gas turbine exhaust is readily available as a heat source for LNG regasification [1].

From a thermodynamic point of view, the most efficient use of LNG is in cooling and as a refrigerant in cryogenic separation processes. While locating cryogenic separation plants close to a LNG regasification terminal may be costly and unrealistic, a LPG separation unit integrated with LNG regasification for production of LPG fuel and butane plus components for gasoline blending can be economically attractive.

LNG Regasification

LNG vaporization is an energy intensive process. Vaporizing 1,200 MMscfd of LNG requires about 750 MM Btu/hr of heating duty. This heat duty can be supplied by lowering the temperature of 100,000 gpm of seawater by 15 °F. The large seawater flowrate is required to comply with environmental guidelines. The criteria for seawater effluent specified by the 1998 World Bank [5] is that the seawater effluent should result in a temperature change of no more than 5.4 °F at the edge of the zone where initial mixing and dilution take place. For terminals located in warm water, such as along the US Gulf coast and the Caribbean, the use of seawater for LNG regasification would conserve fuel gas and minimize LNG shrinkage. The challenges are the ecological impacts of the cold seawater discharge on marine life; remediation measures must be taken to protect ecological sensitive areas. For terminals located in cold climate areas where the seawater is close to freezing, fuel fired submerged combustion vaporizers are more practical. Operating a 1,200 MMscfd regasification terminal would require about 20 MMscfd of fuel gas (based on 1050 Btu/scf heating value) that must be factored into the life cycle cost of the terminal.

Other LNG regasification methods may also be suitable, including the use of a heat transfer medium to transfer heat from various heat sources, such as waste heat from processing plants or heat from ambient air. The more innovative method is using the condensation duty in a Rankine power cycle for LNG regasification.

Rankine Cycle and Working Fluids

The Rankine power cycle is used in today's steam power plants. The ideal Rankine cycle efficiency (or Carnot Cycle efficiency) can be defined as (T2-T1)/ T2, T2 is the absolute temperature of the heat source and T1 is the absolute temperature of the heat sink. When LNG is used as the heat sink, the temperature difference (T1-T2) is increased, resulting in higher power generation efficiency. The selected working fluid must be thermally stable at high temperatures and condense at low temperatures without freezing problems. The basic configuration of a Rankine cycle using LNG as the heat sink is depicted in **Figure 1**.

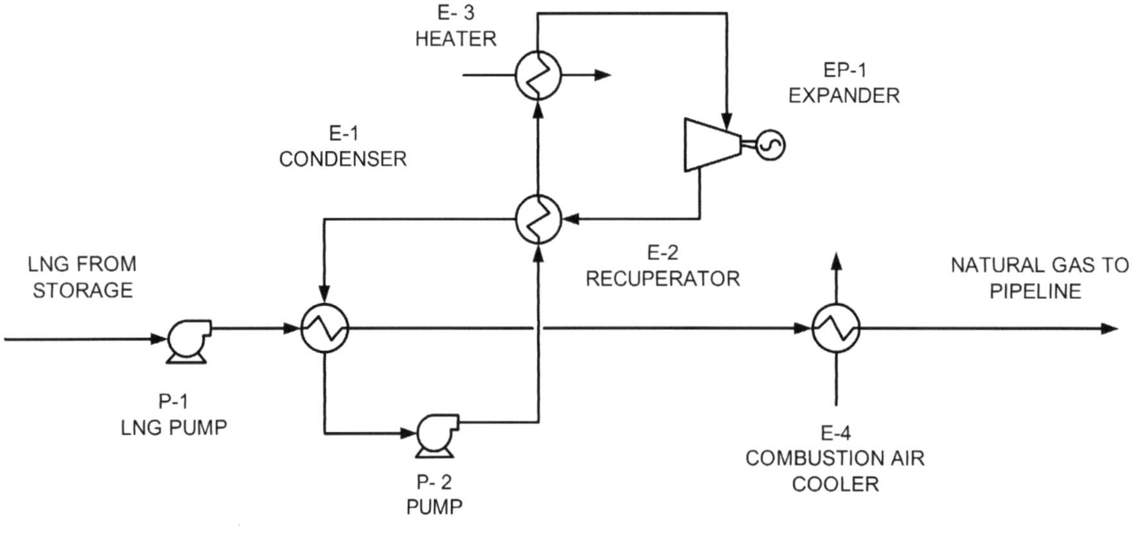

Figure 1 – Rankine Cycle for LNG Power Generation

Per **Figure 1**, the working fluid is pumped by Pump P-2 to supercritical pressure, typically 1500 psig. The high-pressure fluid is first heated with the hot expander discharge in the Recuperator, E-2. The high pressure fluid is further heated in Heater, E-3, to about 600 °F. Heat can be supplied by fired heaters or waste heat exhausted from a gas turbine generator. The high temperature supercritical fluid is then expanded to atmospheric pressure in the Expander, EP-1, generating power. The low pressure vapor is then cooled in the Recuperator, E-2, and condensed in the Condenser, E-1. LNG is vaporized in the condenser using the condensation duty from the working fluid.

Pure components, such as butane or propane, have been used as a working fluid in power generation. The power generation efficiency using butane is fairly low due to its high condensation temperature. Butane condenses at 30°F under atmospheric pressure. Therefore, with butane as the working fluid, the very low temperature of LNG cannot be effectively utilized. Propane is a more efficient working fluid as it condenses at a much lower temperature (e.g., -44°F at atmospheric pressure). Other lower boiling fluids, such as ethane or ethylene, are not suitable for LNG vaporization, as their condensation temperatures are too low to heat the LNG to the required pipeline temperature.

Multi-component working fluid is commonly used in refrigeration cycle for LNG liquefaction due to its high thermal efficiency. The same principle can be applied to power generation cycle where LNG is used as the heat sink. The advantage of a multi-component working fluid is that it condenses at varying temperatures and its composition can be tailored to match the LNG vaporization curve. In today's plate and fin exchanger design, a close temperature approach is feasible. The working fluid condensation curve can be adjusted to parallel the LNG vaporization curve, avoiding the temperature pinch of a pure component system.

There is an optimum mixed fluid composition for each LNG composition. An optimum mixed fluid composition and a corresponding LNG composition are shown in **Table 1**.

Table 1 – Import LNG Composition and Mixed Fluid Composition

Mole %	Import LNG	Mixed Fluid
C_1	90	18
C_2	5	24
C_3	3	33
C_4	2	24

Rankine Power Cycle Performance

Table 2 compares the power cycle performance for three different working fluids: butane, propane and a mixed fluid. The power cycle output for the butane working fluid is 55.7 MW. With a propane working fluid, the power output is increased to 88.6 MW. The Mixed Fluid

Power cycle has the highest power output of 107.1 MW achieving a power generation efficiency of 37.8%.

To compare the thermodynamic efficiencies of the different working fluids, the LNG vaporization curve is plotted against the condensation curves as shown in **Figure 2**. Butane and propane, due to their higher condensation temperatures, must be superheated at the condenser inlet in order to avoid a temperature pinch in the condenser. On the other hand, the mixed fluid at the condenser inlet is in a two-phase region, thus avoiding the temperature pinch. The mixed fluid composition can be tailored to condense at a close temperature approach to LNG, resulting in higher power cycle efficiency.

Table 2 –Rankine Power Cycle Performance with Butane, Propane and Mixed Fluid

	Butane	**Propane**	**Mixed Fluid**
Expander Inlet Temp, °F	600	600	600
Expander Inlet Pressure, psig	1,450	1,450	1,450
Expander, MW	62.5	96.6	115.0
Pump MW	6.8	7.9	7.8
Net Power, kW	55.7	88.6	107.1
Heat Input, MM Btu/h	791	903	967
Condensing Temperature, °F	31	-44	- 10 to -220
Power Generation Efficiency,%	24.1	33.5	37.8

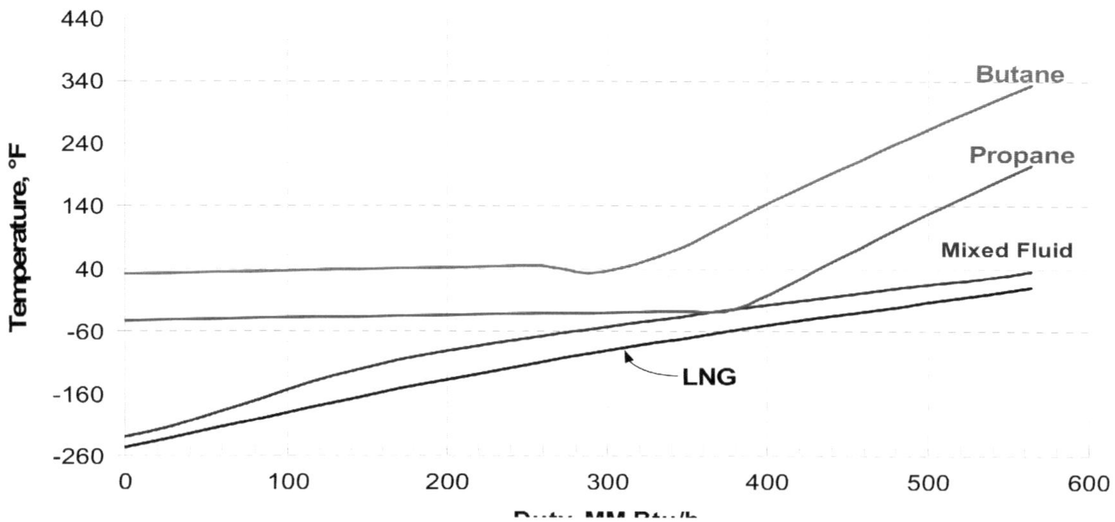

Figure 2 – LNG Vaporization vs Butane, Propane and Mixed Fluid Condensation

Gas Turbine Inlet Cooling

Gas turbine inlet cooling with the use of refrigeration to increase power output is a common practice in the power industries. LNG cold can be utilized for chilling gas turbine inlet air. One of the first plants using this concept with LNG is the Dominican Republic LNG receiving terminal which is thermally integrated with the power plant. This plant uses a glycol-water solution in a chiller coil at the gas turbine inlet. The colder air temperature allows higher mass flow through the gas turbine, resulting in higher gas turbine power output and efficiency. **Figure 3** shows the impact of ambient air temperature on power generation for a typical combined cycle power plant.

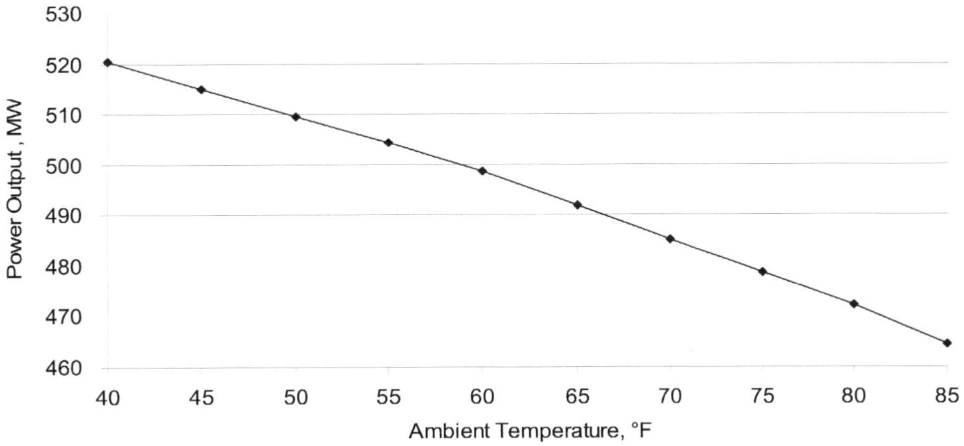

Figure 3 - Combined Cycle Power Plant Outputs vs Ambient Temperatures

Gas turbine inlet air-cooling decouples power generation capacity from changes in ambient temperature. Power plants typically experience a drop in power output when ambient temperatures rise. In many cases this drop in power output occurs when power demand peaks and can be sold for a premium. Using LNG to cool the gas turbine inlet air, a constantly high power output can be maintained throughout the year, greatly improving the power plant economics.

Depending on the heat balance between the LNG regasification duty and the chilled air requirement, excessive chilled glycol-water may be available, that can be used to lower the cooling water supply temperature. This will effectively lower the steam turbine surface condenser pressure resulting in a lower steam turbine back-pressure and a higher steam turbine output.

Integrated LNG Regasification/Power Cogeneration

The conceptual design of an integrated plant is depicted in **Figure 4**. Power is produced in three steps:

- Mixed Fluid Power cycle
- Gas turbine combustion air cooling

- Fuel gas saturator using condensate recovered from the gas turbine inlet

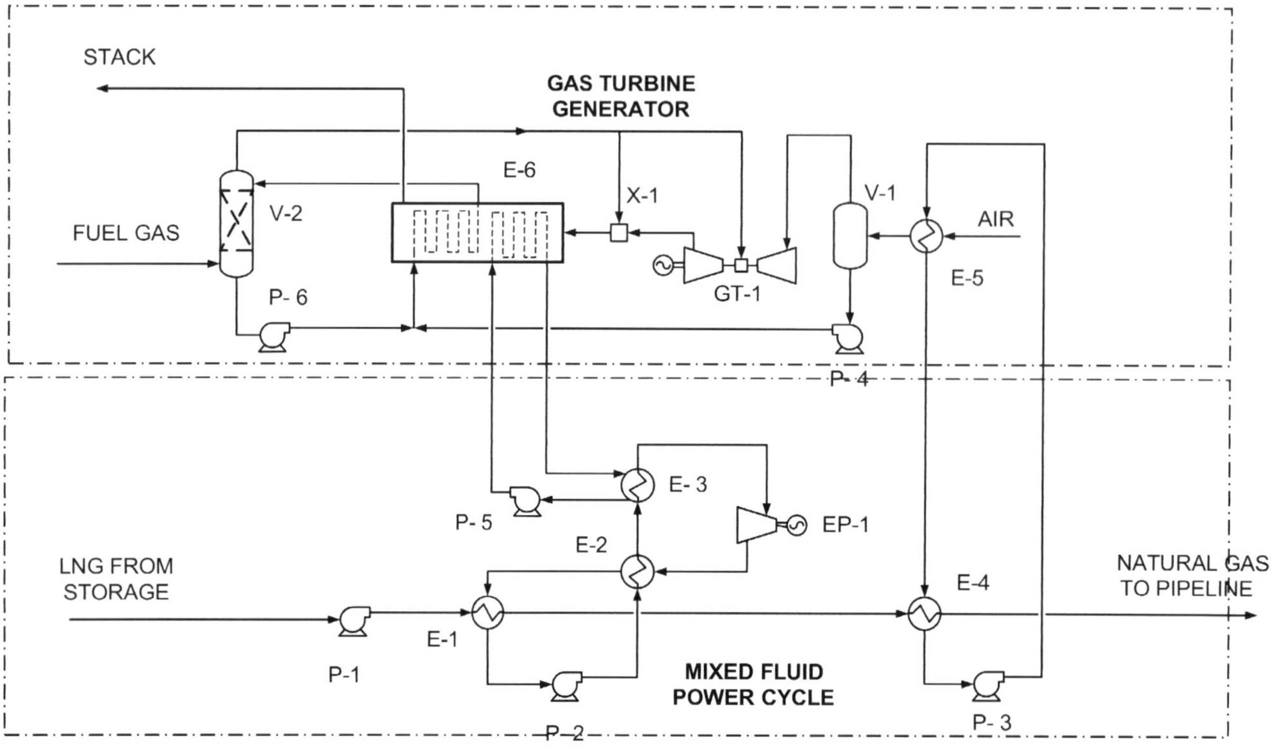

Figure 4 – Integrated LNG Regasification/Power Cogeneration Facility

The main concerns on integrating a power generation plant with a LNG regasification terminal are safety and reliability. The heat exchange design must consider the potential risks of tube rupture and the hazards of leaking hydrocarbons to the gas turbine inlet or exhaust ducting. One of the solutions is using an intermediate heat transfer medium to isolate the hydrocarbons and LNG from the gas turbine.

A stable heat transfer fluid can be used to supply heat to the Rankine power cycle. Per **Figure 4**, the heat medium pump, P-5, is used to transfer heat from the heat recovery exchanger, E-6, to the heat exchanger, E-3. In a similar manner, LNG indirectly cools the gas turbine inlet air with the use of a glycol-water mixture. The glycol mixture is pumped by the glycol pump, P-3, circulating between the LNG vaporizer, E-4, and the gas turbine inlet chiller, E-5.

When ambient air is chilled to 40°F, almost all of its moisture content is condensed. The condensate can be used as makeup water to a steam cycle power plant, reducing its demineralized water makeup requirement. Alternatively, when a simple gas turbine cycle is used, the condensate can be vaporized into the fuel gas to the gas turbine increasing mass flow to the gas turbine, hence increasing its power output. Fuel gas moisturization has the side benefit of reducing NOx emissions from the gas turbines. In this process, the water condensate is recovered and is pumped to the fuel gas saturator, V-2, where water is vaporized by contacting the fuel gas using heat from the turbine exhaust.

Reliability of the gas turbine exhaust as a heat source is a concern in an integrated facility. The gas turbine power output varies with consumer demands, resulting in fluctuations in available waste heat, while the heat input to a LNG regasification plant must be maintained constant. Additional heat sources, such as duct firing and backup steam boiler must be used to supply the LNG heating duty in case the power plant is down for maintenance.

Integrated LNG Regasification/ Power Cogeneration Performance

Table 3 compares the performance of an integrated 1,200 MMscfd LNG regasification terminal/power plant with a standalone 1,200 MMscfd LNG regasification terminal with independent power generation. Both seawater regasification and submerged combustion type regasification plants are compared. Power production is based on a combined cycle power plant using a 141 MW gas turbine and a 68.4 MW steam turbine generator.

Table 3 – Comparison of Conventional LNG Regas with Integrated LNG Regas Cogeneration

	Seawater LNG Regasification	Submerged Combustion LNG Regasification	Integrated LNG Cogeneration
LNG Regasification:			
Seawater Flow rate, GPM	100,000	None	None
LNG Pumps, MW	8.0	8.0	8.0
Seawater Pumps, MW	6.0	0.0	0.0
Fuel fired, MM Btu/hr	0.0	800.0	0.0
Mixed Fluid Cycle, MW	0.0	0.0	111.0
Power Production, MW	-14.0	-8.0	103.0
Power Plant:			
Combustion Air Inlet, °F	90	90	40
Gas Turbine, MW	141.0	141.0	191.8
Steam Cycle MW	68.4	68.4	None
Cooling Water, GPM	40,000	40,000	None
Total Power Plant, MW	209	209	191.8
Power Plant Fuel, MM Btu/hr LHV	1,463	1,463	1,808
Total Fuel Gas, MM Btu/hr LHV	1,463	2,263	1,808
Net Power Output, MW	195.5	201.5	294.8
Power Generation Efficiency, Btu/kW-hr	7,487	11,235	6,133

A conventional 1200 MMscfd LNG regasification terminal requires about 100,000 gpm of seawater or 800 MM Btu/hr of fuel gas consumption. The integrated LNG regasification/power generation facility eliminates these requirements and used the condensation duty from the mixed fluid Rankine cycle for heating. The Mixed Fluid Rankine cycle produces 111 MW and a portion of this power can be used to supply the LNG pumping requirement.

The bottoming steam cycle is not required in the integrated facility since waste heat from the gas turbine is fully utilized in the Mixed Fluid Rankine cycle and in turbine inlet air cooling. The integrated facility also does not require a cooling water system. The capital, operating and maintenance costs of the steam generation system and the cooling system are no longer needed. The power output from the gas turbine is increased to 192 MW with turbine inlet air cooling and fuel gas water saturation.

The fuel gas consumption of the LNG regasification/ power generation integrated plant is 1,808 MM Btu/hr. This is higher than a LNG regasification terminal using seawater, but less than a LNG regasification terminal using submerged combustion vaporizers. The integrated complex achieves an overall thermal efficiency of 6,133 Btu/kW-hr.

LNG Heating Values

One of the problems in LNG receiving terminals is accepting rich LNG with non-compliant heating values. As the LNG import market grows, spot LNG trades will become common, as in today's crude oil trade market. The LNG terminals may be required to import spot-cargo LNG with various compositions and heating values. In some markets, the rich LNG is advantageous as its propane content can be extracted to supply the local LPG market and the butane plus liquid can be produced for gasoline blending. Processing steps for extraction of the heavier components from the rich LNG will be necessary to meet pipeline specifications and/or the liquid fuel market demands.

In upstream LNG liquefaction plants, removal of the pentane, hexane and heavier hydrocarbons is required only to avoid wax formation inside the cryogenic liquefaction exchanger. The LPG components (C_2, C_3 and C_4+) are not removed and liquefied together with the methane component, resulting in LNG with a fairly high gross heating value. The heating values of LNG from a number of LNG export plants in the Atlantic, Pacific Ocean and Middle East LNG plants are shown in **Figure 5**. The higher heating values indicate a higher proportion of the non-methane components. The compositions of the ethane, propane, and butane plus components for these LNG are shown in **Figure 6**.

In North America, many pipeline operators require very lean gas for transmission, and in some mid-west regions, natural gas gross heating value ranges between 960 and 1050 Btu/scf. In California, the acceptable gross heating value is between 970 and 1150 Btu/scf. California also imposes constraints on specific gas components. Currently, acceptable LNG that meets the California specification is limited to sources such as the Kenai, Alaska LNG [4], or the Atlantic LNG from Trinidad.

To meet the North American natural gas specifications, these regasification terminals must have facilities that are capable of processing these non-compliant LNG [3]. The common method of controlling the LNG heating value and Wobbe Index is by dilution with nitrogen or blending with a leaner natural gas. However, there are also limits on the maximum amount of inerts that can be introduced to the pipeline gas. Dilution with nitrogen is also very costly and may not be acceptable if the maximum amount of nitrogen for pipeline distribution is exceeded.

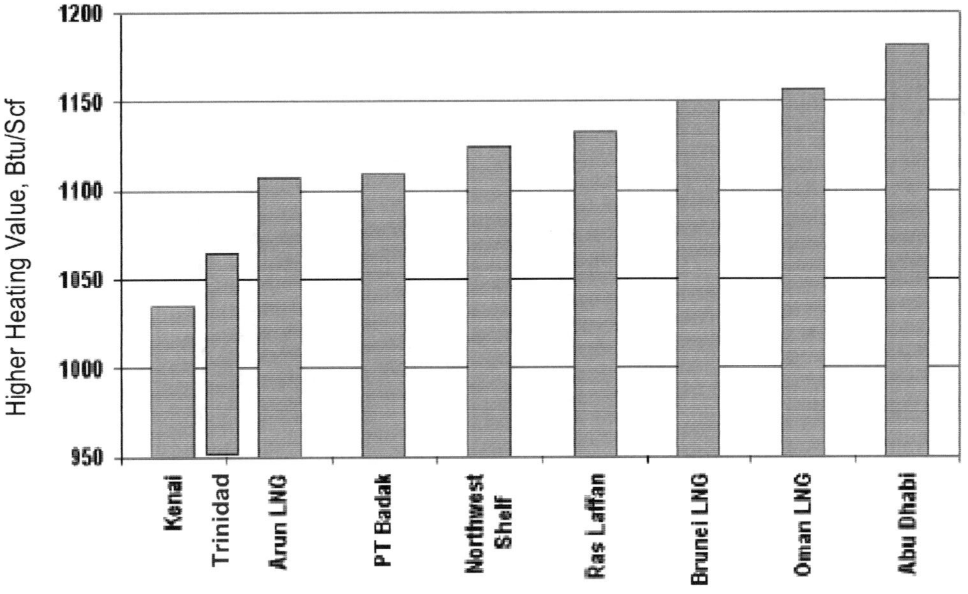

Figure 5- LNG Energy Content for Selected Atlantic, Pacific and Middle East Plants [4]
(Data source: GTI's World LNG Source Book, 2001; ZDC Analysis)

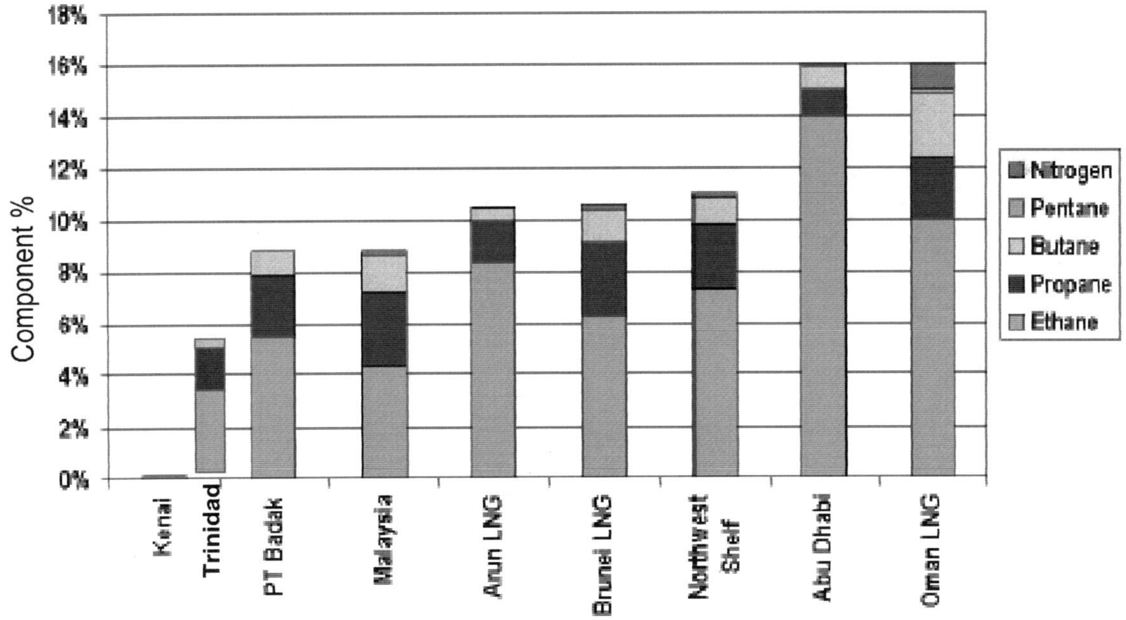

Figure 6– Non-Methane Constituents for Selected Atlantic, Pacific and Middle East Plants [4]
(Data source: GTI's World LNG Source Book, 2001; ZDC Analysis)

LPG Fractionation with LNG Cogeneration

The LPG components in LNG are of high quality as all the contaminants, water and heavy hydrocarbons have been removed in the liquefaction process. LPG from LNG can be sold as a clean liquid fuel and can command a higher price than natural gas.

The use of LNG cold to recover LPG requires significantly less capital and has lower operating costs than conventional NGL recovery processes. Costly gas treating, conditioning and processing units, such as gas treating, dehydration, turbo-expander, refrigeration and residue gas compression are not required. Typically, the power requirement to process 1,200 MMscfd feed gas is 60,000 HP or 44.7 MW, which is mainly used for chilling the feed gas and recompressing the residue lean gas. The LPG fractionation process shown in **Figure 7** eliminates these units and their energy consumption. This process can achieve 99% propane recovery without gas chilling and recompressing requirements.

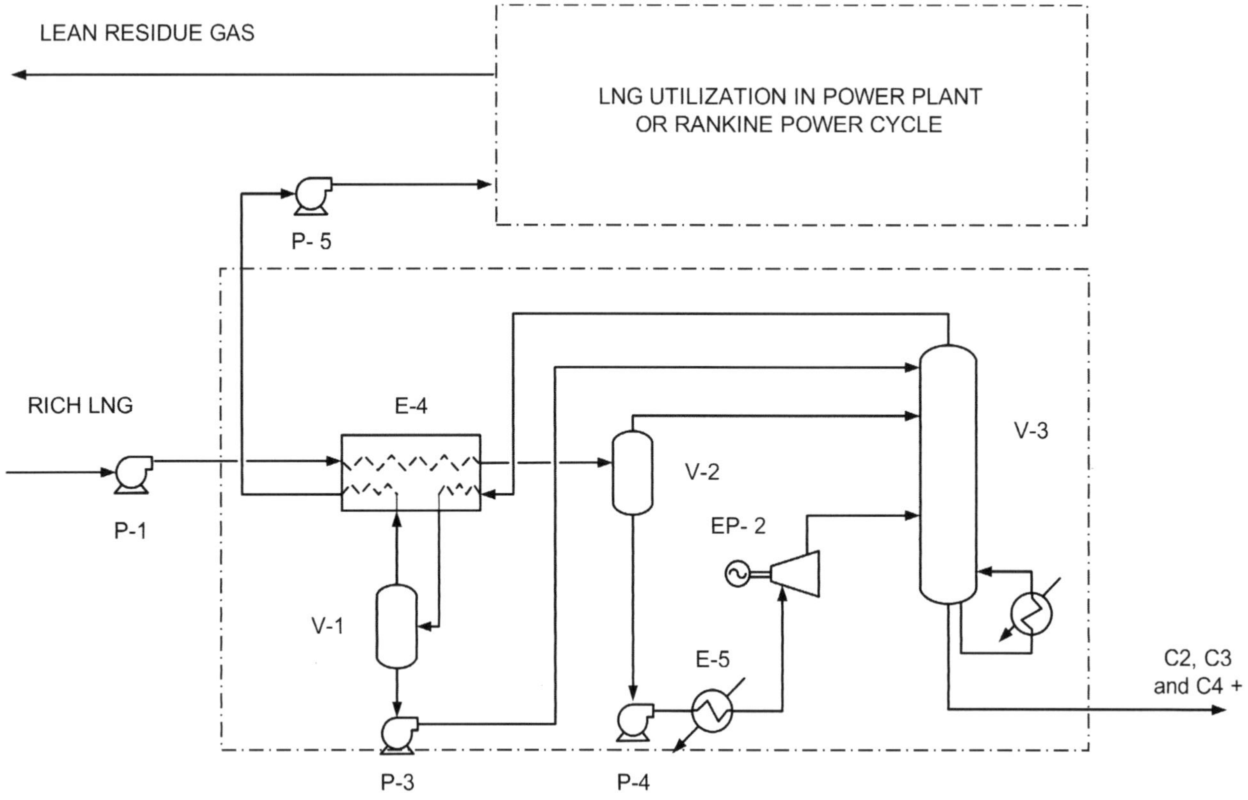

Figure 7 – Integrated LNG Regasification, LPG Fractionation and Power Cogeneration

Per **Figure 7**, LNG send-out rate of 1,200 MMscfd is pressurized by LNG pump P-1 to about 500 psig and fed to the LPG fractionation unit. The LNG cold is utilized in the brazed aluminum type exchanger E-4. This exchanger serves dual functions. First, this exchanger uses LNG cold to condense the deethanizer, V-3 overhead to produce an ethane rich cold reflux to the deethanizer. Secondly, the exchanger condenses the lean residue gas, allowing the condensed liquid to be pumped to the gas pipeline pressure. This eliminates costly gas compression systems.

After heat exchange in E-4, LNG is heated to about -90°F and is partially vaporized. The two-phase stream is separated in vessel V-2. The flashed vapor is fed to the upper rectification trays in the deethanizer, V-3, and the flashed liquid is used for power generation and stripping. The flashed liquid is pumped by pump P-4 to about 1500 psig and is then heated in exchanger E-5 to about 600°F. The high-pressure high-temperature vapor is expanded in expander EP-2 to 460 psig, generating power. The expander discharge vapor, which is quite warm, typically at 300°F, is used as a stripping vapor in the fractionator. The stripping vapor can supply most of the reboiler requirement and can be trim-heated with a bottom reboiler as required.

The deethanizer operates at 450 psig with an overhead temperature of -65°F. The overhead vapor is cooled to about -105°F in exchanger E-4 and partially condensed. The condensed liquid is separated in reflux drum, V-1 and is returned to the deethanizer as reflux using the reflux pump, P-3. The lean residual vapor from V-1 is further cooled and condensed in E-4. The lean residue gas is re-condensed as a lean LNG at -140°F. The lean LNG is then pumped by pump P-5 to about 1100 psig as required for pipeline transmission. This re-condensation and pumping operation is more energy efficient and less costly than the residual gas re-compression in a conventional NGL recovery plant.

The lean LNG at -140°F still contains a significant amount of cold energy that can be utilized. The cold energy can be used for combustion air cooling in a combined cycle power plant or in power generation using the Mixed Fluid power cycle. When used in a Rankine cycle, due to the leaner LNG composition, the composition of the multi-component fluid must be accordingly adjusted to match the heat curve of the leaner LNG.

The Rankine cycle can replace conventional steam cycle which can be justified in a small simple cycle power plant where gas turbine exhaust heat is available at no cost. However, when a large combined cycle power plant is considered, LNG cold is more suitable for gas turbine inlet air cooling. The gas turbine output can be significantly increased with relatively low cost air chillers. The overall balance for a 1,200 MMscfd Integrated LPG Fractionation/LNG Regasification Facility is shown in **Table 4**. This facility produces 40,700 Barrel per day of propane plus liquids and 1,141 MMscfd of 1051 Btu/SCF HHV pipeline gas. The propane plus product can be separated into a LPG liquid fuel for the local market and butane plus for gasoline blending for local refineries.

This process can also be used for ethane recovery. Ethane production would require a second column to fractionate the ethane from the LPG product. The first column would operate as a demethanizer recovering the ethane and heavier components while producing a leaner gas with lower ethane content.

Table 4 – LPG Production Overall Balance

Component, Mol Fraction	LNG Feed	LPG Plus	Pipeline Gas
N_2	0.0034	0.0000	0.0036
C_1	0.8976	0.0028	0.9439
C_2	0.0501	0.0200	0.0517
C_3	0.0316	0.6251	0.0009
IC_4	0.0069	0.1404	0.0000
NC_4	0.0103	0.2096	0.0000
NC_5	0.0001	0.0020	0.0000
Heat Value Btu/SCF (HHV)	1,137	2,829	1,051
MMscfd	1,200	59	1,141
Std Barrels per day		40,742	

Conclusions

LNG regasification terminals should be designed to provide flexibility in processing non-compliant LNG to meet the North American pipeline standards. LNG regasification using the rejected heat from the Rankine power cycle or from the gas turbine waste heat can replace the traditional regasification equipment, while reducing environmental impacts. Gas turbine inlet air cooling can significantly increase the power plant output, and improve the LNG power plant integration economics. Rankine power cycle with a mixed fluid using LNG as a cold sink is very efficient in a simple cycle power plant.

References Cited

1 Mak, J., “Power and LPG Production with LNG Import” 2nd Annual Atlantic Canada Power Summit, November 2004, Saint John, New Brunswick, Canada.

2. Mak, J., Nielsen, D., Schulte D., and Graham C., “LNG Flexibility”, October 2003, Hydrocarbon Engineering.

3. Mak, J., Nielsen, D., Schulte D., and Graham C., “A New and Flexible LNG Regasification Plant”, Gas Processors Association, March 2004, New Orleans, Louisiana.

4. Zeus Development Corporation, “LNG-West East Meet West,” June 2003, Long Beach, California.

5. World Bank Group, “Pollution Prevention and Abatement Handbook,” Effective July 1998.

Biography of Speaker

John Y. Mak is a Technical Director and a Technical Fellow of Fluor's Energy and Chemical Group in the Aliso Viejo, California. He has been working in the Fluor's California office for over 30 years and has been involved in many oil and gas and energy projects. He leads the Fluor technical group in the development of new applications in the oil and gas and power industries. Recently, he has been focusing on LNG regasification technologies, and has developed several patented processes in the area of NGL recoveries and LNG regasification and power generation. He recently published several articles in these technical areas. He attended Oregon State University and received his BS in chemical engineering, and also received a MS degree in mechanical engineering from the University of California. He is a registered professional engineer in California.

Badak's Utility System to Support High Reliability LNG/LPG Plant Operation - Present and Future

Yudi Siswadi, PE
Process Engineer
PT. Badak NGL Bontang Indonesia

Soewignyo
Utilities II Section Head
PT. Badak NGL Bontang Indonesia

AIChE Spring Meeting, April 2005
5th Topical Conference on Natural Gas Utilization
Atlanta April 10-14, 2005

Abstract

Badak's utilities system is divided into two operational modules, Module I to serve primarily Trains A~D/Community and Module II to support Trains E~H. To operate this system reliably to support plant operations and for energy efficiency, the Company applies a combination of "modular" and "integrated" operating modes. For the critical units, the Company provides "N+2" capacity in each module. The Nitrogen and Electrical systems are in integrated mode, while the others are in modular mode. In order to have steam system flexibility, one boiler in each module can be switched to the other module.

This article explains the operation philosophy and operational mode of Badak's utilities system to support LNG/LPG production, the efforts of improvement, and the future plan of 9 operating trains with considerations for safety, reliability, operation flexibility, and maintainability.

I. INTRODUCTION

The Badak LNG/LPG plant is located in Bontang, East Kalimantan, Indonesia, and has eight LNG Trains (A thru H) with a total production capacity of 22.5 million ton/year of LNG and 1.0 million ton/year of LPG. The capacity of the future ninth train (Train I) will be 2.95 million ton/year of LNG.

Trains A/B were commissioned in 1977, and sequentially continued with Trains C/D, E, F, G and H in 1983, 1989, 1993, 1997 and 1999 respectively. The additional utilities facilities were also commissioned and connected to the previous utilities system at the time of their train constructions.

In order to manage the whole plant operation, the plant is divided into 3 areas, i.e.: Process Train, Storage and Loading, and Utilities. There are two modules to operate the process trains and utilities; Module I for Trains A~D and Utilities-I, and Module II for Trains E~H and Utilities II. Utilities I primarily serves Trains A~D and the community, while Utilities II serves Trains E~H.

This article explains the operation philosophy and operational mode of the utilities system to support LNG/LPG production, the efforts of improvement, and the future plan of 9 operating trains with considerations for safety, reliability, operation flexibility, and maintainability.

II. BADAK'S OVERALL UTILITY SYSTEM

The Badak LNG/LPG plant operates a stand-alone utilities system to meet the requirements of the plant as well as the community. The facility of the overall utilities system consists of A Water Treatment System, Compressed Air System, Nitrogen Generation System, Steam Generation System, and Electrical Power Generation System.

In general, the overall utilities system is separated in a "modular" concept considering the safety and reliability aspects. Some of them, however, are operated in an "integrated" mode.

The capacity and the operational mode of the overall utilities system can be seen on following table.

Utilites System	Rated Capacity		Spare Capacity	Operation Mode
	Module I	Module II		
Ground water system, m^3/h	8 @230		N + 2.5	Integrated
Water treatment system, m^3/h	228	342	N + 2	Modular
Compressed air system, Nm^3/h	4 @5,100 and 850 for black start	4 @5,100	N + 2	Modular
Nitrogen generation, hourly	3,100 Nm^3 GAN and 150 liters LIN	600 Nm^3 GAN	N + 1	Integrated
Steam generation, T/h	11 @295	10 @379	N + 2	Modular-Integrated
Power generation, MW	7 @12.5 and 5 for black start	7 @12.5	N + 4	Integrated

1. Water Treatment System

Ground water as raw water from eight wells "integrated" in a piping network is distributed to the utilities and community (potable) water treatment plants. Currently, five of eight water wells are operated to fulfill the water demand of 1,000 m^3/h. Raw water is distributed to the plants through three underground pipes of 8-inch, 10-inch, and 14-inch diameter. The piping network is adequate to support maximum LNG production and high water demand.

Some wells cannot be continuously operated due to their water level decreasing down below the pump suction. Therefore, in order to maintain the reliability of the water supply and for operation flexibility, the well water system is operated in an "integrated" mode.

The typical make-up rate of demineralized water is 4% of the boiler feed water flow. The peak make-up occurs when seawater leaks into the boilers from the surface condensor causing increased blowdown rates. Each module has two 2,460 m^3 demineralized water storage tanks that can back up 5½ days for continued operation concurrent with disposal of contaminated condensate. A 10-inch crossover line between these tanks is provided as an export/import facility between both modules for operation flexibility and maintainability.

About 96% of the steam condensate is recovered or recycled as boiler feed water. It is processed through resin polishers to remove cationic contaminant. Module I has seven larger polisher vessels (6 operating, 1 standby) with a capacity of 3,465 m^3/h, sufficient to handle the peak flow of 2,800 m^3/h. Module II has eight smaller polishers with a capacity of 3,178 m^3/h (7 operating, 1 standby).

Future plan. The additional raw water demand is estimated at 115 m^3/h for the ninth train operation. This requirement still meets with the existing five wells in operation. However, to ensure a reliable water supply, an additional well of 240-m^3/h capacity shall be provided.

No aerator package and demineralizers are required. The peak consumption of 325 m^3/h can still be supplied by the existing units. Taking into account the steam condensate load from the new train, three 454-m^3/h capacity polishers *identical* to the existing ones are required (2 in operation and 1 standby). No additional storage tank for filtered water and condensate is required as it will reduce the storage residence time from 5½ to 4½ days for a continuous operation concurrent with disposal of contaminated condensate.

2. Compressed Air System

Atmospheric air is compressed to 9.1 kg/cm^2g and 30°C, and delivered to users as general utility air, feed to the nitrogen plants and instrument air. There are eight centrifugal air compressors rated at 5,100 Nm^3/h each. For black start purposes, a 850-Nm^3/h diesel compressor is installed in Module-I. Four compressors are provided in Module I to supply compressed air for Trains A~D, Utilities-I, Storage and Loading, and Nitrogen plants; while the others in Module II for Trains E~H, Utilities-II, and the other Nitrogen plant.

The previous operating philosophy was to operate 2 compressors in each module while importing compressed air from Module II. The compressed air facilities were operated in an "integrated" mode, while the instrument air on "modular" operation. Currently, 2 of 4 air compressors in each module are operated to support 8 trains operation with "modular" operation. One of Module-II compressors is dedicated to supply compressed air to the Nitrogen plant.

The control of compressed air distribution is to prevent plant upset and shutdown due to loss of instrument air. The air compressors are operated by controlling the discharge pressure, while the distribution system is controlled by instrument air header pressure. If the header pressure of instrument air drops to 7.4 kg/cm^2g, from normal 8.4 kg/cm^2g, utility air flow is automatically stopped. If this pressure falls to 7.2 kg/cm^2g, the feed air flow to the Nitrogen plant is also automatically stopped.

In order to improve the reliability of the compressed air system, the improvements have been performed by: (i) balancing the load of compressed air, a portion of the compressed air supply to the Storage and Loading area is changed from Module I to Module-II, and (ii) separating the electrical supply for the Module-II compressors from one to two buses/feeders.

Future plan. Train I will add an estimated air consumption of 770 Nm^3/h in normal operation and 2,540 Nm^3/h during train shutdown. The average consumption of nine trains will be about 18,300 Nm^3/h, which can still be covered by the existing four air compressors. In order to accommodate Train I requirement on Nitrogen and to maintain reliable operation, an additional air compressor shall be dedicated to the existing Nitrogen plant.

3. Nitrogen Generation System

Nitrogen is used for refrigerant makeup, purging of loading arms, blanketing of chemical storage tanks and purging of equipment during shutdown. The operating conditions are 6.5 kg/cm^{2}g and 25°C, while the mechanical design conditions are 8.8 kg/cm^{2}g and 38°C. The Nitrogen distribution system is in an "integrated" operation.

The feed to the Nitrogen plants is compressed air supplied by the air compressors. Gaseous Nitrogen is distributed to users via an "integrated" piping network by maintaining the header pressure at 7.3 kg/cm^{2}g. In the event Nitrogen demand exceeds the production, the pressure controller valve opens to supply liquid Nitrogen from storage through the vaporizer into the header system.

The normal consumption is 2,500 Nm^3/h. The additional consumption is 500 Nm^3/h during one train shutdown and 550 Nm^3/h during one LNG ship loading. These consumptions can be supplied by five Nitrogen generation units with the total capacity of 3,700 Nm^3/h gaseous Nitrogen (GAN) plus 150 liter/h liquid Nitrogen (LIN), or 3,200 Nm^3/h GAN plus 238 liter/h LIN. To maintain the reliability of supply, one of the Nitrogen plants is provided with a LIN storage of 500 m^3 in five tanks with the vaporizer capacity of 5,680 Nm^3/h to meet peak Nitrogen demand such as purging of a LNG storage tank.

Future plan. The additional Nitrogen consumption of Train I is estimated at approximately 200 Nm^3/h, which brings the total consumption for nine trains to about 2,700 Nm^3/h. It requires the installation of a new Nitrogen generation unit to maintain the spare capacity. No additional liquid storage nor vaporizer is required.

4. Steam Generation System

All motive power and heat in the plant arise from steam, which is generated at high pressure (HP) in gas-fired boilers. Steam is delivered at four pressure levels to satisfy the various requirements:

- High Pressure (HP) steam at 59.8 kg/cm^{2}g (850 psig) and 445°C,
- Medium Pressure (MP) steam at 17.6 kg/cm^{2}g (250 psig) and 335°C,
- Medium Low Pressure (MLP) steam at 10.5 kg/cm^{2}g (150 psig) saturated, and
- Low Pressure (LP) steam at 3.5 kg/cm^{2}g (50 psig) saturated.

There are eleven 295-T/h boilers in Module-I and ten 379-T/h boilers in Module-II to produce HP steam. The steam requirement of Module-I and II is 2,620 and 2,800 T/h respectively. All available boilers are normally operated. So, in each module, two of the boilers serve as hot spares (applying "N+2" philosophy-N is the number of machines required to meet the power demand). One extra boiler is required to allow continued operation during boiler inspection, to ensure sufficient reserve in case one boiler trips.

The steam distribution system is split into two modules. Cross connections are provided between both modules for HP, MP and LP steam, as well as condensate water to provide more operation flexibility. HP steam distribution is operated on "modular" operation. To improve the operation flexibility, one boiler from each module can be operated or exported to the other module through a 24-inch crossover line.

MP steam is extracted from the exhaust of the backpressure turbines. The balance of MP steam is controlled by adjusting the load of these turbines. HP to MP letdown stations are provided for each pair of trains. Most LP steam is produced from the exhaust of the process and utility drivers. Less than 15% of LP steam is supplied by extraction from the condensing turbines. LP steam pressure is controlled by induction/ extraction to/from these turbines. MP to LP letdown stations are also provided for each pair of trains.

The table below shows the steam balance between both modules on different operation modes, "modular" versus "integrated", of the MP and LP steam systems.

Description		Modular Mode		Integrated Mode	
		Module-I	Module-II	Module-I	Module-II
HP Steam	Production, T/h	2778	2966	2733	2891
	Used, T/h	2654	2812	2654	2812
	Letdown HP to MP steam, T/h	123	154	79	79
	Available Boiler Capacity "N+"	1.58	2.18	1.74	2.37
MP Steam	Production, T/h	673	584	1257	
	Used, T/h	796	552	1348	
	Excess (+) or Shortage (-)	-123	+32	-91	
	Letdown MP to LP steam, T/h	0	186	66	
LP Steam	Production, T/h	870	639	1509	
	Used, T/h	750	825	1575	
	Excess (+) or Shortage (-)	+120	-186	-66	

When the MP and LP steam systems are on "modular" operation. Module-I will require additional MP steam of 123 T/h (shortage) and result in an excess of LP steam at 120 T/h, while Module-II will result an excess of MP steam of 32 T/h and require additional LP steam of 186 T/h (shortage). The excess 120-T/h LP steam in Module-I will be *vented* to the atmosphere. Whereas the shortage of LP and MP steam in each module can be supplied by the Letdown Stations. It will reduce operational flexibility of the steam distribution system and reduce the overall plant thermal efficiency.

Operationally, the both MP and LP steam systems can be operated on "modular" mode. However, they have some implications i.e.:

- increase the required HP steam by 123 T/h in Module-I,
- decrease the available boiler capacity of Module-I to "N+1.58",
- reduce the overall plant thermal efficiency proportionally, and
- reduce the operational flexibility of the boilers and power generations.

In order to maintain spare capacity in each module and to maximize energy conservation and improve operation flexibility, the MP and LP steam distribution systems are operated in an "integrated" mode. Block valves are provided on the interconnecting lines between the modules.

Future plan. To avoid all trains trip due to loss of MP and LP steam, the manual block valves on the interconnecting lines between both modules will be improved by replacing them with motor operating valve (MOV) to *automatically* close if their pressures drop.

To expand an additional train, the steam generation system will be operated in an "integrated" with Module II system by connecting all steam levels and the condensate system.

5. Power Generation System

The power generation system consists of 15 generator units, eight 13.8 KV main buses, an integrated 34.5 KV synchronizing ring bus, and associated distribution equipment to serve the plant and community. Power supply from an outside utility (Bontang city) is not available. The facilities comprise 14 generators rated at 12.5 MW each and one 5-MW diesel generator. The total generating capacity is 180 MW, whereas the peak demand is 127 MW. The generator driver types are 7 back-pressure steam turbines, 5 extraction/induction condensing steam turbines, and 2 gas-fired turbines.

The operating philosophy is as follows:

- The diesel and gas turbines are used for emergency in case of "black start".
- The extraction/induction condensing turbines are operated at moderate load (7~9 MW) on "isochronous" or fast response mode, whereas the back pressure turbines on "droop" or slow response mode. This avoids MP steam pressure swings when a large electrical user is started or stopped. The load of the back pressure units is adjusted to meet MP steam demand.
- "N+2" power generators are provided in each module. N+1 machines are needed in operation to provide sufficient spinning reserve and one machine for maintenance shutdown.

Electrical distribution. Separated 34.5 KV ring buses are provided for Module I and II. These two ring buses are linked to permit operation as one integrated ring bus. Synchronizing buses are connected between the two modular ring buses. Current limit reactors are connected between the 34.5 kV Trains A~D and Trains E~H synchronizing buses, thus restricting short circuit current.

The main features of the distribution system design philosophy are:

- Module I and II can be operated independently ("modular"). However, the design allows for integrated operation using the inter-tied 34.5 kV synchronizing ring bus. Due to simplicity of operation, reliability, operation flexibility, maintainability, and safety, the electrical distribution system is operated on "integrated" mode.
- One gas turbine generator can be connected either to Module-II or Module-I.
- Each load area in the plant has power supplied over two primary (13.8 kV) feeders through two transformers. The capacity of each feeder cable and transformer can safely carry the entire load in the area served by the double-ended substation.

Future plan. The power demand for 8 trains is 120.8 MW. The demand for 9 trains will be 140.9 MW and assumes that four cooling water pumps of about 3.3 MW each pump are required. Train I will be integrated to Module II.

III. CONCLUSIONS

1. Badak's LNG/LPG plant provides "N+2" spare capacity for the critical equipment and applies a combination of "modular" and "integrated" operation mode for the utilities system in order to support high reliability plant operation with considerations of safety, operation flexibility and maintainability aspects. Some improvements were performed while maintaining these aspects.

2. The steam distribution system is operated on "modular - integrated" mode. The HP steam distribution system is operated on "modular", while the MP and LP steam distribution systems on "integrated" considering energy conservation and operation flexibility. One boiler from each module can be operated/exported to the other module to improve the operation flexibility. Motor operated valves will be installed on the interconnecting lines of the MP and LP steam systems to prevent the upset of all trains due to the loss of their steam pressure.

3. For installation of an additional train, the design and operating philosophy will be applied to ensure high reliability of the utilities supply and to determine additional utilities requirements.

REFERENCES

1. Balance of Medium and Low Pressure Steam on "Modular" Operation, PTB Process Engineering Section, 2002, Bontang, Indonesia
2. Front End Engineering Design for Bontang LNG Expansion Project Train H, PKP Pertamina, 1996, Jakarta, Indonesia
3. Follow-up Items for Train "I" Project, PTB Process Engineering Section, 2000, Bontang, Indonesia
4. Statement of Requirement (SOR) Train I Project, PKP Pertamina, 1999, Jakarta, Indonesia

World Largest Acid Gas Incinerator in LNG Plant: Startup and Operational Experiences

S. Jamaludin
Process Control Engineer Malaysia LNG Tiga Startup Team
Sarawak, Malaysia

S. Manaf
Senior Manager Malaysia LNG Tiga Startup Team
Sarawak, Malaysia

AIChE Spring Meeting, April 2005
5th Topical Conference on Natural Gas Utilization
Atlanta April 10-14, 2005

ABSTRACT

Environment is fast becoming a major aspect of oil and gas industry alongside safety and production. With increasing plant capacity, tighter government and international regulation as well as increasing sensitivity and awareness of the public; meeting environmental limits has never been as important and challenging as what we are facing today. Demand for gas and specifically LNG has been steadily growing and is seeing its best growth since past years. Even though known for its clean combustion property, gas processing still has to deal with discharging harmful byproduct waste to the environment safely. In most LNG plant, due to the presence of CO_2, H_2S and BTX in the feed gas, continuous stream of acid gas from the process has to be discharge to the air. The best way to do this safely and economically is by acid gas incineration. Malaysia's PETRONAS LNG complex in Bintulu, Sarawak, comprises of MLNG Satu (1983), MLNG Dua plants (1995) and MLNG Tiga (2003). MLNG Tiga has 4 units of acid gas incinerators in operation capable of incinerating 4 x 1300 t/d of acid gas. The designer faced tremendous challenge of building an even larger acid gas incinerator than the problematic MLNG Dua incinerators, which were the world's largest before MLNG Tiga came online. The paper is about the accomplishment of the new design and more importantly the lessons learnt from the startup and operational experience of the unit from process and process control point of view. A story of success with a lot to be learns for an even greater future success.

1 INTRODUCTION

1.1 PETRONAS LNG COMPLEX

Situated in a small town of Bintulu in the island of Borneo, Malaysia; Petronas LNG complex consist of 3 LNG plants namely Malaysia LNG (MLNG), Malaysia LNG Dua (MLNG Dua) and Malaysia LNG Tiga (MLNG Tiga). MLNG, which consist of 3 LNG trains was incorporated in 1978 and delivered its maiden cargo to Japan in January of 1983. Growing demand of LNG and vast reserve of natural gas from the central Luconia basin open ways for further establishment of MLNG Dua Train 4, 5 and 6 in 1994 and MLNG Tiga Train 7 and 8 in 2003. With the completion of MLNG Tiga, Petronas LNG complex became the world largest LNG plant in a single location with total capacity of 23 million tonne per annum. To date,

Petronas LNG complex had successfully delivered more than 4000 cargoes of LNG to its buyers without a single failure. With impeccable delivery record and world's class safety achievement; Petronas LNG complex sets the industry standard.

1.2 PROCESS DESCRIPTION

The plant can be divided into 4 main sections, which are the upstream facilities, gas treating, liquefaction and storage/terminal. The upstream facilities receive gas via 125km pipeline from the sea in the Central Luconia basin and separate the condensate from the gas. The gas is metered and enters the individual train gas treating section. The first part of the gas treating section deals with the acid gas in the feed. CO_2 and H_2S need to be removed to meet product specification and to prevent freezing in the liquefaction section. Gas from Central Luconia can have up to 6.8 mole % of CO_2 and 400 ppm of H_2S. Amine-based solvent is used in the acid gas removal process. Solvent with absorbed acid gases are regenerated in regenerator column at high temperature and low pressure. The acid gases are then incinerated before being discharge safely to the environment. In MLNG, acid gases are incinerated in steam boilers while in MLNG Dua and MLNG Tiga, acid gas thermal incinerators are utilised. The final process of gas treating is the removal of water and mercury up to 1 ppm and 10ng/sm^3 respectively.

The natural gas is then ready to be liquefied by using Air Products and Chemicals Inc. (APCI) license process. APCI process uses 2 types of refrigerants, which are propane and mixed component refrigerant (mixture of nitrogen, methane, ethane, propane and butane). Natural gas is first pre-cooled by propane before it is finally chilled to –151 deg C in the Main Cryogenic Heat Exchanger (MCHE) by the mixed component refrigerant. The final cooling down to –161 deg C is by Joule-Thompson expansion through control valve. Finally, LNG is piped to the tank farm for storage at atmospheric pressure. Once the LNG ship is ready, LNG is loaded to the ship from the tank through loading arms at the jetty head. That completes the process, and the LNG is now ready to be used as fuel for power generation upon gasification at the customer's LNG receiving terminal.

2 CHALLENGES FACED BY THE DESIGNERS OF MLNG TIGA ACID GAS INCINERATORS

2.1 DESIGN OF MLNG DUA ACID GAS INCINERATOR

2 of MLNG Dua acid gas incinerators started its operation in 1996 with design capacity of 3125 t/d. However, the operation was highly unstable due to instability in controls and combustion. Based on maximum capacity test, the maximum load achieve temporarily in each acid gas incinerator is 1500 t/d, which is only about 50% of the design capacity.

MLNG Dua acid gas incinerators are vertically fired main burner with 4 auxiliary burners for the acid gas. The combustion chamber was severely undersized and design consideration was not given to the importance of mixing. Both had resulted in unstable flame condition due to insufficient residence time and insufficient mixing. To make things worst, the control system was unreliable and highly unstable control with fluctuating temperature control. Incomplete burning was apparent with continuous visible flame on the stack. The acids gas incinerators frequently trip on high stack temperature or low O_2 in the stack and the trip of one incinerator will trip the other incinerator on sudden acid gas load increase.

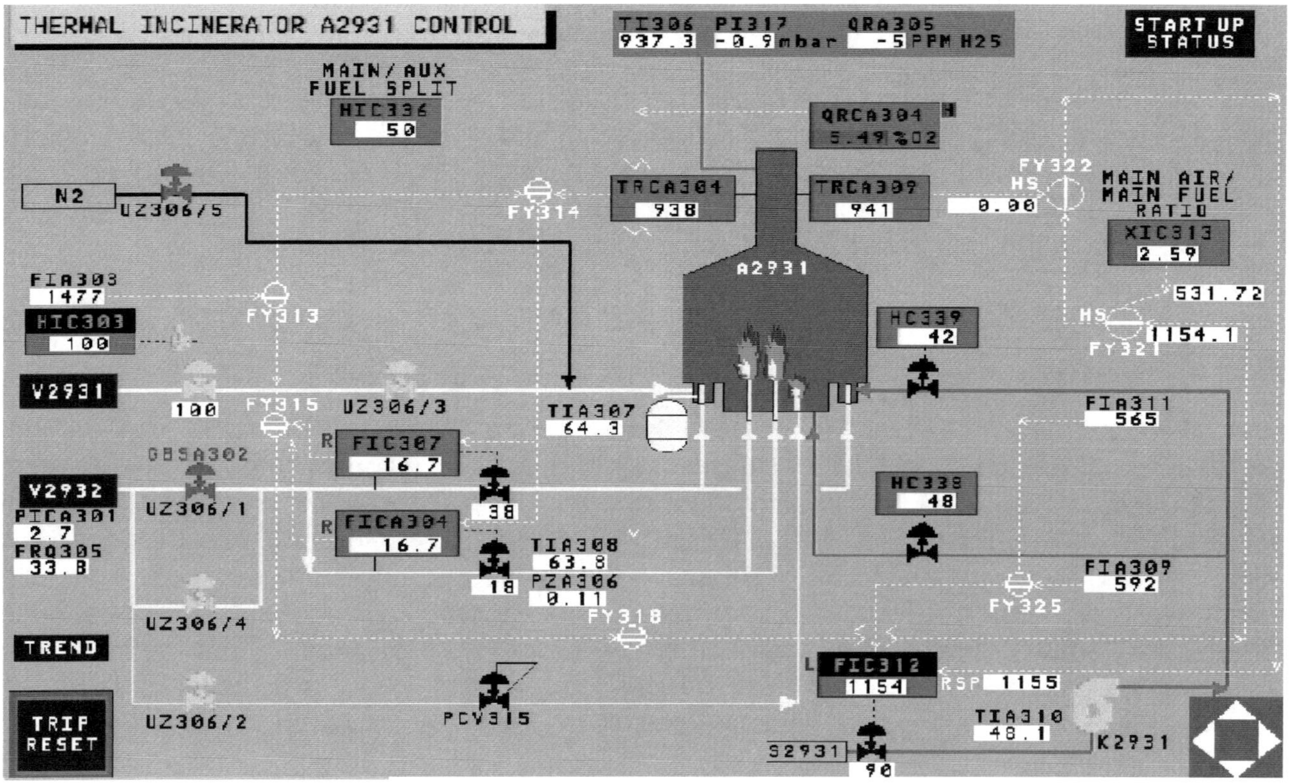

Figure 1: MLNG Dua Vertical Firing Burner

In 2001, the problem with MLNG Dua acid gas incinerator became a major concern when latest study shows that due to changes in offshore gas supply, sour gas content to MLNG Dua is expected to increase to 1815 t/d per module by 2003. In 2003, a third unit of acid gas incinerator was installed with a design capacity of 1815 t/d. Larger combustion chamber was build and swirlers were introduced at sour gas and air supply counter current to each other at the auxiliary burners. Sour gas nozzles were also modified to include 6 sour gas nozzles to the main burner.

Control scheme of the new and existing acid gas incinerators were modified to include sour gas measurement as feed forward signal. Automatic combustion air damper was installed, temperature override control was added and main and auxiliary fuel gas split to the burners was optimised to have more fuel going to the auxiliary burners. The new control scheme and the improved design of the new acid gas incinerator worked well. The unit has since been operating without any problem and able to achieve its design capacity.

Physical modifications were also done to the existing acid gas incinerators to increase the supply of combustion air by enlargement of the air duct. In an effort to improve mixing and increase the residence time in the combustion chamber, swirlers were added to the 4 auxiliary burners. In one of the acid gas incinerator, the swirlers were installed at sour gas and air supply co-current with each other instead of counter current. This was purposely done in order to find the best arrangement.

With all the modifications, the existing acid gas incinerator is able to achieve the target sour gas load of 1815 t/d except for the acid gas incinerator installed with co-current swirlers. However, the operation is still not fully stable with occasional after burner still visible and tight temperature control still not possible. To some extent, poor mixing and insufficient residence time still persist in the improved design of the existing acid gas incinerators.

2.2 INCORPORATED PHYSICAL DESIGN IMPROVEMENT IN MLNG TIGA ACID GAS INCINERATOR

With vast experiences and invaluable lessons learnt from MLNG Dua, the same vendor was selected to design MLNG Tiga acid gas incinerator. The design work goes hand in hand with the design of MLNG Dua new acid gas incinerator. While it was decided that MLNG Dua new acid gas incinerator will be designed similar to the initial design but with the necessary refinement and improvement, MLNG Tiga acid gas incinerator design took a radical approach. The old vertically fired main and auxiliary burners design was replaced with horizontal firing burners, tangentially arranged at the incinerator wall. In anticipation of the much higher incinerator load of 3800 t/d of acid gas, the new design is believed to be able so solve the problem of insufficient mixing and residence time in the combustion chamber.

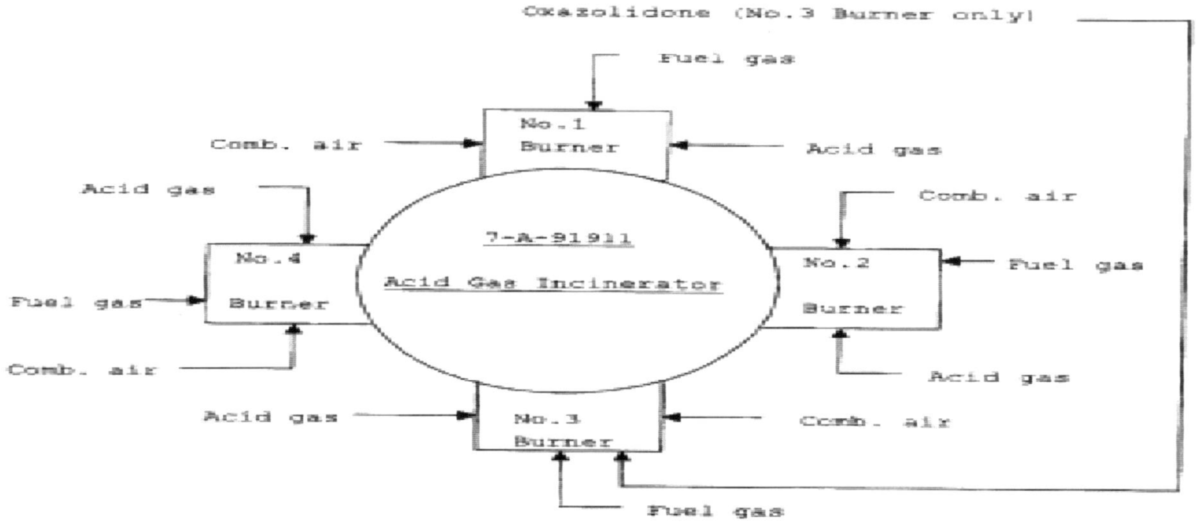

Figure 2: MLNG Tiga Horizontal Firing Burner

The design capacity of the acid gas incinerator was however later reviewed and it was decided that 2 acid gas incinerators per train instead of a single installation would improve the availability of the acid gas incinerator. The design capacity philosophy was also changed to only 100% x 1.1 instead of 150% x 1.1 in the original design. The revised design capacity is justified as that even if one incineration train unit trips, the LNG train production is still possible with turndown throughput to 60%. Although the incineration unit is designed at 55%, it will be able to treat the acid gas correspond to 60% of the feed gas into the LNG train since the unit is designed at highest CO_2 and H_2S concentration and assuming all the aromatic components are absorbed in acid gas removal solvent. In the end, the final design capacity of the acid gas incinerator is 1300 t/d. Even with a lower capacity, the vendor decided to go ahead with the initial design proposal of horizontal tangentially fired burners but with only 4 numbers of burners instead of 6.

The risk with tangentially fired burners is that there is a possibility for the flame detector to detect adjacent flame. Furthermore, the flame will be closer to the combustion chamber wall, hence higher stress to the refractory. This could shorten the life of the refractory as it might see higher temperature than the design adiabatic flame temperature. MLNG Dua burners fire vertically, and as a consequence all of the combustion was taking place in a narrow cylinder of approximately the same diameter as the exhaust stack. The available residence time in the combustion chamber is not being utilized since most of the volume was dead space. MLNG Tiga burners are oriented to give better utilization of the volume, hence increasing residence time and allowing more complete combustion. Furthermore, the proportional volume of combustion chamber to gases was increased to further increase the residence time.

Auxiliary burners for the acid gas are replaced with scroll type burners wrapped around the fuel gas lance burners. Scroll type burner is generally applied to burn low calorific fuel gas in incinerator and boiler. In this case, acid gas is supplied through the scroll gas burner and fuel gas is supplied through multi lance burner.

Combustion air is fed through primary and secondary swirl vane of air register. So as combustion air has swirl momentum, mixing with both straight flow acid gas and fuel gas is made effectively. Primary air register provides 100% air for combustion of fuel gas at the lance burner, 50% air fir combustion of acid gas at the scroll burner and excess air for cooling. Secondary air register on the other hand provides 100% air for combustion of acid gas and excess air for cooling. High adiabatic temperature of the fuel gas flame causes the lower adiabatic temperature of the acid gas flame to mix well with it. Acid gas flame that is heavier is directed towards the center of the fuel gas flame. Secondary air register can be used for mixing of the acid gas and fuel gas by changing the flame temperature and recirculating the flame.

Further, these burners are installed horizontally and tangentially, and so, combustion fuel gas from burners is effectively mixed. Thermal oxidation will also take place efficiently due to the long combustion flue gas trace of horizontal tangential burner installation compared to vertically upper firing incinerator.

MLNG Dua acid gas incinerator was designed for normal combustion temperature of 800 deg C, with internal design temperature of 1000 deg C to allow for local hotspots. MLNG acid gas incinerator normal operating temperature on the other hand is 900-1000 deg C, which considerably improved the operation. 800 deg C operating temperature is normal for burning H_2S but is insufficient for complete burning of aromatics in the acid gas. Toluene for e.g. typically requires 925-950 deg C for complete combustion in reasonable time period. The higher operating temperature also increases the reaction rate, thereby reducing the required residence time. It also reduces the requirement for cooling air.

Actual heating value of the acid gas in MLNG Dua is considerably higher than allowed for in the design due to higher than design pickup of hydrocarbons in the acid gas removal absorber column. The higher pickup of combustibles increased the stoichiometric demand for oxygen and consequently MLNG Dua acid gas incinerator was limited by the amount of available air from the blower. For MLNG Tiga, the licensor of the acid gas removal unit was consulted repeatedly to ensure that the actual absorbed hydrocarbon in the acid gas solvent is as expected in the design of the acid gas incinerator. The number of design cases was significantly increased to address all the intended modes of operation and the combustion air blower was sized accordingly.

2.3 INCORPORATED CONTROL DESIGN IMPROVEMENT IN MLNG TIGA ACID GAS INCINERATOR

When there is an upset in the LNG train, operator attention will be devoted to the process train. The acid gas incinerator largely has to look after itself through the control system. Hence, in order to ensure stable continuous operation of the acid gas incinerator during steady operation and during upset and disturbances in the natural gas feed composition and the acid gas removal unit, the control system need to be able to function well. With 2 acid gas incinerators per train, there is the possibility of one acid gas incinerator tripping during normal operation. The acid gas flow will be automatically diverted to the remaining acid gas incinerator and the control system needs to react to this fast disturbance. Furthermore, during turndown, there will normally be a lag between cut back of feed gas rate and acid gas removal solvent circulation rate. There will be an increase in acid gas heating value during the lag and the required response time for the incinerator controls need to cater for the consequent step changes.

MLNG Tiga acid gas incinerator control scheme adopted the double ratio cross limiting system. The incinerator stack temperature acts on the fuel and combustion airflow controller setpoint in parallel, with limits (maximum for fuel, minimum for air) to avoid sub-stoichiometric combustion. In comparison to MLNG Dua control scheme, in parallel to adjusting the fuel flow, the total fuel demand signal passes via the air/fuel ratio controller to adjust the setpoint of the combustion airflow controller. Hence, there will be minimum disturbances to the air/fuel ratio on load increase and decrease. The required air/fuel ratio can either be manually set by panel operator, or automatically set by the closed-loop stack oxygen controller.

Acid gas flow is included as feedforward signal in MLNG Tiga acid gas incinerator control scheme. The measured fuel and acid gas flow are converted to Standard Refinery Fuel (SRF) and added together in order to derive the total measured fuel flow. SRF is a Shell's standard method of measuring furnace fuel mass flow where the fuel mass flow is in referenced to a standard heating value. SRF is defined as a fuel with Lower Heating Value (LHV) of 40.5 MJ/kg and Standard Air Requirement (SAR) of 13.66 kg/kg.

In addition to the limit imposed on the range of the air/fuel ratio controller, limits are also sets on the adjustment of the setpoint of fuel and combustion air as follows. The total fuel flow is multiplied by a factor of 0.9 and provides a minimum limit via a high selector to the total fuel demand signal being sent to the combustion airflow controller setpoint. If the total fuel demand decreases and the actual fuel flows do not react, this signal will limit the decrease in the combustion airflow to prevent sub-stoichiometric combustion. The control system changes from a parallel control system to a "fuel-leading" system (fuel decrease leads air decrease) after the high selector had limited the decrease in combustion airflow.

A similar system applies to the fuel flows. The measured combustion air flow passes through a minimum air/fuel ratio block with setting of 10% lower than the air/fuel ratio, and the signal provides a maximum limit via a low selector to the total fuel demand signal to be sent to the fuel flow controller setpoint. If the total fuel demand increases and actual combustion airflow does not follow, this signal will limit the increase in fuel flow demand. The control system changes from a parallel control system to an air leading system (air increase leads fuel increase) after the low selector has limited the increase in fuel flow.

If not all burners are in operation, the air that is supplied to the burners not in operation will not be taken into account in the air/fuel ratio control. For this reason, the measured fuel flow signal used for the air/fuel ratio control is multiplied by the ratio of the number of burners installed to the number of burners in operation. Furthermore, if not all burners are in operation, the oxygen measurement will not give proper information on the actual air/fuel ratio at the burner. Therefore the oxygen controller shall be switched automatically to tracking if not all burners are in operation.

Similar to the new control scheme of MLNG Dua acid gas incinerator, stack temperature override control is added which will increases the amount of combustion air via a high selector to meets its temperature setpoint. It is particularly useful during large variation of acid gas heating value, where the constant used in calculating the acid gas SRF will be off by a large amount that the amount of air requested by the air/fuel ratio controller will cause too high temperature in the stack.

Density compensation is added for compensation of the fuel gas flow measurement and to sets the constant used to convert the fuel gas flow from actual mass to SRF. As density increases, the mass flow will be higher than the uncompensated value, but at the same time, the LHV decreases. Using both compensations gives about 80% of the correction based on mass flow alone. So the major compensation is for the mass flow correction of the flow meter. For a head type flow meter, density changes of +/- 20% will results in uncorrected flow to be incorrect by +/- 10%. This is an accepted range of error for air/fuel ratio control. However, other variations in acid gas composition and CO_2 content of the fuel gas may give variations in air/fuel ratio of +/- 15%. So the total variations without fuel gas density compensation could be up to +/- 25% and this would be too excessive.

3 ACID GAS INCINERATOR CONTROL VS. FURNACE CONTROL

MLNG Tiga acid gas incinerator control scheme is a vast improvement from MLNG Dua. However, it is still based on standard fired heater control scheme. The scheme is premised on varying the combustion heat input to meet an external process demand; i.e. process outlet temperature or steam production rate. In this scheme the acid gas is treated as a waste gas flow, supplying heat input to the system.

An incinerator is not a heater. It is an adiabatic reactor, with the incineration (acid) gas, combustion air and supplemental fuel gas as reactants. The operational objective of maintaining target combustion gas (stack) temperature and excess 0_2 concentrations are not obtained by manipulating the total heating value of the fuel supplied, but rather by manipulating the net composition of the reactants (i.e. combustion stoichiometric). Supplemental fuel gas (with a higher heating value) is necessary to produce the required combustion gas temperature. For the incinerator, acid gas is a contributor of combustion heat but it is more significant load on the system (mass of gas/air to be heated to combustion temperature). For an increase in acid gas flow, a proportional increase in fuel gas and combustion air will be required, and vice-versa for a decrease in acid gas flow.

MLNG Tiga acid gas incinerator control scheme attempts to maintain the total combustion heat input at the value called for by the stack temperature controller. For an increase in acid gas flow, this scheme will decrease the fuel gas flow, a move in the opposite direction from the required steady state value. In order to correct this situation, the control scheme will wait for the incinerator stack temperature to deviate from setpoint, only then it would proceed adjusting the fuel gas flow until the correct steady state value is reached.

The acid gas incinerator is expected to display relatively fast response to manipulations, on the order of 5 minutes time constant. Relative to this process dynamic, changes in acid gas or fuel gas compositions are expected to be slow. It is reasonable expectation that the feedback controllers for incinerator temperature and excess O_2 will adequately respond to composition (heating value) changes. However, the disturbance of acid gas flow changes can occur suddenly and may be large. These disturbances may be due to an upstream disturbance or the parallel incinerator tripping, among others. The control scheme should take this fast disturbance into account.

4 STARTUP OF TRAIN 7 ACID GAS INCINERATORS

Train 7 acid gas incinerators were started up in March 2003. Various implementation problems were found during dynamic simulation, mainly related to input and output ranges of calculation relays, and tracking and initialization to achieve bumpless transfer. Control system was further checked onsite by applying "dummy" input values prior to commissioning itself, and various modifications were made to the detailed configurations. As a result, there were no further configuration problems during commissioning.

Series of inspection were carried out on the burners and its accessories assembly prior to final box up. Main findings were sticky air register, loose ionization rod, wrongly positioned air damper, plugged burner tips and obstructed flame detector pathways. Position of peepholes and flame detector were also checked to ensure correct line of sight.

Prior to the first attempt of lighting up the burners the pressure in the wind box of each burners was check with the combustion air blower running on maximum inlet damper opening. It is important that the pressures are similar in order to ensure symmetrical firing. Air damper for each main burners were then locked at the position where all wind box pressures are equalized.

For each of the 4 burners, there are 2 ultraviolet (UV) radiation semi conductor type detection cell flame detectors. Initially it was assumed that these flame detectors come with initial factory settings. However, as it was later found out, the flame detectors need to see the real flame before it will be able to be configured. Hence, the vendor was called in and each of the flame detectors was configured at site during startup.

When the burners were finally lighted up, it can be seen clearly that the combustion was complete without any soot formation and the flame was very stable. Without acid gas, the flame was bright yellowish and due to the tangentially positioned burners, it was circling the combustion chamber.

Burner was found to exhibit flame instabilities and flame out with burner pressures below 0.01 barg. Consequently, active minimum stop settings of 0.02 - 0.05 barg were implemented. A minimum stop value of at least 0.04 barg was required to avoid flameout during ignition of a second burner. Mechanical minimum stop for the fuel gas valve was set at 15% opening. This setting was found to be sufficient for combustion of fuel gas at all 4 burners. The minimum stop is needed to ensure continuous firing on all burners when the acid gas incinerator trip to minimum firing where acid gas is vented and the fuel flow controller is switched to manual with 0% output. The setting has to be low enough that it will not cause too high pressure during startup of the first burner.

With the introduction of acid gas, the flame turns blue and becomes much shorter. The change in the color of the flame indicates the increase in UV radiated from the flame and the reduction of infrared radiation. The flame is still very stable and not seems the stability seems not to be affected by further increase in the amount of acid gas.

As the load increases, flame detectors were continually programmed to capture all possible combination of different flame signatures from zero acid gas to maximum acid gas flow. At each burner, primary and secondary air registers were provided to change the ratio of air flowing to the acid gas scroll burner and the fuel gas lance burner. These registers are useful in balancing the air split and helps to obtain the best color of flame, indicating sufficiently good mixing and combustion. The split of air also affects the flame detection. If there are too much air going through the secondary air register towards the acid gas scroll burner, then the acid gas flow will be cooled off and the aromatics in the stream will not be combusted. The uncombusted aromatics at the burners will absorb most the UV from the flame and the flame detector will have difficulty in detecting the flame.

It was found that with acid gas it was not possible to achieve 3% O_2 levels (1.2 air/fuel ratio) at the stack, and the protective temperature controller was continuously active to limit stack temperatures. This is attributed to the higher than design hydrocarbon content of the acid gas from the acid gas removal solvent regenerator column.

5 STEADY OPERATIONS OF TRAIN 7 ACID GAS INCINERATOR

Train 7 acid gas incinerator has been in operation for about 2 years now. It is taking on average of about 1100 t/d of acid gas daily, making it the world largest acid gas incinerator in terms of actual operating capacity. This is larger in comparison with MLNG Dua acid gas incinerator, which only incinerating on average of 700 t/d of acid gas daily. There had never been any problem of insufficient residence time or mixing. The flame continues to be stable at all operating cases and varying acid gas load.

The only problem with the unit is the instability of the flame detector at high acid gas load of 1200 t/d. After rigorous checking and testing done on the flame detector, it is believe that the root cause of the problem is the viewing angle of the flame detector. The position of flame detector is fixed, hence the line of sight of the flame detector cannot be adjusted. As acid gas load on the burner increases, the flame shortens. It is likely that the flame shortens to the extent that the line of sight of the flame detector does not cross the flame anymore. This is supported by the fact that the flame can be seen to be visually stable at 1200 t/d of acid gas, radiating a lot of UV radiation as the indicated by the bluish color of the flame. Hence, the maximum flow of acid gas to the acid gas incinerator is limited to 1200 t/d only, which is about 92% of design capacity.

At steady state, the control system is able to maintain all control variables at setpoint without any instability. Constant stack temperature and oxygen content suggests good mixing and combustion. Heating value of acid gas continues to be high, therefore the override temperature control continuously selected to control the combustion air. However, as expected, on sudden large changes of acid gas load to the acid gas incinerator, the disturbance on the stack temperature is large. Due to the fact that the control scheme is based on standard fired heater instead of adiabatic reactor, increase of acid gas flow will decrease the fuel gas flow, which is a move on the opposite direction. Fuel gas flow will only be increased from the action of the stack temperature controller once the measured temperature starts to deviate from the setpoint. Nevertheless, in most operational upsets, the swing in the stack temperature is recovered relatively fast and is still within operational margin, hence the acid gas incinerator is still able to stay online on changes of acid gas load.

In June 2003, fire broke out inside the gas turbines and incinerators combined stack for about 3 minutes. The root cause of the fire was the presence of unburned hydrocarbon from the acid gas stream in combustible environment in the combined stack. Rich amount of hydrocarbon in the acid gas was due to the high amount of hydrocarbon being released from acid gas removal solvent when heat was restored to the acid gas removal regenerator column after an earlier trip of the heat transfer fluid pumps on loss of power to Train 7. As a consequence, some changes are made to the safeguarding system to trip incinerator to minimum firing on trip of incinerator air blower and on MCHE trip. Also, it was included in the Operational Procedures Guide that the acid gas removal unit should be started up with acid gas routed to vent, and only diverted to incinerator after stable operation has been established. This is to ensure that during startup and operational upsets in the acid gas removal unit, acid gas will not be routed to the acid gas incinerator as the heating value of the acid gas can be considerably higher than what the acid gas incinerator is designed for.

6 FUTURE EXPECTATIONS AND CONCLUDING REMARKS

In conclusion, the design of MLNG Tiga acid gas incinerator was a story of success. Benefited from vast lessons learnt of MLNG Dua acid gas incinerator operations, large improvements were made in both the physical design and control scheme of the incinerator. Startup and steady operations were not troubled by major difficulties and constraints. Although the maximum capacity of the incinerator is currently limited to only 92%, this is not attributed to the stability of the flame or operating conditions that is beyond the design envelope. It was simply due to the limited line of sight of the flame detector. Even though more are understood of the process now, there are yet many unknowns to be mastered. Design of large-scale acid gas incinerator is still an uncharted territory. With MLNG Tiga acid gas incinerator capable of continuously steadily incinerating 1200 t/d of acid gas, the windows are open wide for future design of a much larger horizontal firing acid gas incinerator.

MLNG Tiga acid gas incinerator design not only is the largest horizontal firing acid gas incinerator but it is also the only one that shares its stack with a gas turbine exhaust. A brilliant solution to the regulatory emission limitation, but poses significant challenges and risks due to the chances of delayed combustion of unburned hydrocarbon in the hot stack. A simple stack design were made complicated with numbers of safeguarding system that had to be in place along with some operational limitations. Still a major cost saving in comparison to its sulfur recovery alternative, it is yet too early to be concluded that combining the gas turbine and the acid gas incinerator stack was the best decision. Though operations are only slightly affected by the design with integrity not seems to be affected, only time will tell.

6.1 PHYSICAL DESIGN OF ACID GAS INCINERATOR

Startup and operational experience of acid gas incinerator in MLNG Tiga had proven that the theory behind horizontal firing burners with tangential arrangement at the incinerator wall does works. Mixing started at the point the reactants (acid gas, fuel gas and combustion air) are introduced into the acid gas incinerator. The combination of the acid gas scroll burner and the fuel gas lance burner with combustion air that comes through primary and secondary registers ensures that the reactants are sufficiently mix as it is combusted at the burners. The well-mixed hot combustion gases from each burner are then thermally oxidized as it mixes with each other along the spiral path from the combustion chamber to the top of the stack.

The horizontal firing and tangentially arranged burner had proved itself to be a superior design that solves the problem of improper mixing ant insufficient residence time. Maximum utilization of the incinerator volume means that capital cost can be saved in size reduction of the acid gas incinerator. It is a clear way forward in future acid gas incinerator physical design.

6.2 CONTROL SCHEME OF ACID GAS INCINERATOR

The standard fired heater control scheme that was implemented worked well in steady state but less so in transient conditions. Though the disturbances is normally not large enough to cause a trip during large load changes, the resulting transient is not desirable and can cause problem for a larger design capacity of acid gas incinerator. Hence to obtain the control objective of maintaining the stack temperature and O_2 concentration, the net composition of the reactants need to be varied rather than varying the total heating value of the fuel supplied. For an increase in acid gas flow, the amount of fuel gas supplied and combustion air need to be increased and vise versa.

7 REFERENCES

Prakash, K.K and Poo H.C. (2004). ‘Presentation Material MLNG Dua Incinerator Control’.

Nippon Furnace Kogyo Kaisha Ltd. (2000). ‘Minutes of Meeting M-YPC-NFK-001 Technical Clarification Meeting for Acid Gas Incinerator’.

Shell Design and Engineering Practice 32.24.20.38-Gen. (1997). ‘Control System and Instrumented Protective Functions For Fired Equipment’.

Shell Global Solutions International, B.V. (2001). ‘M149 Combustion Technique for Gas Plant Course Manual’.

Brandt, R. R. (2003). ‘Email Correspondence MLNG Tiga Incinerator’.

Beute, N. N., Dam W., Habets G. and Brandt R. R. (2003). ‘Email Correspondence Safety Issues Related To The Combined GT/Incinerator Stack Design As Applied In MLNG Tiga’.

Moody, A. (2001). 'Email Correspondence Acid Gas Combustion Control Revision'.

Shell Global Solution International, B.V. (2001). 'PH-2037 SGS Review On The Combined GT And Incinerator Stacks Design'.

JMSB-KMSB-SIME. (2001). 'HP-3545 SGS Review On The Combined GT And Incinerator Stacks Design'.

JMSB-KMSB-SIME. (2000). Minutes of Meetings 'M-YOC-PMT-0248 Brainstorming For Sulphur Treating'.

Nippon Furnace Kogyo Kaisha Ltd. (2002). 'Installation, Operation and Maintenance Instruction of Burner and Accessories for 7/8-A-91911 and 7/8-A-91921 Acid Gas Incinerator'.

8 BIOGRAPHY OF SPEAKER

Mr. Saifullah Jamaludin is an Associate Member of Institution of Chemical Engineering UK. He graduated from University of Edinburgh, Scotland with BENG (HONS) Chemical Engineering in 1999 and joined Malaysia LNG Sdn. Bhd., a subsidiary company of PETRONAS.

Saifullah first assignment was as a Process Technologist in the Technical Services Department with responsibility of troubleshooting and monitoring of MLNG and MLNG Dua process including storage and loading.

In 2001, Saifullah was transferred to Operations Department to join a newly formed section of MLNG Tiga Startup Team as Process Control Engineer. At that point in time, construction of MLNG Tiga plant had just begun. MLNG Tiga Startup Team was formed early in the construction stage of the project with the intention of integrating it with the MLNG Tiga Project Team to ensure safe, timely and efficient startup of the plant in 2003.

At the end of 2001, 2 Senior Process Control Engineers from Shell joined the group and together they formed Process Control Group of the MLNG Tiga Startup Team. Early in the project, Process Control Group was deeply involved with review of Process Engineering Flow Scheme, Operating Manual, Operational Procedure Guide, Instrumented Protective Function, Process Control Narrative and Operational Functional Test Procedure. During commissioning, all control parameters and especially complex control loops in Distributed Control System (DCS) and compressor anti surge control was checked and simulated. When MLNG Tiga Train 7 and 8 were finally started up in March 2003 and August 2003 respectively, fine-tuning and minor troubleshooting was done. Both of the trains were successfully started up safely and smoothly.

With the end of the MLNG Tiga project, Saifullah is currently transferred to MLNG Process Control Group responsible for monitoring and troubleshooting of process control issues in MLNG, MLNG Dua and MLNG Tiga. Saifullah is also currently involved with the planning and implementation of Advanced Process Control in MLNG Tiga.

The Shell PMR Process for Large Capacity LNG Trains

Jolinde M. van de Graaf
Barend Pek

Shell Global Solutions International B.V.
The Netherlands

AIChE Spring Meeting, April 2005
5th Topical Conference on Natural Gas Utilization
Atlanta April 10-14, 2005

Abstract

World LNG trade has doubled over the last decade and is forecasted to continue to grow rapidly in the coming decades. Large capacity LNG trains – around 8 Mtpa - can help meet this high market demand in the most cost effective way when large gas reserves are present, due to economies of scale.

Shell Global Solutions has developed the Shell Parallel Mixed Refrigerant (PMR) process for LNG trains in the large capacity range. The technology is based on an alternative configuration of well-proven equipment and employs a common pre-cooling circuit and two parallel main liquefaction circuits, thus having only two cooling cycles in series. This improves the overall availability compared to three-cycles in series, as found in most other large train concepts. Moreover, the specific power (power required per unit of LNG produced) of the Shell PMR process is highly competitive (up to 10% lower than alternative processes on a like-for-like basis).

This paper presents the development of the Shell PMR process, with a case study for a large LNG liquefaction train with integrated NGL extraction facilities in a tropical climate.

Introduction

World LNG trade has doubled over the last decade (Figure 1) and is forecasted to continue to grow rapidly in the coming decades. In order to meet this demand, the LNG industry is rapidly expanding its LNG export facilities.

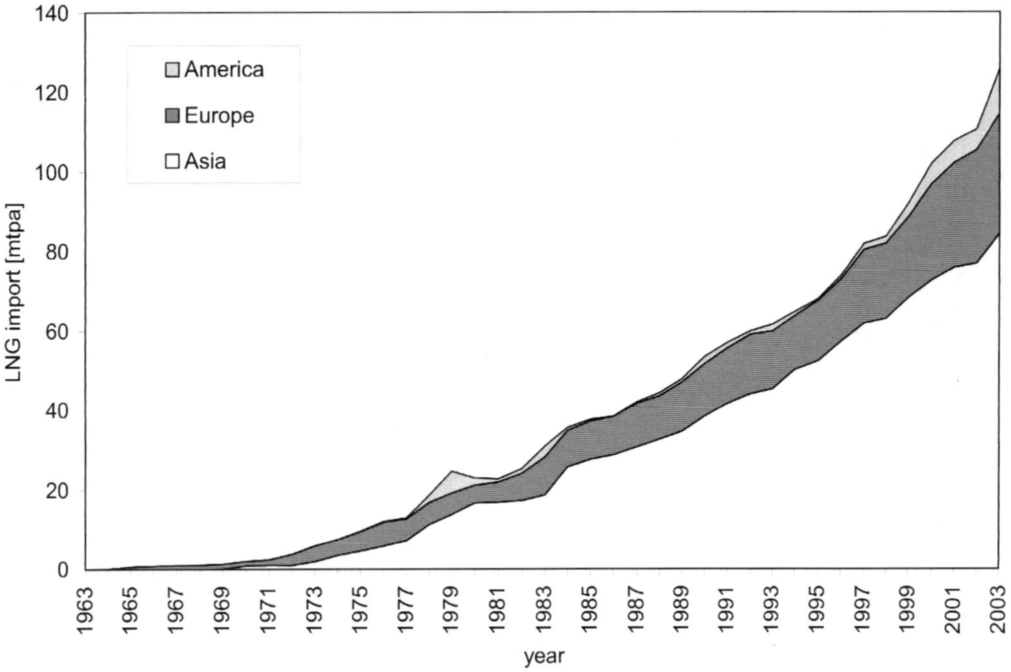

Figure 1: LNG import as a function of time

Figure 2 shows that up to 2000 the trend was to increase liquefaction train capacity to fully benefit from the economies of scale. However, from 2000 onwards there is more spread in train capacity ranging between 3 and 8 Mtpa per train. High capacity trains are the most economic when reserves are abundant and relatively easy to produce, assuming that the market can readily absorb large volume in one go. If not all of these conditions are fulfilled, projects with lower capacity trains can also be economically attractive.

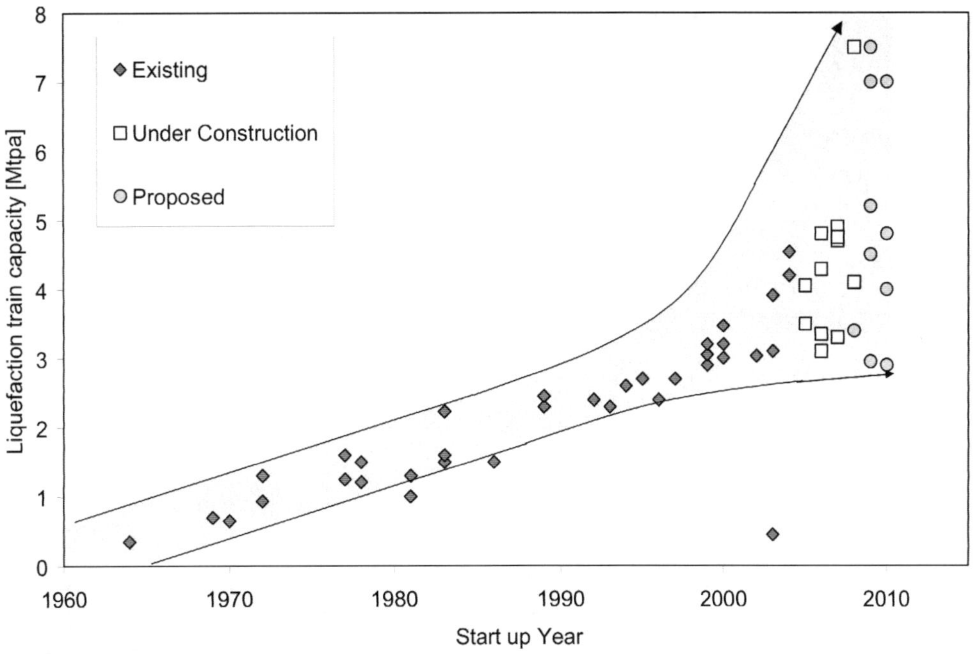

Figure 2: LNG train capacity over the years

Figure 3 shows that there is now not only more variation in train capacity, but also in the liquefaction technology applied. Up to 2003 85% of the worldwide LNG was produced using Propane/Mixed Refrigerant (C3/MR) technology. Based on public information on newly built and proposed capacity for 2004 to 2010, it is estimated that the portion of LNG produced with classical C3/MR technology, will have been reduced to some 50% and new technologies will have been implemented.

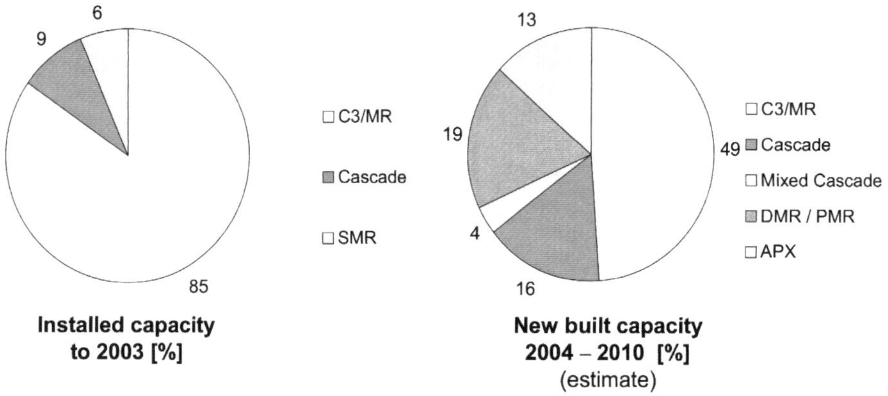

Figure 3: Evolution of LNG technology; world wide applied LNG liquefaction technologies.

Different specifications are required for different LNG Markets. These specifications are mainly related to the combustion properties of the LNG that need to be compatible with the combustion properties of the gas in the distribution network. For example, the USA market requires a relatively lean LNG with a Higher Heating Value of 1050 – 1070 Btu/scf, whereas in Asia end-consumers require a higher heating value up to 1160 Btu/scf. To a limited extent LNG combustion properties can be adjusted at the import terminal by the addition or removal of LPG's or the injection of inerts, like nitrogen. Depending on the market or markets served, the LNG plant needs to configure its process in a way that meets the LNG specifications, i.e. by the extraction of LPG. Apart from market requirements for the LNG specification, the extraction of LPG may be a very attractive option to increase the overall project value.

In the design of an LNG export plant gas availability, local operating conditions and targeted markets need to be considered to come to an optimal plant capacity and process configuration. Against this background Shell has developed a portfolio of cost effective, proven technology liquefaction process options for 3-10 Mtpa train sizes (Pek et al. 2004, Jager et al., 2004). This paper will address our approach to large train designs. Moreover, a case study is presented that will compare two process options for an 8 Mtpa train, that will produce LNG of different quality.

Developments in LNG train capacity

The process options in the Shell portfolio in the production range from 3-10 Mtpa are all based on two refrigeration cycles in series, as schematically depicted in Figure 4. Depending on the required capacity and local circumstances, different choices are made with respect to the line-up of rotating equipment, type of refrigerant, ambient cooling medium and types of cryogenic heat exchangers.

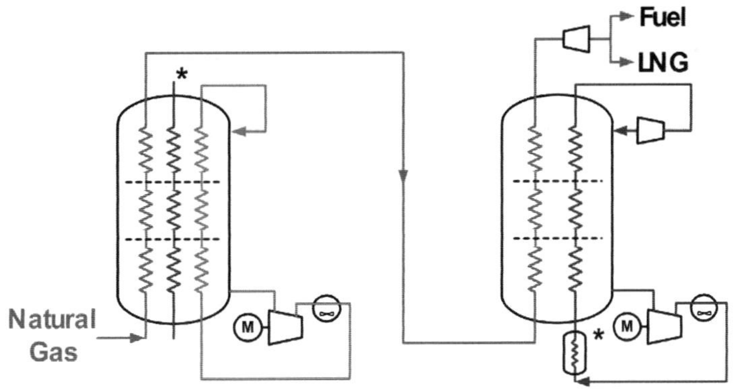

	Precool	Liquefaction
C3/MR =	Propane (**C3**)	**M**ixed **R**efrigerant
DMR =	**M**ixed **R**efrigerant	**M**ixed **R**efrigerant
PMR =	**MR** or **C3**	2 parallel **MR** cycles

Figure 4: Schematic representation of process options in Shell Portfolio of liquefaction processes

The Propane/Mixed Refrigerant (C3/MR) designs were traditionally the industry standard. Modern C3/MR designs typically have an arrangement of two large mechanically driven compressors, one for the pre-cooling refrigerant (C3) and one for the liquefying and sub-cooling refrigerant (the Mixed Refrigerant). Technology developments have resulted in a gradual increase in capacity for the C3/MR process to about 5 Mtpa.

The use of a mixed refrigerant in pre-cooling while keeping the compressor configuration the same results in a more flexible design. The Shell Double Mixed Refrigerant (DMR) process (Grootjans et al., 2002) overcomes the inherent limitations of using a single component refrigerant in pre-cooling in the C3/MR design; the additional degree of freedom resulting from the use of two mixed refrigerant cycles allows full utilization of power in a design with two mechanically driven compressors. Furthermore, it allows keeping the compressors at their best efficiency points over a very wide range (up to 50 °C) of ambient temperature variations and changes in feed gas composition. This design has been selected for the Sakhalin LNG project (Dam and Ho, 2001) currently under construction, with an annual design capacity of 9.6 Mtpa (two trains). Developments of the DMR process have further improved process efficiency. An example is the application of a three stage pre-cooling cycle. A number of publications have confirmed the high efficiency of the DMR process (Nibbelke et al., 2002, Paradovski and Hagyard, 2003).

A further development of the Double Mixed Refrigerant Process is the electrically-driven DMR (LNG GameChanger^{TM}) design (Heyman et al. 2002). The Gamechanger^{TM} concept is based on parallel line-up of electrically driven refrigerant compressors around a common set of cryogenic spool wound exchangers. Electric motors of 65 MW have already been constructed for LNG service. The current electrically-driven DMR design is particularly attractive in the 5-7 Mtpa capacity range (Del Vecchio and De Jong, 2004). Motors up to 80 MW are considered feasible (Kleiner and Kauffman, 2005), enlarging the viable capacity range of this process line-up further.

Electrically-driven LNG trains can compete with mechanically-driven trains because the increase in cost is compensated for by the increased availability. Other benefits of the electric option are the variable size and speed of the driver, the increased vendor base, and the potential to make a step change reduction in overall plant CO_2 emissions, by using combined cycle electric power generation.

Shell has developed the Parallel Mixed Refrigerant (PMR) process to meet the current challenge of the industry for larger train sizes (Klein Nagelvoort, 2002). With a single pre-cooling cycle and two parallel mixed refrigerant cycles the capacity can be boosted up to 8 Mtpa with 3 GE-F7 compressors in a tropical climate. The process can either use propane or a mixed refrigerant in pre-cooling. Proven refrigerant cycles can be used and the design can be applied today, without step changes in technology. The capacity can be increased further with different (larger) drivers.

Case study

A case study has been carried out to determine the potential of the Shell PMR process for an LNG plant in tropical conditions. This plant will process associated gas that is heavy in LPG's and lean in Nitrogen. The envisaged capacity of the plant is around 8 Mtpa. For this case study, only the use of GE-F7 was considered as drivers for the refrigerant circuits, and only propane as pre-cooling refrigerant.

In order to deliver the LNG to the envisaged markets, LPG and condensate need to be extracted from the natural gas feed to obtain a product of suitable quality. Different options exist, ranging from modest in-train LPG extraction to almost complete LPG recovery via a dedicated NGL recovery unit upstream the liquefaction plant. Extraction of LPG's can be limited to a moderate level if only the European market is considered, or if the North American import terminals have facilities to reduce the heating value of the LNG. For these cases, LPG recovery with a high pressure scrub-column is sufficient. Deep extraction of LPG is required in order to have the LNG quality that is directly suitable for North American import terminals. Deep LPG extraction can be achieved with a dedicated NGL extraction unit, operating at lower pressure. Therefore two cases are considered:

Case 1: Moderate LPG extraction, using a high-pressure scrub-column. The targeted heating value of the LNG for this case was 1110 Btu/scf.

Case 2: Deep LPG extraction, using an up-stream NGL extraction unit. The targeted heating value of the LNG for this case was 1070 Btu/scf.

The premises of this study are detailed in Table 1.

The focus of the study is to determine whether production capacities of about 8 Mtpa could be achieved within the constraints set out above.

Table 1: Premises for the case study

Gas Composition [mol%] (ex treating unit)	Methane 86 Ethane 6.4 Propane 4.3 Butane 2.4 C5+ 0.7 N2 0.1
Ambient Temperature	28.5 °C
Feed Gas Pressure (to train)	62 bara
Liquefaction train drivers	3 x GE-F7 + helper motors
Pre-Cooling refrigerant	Propane
Liquefaction and sub-cooling refrigerant	Mixture
Cooling medium type	Air

Case 1: The Shell PMR process with in-train NGL extraction

The process flow scheme of the Shell PMR process with in-train NGL extraction is given in Figure 5. Gas receipt and natural gas treating is followed by a single propane pre-cooling cycle. After pre-cooling, the flow is distributed over two parallel strings each having a scrub column for NGL extraction and a mixed refrigerant cycle for liquefaction and sub-cooling of the natural gas. The scrub column overheads are cooled by the mixed refrigerant to create reflux and ensure the required extraction level. Each liquefaction cycle has its own mixed refrigerant circuit, driven by a GE F7 gas turbine. The sub-cooled liquid from the two liquefaction cycles is combined in an end-flash system, where fuel gas is flashed off and LNG is sent to storage at atmospheric pressure.

The Splitpropane technology (Grootjans, 2003) is applied to limit the propane suction flows to acceptable levels. In this arrangement the four stage propane compressor is split over two casings: the first machine compresses the Low Pressure (LP) and High Pressure (HP) propane to discharge pressure whereas the second machine handles the medium Pressure (MP) and High High Pressure (HHP) flows. This split arrangement results in a lower volumetric flow per stage, for the same compression duty.

In order to achieve the targeted 8 Mtpa production capacity with the driver configuration chosen, an extended end-flash system is required for the conditions prevalent in this study. The application of an extended end-flash system allows an increase in the rundown temperature from the Main Cryogenic Heat Exchanger liquefaction cycle. The additional flash gas that is generated as a result of this, and that cannot be accommodated in the fuel gas system, is compressed, condensed and recycled back into the end flash gas system. In this way, power in the MR circuit is freed-up and replaced by power in the end-flash gas compressor. This relieves the power constraint in the MR circuit and enables higher LNG production capacity. The extended line-up as chosen here results in some 4-5% additional LNG capacity.

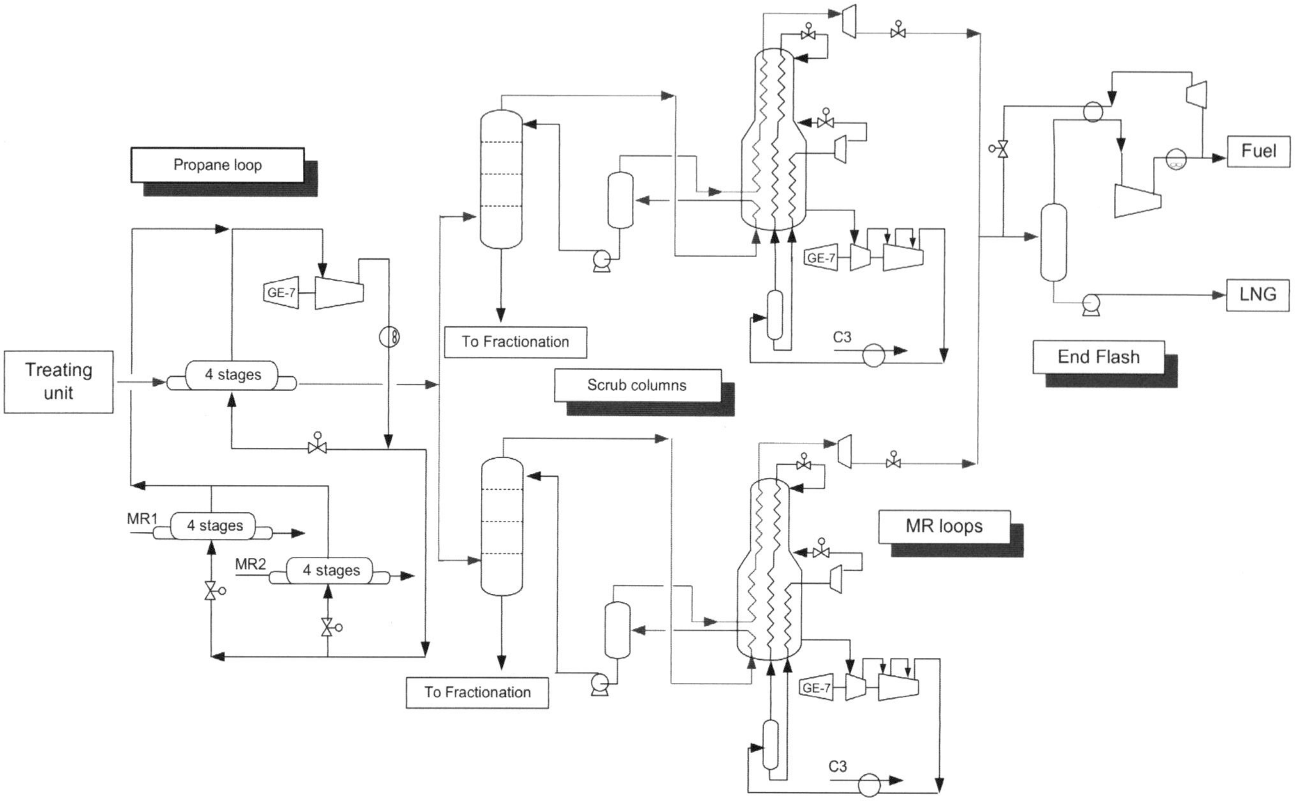

Figure 5: Simplified process flow scheme of the Shell PMR process, with in-train NGL extraction (case 1).

As an alternative for the extended end-flash system, larger compressor drivers like GE-F9 or Siemens V84.2 can be used to achieve a production capacity of 8 Mtpa or higher.

Two of the key parameters in this process design are the scrub column overheads temperature and the cut point temperature between the propane and mixed refrigerant cooling cycles.

The scrub column overheads temperature sets the level of propane recovery. In order to achieve the specification of an LNG product quality of 1110 Btu/scf, a propane recovery level of ~ 40-45% is required. At the pressures prevalent in the scrub column, this is accomplished by using mixed refrigerant cooling to ~ minus 45 °C.

The lowest propane cooling level is the cut point temperature between pre-cooling and liquefaction. This temperature is optimized so that the utilized power is balanced to a 1:2 ratio between the pre-cooling and the liquefaction cycles. The ability to tune the power balance exactly to the installed mechanical refrigeration capacity in the PMR process is an advantage over the conventional C3/MR process where, due to other constraints, the pre-cooling cycle cannot be loaded fully.

Case 2: The Shell PMR process with front end NGL extraction

The process flow scheme of the Shell PMR process with front end NGL extraction is given in Figure 6. The liquefaction process is similar to the one described for case 1, only the scrub column and related equipment is absent in this line-up. Instead, an NGL extraction unit is added downstream the natural gas treating unit. The NGL recovery process is derived from the Gas Sub Cooled (GSC) process. The lean gas from the NGL extraction unit is pre-cooled with propane in a single refrigerant cycle and next liquefied in two parallel mixed refrigerant cycles. The sub-cooled liquid from the two liquefaction cycles is combined in an end-flash system where fuel gas is flashed off and LNG is sent to storage at atmospheric pressure. In this case the amount of flash gas generated is in balance with the fuel gas requirements and there is no recycle of end flash gas. As for case 1, the Splitpropane technology is applied to limit the propane suction flows in the pre-cooling cycle to acceptable levels.

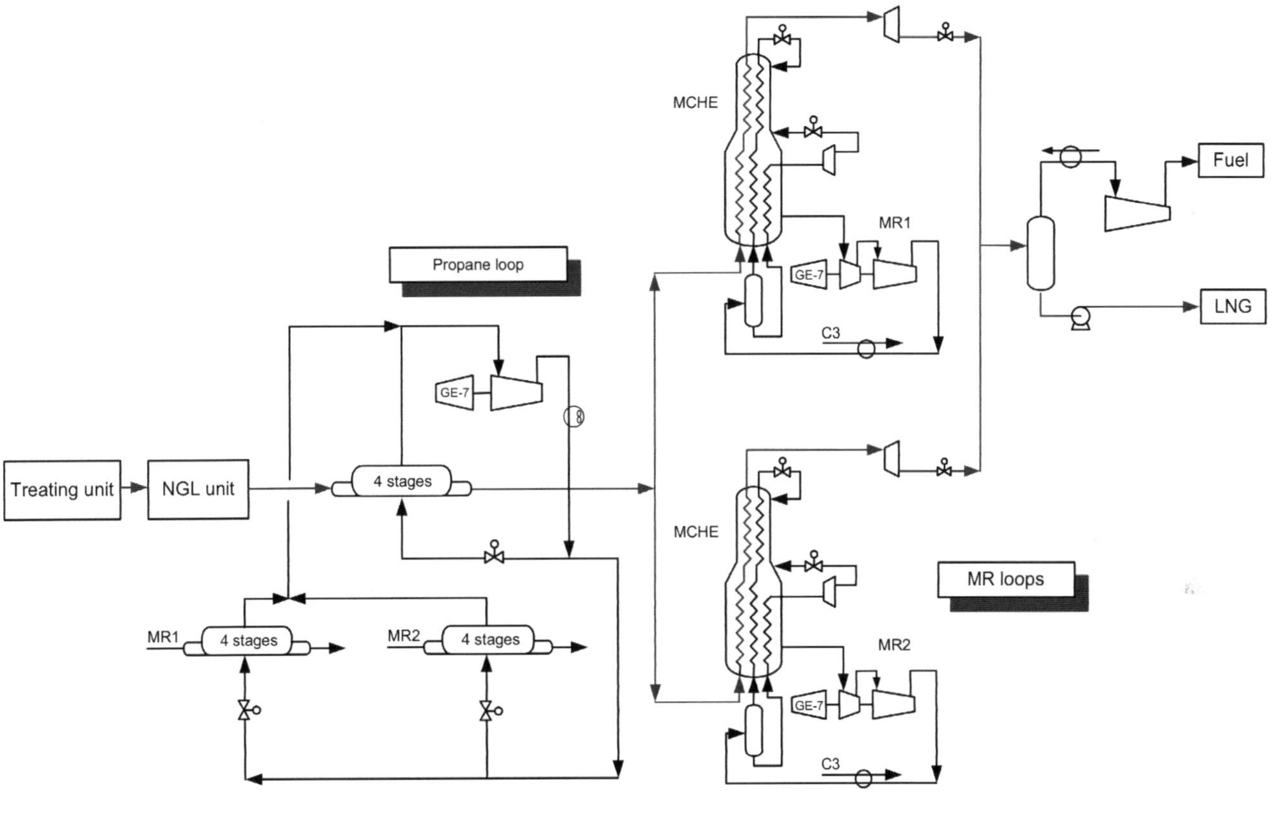

Figure 6: Simplified process flow scheme of the Shell PMR process, with front end NGL extraction (case 2).

The requirements for the NGL extraction unit are set by the targeted LNG product quality. To produce LNG with a heating value of 1070 Btu/scf, a propane recovery of > 90% is required. This high recovery level cannot be achieved in a scrub column line-up due to the poor separation characteristics at higher pressure. In the NGL extraction unit the pressure is reduced providing better separation. After the NGL extraction, recompression of the lean natural gas is done in two steps. The first compression step is coupled to the natural gas expander. Next a booster compressor is added to raise the pressure to the liquefaction pressure.

The production of 8 Mtpa per annum can be achieved with 3 GE F7 machines and a small propane cycle in the up-front NGL extraction unit. Heat integration within the NGL extraction unit and booster compressor system can be applied to off-load the main propane cycle and boost the capacity. Alternatively, application of larger compressors driven by for example GE-F9 or Siemens V84.2 can achieve production capacity well above 8 Mtpa.

As for Case 1, the cut point temperature between the lean gas pre-cooling cycle and the liquefaction cycles enables balancing the available power in pre-cooling and liquefaction cycles.

Synergies between the NGL extraction unit and the PMR process lead to an overall cost effective scheme.

- The scrub column that is normally required to meet the C5+ specification in LNG has been replaced by the NGL demethanizer that operates at lower pressure than the original scrub column. The NGL fractionation section will also produce ethane, propane and butane for refrigerant make-up, if required. No separate fractionation unit for refrigerant production is therefore required.
- Due to the higher quantities of propane and butane produced compared to a regular refrigerant make-up section, the required heat for the re-boilers has increased accordingly. Waste heat from the gas turbine exhaust can be used as re-boiling medium.
- Maximizing the power input to the NGL extraction unit propane cycle by transferring more cooling duty from the lean gas pre-cooling circuit to the NGL pre-cooling circuit can enhance the LNG production to well above 8 Mtpa.

Advantages of the Shell PMR process

The Shell Parallel Mixed Refrigerant technology for large LNG trains has a number of appealing advantages.

The processes as described in the case study are robust through the application of well-proven equipment (e.g. spool wound heat exchangers, proven rotating equipment) without requiring further scale-up. In fact the main equipment in the PMR process is already in operation in plants like Nigeria LNG, Australia LNG and Malaysia LNG.

The parallel line-up of the liquefaction cycles improves the reliability of the train since the LNG production can be designed to continue at 60% of the train capacity when one of the liquefaction cycles trips. Moreover, it allows high production capacity with only two refrigeration cycles in series, compared to three in most other large train concepts. Table 2 illustrates this; The reliability (fraction of time that the unit is working when required) of a single compressor/driver combination is taken as 99.5%. Note that this is a statistical value, used here for a theoretical analysis. Actual reliability is determined by trip frequency and time to restart and depends on factors like design, operating experience, type of compressor/turbine. The reliability/availability of other equipment than the compressor/driver combinations is not considered for the calculation shown in this table. The reliability (expressed as full production equivalent) of the parallel strings in the PMR process is slightly higher than that in the single string processes, since production is assumed to continue at 60% of the train capacity when one of the strings trips. Overall the unscheduled down time of the PMR process, expressed as full production equivalent, is up to 0.6 % less than processes that have three mechanically driven refrigerant compressors in series, when only the reliability of the compressor/driver systems are considered (Table 2). In practice reliability of other equipment will also influence train availability, and detailed analysis is required to determine the unscheduled downtime of the train. The time required for scheduled maintenance also affects the overall train availability and varies depending on the equipment installed.

Table 2: Theoretical reliability of compressor/driver systems in series

Line-up	Reliability of compressor/driver system [% full production equivalent]			Unscheduled down time [% full production equivalent]
	Cycle 1	Cycle 2	Cycle 3	
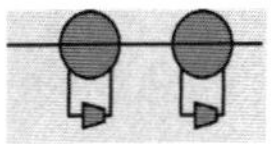	99.5	99.5	n.a.	1 % (~ 3.6 days)
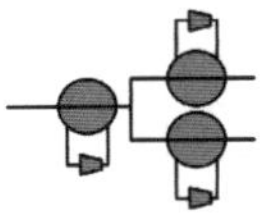	99.5	99.6	n.a.	0.9% (~3.3 days)
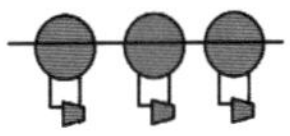	99.5	99.5	99.5	1.5% (~5.5 days)

Because of the PMR line-up and the application of Splitpropane technology the installed power of three GE-F7 drivers can be utilized fully. Since these machines form a large part of the liquefaction unit cost, this makes PMR designs very cost-effective. In the conventional C3/MR process this cannot be achieved as the inherent limitation in the lowest cut-point temperature of the propane cycle (close to the atmospheric boiling point of propane), limits the amount of power that can be put into this cycle efficiently.

The Shell PMR process has a high efficiency through the use of two very efficient refrigeration cycles. The parallel line-up reduces the pressure drop in the system, which also helps improve efficiency. Another advantage is the absence of a third cycle with associated efficiency losses due to temperature approaches in the cooling of the third cycle refrigerant. Internal comparison of different large-train processes for similar conditions and design premises has shown that the Shell PMR process has an efficiency that is up to 10% better than alternative processes.

Conclusions

For the premises of this study, a tropical climate, air-cooling, and the use of GE Frame 7 drivers, an LNG production capacity of 8 Mtpa can be achieved with the Shell PMR process. This production capacity can be achieved without step-outs in main equipment or process technology. The Shell PMR process can be designed to produce different grades of LNG.

Moderate, in-train extraction of LPG's is sufficient to produce LNG that is suitable for the European and North American Market (in the latter case some processing at the import terminal may be necessary) from associated gas that is rich in LPG. The installed driver power can be utilized fully by changing the cut-point temperature between the pre-cooling and liquefaction refrigerant cycles. An extended end-flash enables the production of the target capacity of 8 Mtpa.

Deep extraction of LPG's is required to make an LNG product directly suitable for North American import terminals. This is achieved through front end LPG extraction unit with a second propane cooling cycle. Maximizing heat integration between the NGL extraction unit and the Natural Gas pre-cooling cycle can enhance the LNG production to well above 8 Mtpa.

Application of larger gas turbines, such as GE-F7 or Siemens V84.2, is considered for both schemes, to achieve production capacities above 8 Mtpa.

References

Dam, W., Ho, S.-M., "Unusual design considerations drive selection of Sakhalin LNG plant facilities", Oil & Gas Journal, 58-69, October 1, 2001

Del Vecchio, J, De Jong, E, "The Mariscal Sucre LNG Project – a downstream perspective", 4th Topical conference on Natural Gas Utlization, AIChE Spring National Meeting, New Orleans, 2004

Grootjans, H.F., Klein Nagelvoort, R., Vink, K.J., "Liquefying a stream enriched in methane", US Patent, No. 6,370,910, 2002

Grootjans, H.F., "Compression apparatus for gaseous refrigerant", US Patent, No. 6,637,238, 2003

Heyman, E., Bliault, A., Pek, B., "The LNG GameChanger^{TM} Technology", GASEX 2002

Jager, M., Mahony, S. Van Driel, A, "The key to success", Hydrocarbon Engineering, December 2004.

F. Kleiner, S. Kauffman, "All Electric Driven Refrigeration Compressors in LNG Plants Offer Advantages", Gastech March 2005, Bilbao, Spain

Klein Nagelvoort, R., "Plant for liquefying natural gas", United States Patent 6,389,844, 2002

Nibbelke, R., Kauffman, S., Pek, B., "Liquefaction process comparison of C3MR and DMR under tropical conditions", GPA March 2002, Dallas, Texas.

Paradowski, H. and Hagyard, P., "Comparing five LNG processes", Hydrocarbon Engineering October 2003.

A Technical and Economic Comparison between 9% Nickel and Cryogenic Concrete for Primary LNG Containment.

Paul Vincent Sullivan
Whessoe Oil & Gas
Houston, Texas

AIChE Spring Meeting, April 2005
5th Topical Conference on Natural Gas Utilization
Atlanta April 10-14, 2005

ABSTRACT

Whessoe Oil & Gas is the licensor of the Cryocrete system for LNG containment. Whessoe's Cryocrete system is the only concrete LNG primary containment system with a proven track record in the industry.

Previously primary concrete containment appeared to be uncompetitive with 9%Nickel steel in tank sizes under 100,000m3 (2bcf). With the staggering rise in steel and Nickel prices over the last 18 months, this assumption may no longer be true. Additionally the shortage of suppliers and the extending delivery times for 9%Ni may be further eroding it's position as the material of choice for LNG primary containment.

The paper will consider the following issues:

- A technical comparison of the two materials.
- An engineering comparison of the strengths and weaknesses of the two materials.
- An economic comparison of the two materials for primary containment at various volumes, to establish a basis for selection.
- A schedule comparison for tanks using both materials.
- Constructability issues.

Modeling LNG Pool Spreading and Vaporization

Paper Prepared for AIChE Spring Meeting, Atlanta, GA, April 10 – 14, 2005

Paper number: 102g

Authors: S. R. Saraf and G. A. Melhem
ioMosaic Corporation
93 Stiles Road
Salem, New Hampshire 03079

Modeling LNG Pool Spreading and Vaporization

S. R. Saraf[1] and G. A. Melhem
ioMosaic Corporation
93 Stiles Road
Salem, New Hampshire 03079

Abstract

Liquefied Natural Gas (LNG) vapor dispersion analysis is heavily influenced by the estimation of the source term: (a) the LNG (liquid) leak rate and duration, and (b) the pool spreading and vaporization. A sophisticated dispersion model will produce the wrong answer if the source term used is in error.

This paper discusses the validation of SuperChems™ Expert's dynamic liquid pool model for simultaneous spreading and vaporization. The model is capable of performing multi-component analysis to account for differences in LNG composition. LNG spills can be modeled on various surfaces including water.

SuperChems Expert's dynamic pool model is shown to accurately reproduce field data for LNG as well as other liquids.

---

[1] Author to whom correspondence should be addressed
saraf@iomosaic.com
Ph. 603-893-7009, Fax 603-893-7885

Introduction

Consequence analysis of LNG hazards requires modeling of liquid spills on land and water surfaces. The liquid spill duration, release rate, and rate of vaporization are important factors for the estimation of safe separation distances for flammable vapor dispersion and thermal radiation from pool fires. Dispersion distances calculated to ½ LFL are required for the siting of LNG facilities.

For an LNG spill, NFPA 59A (2001) requires facilities to have impoundments be sited such that the average concentration of methane in air does not exceed 50% of its lower flammability limit (LFL) beyond the property line[1]. The federal code of regulations for siting LNG facilities in 49 CFR Part 193 allows the use of both DEGADIS and FEM3A[2] for dispersion analysis. DEGADIS is a refined box model while FEM3A is a computational fluid dynamics (CFD) model. There is no defined model for calculating the source term.

LNG dispersion analysis is heavily influenced by the estimation of the source term: (a) the leak rate and duration, and (b) the pool spreading and vaporization. LNG vapors are heavier than air. They usually form low, pancake shaped clouds which spread due to their own density even in the absence of wind. Since the heavy gas dispersion behavior of LNG is different from buoyant clouds, numerous models have been proposed to mathematically represent heavy gas dispersion. The proliferation of dispersion models has resulted in several good dispersion models. DEGADIS and FEM3A are two such heavy gas dispersion computer programs recommended for calculating flammable vapor dispersion distances.

The applicability of a mathematical model generally depends on the degree to which important physical phenomenon are represented. In that regards, FEM3A is least limited by various approximations but at the same time needs significantly more computational time. FEM3A is free of surface geometry constraints and can treat obstacles.

However, in general, the quality of predictions of dispersion models is very similar, as demonstrated in Table 1 and Table 2. The selection of dispersion models does not appear to significantly affect the quality of results.

A sophisticated dispersion model will produce the wrong answer if the source term (vaporization rates and duration) used are in error. We need to establish realistic, accurate, and prudent estimates of the vaporization rates. Fortunately, and despite several factors that are difficult to quantify such as wave action for spills on water, we can use existing methods to establish safe estimates of pool spreading and dynamic vaporization rates.

This paper discusses the theoretical basis for SuperChems' dynamic pool spreading model and details the performance of SuperChems against several field data sets for a wide variety of fluids including LNG.

Table 1. Comparison of measured field peak concentrations to model predictions[3]

	Maximum downwind extent of the LFL. (in meters)			
	Field Data	Germeles-Drake (Box)	SLAB (refined box)	FEM3 (3D)
Burro 3	255	126	215	199
Burro 7	200	150	264	210
Burro 8	420	661	418	630
Burro 9	325	235	315	330

Table 2. Comparison of measured field peak concentrations to model predictions [4]

	Ratio of Predicted/Experimental Peak Concentrations			
	FEM3A (3D)	SuperChems (Box)	DEGADIS (Refined Box)	Gaussian
Falcon 4	0.7 – 1.35	N/A	N/A	N/A
Burro 8	0.6 – 1.14	1.1 – 1.8	0.63 – 2.55	1.14 – 2.65
Burro 9	1.16 – 1.34	0.8 – 1.84	0.8 – 3.54	0.9 – 1.45

SuperChems Overview

SuperChems contains advanced software for the estimation of single and multi-phase flows, dispersion, chemical reaction systems, fire, and explosion dynamics. SuperChems was developed in 1988 by Arthur D. Little Inc. Its models have been the subject of many publications and validation studies with field and incident data. SuperChems is used by more than 250 users worldwide.

The development of flow models in SuperChems is overseen by a Technical Steering Committee of the American Institute of Chemical Engineers[2]. In July 2002, the American Institute of Chemical Engineers (AIChE) Design Institute for Emergency Relief Systems Users Group (DIERS) adopted SuperChems to replace DIERS' existing software for designing emergency relief systems. AIChE currently markets its own version of SuperChems, called SuperChems for DIERS which is a subset of SuperChems Expert that is primarily focused on single and multiphase flow and relief systems dynamics.

As part of a New Jersey Toxic Catastrophe Prevention Act (TCPA) program, the New Jersey Bureau of Release Prevention determined that SuperChems constitutes a set of models which meet the TCPA regulations in complying with risk assessment requirements.

---

[2] http://www.iomosaic.com/iomosaic/products/software/superchems_user_group.html

SuperChems has also received favorable review and acceptance from several local and regional regulatory agencies in California. SuperChems Expert is currently used by the California South Coast Air Quality Management District. SuperChems is also used by agencies internationally such as Transport Canada, and most of the leading universities in the United States and Canada with active process safety programs in Chemical Engineering.

Details of SuperChems' Dynamic Pool Spreading and Vaporization Model

SuperChems' dynamic pool model is based on numerical solutions of the conservation equations for a liquid pool[5], as illustrated in Figure 1, for a liquid pool of radius "r" and thickness "z". A mass balance is written for the ith component in the liquid pool mixture such that:

$$Accumulation = Input \; - \; Evaporation \; - \; Percolation / Dissolution$$

$$\frac{dn_i}{dt} = \frac{dn_{i,in}}{dt} - \frac{dN_{i,out}}{dt} - \frac{dn_{i,out}}{dt}$$

$$\frac{dN_{i,out}}{dt} = \pi \; r^2 \; w_i$$

Where n is the liquid number of moles (kmol), N is the vapor number of moles (kmol), and w is the liquid regression rate (vaporization rate) in kmol/m2/s. Usually, measured evaporation/ vaporization data is reported as liquid regression rate. The liquid regression rate is a function of composition, the spill surface type, the spill surface thermal properties, and time.

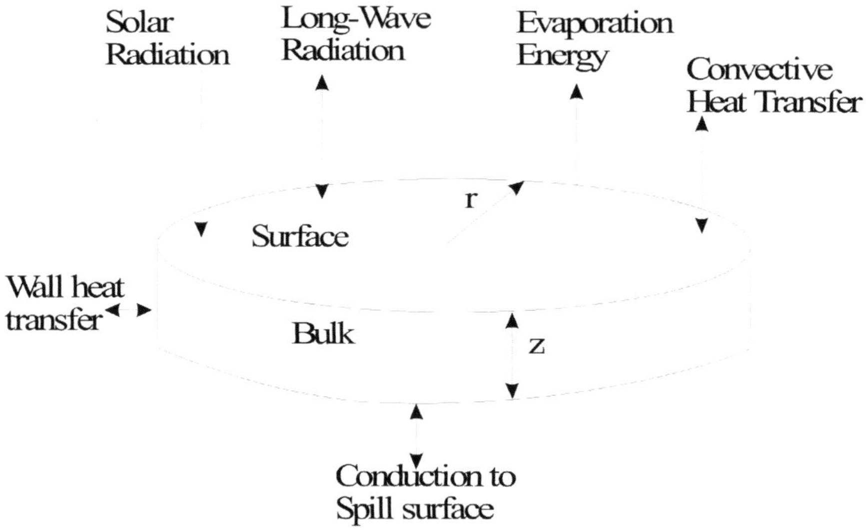

Figure 1. Liquid pool representation used in SuperChems

SuperChems uses two energy balances to dynamically solve (as a function of time) for the liquid pool surface temperature, bulk liquid temperature, and the underlying surface temperature. All excess energy is converted into boiling energy when the bulk liquid temperature reaches the composition-averaged bubble point of the liquid pool mixture. Note that the bubble point and liquid pool composition are changing with time as light ends are depleted, i.e. vaporize first. The energy balances account for all possible modes of heat transfer for a liquid pool. For cryogens, conductive heat transfer from the substrate usually dominates. SuperChems uses a spill / spill surface correction factor, ϕ[3], to account for surface irregularities for spills on land, and for interfacial distortions leading to increased surface area for spills on water. ϕ is determined based on field data for a variety of spill surfaces including water. A porous dry soil will have a higher effective heat transfer area than highly compacted soil or moist soil.

The extent of spreading of a liquid pool is a key factor which affects the overall emission rate. Larger pools yield higher emission rates. The spreading of liquid pool in SuperChems is described using numerical solution of mass and momentum conservation equations based on Weber's[6] formulation of the shallow water equations. For a liquid spill spreading symmetrically on a flat surface, Weber simplified and integrated shallow water equations to describe the bulk properties of liquid pool as follows:

$$\frac{dn}{dt} = \frac{dn_{in}}{dt} - \frac{dN_{out}}{dt} - \frac{dn_{out}}{dt}$$

$$\frac{du}{dt} = \frac{4\Delta(1-s)z}{r} - F$$

$$\frac{dr}{dt} = u$$

where

u – liquid radial spreading velocity

F – viscous effects resistance

$\Delta = g =$ acceleration due to gravity due to spills on land

$\Delta = g\left[1 - \frac{\rho_l}{\rho_w}\right] =$ acceleration due to gravity reduced by the relative density differecne for spills on water

As the liquid pool spreads three flow regimes are recognized:

1. Gravity – inertia
2. Gravity – viscous
3. Surface tension – viscous

In regime 1, the gravity force and liquid inertia are dominant. In regime 2, the viscous drag becomes important and must be taken into account. In regime 3, the surface tension replaces gravity as the driving force. Cryogenic liquids spills rarely reach the second

[3] A heat transfer area enhancement factor

regime because they vaporize quickly. The third regime is important for spills of heavy hydrocarbons on water.

The dynamic simultaneous liquid pool spreading and vaporization equations are solved for every time step leading to concentric annular rings of the spilled liquid to be in contact with the substrate for different periods. For cryogenic spills, the heat exchanged between the bulk liquid and the substrate results in cooler surface temperatures and consequently reduced vaporization rates. Furthermore, SuperChems accounts for the film boiling and the transition from film to nucleate/pool boiling.

SuperChems successfully captures the underlying physics involved in pool spreading and vaporization for cryogenic and non-cryogenic spills alike. Technical details of the pool model equations can be found in the SuperChems reference manual[5].

Model Validation

The following sections detail the performance of SuperChems against several field data sets including a wide variety of liquids as well as LNG.

1. Drake and Reid's LNG Experiments[7]

Drake and Reid conducted an experimental program to study the effect of soil type, moisture content, and LNG composition on boiling rates of LNG. A cylindrical well, 0.18 m in diameter and 0.22 m deep, was cut into a 0.31 m Styrofoam cube. Soil samples were packed to a depth of 0.05 – 0.06 m at the bottom of the well. Approximately 0.27 – 0.4 kg of cryogen was spilled in 2 – 3 seconds. The experiments were performed with the following cryogens:

1. Liquid nitrogen
2. Liquid methane
3. A mixture of 93 % methane and 7 % ethane

The entire test apparatus was mounted on a load cell to monitor changes in the system mass. The boiling rate of cryogens appears to increase with lower moisture content because when water freezes it plugs the pores thereby reducing the surface area available for heat transfer. Reduced boiling rates were observed when a plastic barrier was used to prevent percolation in the soil. The boiling rates observed for LNG were higher than pure methane. Figure 2 and Figure 3 provide a comparison of experimental and predicted data for liquid methane and LNG spills respectively.

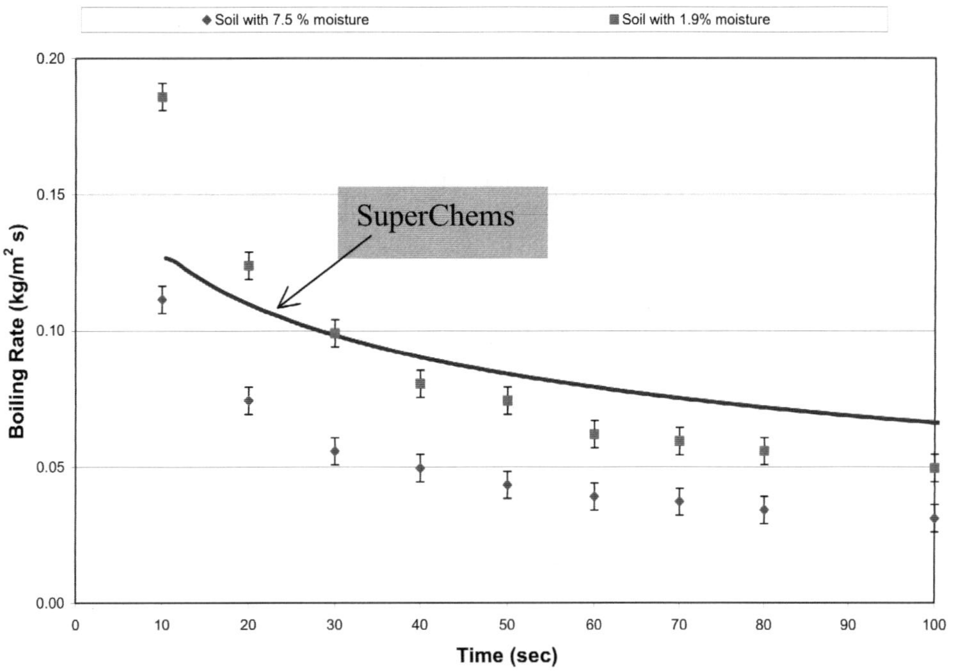

Figure 2. Liquid methane vaporization data

Simulation details: Spill rate – 0.04 kg/s; Spill duration – 10 sec.; Spill area – 0.025 m^2; Spill composition – Liq. Methane

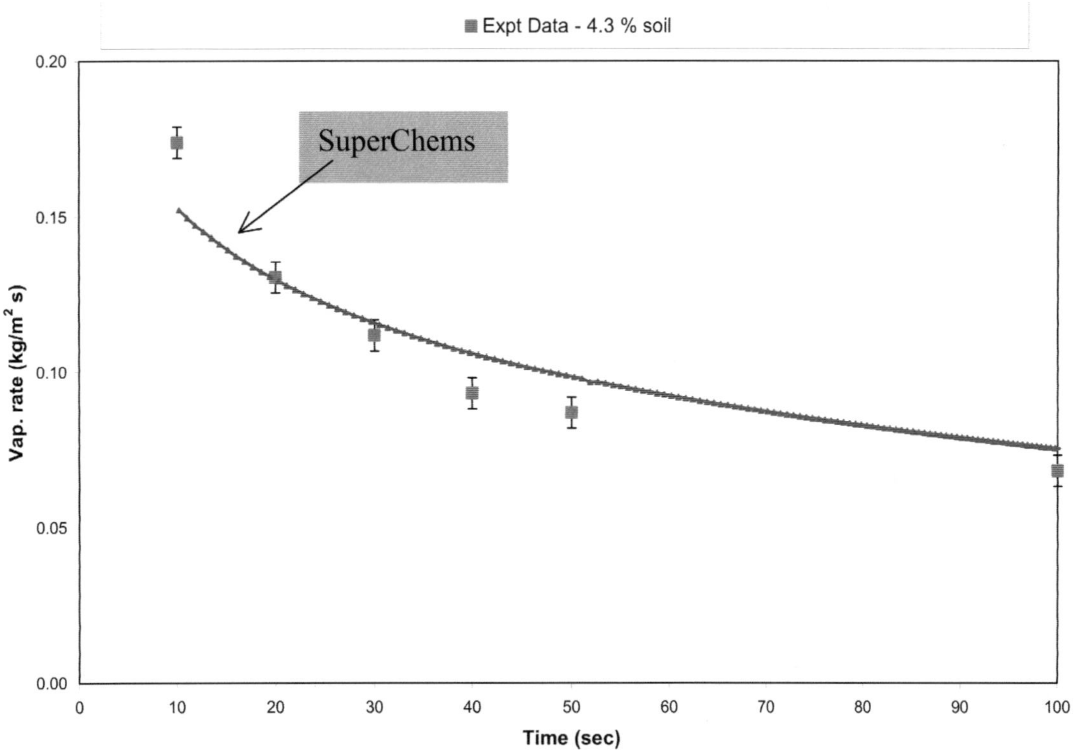

Figure 3. LNG (93 % methane, 7 % ethane) vaporization data

Simulation details: Spill rate – 0.04 kg/s; Spill duration – 10 sec.; Spill area – 0.0254 m2; Spill composition – 93% Methane, 7 % ethane.

2. MIT LNG Research Center Data[8]

Boiling rates of LNG and liquid methane on various dike floors were measured at the MIT LNG research center for GRI study. In this report, tests 38 and 40 are chosen for data validation, and details of the tests are summarized below. A comparison of predicted and experimental data is shown in Figure 4 and Figure 5.

Test no.	Test 40	Test 38
Spill mass (kg)	**0.663**	0.875
Spill composition (mol %)	**99 % methane, 1 % ethane**	99 % methane, 1 % ethane
Spill area (m^2)	**0.1**	0.1
Substrate	**Low compact soil**	Sand

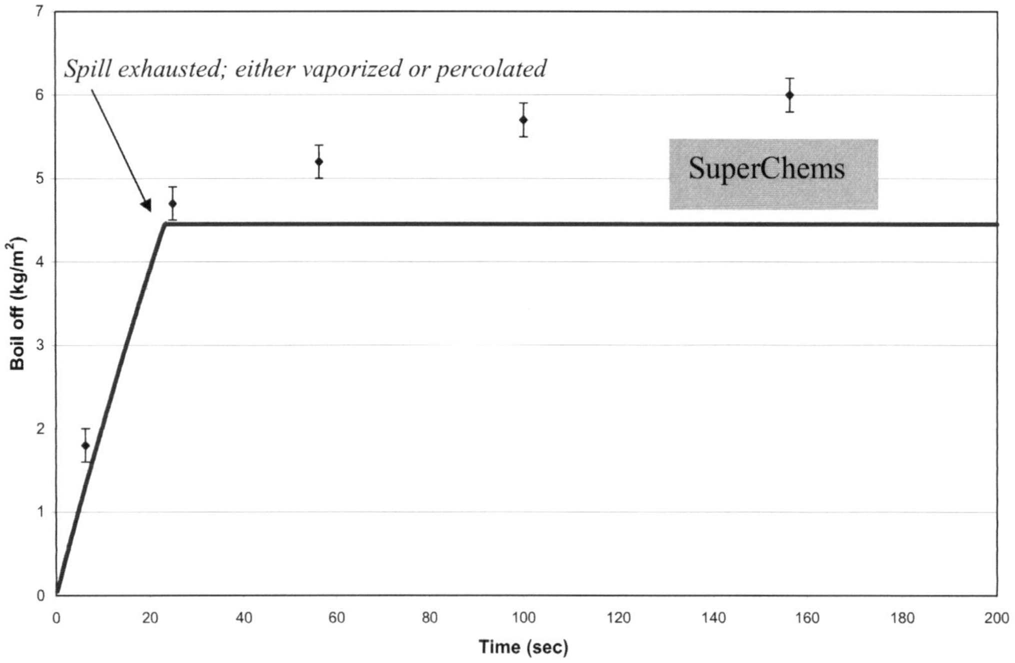

Figure 4. Test 40

Note: There is an inconsistency among the reported experimental data since the reported vaporized mass is greater than the reported spill mass.

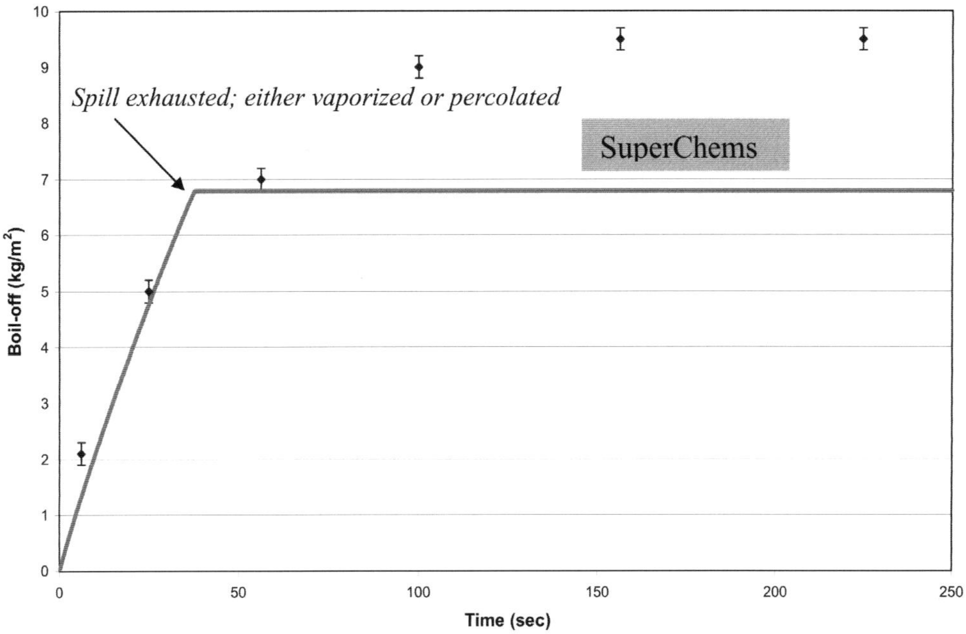

Figure 5. Test 38

Note: Although sand was used, rates similar to soil were obtained. Soil heat transfer properties were used during simulation. There is an inconsistency among the reported experimental data since the reported vaporized mass is greater than the reported spill mass

3. Bureau of Mines LNG Spills on Water[9]

The U.S. Bureau of Mines conducted 13 LNG spills onto water in strip mine lake near Florence, PA. For measuring vaporization rates, up to 2 x 10^{-3} m^3 of LNG was poured from a height of 0.46 m (1.5 ft) into a bucket containing 22.7 m^3 of water. Based on the weight loss and a heat of vaporization value of 577.4 kJ/kg, a 20-second average vaporization rate of 0.155 kg/m^2 s is reported.

Table 3. Liquid regression rates for LNG spills on water[4]

Surface	*Regression rate (m/s)*
Water	4.34 x 10^{-4}
Water	3.72 x 10^{-4}
Aluminum	1.74 x 10^{-4}
Water	4.68 x 10^{-4}
Water	0.57 x 10^{-4}
Water	4.68 x 10^{-4}

SuperChems predicts a 20 second average vaporization rate of 0.155 ± 0.016 (2 σ) kg/m^2 s. The model predicts a steady state mass flux of around 0.1 kg/m^2s, which is consistent with literature values.

SuperChems predicts an average regression rate of 3 x 10^{-4} m/s and a maximum regression rate of 6.3 x 10^{-4} m/s for the Bureau of Mines tests. This is in agreement with the reported values in Table 3, which summarizes regression rates for LNG spills on water from different experiments.

1. Welker[10] LPG Data

Welker conducted numerous experiments to quantify the vaporization rate of propane pools. In test P-15, ~ 18.2 kg of propane was spilled into a 0.47 m^2 (5 ft^2) perlite concrete pit in 15 seconds. Ambient wind speed was 5.1 m/s. Figure 6 presents a comparison of the mass of propane remaining in the pit as a function of time.

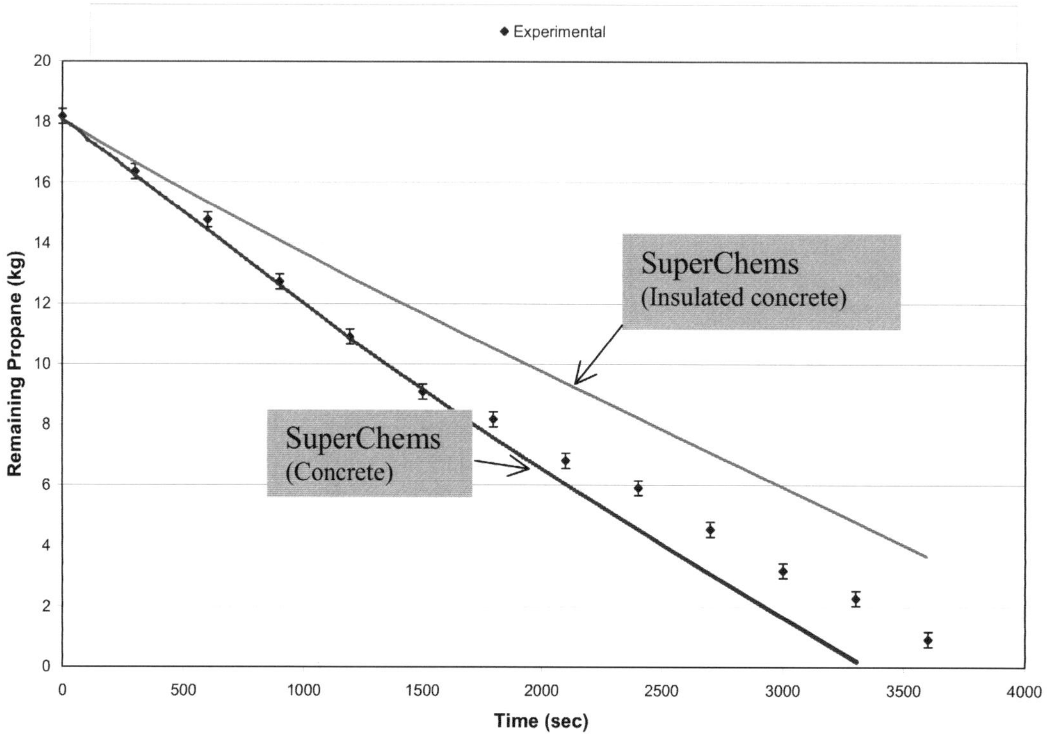

Figure 6. Weight of propane remaining in the pit

1. Mikesell et al[11] Ammonia-Water Non-Ideal System

2240 grams of 28.8 % by weight aqueous ammonia was poured into a 18.8 x 28.8 x 4.76 cm plastic pan. The ambient temperature was 25 °C, with a prevailing wind of 1.59 m/s and 59 % relative humidity. Figure 7 compares the predicted values for ammonia weight fraction in the pool and the total amount of ammonia evaporated from the pool as a function of time to experimentally measured values. SuperChems also successfully predicts the bulk and surface temperatures measured for the pool as shown in Figure 8.

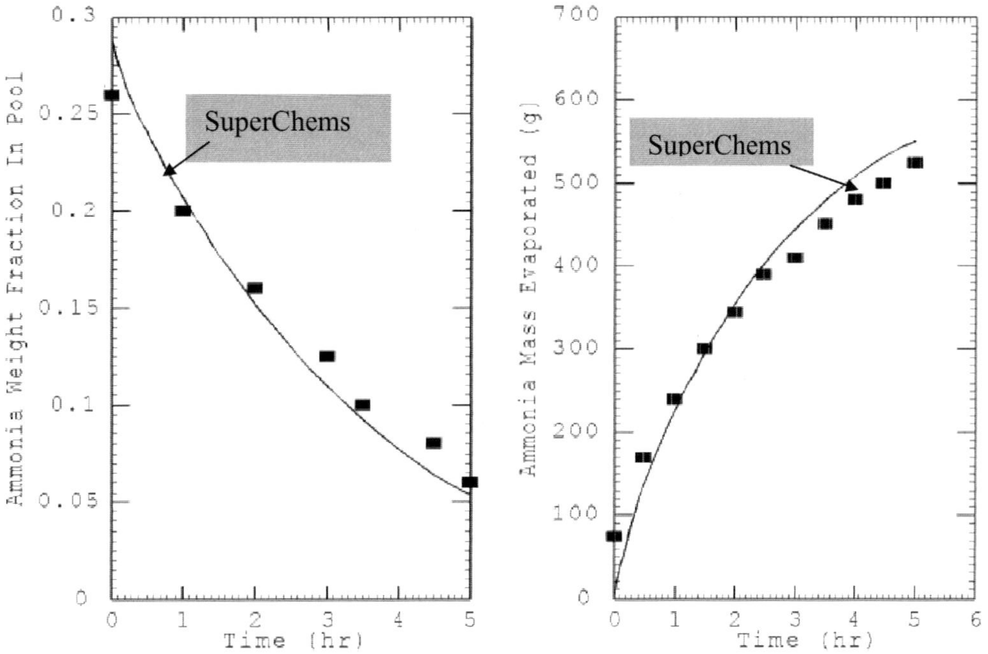

Figure 7. Ammonia weight fraction in pool and total mass evaporated

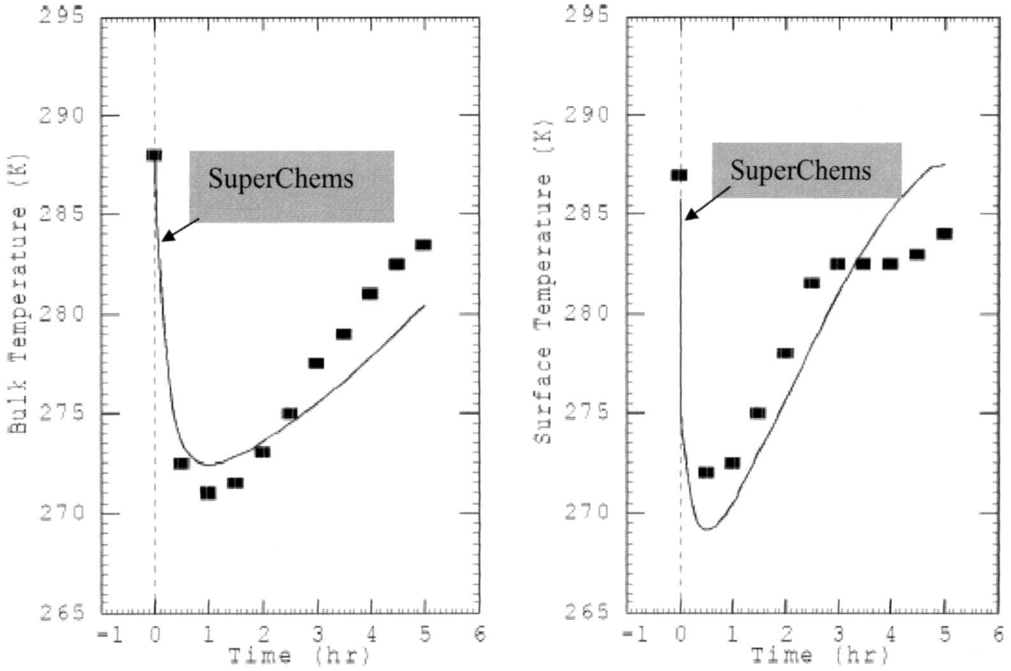

Figure 8. Liquid pool temperature histories

1. Merck[12] Organic Solvents (Methanol, Acetonitrile, and Toluene)

A glass cylinder 5 cm in diameter and 1.8 cm deep was used for the Merck study. The cylinder was placed in a chamber to control temperature and humidity. The effect of wind speed was studied by changing the fan speed in the chamber. The predicted results are in excellent agreement with test data as shown in Figure 9.

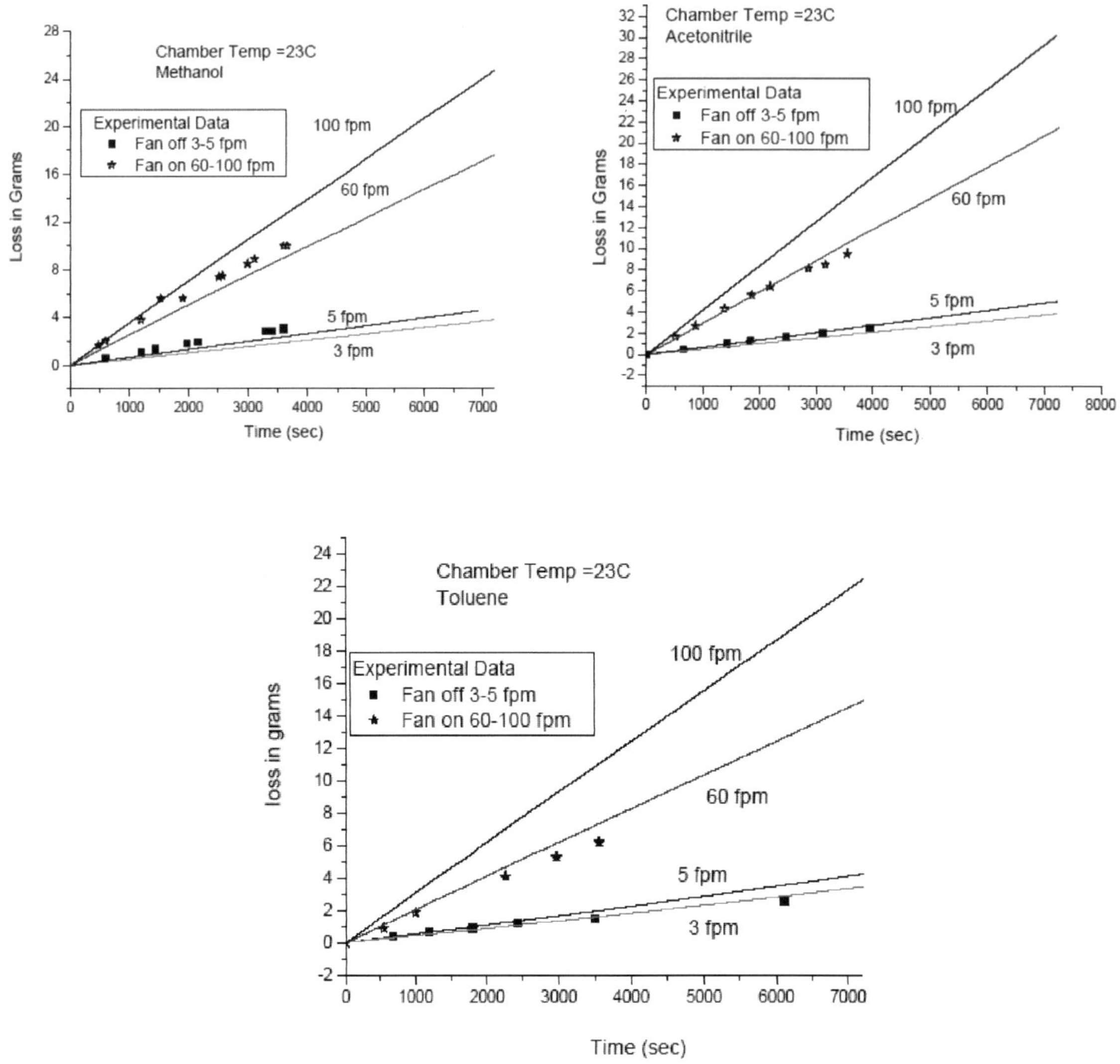

Figure 9. SuperChems Predicted vaporization data for organic solvents

Note: The solid lines represent results from SuperChems^{TM}.

Conclusions

A prudent estimation of hazard zones requires a validated pool spreading and vaporization model. The SuperChems dynamic simultaneous spreading and vaporization liquid pool model represents all significant liquid pool heat and mass transfer mechanisms, vapor-liquid equilibrium, and the radial spreading of the liquid. It successfully captures the underlying physics involved in pool spreading and vaporization as evident by its performance against measured data for a wide variety of fluids. The pool model can be used to study liquid spills on different surfaces including water. The SuperChems model has been successfully validated against experimental data for cryogens and non-cryogens, and is commercially available through ioMosaic Corporation.

References

1. NFPA 59A: Standard for the Production, Storage and Handling of Liquefied Natural Gas (LNG), 2001 Edition.
2. 49CFR 193 – Liquefied Natural Gas Facilities: Federal Safety Standards, Sept. 2004
3. D. L. Ermak, "LNG Vapor Dispersion Research at LLNL: Remaining Questions", UCRL-87668, May 1982.
4. G. A. Melhem, "LNG Release Assessment", Arthur D. Little Draft Final Report to the U.S. Department of Transportation (DOT), Ref. 61230-30, February 1991.
5. SuperChems Expert, Version 5.81, Reference Manual, ioMosaic Corporation, 2005.
6. D. M. Weber and S. J. Jones, "A model of spreading vaporizing pools", International Conference on Vapor Cloud Modeling, Center for Chemical Process Safety (CCPS), AIChE, 1987, pages 226–250.
7. E. M. Drake and R.C. Reid, "How LNG Boils on Soils", Hydrocarbon Processing, 54 (5), 191 - 194, May 1975.
8. R. C. Reid, "Boiling of LNG on typical dike floor materials", GRI-79/0026, Mar. 1980.
9. D. Burgess, J. Biordi, and J. Murphy, "Hazards of Spillage of LNG into Water", U.S. Dept. of the Interior, Bureau of Mines, Pittsburgh, PMSRC report no. 4177, 1972.
10. J. R. Welker and W.D. Cavin, "Vaporization, dispersion, and radiant flux from LPG spills", DOE/EV/66-020-T7, 1981.
11. J. L. Mikesell, A. C. Buckland, V. Diaz, and J. J. Kives, "Evaporation of contained spills of multicomponent nonideal solutions", International Conference and Workshop on Modeling and Mitigating the Consequences of Accidental Releases of Hazardous Materials, CCPS, AIChE, 1991, pg. 103–125.
12. D. Muzzio, "Modeling Experimental Evaporation Rates of Organic Solvent", Merck and Co., Presented at DIERS User Group Meeting, April 2003.

About the Authors:

Georges Melhem

Dr. Georges A. Melhem is president and chief executive officer of ioMosaic Corporation. He is an internationally known pressure relief design, chemical reaction systems, and fire and explosion dynamics expert. In this regard he has provided consulting and design services, expert testimony, and incident investigation support and reconstruction for a large number of clients. Dr. Melhem has participated in large scale LNG studies, LNG risk assessment projects, and LNG public testimony.

Dr. Melhem was president of Pyxsys Corporation; a technology subsidiary of Arthur D. Little Inc. Prior to Pyxsys and during his twelve years tenure at Arthur D. Little, Dr. Melhem was a vice-president and managing director of Arthur D. Little's Global Safety and Risk Management Practice and its Process Safety and Reaction Engineering Laboratory.

Dr. Melhem holds a Ph.D. and an M.S. in Chemical Engineering, as well as a B.S. in Chemical Engineering with a minor in Industrial Engineering, all from Northeastern University.

Dr. Melhem is the author/co-author of four books, more than seventy technical papers/presentation, and more than 300 technical reports. He is a member of the American Institute of Chemical Engineers and the American Chemical Society.

Sanjeev Saraf

Dr. Saraf is currently a partner at ioMosaic Corporation. Prior to joining ioMosaic, he graduated with a Ph.D. in Chemical Engineering from Texas A&M University. Dr. Saraf's areas of specialty include emergency relief system design, process safety management, consequence modeling, reactivity hazard assessment, calorimetric studies, and molecular modeling. He is a member of the American Institute of Chemical Engineers (AIChE), American Chemical Society (ACS), North American Thermal Analysis Society (NATAS), and the DIERS Users Group.

Improved Design of Mixed Refrigerant Cycles using Mathematical Programming

Del Nogal, F.L., Kim, J., Perry, S.J. and Smith, R.
Centre for Process Integration
School of Chemical Engineering and Analytical Science
The University of Manchester
Manchester, U.K.

AIChE Spring Meeting, April 2005
5th Topical Conference on Natural Gas Utilization
Atlanta April 10-14, 2005

ABSTRACT

The advantage of mixed refrigerants over pure refrigerants is that they undergo isobaric phase change along a range of temperatures within the dew and bubble temperatures of the mixture. Given the correct refrigerant pressures, flowrate and composition, this allows a good match between the process and refrigerant temperature profiles (an indicator of cycle efficiency) with a simpler configuration than multilevel pure refrigerant cycles. Refrigerant composition becomes an important degree of freedom both at the design stage and during operation.

A new approach for the optimal design of mixed refrigerant cycles is presented. It is based on mathematical programming and offers significant improvements in relation to previous approaches. It includes multistage refrigerant compression, full enforcement of the minimum temperature difference along the temperature profiles, simultaneous optimisation of variables, incorporation of capital costs in the objective function and the use of stochastic optimisation (genetic algorithm) to overcome local optima. The effectiveness of the method is illustrated by revisiting previously published LNG case studies, for which better and feasible solutions are produced. It has also proved the importance of considering multistage compression and capital costs during optimisation. The application of genetic algorithms in the design of mixed refrigerant cycles permits having greater confidence in the optimality of the results.

INTRODUCTION

A major limitation of vapour compression cycles that make use of pure components as refrigerants is that refrigeration is provided at a constant temperature while the cold refrigerant is evaporating. For a chosen refrigerant, the temperature at which cooling is provided is a consequence of the evaporator pressure, i.e. the saturation temperature.

If the hot streams (streams that need cooling) demand the cooling task to be carried out along a wide temperature range, a system providing all the refrigeration at a single level is likely to have a poor performance. This is because large temperature differences would exist in the heat exchanger(s), moving the system away from thermodynamic reversibility, and hence, from thermodynamic efficiency. Being that the case, a multilevel pure refrigerant system is likely to be implemented, in an attempt to reduce the temperature differences in the system, as seen is Figure 1, in which three pressure levels are used to provide refrigeration at three temperature levels. However, both heat transfer area and complexity would increase as a consequence. The refrigerant compressor would need as many stages as refrigeration levels.

Multicomponent, or mixed refrigerants, unlike pure refrigerants, undergo isobaric phase change through a range of temperatures contained within the dew and bubble temperatures of the mixture. Given the right pressures and compositions, a good match between the process and refrigerant temperature profiles can be obtained with a simple configuration, as shown in Figure 2. Although refrigerant compression can take place in several compression stages with intercooling, only one stage is shown in the illustration for simplicity. A practical maximum pressure ratio per compression stage (usually 4-5 for industrial applications) is likely to define the number of compression stages, rather than the number of pressure levels, since in this case the cold refrigerant evaporates at a unique pressure (or through a small pressure range, if friction pressure drop is accounted for in the evaporator). Care should be taken, however, since the reduced average temperature difference would lead to an increased heat transfer area. A minimum temperature difference (ΔT_{min}) should be established for practical purposes.

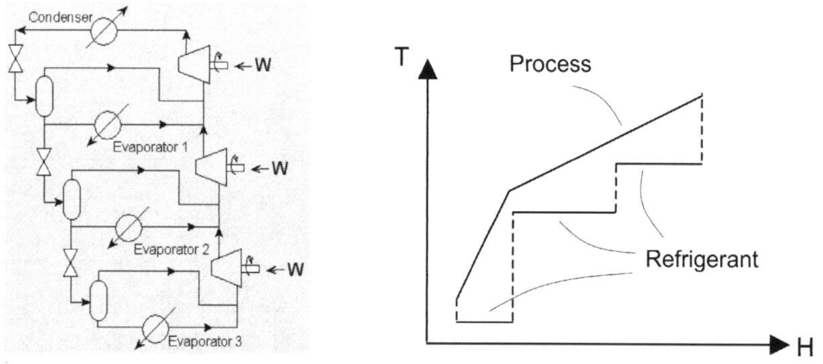

a) Flowsheet. b) Temperature profiles.

Figure 1. A three-level pure refrigerant cycle.

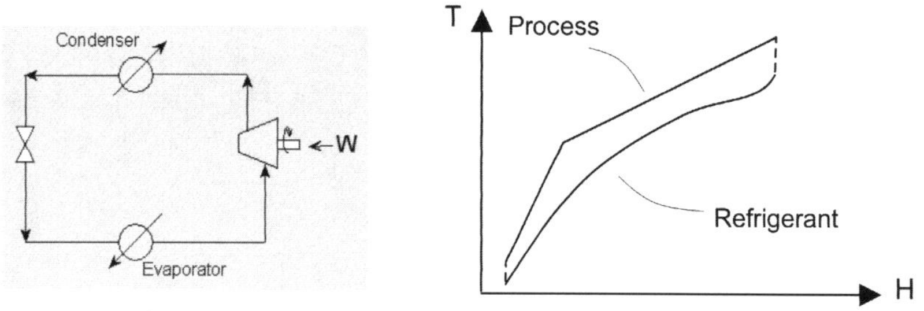

a) Flowsheet. b) Temperature profiles.

Figure 2. A mixed refrigerant cycle.

Since the benefits of using mixed refrigerant cycles are highlighted in situations where refrigeration is to be provided along wide temperature ranges, these types of cycle find a natural application in the LNG industry, where refrigeration is required continuously from ambient temperature to around -160 °C.

Flowsheet variations of mixed refrigerant cycles

An interesting flowsheet variation in mixed refrigerant (MR) cycles is the one shown in Figure 3, where the high pressure refrigerant rejects heat not only to an external heat sink but also to itself, once expanded. In this way the hot refrigerant is further cooled down, probably subcooled, before expanding, allowing the cold refrigerant to reach a lower temperature and/or a lower vapour fraction after the expansion. Such a benefit is achieved at the expense of a higher heat transfer area, since the total duty in the heat exchanger increases significantly. A multistream heat exchanger can be used to handle all hot streams and the cold stream in a single piece of equipment. This process is known as the Pritchard cycle and is described in more detail by Walsh (1993).

Repeated partial condensation and separation of the refrigerant stream has been reported to achieve a better match between the temperature profiles (Finn *et al.*, 1999). Figure 4 shows a mixed refrigerant cycle with three refrigeration stages, as an extension of Figure 3a. Although increasing the number of stages apparently reduces the power consumption, the design also grows in complexity and probably in capital cost. This results in an important trade-off for designers.

Multistage compression

As compressors used in industry have a practical maximum stage pressure ratio of around 4 to 5, it is very common to find compression tasks performed in multiple stages. Further, if there is an appropriate cold utility, the partially compressed gas originating from a given stage may be cooled down before entering the next compression stage (intercooling). The lower temperature of the partially compressed gas reduces the volumetric flowrate and in consequence reduces the compression power of the next stage. If part of the gas condenses, then it is necessary to remove the liquid so that a dry vapour stream enters the next compression stage. The liquid removed is then pumped and mixed with the fully compressed gas and with any liquid streams from other intercoolers at the outlet of the last compression stage at same pressure. Condensate removal and pumping also helps in saving power, as increasing the pressure of liquids is much cheaper in power (and capital) than doing so with gases.

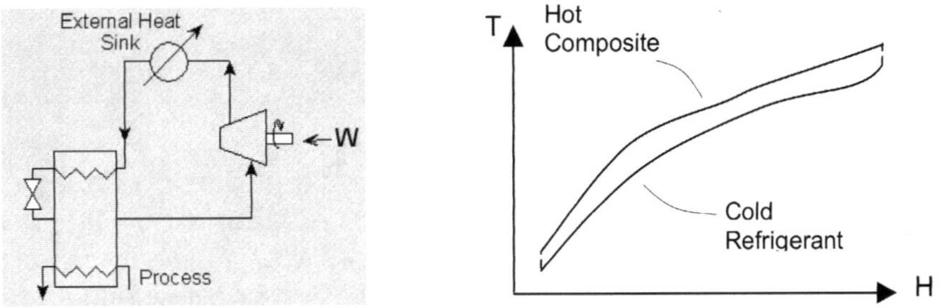

a) Flowsheet. b) Temperature profiles.

Figure 3. A self-cooling mixed refrigerant cycle.

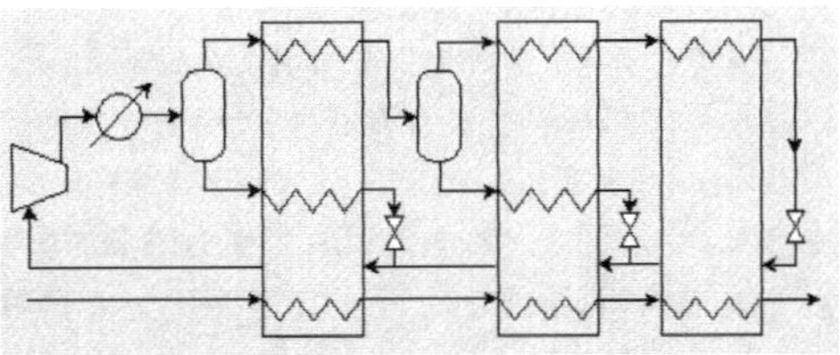

Figure 4. A three-stage mixed refrigerant cycle.

The impact of multistage compression with intercooling is illustrated in Figure 5. A stream, consisting of a mixture of nitrogen and hydrocarbons C1 to C4, is compressed from and to the same conditions both in a single compression stage (Figure 5a) and in three stages with intercooling (Figure 5b). A compressor isentropic efficiency of 80% is assumed. Since the total compression ratio is 26.25 (42 bar/1.6 bar), three stages were necessary in the multistage case so that the stage pressure ratios were reasonably practical (2.97, assuming identical pressure ratios). Condensates are present after the second intercooler.

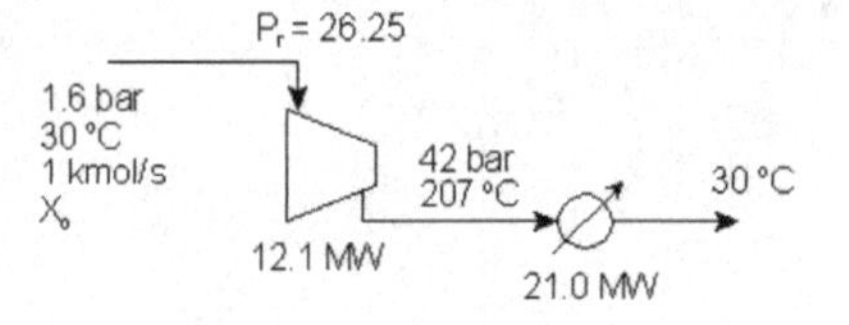

a) Single stage.

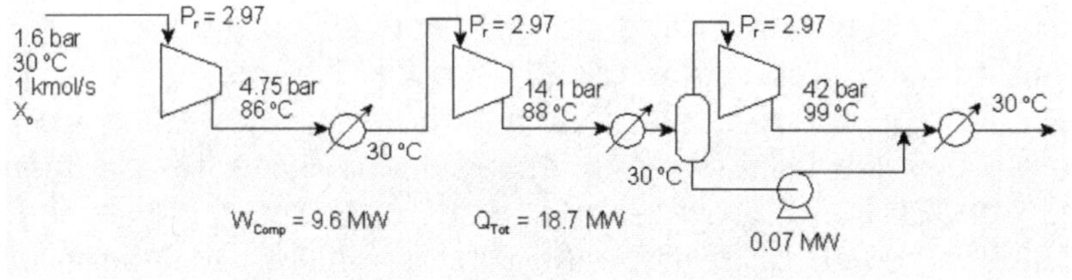

b) Several stages with intercooling.

Figure 5. A compression example.

In Figure 5b, the total compression power was reduced by 21% with respect to Figure 5a. A significant reduction in the total cooling duty (11%) was also achieved. It is worth noting that the total cooling duty will always move in the same direction as the total power for a fixed compression task according to the overall energy balance, since the inlet and outlet enthalpies are fixed.

PREVIOUS WORK

Although there are many publications on the analysis of mixed refrigerant systems, only a few exist on their optimal design using mathematical programming. One of the earliest suggestions on how to synthesise mixed refrigerant cycles was made by Cheng and Mah (1980). Although their work was on pure refrigerant cycles (and heavily based on heuristics), they suggested that their approach could be extended to mixed refrigerant cycles if they were seen as an abstract equivalent of cascaded pure refrigerant cycles. However, the underlying assumptions of such equivalence are very likely to lead to either temperature infeasibilities or non-optimal solutions, as Cheng and Mah themselves warned.

Lee (2001) (part of this work was also published as Lee *et al.*, 2002) worked on the optimal design of multistage mixed refrigerant cycles. Given a refrigeration task in the form of a hot composite curve (a combined temperature-enthalpy profile of all hot streams) and the number of refrigeration stages, his approach allows for the optimisation of key variables in a process flowsheet of the type shown in Figure 4. In principle, the optimisation variables are the refrigerant composition and flowrate (at compressor inlet), the compressor inlet and outlet pressures and the intermediate temperatures (at which refrigerant is separated into vapour and liquid streams). However, the decision variables are not optimised simultaneously. Instead, the refrigerant compositions are optimised at fixed refrigerant flowrate and pressures using non-linear programming (NLP). Once the optimal composition is obtained, the hot and cold temperature profiles are checked for feasibility. If they do not cross, then new refrigerant flowrate and pressures are proposed based on heuristics, judgement or optimisation and the procedure is repeated until no further improvement is possible without incurring temperature infeasibilities.

Lee (2001) also used three different types of objective function. Two of them tried to match the temperature profile of the cold refrigerant to an ideal profile (the hot composite curve shifted down by ΔT_{min}) by minimising either the maximum violation of the minimum temperature difference or the sum of such violations along the profile. The third possible objective function was the minimisation of compressor power, but its use was not recommended until the final part of the overall optimisation. That is because, since there is no constraint on the vapour fraction at the inlet of the compressor, it is less likely to have wetness at the compressor inlet towards the end of the overall optimisation due to a probable lower value of the refrigerant flowrate.

The design of mixed refrigerant cycles is a highly nonlinear problem with many local optima. One of the main drawbacks of Lee's approach is that the non-simultaneous optimisation of variables coupled with the dependence of the final solution on the initial guess, on the non-systematic selection of refrigerant flowrate and pressures (especially if updated using heuristics or judgement) and on the switching of the objective function, makes reaching the global optimal solution (or a good near-optimal solution, for practical purposes) very unlikely.

In Lee's method, the minimum temperature difference between the hot and cold temperature profiles is not fully enforced, which often leads to very tight profiles, apart from relying on human intervention at each iteration to check for temperature feasibility. It is assumed that the refrigerant compression takes place in one stage with no intercooling, regardless of the compression ratio. No capital costs are considered during the optimisation procedure.

Another relevant precedent in the synthesis of multistage mixed refrigerant cycles is that by Vaidyaraman and Maranas (2002), which is also based on NLP. The cycle topology used by them is the one shown in Figure 6a. Although in principle it might look very different, it is just a slight variation of the type of structure in Figure 4, the only difference being that the liquid refrigerant stream is not subcooled after each flash step, as shown in Figure 6b as an alternative representation. The variables being optimised in this case were the refrigerant composition (at the stream being expanded in the last stage), the compressor inlet and outlet pressures and the vapour fraction at flash drums 2 to N. All the design variables were optimised simultaneously.

Two key assumptions were made by Vaidyaraman and Maranas (2002). The first one was that the hot refrigerant leaving stage N was at its bubble point and the second one that the cold refrigerant streams leaving each stage were at their dew point. These assumptions, although not unreasonable, constrain the solution space unnecessarily and could lead to good opportunities missed by the optimiser.

Another major shortcoming of the method proposed by Vaidyaraman and Maranas (2002) is that temperature feasibility is only enforced at the ends of the heat exchangers. This opens the possibility of not only minimum temperature difference violations but also of temperature crossovers not being identified during the optimisation. Although the authors suggest that, if violated, feasibility might be regained after the optimisation by correcting the refrigerant pressures, doing so could lead the solution to lose its optimality.

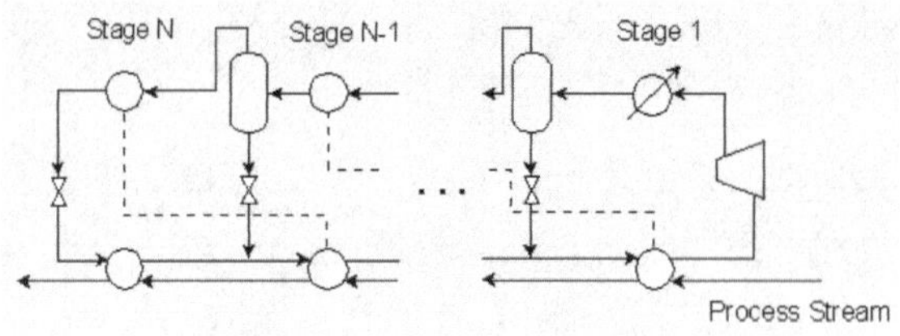

a) Original representation.

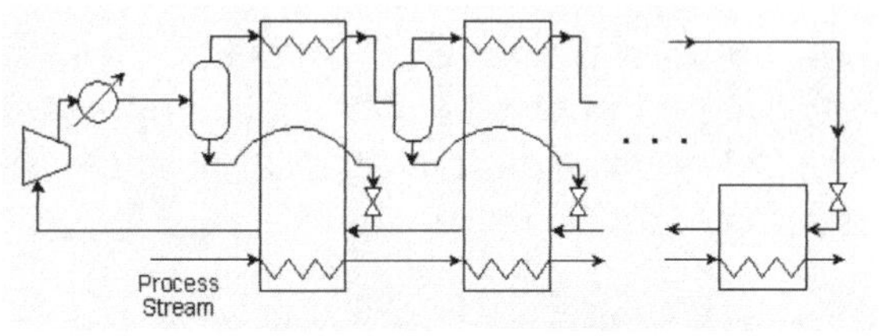

b) Alternative representation.

Figure 6. MR structure used by Vaidyaraman and Maranas (2002).

As in Lee (2001), Vaidyaraman and Maranas (2002) assumed single stage compression without intercooling and did not consider capital costs. However, they made a further effort in trying to overcome local optima by performing a number of optimisations with different starting points. The effect of refrigerant pressure drops was not covered in their formulation.

PROCESS MODEL

A detailed description of the equations involved in the process model is omitted from this paper as it would increase its length significantly. However, some of its main features and assumptions are discussed below.

The optimality criteria must be defined by an objective function as a generic function of the main process variables. This objective function may well change from case to case according to the purpose of the designer (e.g. minimum compressor power, minimum capital investment, minimum total cost, etc.) and to the economic models employed (e.g. capital cost of cryogenic heat exchangers).

Accurate pressure drop predictions across heat exchangers, especially of the compact type, can result in a demanding task, since details of the actual exchanger geometry and internals (e.g. type of fins), besides a set of additional fluid transport properties, are required for such a purpose (Pua and Zhu, 2002). If a rigorous pressure drop prediction of such kind was implemented in the process model it would increase the mathematical complexity considerably, taking the focus away form the main design variables. On the other hand, obviating the pressure drops in the cycle could lead to unidentified infeasibilities because of inaccurate prediction of temperature profiles and/or to a non optimal and sub-designed system. An intermediate approach was adopted in this work. The total hot refrigerant pressure drop has a fixed value defined by the user before the optimisation, which allows the estimation of intermediate hot refrigerant pressures by approximating them as a linear function of the respective temperature. A similar approximation is performed to obtain the intermediate cold refrigerant pressures. However, these are assumed as a linear function of the temperature of the hot refrigerant instead of that of the cold refrigerant for convenience.

For a given refrigerant composition and inlet and outlet temperatures, the temperature-enthalpy profiles at each refrigeration stage (hot vapour refrigerant, hot liquid refrigerant and cold refrigerant) are obtained by evaluating the stream enthalpy repeatedly at a number of intermediate points within the respective temperature range. A linear behaviour of refrigerant pressure with temperature is also assumed during the calculation of the temperature-enthalpy profiles. Cold refrigerant pressures, this time, are estimated according to the cold refrigerant temperatures and not those of the hot refrigerant as done previously. Extra evaluation points are added at the exact dew and/or bubble points of the mixture if found within the temperature and pressure range. Since the process composite curve is an input to the problem, the process profiles at each stage are obtained by just extracting the corresponding data within each respective range of temperatures. The overall process profile input to the problem should consider the effect of pressure drop. If the process composite curve was not an input to the problem, the process profiles could be constructed similarly to the refrigerant profiles. All the hot profiles at each stage are combined to form a single hot composite curve.

Once the hot and cold composite curves are known at each stage, the temperature feasibility can be assessed by comparing the temperature of the hot composite curve to that of the cold one for a given enthalpy at a number of sample points. If the temperature of the cold composite evaluated at a given enthalpy is greater than that of the hot composite then a temperature cross (or crossover) occurs. If the temperature of the cold composite is lower than that of the hot composite by less than the minimum temperature difference (ΔT_{min}) then a ΔT_{min} violation occurs. If the temperature of the cold composite is lower than that of the hot composite by more than ΔT_{min} then the point is perfectly feasible. This temperature feasibility check is further illustrated in Figure 7.

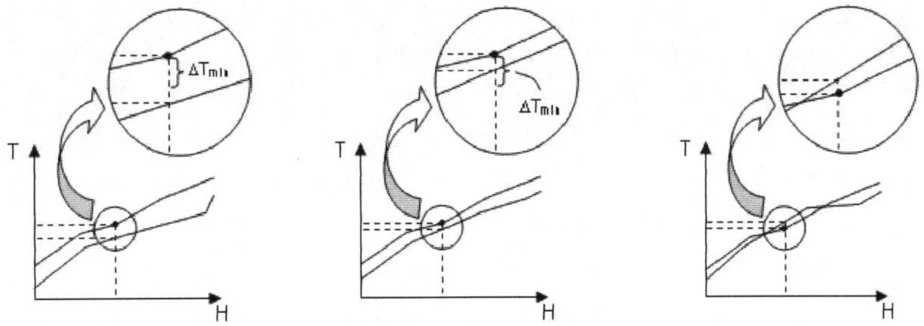

Figure 7. Temperature feasibility check.

In practice, stages belonging to the same compressor normally feature similar pressure ratios. The process model assumes that all stage pressure ratios are identical and ensures that such value is consistent with the number of stages and with the compressor inlet and outlet pressures. In case the refrigerant pressure levels imply a total pressure ratio greater than a previously defined maximum value, the model has the ability to automatically increase the number of compression stages so that the resulting stage pressure ratios do not exceed the maximum allowed. The resulting number of stages must be the minimum possible without violating the limit imposed on the stage pressure ratio. As an alternative, the problem could also be solved using a fixed number of stages as input. However, stage pressure ratios not within a practical range may result.

Mixed refrigerant cycles without hot liquid refrigerant subcooling, as in Figure 6, are supported by the model as well and are handled as a relatively minor flowsheet variation indicated by a flag parameter.

SOLVING THE PROBLEM

The optimisation problem is of the NLP type. However, not all the formulation is explicit (e.g. phase equilibrium and physical property calculations) and in practice part of the calculations may take place in external subroutines (e.g. interfacing with commercial process simulators). Also, as stated before, the problem is nonlinear and features many local optima. Traditional deterministic optimisation methods would become trapped in one of those quite easily. On the other hand, stochastic methods (e.g. genetic algorithm, simulated annealing) may offer more confidence on the optimality of the final solution at the expense of computational time. A genetic algorithm (GA) has been chosen as the main optimiser in this work. Genetic algorithms perform iterations not on a single candidate solution but on a population of candidate solutions (individuals) instead. The path towards optimality is built by sharing information between individuals (reproduction) and by evaluating and ranking the resulting offspring according to their relative optimality (fitness) in a given number of major iterations (generations). The information defining a particular individual (chromosome) consists of a set of values of the chosen independent variables (genes), over a discretised solution space, encoded as a string (e.g. binary encoding, real encoding). Another key advantage of GA is that no initial guess is needed to start up the optimisation process. Further details on how genetic algorithms work can be found in books such as Goldberg (1989) and Stender *et al.* (1994).

Release 1.2 of Pikaia (Charbonneau, 2002), a genetic algorithm in the public domain by the US National Center for Atmospheric Research (www.hao.ucar.edu/public/research/si/pikaia/pikaia.html), has been adopted in this work and modified in order to improve the quality of the initial generation. Randomly generated individuals are now evaluated before being allowed into the initial generation. They qualify as a member of the initial generation only if excessive infeasibilities are not incurred (e.g. average crossovers of no more than 3 ^{o}C and a compressor inlet temperature at most 2 ^{o}C below dew point), otherwise they are rejected. This filter, although it increases the computational time for the first generation considerably, ensures a departure from a better set of candidate solutions, shortening the path to optimality.

Figure 8 illustrates the flow of information within the general optimisation framework. The optimiser (GA in this case) proposes a series of candidate solutions and relies on the simulator for their assessment. How the optimisation task evolves is decided by the GA based on the assessment of the candidate solutions. The interactions between the GA and the simulator result in a set of the best solutions found over a discretised solution space. Standard NLP optimisation(s) can be carried out afterwards having the best discretised solution(s) as initial guess in order to fine-tune and finally report the optimal solution on the basis of a continuous solution space.

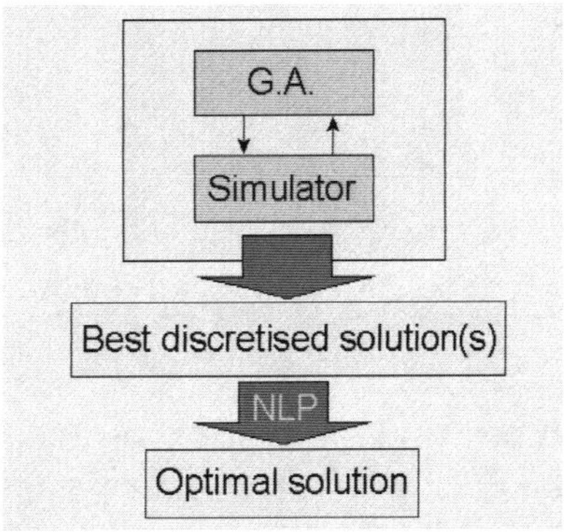

Figure 8. Optimisation strategy.

The chosen independent variables for this application are the refrigerant flowrate and composition, the compressor inlet and outlet pressures and the intermediate temperatures (for cycles with two stages or more). Given the values for all these variables, the simulator is responsible for calculating the remaining variables and the magnitude of infeasibilities (e.g. temperature profiles and compressor inlet wetness) and evaluating the objective function. Penalties may be applied to the objective function according to the magnitude of the infeasibilities as a disincentive in the ranking of solutions incurring violations. However, care should be taken in avoiding excessive penalties since infeasible solutions (especially the slightly and moderately infeasible) may still contain features worth to inherit and evolve by offspring candidate solutions.

CASE STUDY

The proposed process model and optimisation strategy are put into practice using the same LNG case study published by Lee (2001). A pre-treated natural gas stream is to be cooled from 25 °C to -163 °C using a mixture of nitrogen and hydrocarbons C1 to C4 as refrigerant in a single stage cycle. Subcooling of liquid refrigerant is allowed and compression takes place in one stage. External cold utility is available to cool hot streams down to 30 °C. The minimum temperature difference is 5 °C, the isentropic compression efficiency is 80% and the physical properties calculations are performed based on the Peng-Robinson equation of state. The temperature-enthalpy data of the natural gas are given in Table 1. The objective is to minimise the power consumption.

Table 1. Temperature-enthalpy data of the natural gas stream to be liquefied.

Temperature (°C)	Enthalpy (MW)
25.00	20178.8
-6.03	18317.0
-34.09	16352.8
-57.65	14468.0
-70.10	11978.0
-74.55	10198.0
-82.26	7114.0
-96.50	5690.0
-115.00	3840.0
-163.00	0.0

Base case solution

The application of the new methodology results in an optimal solution with a compression power of 33.39 MW. The composite curves for this solution are given in Figure 9.

The optimal solution reported by Lee (2001), on the other hand, had a compression power of 26.60 MW, which is 20.6% lower than the one found with the new methodology. However, it must be considered that because the minimum temperature difference was not fully enforced, Lee's solution ended up with an effective ΔT_{min} of approximately 1.2 °C. Hence, for the comparison to be really meaningful, the base case was re-optimised with the new methodology and ΔT_{min} of 1.2 °C, finding and optimal solution that features 24.53 MW of power. That is, 7.8% lower than in Lee's solution. The three solutions discussed are reported in Table 2.

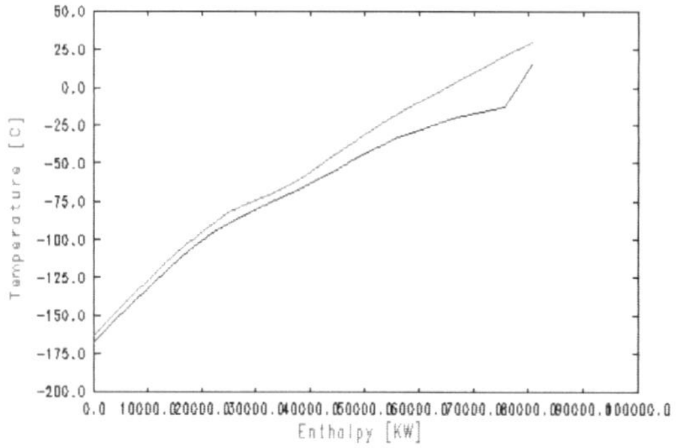

Figure 9. Composite curves for the base case optimal solution.

Table 2. The three solutions for the base case.

	New Method	Lee's Method	New Method
ΔT_{min} (°C)	5.0	1.2	1.2
Power (MW)	33.49	26.60	24.53
Refrigerant flowrate (kmol/s)	3.47	3.2	3.53
Suction pressure (bar)	2.40	3.7	4.84
Discharge pressure (bar)	36.95	40.0	43.87
Refrigerant composition (mol %)			
N_2	15.32	11.0	10.08
CH_4	17.79	27.3	27.12
C_2H_6	40.85	35.6	37.21
C_3H_8	0.41	5.20	0.27
n-C_4H_{10}	25.62	20.9	25.31

Effect of multistage compression

It is worth noting that all three solutions presented in the previous section have a compressor outlet temperature of no less than 140 °C. A gas undergoing compression at such temperatures certainly occupies a relatively high specific volume and hence demands a relatively high specific compression power. To illustrate the effect of multistage compression, another design task was performed. This time the maximum stage pressure ratio was set to 5 and the minimum temperature difference back to 5 °C. The resulting minimum power solution has total compression power of 27.87 MW, which is 16.8% lower than the minimum power found when optimising for a single compression stage (33.49 MW). The compression train features three stages with a pressure ratio of 2.95 each. The total intercooler duty was also reduced from 53.67 to 48.15 MW. Although the refrigerant composition is quite similar to that of the base case optimal solution, meaningful differences are observed in the refrigerant flowrates (2.81 against 3.47 kmol/s before) and in the compressor inlet and outlet pressures (1.76 against 2.40 bar and 45.06 against 36.95 bar). It seems that now that the compression process is more efficient, the system can afford a larger total pressure ratio (25.6 against 15.4) in exchange for a reduction in the refrigerant flowrate, the net effect being the observed reduction in total power whilst still keeping good care of temperature feasibility in the main heat exchanger.

It is tempting to consider that multistage compression is not needed at the synthesis stage, because the power consumption can be reduced after optimising the cycle for a single compression stage by just increasing the number of compression stages and inserting intercooling afterwards. However, doing so in the case study reduces the total power only to 30.18 MW; still 2.31 MW above the 27.87 MW found when considering the effect of multiple stages during optimisation. One possible reason for this could be the unattractiveness of exploring high total pressure ratios when minimising power in a cycle with a single compression stage, as discussed previously.

Two-stage refrigeration and capital cost effect

Several two-stage MR cycles were also optimised in a sensitivity analysis with a maximum stage pressure ratio of 5. The goal was to evaluate the effect that the pre-defined minimum temperature difference has on the minimum power solutions. The effect on the total compression power can be seen on Figure 10a, where the resulting power of each solution is plotted against the respective value of ΔT_{min}. The monotonic trend was to be expected. The closer the hot and cold composite curves are allowed, the lesser the irreversibilities in the process and hence the lesser the total compression power. It is worth noting that the optimisation for a ΔT_{min} of 5 °C gave a solution with a power consumption of 26.58 MW. This is 1.29 MW lower than the solution reported in the previous section and due to the fact that the refrigeration stages have been increased to two.

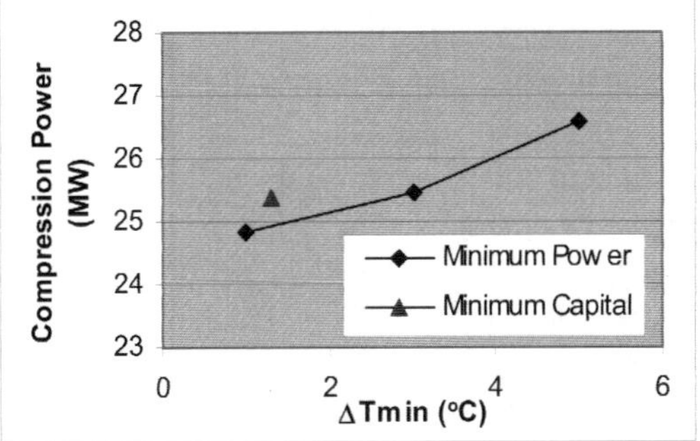

a) Power.

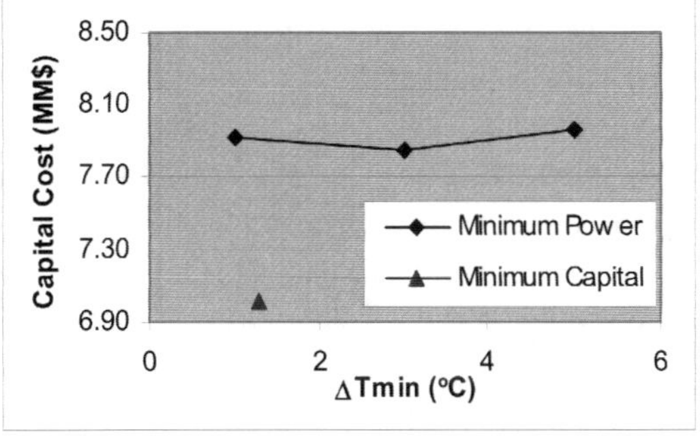

b) Capital cost.

Figure 10. ΔT_{min} sensitivity for two-staged refrigeration cycles.

The capital costs of the solutions appearing in Figure 10a are plotted in Figure 10b also against the minimum temperature difference. The capital cost considered for any given solution includes the cost of the compressor, estimated from vendor quotations, and the cost of the cryogenic heat exchangers, estimated using the method by ESDU (2003) for aluminium-brazed plate-fin heat exchangers (PFHE). This time the trend is not monotonic. A minimum capital cost seems to exist somewhere between 1 and 5 ^{o}C. This is also not surprising behaviour. As the total power decreases with decreasing ΔT_{min}, so does the compressor cost. However, ΔT_{min} has the reverse effect on the volume (and cost) of the PFHEs. The closer the composite curves are to each other, the smaller the temperature driving forces and hence the larger the heat transfer area, volume and cost.

With the purpose of further investigating this compressor/PFHE capital trade-off, the minimum temperature difference allowed during optimisation was decreased to a negligible amount and the objective of the optimisation was changed to minimum capital cost. In this way there would be no restrictions on the temperature differences, apart from avoiding crossovers, as long as the capital is minimised. The resulting solution is also shown in Figure 10. In the point of closest approach between hot and cold composite curves the temperature difference is 1.3 ^{o}C. The total compression power is located slightly above the power curve of minimum power solutions. However, the capital cost is considerably below the cost curve of minimum power solutions. The reason for this is that, apart from exploiting the compressor/PFHE capital trade-off, the resulting pressure levels of the minimum capital solution are such that the compression task can be accommodated in just two stages, against the three needed in all of the minimum power solutions. This reduction in the number of compression stages, albeit the slightly increased total power, provided an additional contribution towards the reduction of capital.

CONCLUSIONS

The design of mixed refrigerant cycles is a challenging task. Also, local optima are abundant. This paper has introduced and illustrated the application of a new formulation and optimisation strategy for the synthesis of mixed refrigerant cycles. It considers multistage refrigerant compression, full enforcement of the minimum temperature difference along the temperature profiles, simultaneous optimisation of variables, incorporation of capital costs in the objective function and the use of stochastic optimisation (genetic algorithm) to overcome local optima. The effectiveness of the method was illustrated by finding improved and feasible solutions for a case study previously published by Lee (2001). The incorporation of multistage compression proved key in the design of more efficient cycles since it not only reduces the power consumption *per se*, but also encourages the full exploration of total pressure ratios, unlike the single stage case. Although only compressor and PFHE capital costs were considered, the final part of the case study demonstrated that it is essential to address all the relevant cost implications during optimisation. Finally, although it makes the solving process time consuming and it still does not guarantee global optima, the application of a genetic algorithm optimiser to this design problem permits accepting the optimality of the results with greater confidence than deterministic methods.

The application of the presented problem formulation to more complex MR cycles is a promising extension of this work. That is, mixed refrigerant cycles in cascade with other mixed refrigerant cycles and with expander cycles. Also, more flowsheet options can be incorporated, including the use of expanders as an alternative to Joule-Thompson valves.

ACKNOWLEDGEMENTS

The main author of this paper would like to thank the Centre for Process Integration at the University of Manchester and the Overseas Research Student Award Scheme for the financial sponsorship of this project.

REFERENCES

Charbonneau, P., Release notes for PIKAIA 1.2, National Center for Atmospheric Research, Technical Note 451+STR, 2002.

Cheng, W.B. and Mah, R.S., Interactive synthesis of cascade refrigeration systems, *Industrial & Engineering Chemistry Process Design and Development*, 19(3), pp 410-20, 1980.

ESDU International Plc., Selection and costing of heat exchangers - Plate-fin type, ESDU Document 97006, July 2003 update.

Finn, A.J., Johnson, G.L. and Tomlinson, T.R., Developments in Natural Gas Liquefaction. *Hydrocarbon Processing*, Vol. 78, No 4, April 1999.

Goldberg, D. E., Genetic algorithms in search, optimization and machine learning, Addison-Wesley Publishing Company Inc., 1989.

Lee, G.C., Optimal Design and analysis of refrigeration systems for low temperature processes, PhD thesis, Department of Process Integration – UMIST, UK, 2001.

Lee, G.C., Smith, R. and Zhu, X.X., Optimal synthesis of mixed-refrigerant systems for low temperature processes, *Ind. Eng. Chem. Res.*, Vol. 41, 5016-5028, 2002.

Pua, L. M. and Zhu, X. X., Integrated heat exchanger network and equipment design using compact heat exchangers, *Heat Transfer Engineering*, 23(6), pp. 18-35, 2002.

Stender, J., Hillebrand, E. And Kingdon, J., Genetic algorithms in optimisation, simulation and modeling, IOS Press, 1994.

Vaidyaraman, S. and Maranas, C.D., Synthesis of mixed refrigerant cascade cycles, *Chemical Engineering Communications*, Vol. 189, No. 8, pp 1057-1078, 2002.

Walsh, B. W., Mixed refrigerant process design, *Off. Proc. Comb. Conf.,* 6th Conf. Asia Pac. Confed. Chem Eng., 21st Australas. Chem. Eng. Conf. (1993), (1), 59/1-64/1.

BIOGRAPHY OF SPEAKER

Frank Del Nogal had a BSc in Chemical Engineering from the Universidad Simón Bolívar in Venezuela (1996). Worked five years as Technical Consultant on process simulation with Aspen Technology, being assigned to a main Venezuelan Petrochemical site. During the last two of those years also performed as part-time lecturer at La Universidad del Zulia. He joined the Department of Process Integration at UMIST (now The University of Manchester) in 2001, where an MSc degree with distinction was obtained in 2002. Immediately started further research as PhD student, working on the synthesis of refrigeration systems and utility systems with special interest in the liquefaction of natural gas.

refrigerant cycle, which requires only the components nitrogen, methane, ethylene, propane and pentane.

Basic Data for the Process Design of the Plant

The design of the LNG plant for the Xinjiang project is based on a state-of-the-art natural gas liquefaction technology. The LNG production capacity of the plant is equivalent to 1 500 000 m3(n)/d with an expected on-stream time of 330 days per year. Design hourly liquefaction capacity is 54 t/h. Storage capacity is 30,000 m3 in liquid form of LNG, which is the equivalent of 12 days production. The LNG send-out and distribution system capacity meets the requirement of loading the 100 trucks/movable containers within 16 hours. The split is 30 % in trucks and 70 % in movable containers.

Specification of the Product LNG:

Composition (mole %): Nitrogen 0.8 (max 1.0), Methane 82.4, Ethane 11.1, Propane 4.6, Others 1.1. Pressure and Temperature at LNG tank: 0.01MPaG, –163 °C. The design LNG has a density of about 490 kg/m3 in the LNG Tank.

Feed gas conditions at Plant Battery Limits:

Composition (mole %): Nitrogen3.81, Methane 81.02, Ethane 9.99, Propane 4.10, Butanes 0.93, i&n-Pentane 0.05, C6+ < 0.0021. In addition CO_2 and traces of H_2S and sulfur are present in the feed gas. The feed gas operating pressure can range from about 0.6 MPaG to 1.1 MPaG. The design pressure is 0.7 MPaG. The feed gas operating temperature can range from -15°C to 40°C. The design temperature is 28 °C.

Process Features

The main process and utility units are illustrated in the block diagram in Fig1. The liquefaction process with a closed mixed refrigerant cycle requires the components nitrogen, methane, ethylene, propane, pentane, which except methane and nitrogen have to be purchased from external sources. Refrigerant nitrogen and purge nitrogen are identical and will be generated in a nitrogen package.

Make-up water for closed cooling water cycle for machinery cooling and demineralized water as make-up water for the MEA in the CO_2 wash unit will be provided from outside the plant. A mixture of compressed LNG tank return gas and feed gas is used as normal fuel gas; start-up fuel gas is feed gas. A closed hot oil cycle is used as heating medium. An MEA (monoethanolamine)-water solution is used as solvent for the CO_2 Wash Unit.

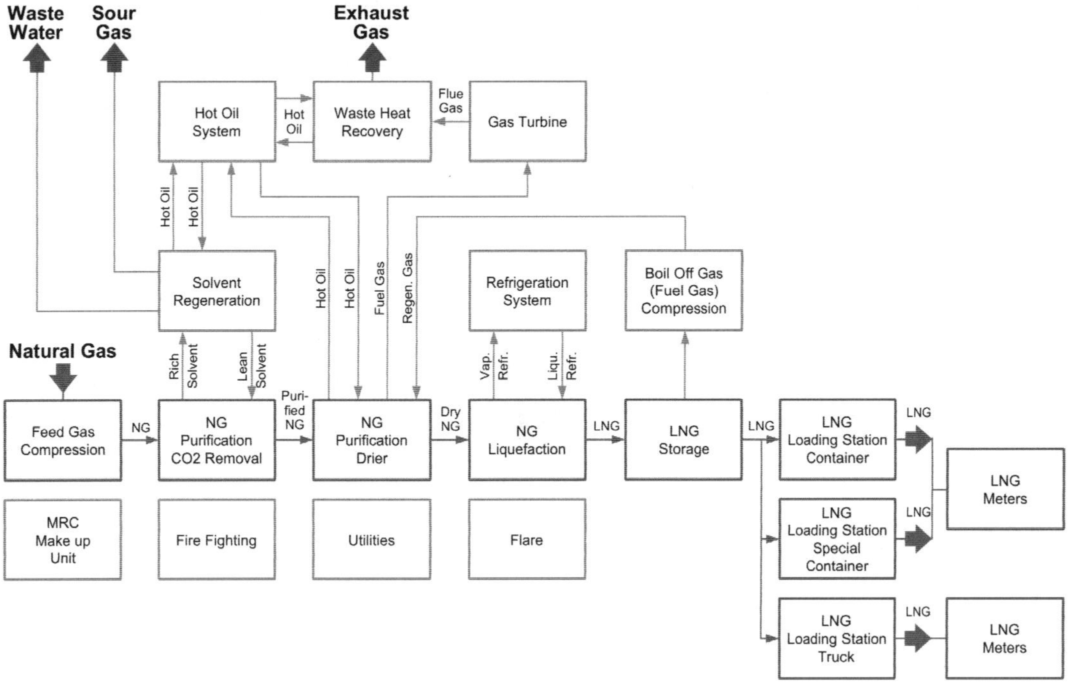

Fig1: Block diagram of the Shan Shan LNG plant with process and utility units

Ambient Conditions at the Site

The average ambient temperatures range from 37.1 °C in the warmest month to -15.6 °C in the coldest month. The design temperature for gas turbine air inlet and for air-cooling is 30 °C, the average ground temperature in the hottest month is about 37 °C, the extreme maximum ground temperature is about 75 °C and the plant elevation above sea level is about 790 m.

Overall Process and Utility Description

The Shan Shan LNG plant has a medium size production capacity in between the two principle types of LNG plants which are currently in operation world-wide. LNG peakshaving or back-up plants with intermittent operation and production have capacities up to about 500 000 Nm3/day.

LNG baseload plants with continuous operation and production have capacities between 5 000 000 m3(n)/day and 17 000 000 m3(n)/day. With the 1 500 000 m3(n)/day LNG production capacity the Shan Shan LNG plant is about 3 times larger than the largest existing peakshaving plants, but about 3 times smaller than existing small baseload plants. The liquefaction process is based on a high efficient single closed mixed refrigerant cycle. The feed natural gas has a low pressure at battery limit, which is too low for an efficient liquefaction process. Therefore, the natural gas is compressed in 3 compressor stages after removal of solid and liquid particles in a separator. The natural gas is cooled, liquefied and subcooled in a spiral wound heat exchanger by a single closed mixed refrigerant cycle. This cycle provides cold temperatures by Joule-Thomson expansion at 3 different pressure levels. The refrigerant cycle is recompressed in a 3-stage turbo-compressor, which is driven by a gas turbine. In order to enhance plant efficiency, the waste heat from the gas turbine is recovered by heating a hot oil cycle, which covers the heating requirements of the process plant.

Natural Gas Treatment

Natural gas (feed gas) has a low pressure at the battery limit. Solid and liquid particles are removed by the Feed Gas Filter Separator before it is compressed in a 3 stage feed gas compressor. After the 1st stage of the feed gas compressor the gas is cooled in an inter-cooler against ambient air to about 40 °C. Potential water condensed in the inter-cooler is separated in the feed gas compressor inter-stage drum and is fed to the wash unit. After this first compression step the feed gas is further compressed in the next two compressor stages with inter- and after-cooling in air-coolers. The Feed Gas is routed to the CO_2 Wash Unit for removal of CO_2. The sweet feed gas leaving the CO_2 wash column is then routed to the drier station.

Natural Gas Liquefaction (Fig. 2)

After the CO_2 and H_2O removal the natural gas is routed to the cold part of the process, which contains three spiral wound heat exchangers integrated in one shell ("rocket"), as well as several separation vessels. The natural gas is first cooled in the Feed Gas Precooler E1, potential off-spec heavy hydrocarbons are separated in the feed gas heavy hydrocarbon separator D3, where only marginal liquids during design feed gas operation are expected. The gas is then condensed in Feedgas Liquefier E2 and subcooled in Feedgas Subcooler E3. The subcooling temperature is controlled by the amount of tank return gas required as fuel gas for the operation of the gas turbine. The cooling is provided by the closed multi component mixed refrigerant cycle.

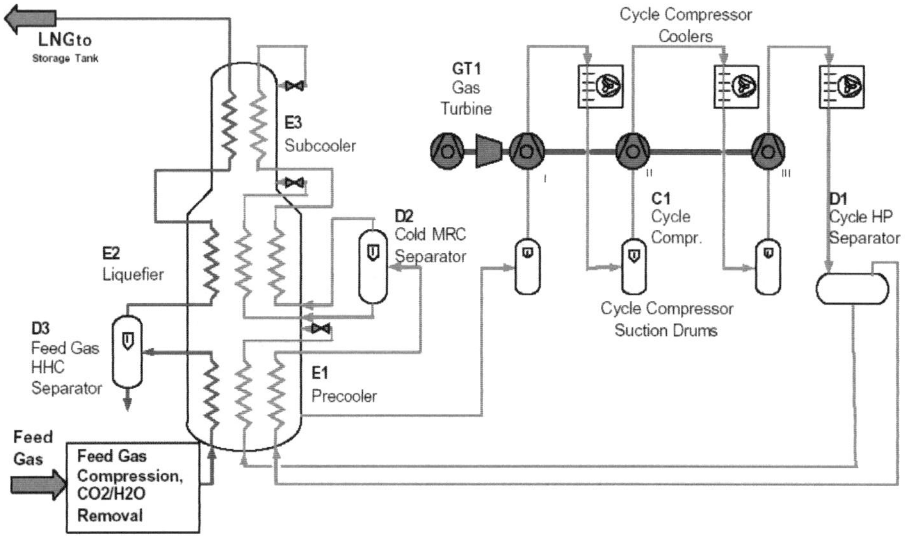

Fig 2: Natural gas liquefaction process of the Shan Shan LNG plant

Refrigerant System

The refrigerant gas stream is withdrawn from the shell side of precooling section E1 of the cryogenic spiral wound heat exchanger set. The refrigerant is slightly super-heated over the dew point condition. It is compressed in the first stage of the 3 stage refrigerant cycle compressor, inter and after cooled against air, where it is partly condensed. Liquid formed in the after-cooler is separated in Cycle Compressor Discharge Drum D1. The liquid from the Discharge Drum D1 is routed to the cryogenic heat exchanger E1, where it is subcooled and then used for the precooling of the natural gas after expansion in a Joule-Thomson expansion valve. The cycle gas from the Suction Drum D1 is cooled in E1 to the same temperature and partly condensed and fed to the Cold Refrigerant Separator D2. The liquid from this separator is subcooled in the cryogenic heat exchanger section E2 to a low temperature, so that it can be used as a refrigerant in E2 after expansion in a Joule-Thomson expansion valve. The vapor from the Cold Refrigerant Separator D2 is condensed in E2 and subcooled in the cryogenic heat exchanger section E3 to a sufficiently low temperature and provides the final cold for the natural gas subcooling after throttling in a

Joule-Thomson valve. After expansion to the lower pressure the cycle gas streams are warmed up in the common shell side of the cryogenic spiral wound heat exchangers E3, E2 and E1 and return jointly to the suction side of the 1st stage of the refrigerant cycle compressor.

Gas Turbine

Gas Turbine GT1 is used as prime driver for the cycle gas compressor C1. Design temperature for gas turbine rating is an ambient air temperature of 30°C. The same design temperature applies for air-cooling. The compressed boil-off, flash and displacement gas from the LNG storage tank is used as regeneration gas and then as fuel gas for the gas turbine.

LNG Storage and Loading System

The LNG from the liquefaction unit with the cryogenic heat exchanger set E1, E2 and E3 is sent to the Storage Tank D-411 via the tank filling line, Fig 3.

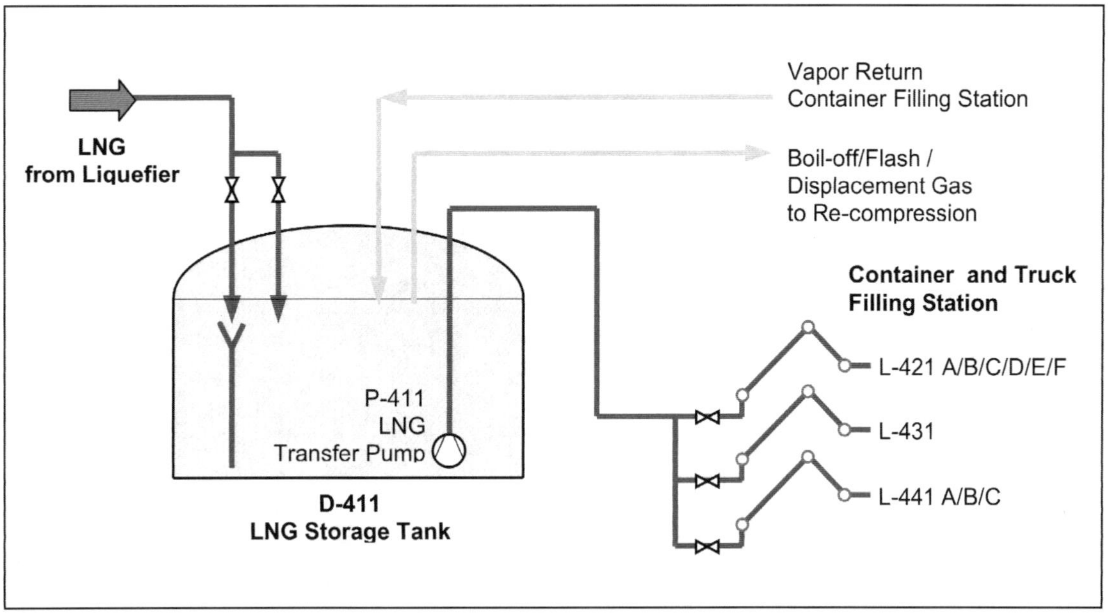

Fig 3: LNG storage tank and loading system of the Shan Shan LNG plant

Fig 4: 30 000 m3 LNG storage tank

The 30 000 m3 LNG tank (Fig 4) is of the flat bottom, double wall, perlite insulated type installed in an endiked area. The tank will be filled continuously during operation of the liquefaction system at a filling rate of about 111 m3/h. During 16 hours per day a discontinuous send out operation to the truck and container filling is scheduled. For send-out operation two submerged in-tank pumps are installed, each designed for 320 m3/h capacity, suitable for 100 % of send-out capacity. One pump is installed as a spare. The pumps are installed in pump columns inside the tank and equipped with foot valves. Each pump is equipped with a kickback line to the tank to control the minimum flow of the pump during the period when no filling operation takes place.

The send out lines to the truck and container filling station are permanently filled with LNG and a small circulation flow keeps the system at cold temperatures. The trucks are weighted ahead of filling. They are connected manually to the loading arm filling and vapor return lines. The initial LNG into the "warm" tank truck evaporates and the created vapor will return to the storage tank. After cooling the truck tank the filling rate increases to the maximum filling rate. The flow counter stops the filling operation automatically via the automatic control valve at the loading station. The truck leaves the plant via the weighbridge after disconnection from the loading arm. Fig 5 shows the LNG truck loading station with 4 trucks each with a storage capacity of 44 m3. The loading capacity of the stations is sufficient to match the send out capacity within 16 hours a day.

Fig. 5: LNG truck loading station

The same operation applies to the container filling system. The only difference is that trucks are mobile by themselves and the container must be moved by gantry crane and trailers. The container will be fixed on rail-platform cars and transported as train of 40 to 70 cars length. The filling time for one container or one truck is estimated to be about 1.2 hours including connection and disconnection time. The system capacity is designed to fill 100 trucks or containers within 16 hours. The filling system consists of 6 loading stations for containers and 4 loading stations for trucks.

Fuel System

The net flash, boil-off and displacement gas coming from the LNG storage tank is compressed, cooled against ambient air and used as regeneration gas in the dehydration section before it is sent as fuel to the Gas Turbine which drives the cycle compressor. To allow for a good pressure control of the fuel gas, an additional fuel stream is taken from the feed gas after the 2nd stage of the feed gas compressor.

Hot Oil Unit

The hot oil system provides the process heat for the Plant at two temperature levels. In order to keep constant flow rates in the system, two cycles are introduced, a medium temperature cycle and a high temperature cycle. The heat for both cycles is provided by a hot oil heater package, a waste heat recovery unit in the exhaust stack of the cycle gas turbine. The hot oil is heated to approx. 260 °C to supply heat for the regeneration gas heating. To allow for start-up during winter conditions, the system is heat traced.

Main Cryogenic Heat Exchanger

Fig 6: Precooling section of the spiral wound heat exchanger in the Linde workshop

A special feature of the cryogenic section of the process plant is the spiral wound heat exchanger designed and built by Linde. It excels by its robustness for the gas precooling, liquefaction and subcooling process, where the refrigerant cycle and product streams reach temperatures down to -160 °C. Fig. 6 shows the precooling section of the spiral wound heat exchanger in the Linde workshop prior to transport.

The outer dimensions (length x diameter) of the three spiral wound heat exchanger sections are:

Precooler: 15 m x 3 m,
Liquefier: 17 m x 3 m,
Subcooler: 11 m x 2 m

All three heat exchanger sections have been transported separately to the site. After concentric stacking and welding in a steel structure the combined spiral wound heat exchangers have an overall height of 43 m. Fig 7 shows the cryogenic section with the spiral wound heat exchanger together with the separator in the permanent steel frame. The heat exchanger is relatively insensitive against thermal shocks, which may occur during start-up or shut-down or mal-operations.

Fig. 7: Three stage 43 m high spiral wound heat exchanger with separator in the steel frame.

Project Execution

The plant execution is an example for the good cooperation between the parties involved:

Xin Jiang Guanghui LNG Development Co Ltd. in Urumqi as client, was also the organizer of the civil and construction works, procurement of local equipment and bulk material.

SPIDI in Shanghai was responsible for the entire plot plan of the plant and detail engineering with utilities. Tractebel Gas Engineering in Bonn was responsible for the design of the LNG storage tank and the loading facilities and for the procurement of the relevant imported equipment and material as well as for the construction and commissioning supervision of the tank and loading units. Linde Engineering in Munich was responsible for the natural gas treatment and liquefaction process design and for the procurement of the imported process related equipment as well as for the supervision of plant construction and commissioning.

Fig 8: Section of the Shan Shan LNG plant with compressor shelter, gas turbine exhaust stack, spiral wound heat exchanger and LNG tank

Fig 8 shows a section of the Shan Shan LNG plant with the compressor shelter building, the gas turbine exhaust stack, the spiral wound heat exchanger in the rack and the LNG tank as well as the air coolers on the pipe rack.

The plant was completed in 2004 followed by commissioning. The equipment and piping was arranged in such a way as to take into account the relevant safety regulations as well as short pipeline lengths. The required plant area is about 58 m x 130 m. The LNG storage tank is connected to the process plant by a pipe – rack, which supports the product and the vapor return line.

Fig: 9: Part of the fleet of LNG trucks for the road transportation of the Shan Shan LNG

A large fleet of LNG trucks, Fig 9, is now permanently transporting the LNG over long distances to the satellite stations in the vicinity of the consumers.

Most of the LNG satellite stations are located in the more densely populated regions in the eastern provinces of China in the vicinity of the natural gas consumers. An overview of the existing and planned

LNG satellite stations as targets of the trucks from the Shan Shan plant is shown in Fig. 10.

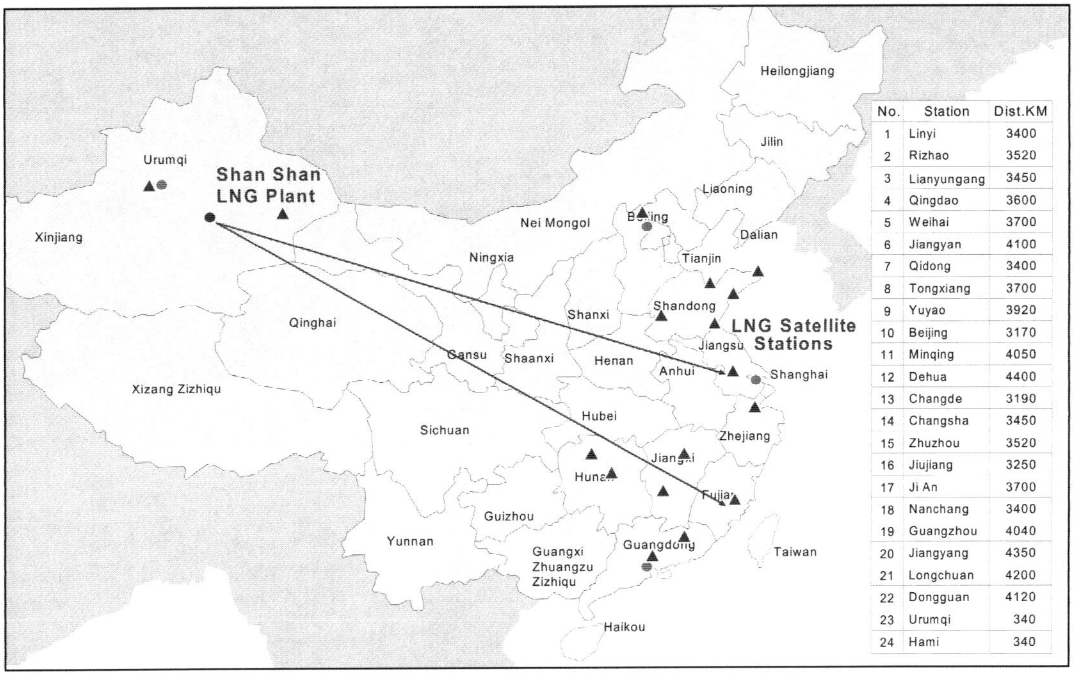

No.	Station	Dist.KM
1	Linyi	3400
2	Rizhao	3520
3	Lianyungang	3450
4	Qingdao	3600
5	Weihai	3700
6	Jiangyan	4100
7	Qidong	3400
8	Tongxiang	3700
9	Yuyao	3920
10	Beijing	3170
11	Minqing	4050
12	Dehua	4400
13	Changde	3190
14	Changsha	3450
15	Zhuzhou	3520
16	Jiujiang	3250
17	Ji An	3700
18	Nanchang	3400
19	Guangzhou	4040
20	Jiangyang	4350
21	Longchuan	4200
22	Dongguan	4120
23	Urumqi	340
24	Hami	340

Fig. 10: LNG transport in China from the Shan Shan plant

As is evident from the table, the one way distances mostly exceed 3 000 km, some are even longer than 4 000 km.

Fig 11: Dehua LNG satellite station

The Dehua satellite station, Fig 11, represents the longest distance from the Shan Shan LNG plant. This station comprises 8 vertically installed cylindrical LNG storage tanks with a capacity of 150 m3 each. They are filled more or less regularly by the LNG trucks. The LNG is vaporized to ambient temperature conditions by blocks of finned heat exchangers using natural convection of ambient air as heat source.

Small LNG containers are very often filled with LNG from these large satellite stations and transported to smaller satellite stations as supply source for a limited residential area, Fig 12.

One of the urgent needs for clean fuel comes from public busses. Therefore, one of the LNG satellite stations supplies the LNG directly to a city bus fleet, Fig 13.

Fig 12: LNG satellite station with containers to supply residential areas with natural gas

Fig 13: LNG fuelling station for busses

Closing Remarks

This Shan Shan LNG plant will open a new era in meeting the increasing demand for natural gas in China, which will consistently rise from the current 3 % in the energy mix in China to the level in industrialized countries of around 20 % or more. With the introduction of such LNG plant types combined with the respective transport infrastructure, natural gas markets can be dynamically introduced and developed. It is evident that natural gas, as a cleaner fuel, will play an increasingly important role in the primary energy mix, as it will serve to reduce the environmental constraints in terms of emission levels. The Shan Shan LNG plant represents a valuable contribution to increase the wealth of the country in a similar way, as is the case in other countries by the commercialization of indigenous natural gas and LNG.

Some of the target regions have not yet been connected to major gas pipelines due to economic reasons, since the initial gas consumption rate would not justify such a large investment. Therefore, the LNG supply will initiate the penetration of these regional markets with environmentally friendly fuel. Thus the LNG from the Shan Shan plant will contribute substantially to the economic development and growth in China. With the LNG from Shan Shan a high degree of flexibility in the energy supply will be made available to the benefit of all natural gas consumers with fluctuating or peak demand profiles.

This LNG scheme is unique in the world with regard to plant type as well as plant and transport capacity. It can be considered as an incentive for the commercialization of remote gas resources with similar market conditions worldwide and particularly in China.

Selected References

1. Eginhard Berger, Linde AG – Linde Engineering Division, Xiang Dong, Xinjiang Guanghui Industry and Commerce Group Co. Ltd., Jin Guo Qiang, Shanghai Pharmaceutical Industry Design Institute of SINOPEC (SPIDI), Albert Meffert, Tractebel Gas Engineering, Lothar Atzinger, Linde AG – Linde Engineering Division: LNG Baseload Plant in Xinjiang, China – commercialization of remote gas resources for an Eco-responsible Future. World Gas Conference, Tokyo, 2003
2. W Förg, W Bach, R Stockmann, Linde and R S Heiersted, P Paurola, A O Fredheim, Statoil: A New LNG Baseload Process and Manufacturing of the Main Heat Exchangers. LNG 12 Conference, Perth, May 1998.
3. E.Berger, Linde: Natural Gas Liquefaction - Technical and Economic Aspects, First Indian LNG Conference, Madras, India, 1996
4. E.Berger, Linde: LNG Satellite Stations in Europe, LNG 10 Conference in Kuala Lumpur, Malaysia, 1992

Biography of Speaker

Eginhard Berger graduated with a M.Sc. degree in mechanical engineering at the Technical University in Munich.

After a short engagement in the space/aviation industry he joined the Linde Engineering Division in 1969.

He was first involved in the computerization of physical properties of natural gas components and mixtures before he calculated natural gas liquefaction plants, mainly LNG peak-shaving plants.

He became project manager for offshore LNG plant developments. In parallel he was serving and pursuing the planned German LNG import terminal.

He was project and sales manager and he negotiated international gas projects, among others:

- the natural gas liquids extraction plant in Zhong Yuan, China,
- the 4.3 MTPA Snoehvit LNG plant in Norway,
- the 0.4 MTPA LNG baseload plant in Xinjiang, China.

He is a member of the German gas association and of the European Technical Committee for the standardization of LNG plants.

Baseload LNG Production in Xin Jiang - a Remote Source of Clean Energy for New Gas Consumers in China

Xiang Dong
President
Xin Jiang Guanghui Liquefied Natural Gas Development Co. Ltd.
Shan Shan, PRC

Eginhard Berger
Senior Sales Manager, Natural Gas Plants
Linde AG-Linde Engineering Division
Hoellriegelskreuth, near Munich, Germany

Albert Meffert
Project Manager
Tractebel Gas Engineering
Bonn, Germany

Li Wei Bin
LNG Chief Representative
SSEC, SINOPEC Shanghai Engineering Co. Ltd.
Pudong New Area, Shanghai, PRC

AIChE Spring Meeting, April 2005
5th Topical Conference on Natural Gas Utilization
Atlanta April 10-14, 2005

Introduction

In 2004 Xin Jiang Guanghui Liquefied Natural Gas Development Co. Ltd. established the operation of a unique LNG chain. The result is that the natural gas – which until recently has been flared at the Tuha oilfields, some 300 km southwest of Urumqi, can now be utilized as a clean primary energy source. This new LNG scheme is a feasible and workable alternative to existing peak shaving and conventional baseload plants. With the LNG production capacity of 0.4 MTPA (million tons per annum) the plant represents a new category of LNG plant types, with which a specific demand can be fulfilled. The gas is treated and liquefied in an LNG plant near Shan Shan in the Xin Jiang Province of China. The plant is operated in a baseload mode and employs intermediate storage of the LNG product in an insulated tank before it is loaded into LNG road tankers. A large number of these trucks then carry the LNG over long distances to satellite and car fuelling stations in various cities of China. After revaporization of the LNG at these stations the natural gas is finally distributed partly via pipeline to a variety of industrial and private consumers.

Since it is considered to be the most environmentally friendly hydrocarbon fuel, it is expected that this domestic natural gas initiative through LNG creates new gas markets and provides a great improvement in the tight energy supply situation in China. This paper describes the Shan Shan LNG facilities ranging from gas treatment, liquefaction with a single mixed refrigerant cycle in spiral wound heat exchangers through storage to unloading and to the distribution of the LNG to various cities in China.

Design Basis

The baseload LNG Plant for the production of LNG equivalent to 1,500,000 m3(n) per day consists of natural gas treatment and gas liquefaction, LNG storage tank and LNG distribution system. The liquefaction process is based on a high efficient single mixed

Optimal Process Location for NGL Recovery in LNG Plant

Douglas A. Attaway
Stanley Huang
Heinz Kotzot
Charles A. Durr

Kellogg Brown & Root, Inc.
Houston, Texas 77002

AIChE Spring Meeting, April 2005
5th Topical Conference on Natural Gas Utilization
Atlanta April 10-14, 2005

ABSTRACT

Natural Gas Liquid (NGL) recovery is an inherent part of most LNG production facilities. The feed gas to a LNG plant contains in most cases ethane and heavier hydrocarbons. Even if the C_2+ components are not desired as a separate product, they often must be recovered to varying degrees to:

- satisfy refrigerant makeup needs
- prevent plugging in the liquefaction section
- meet LNG specifications, primarily heating value

The placement of the NGL recovery system within an LNG process is becoming an area of greater interest. Traditionally the removal of NGL from the feed gas has been integrated into the precooling section, which provides partial condensation for separation. However KBR recently has been involved in cases where front end NGL recovery, which is upstream of the gas cooling section, became the preferred design.

The purpose of this paper is to compare the efficiency of front end versus integrated NGL recovery for an LNG production facility in a neutral approach to liquefaction technology. It will use a generic liquefaction process as a basis for comparison. The analysis is not based on any proprietary liquefaction process.

The NGL recovery technique varies with the placement in the process. Front end NGL recovery will use a conventional expander plant design with full pressure recovery. Integrated NGL recovery can employ different schemes, such as condensation by refrigerant or expander technology. A series of process simulations are used to model the liquefaction process with selected NGL recovery methods at various temperature levels in the feed gas chilling train. A comparison based on specific power is applied to evaluate the options. This will address the impact of process locations on different NGL recovery methods. The results of this work will shed light on trends in this area and serve as a useful guideline for future process design.

INTRODUCTION

Virtually every liquefied natural gas (LNG) production facility has some form of natural gas liquids (NGL) recovery. It is normally required for preventing solidification and plugging in the cryogenic portion of the plant as well as meeting LNG product specifications and providing refrigerant make-up. In many cases it is attractive to recover and market the heavier components as separate products or provide them as feedstock to a nearby chemical plant or refinery [Ref. 4].

The ubiquitous presence of NGL recovery in LNG production facilities has prompted continuing efforts to explore and develop arrangements that might improve the overall performance of LNG facilities. These performance advantages could include higher production rates, higher operating efficiency, and lower emissions [Ref. 3].

LNG production and NGL recovery process configurations can vary substantially, making comparisons and generalizations difficult. KBR has observed a variety of projects and papers where front end versus integrated NGL recovery was scrutinized and seemingly different conclusions drawn. This uncertainty led to an effort to clarify the impact of the placement of the NGL recovery section [Ref. 2, 5].

A key feature of these studies was defining a process configuration that would lend itself to demonstrating trends in the position and arrangement of NGL recovery relative to the liquefaction unit. For this reason it was decided to employ a simple non-proprietary cascade refrigeration system for chilling and liquefaction [Ref. 6]. Although this approach is not as efficient as various licensed processes available, it provides a flexible setup for moving NGL recovery to varying levels in the chilling train while retaining the ability to make comparisons on power requirements.

The configuration of the NGL recovery system by necessity changes somewhat as its placement within the chilling train varies. Two concepts were used as the basic approaches to NGL recovery for all of the cases considered. The first employed a high pressure column with external refrigeration to generate reflux for the required separation. The second used a traditional "expander plant" design utilizing feed gas expansion and recompression to provide the necessary refrigeration for the recovery cycle [Ref. 1]. This approach was further varied by looking at the impact of eliminating recompression.

In all cases the refrigeration power was kept constant while the production rate was allowed to vary. This is consistent with the approach used in many facilities where the refrigeration compressors are driven by a fixed selection of gas turbine drivers. The power consumed by the recompressor, although included in the process specific power, is not considered as part of the refrigeration compression. The significance of this distinction is that it allows additional power to be applied to the process which can have an impact on the overall production rate.

The end use of the recovered NGL will vary from site to site. In some cases it will be exported as a separate product; in others, it is reinjected into the LNG product after satisfying refrigerant makeup requirements. In this report the recovered NGL is not returned to the LNG product stream. This means that the specific power, which is compression power referenced against LNG product rate, will inherently rise with increasing NGL recovery. To recognize this issue and examine the data from a different perspective, an alternate form of specific power referenced against the sum of the LNG and NGL product rates is also tabulated.

STUDY BASIS

The following section describes the assumptions used in developing the data presented in this report.

1) The cascade refrigeration system is based on propane, ethylene, and methane loops. Each refrigeration loop is divided into three stages, chilling the feed stream to the following temperatures
   Propane: 10, -15, -40°C
   Ethylene: -59, -85, -100°C
   Methane: -120, -140, -160°C
2) Natural gas feed is varied to hold the total refrigeration power constant at approximately 143.3 MW.
3) Additional power may be applied to natural gas recompression, which is not included in the limit on refrigeration power.
4) Liquid expansion turbines are not used in refrigeration loops.
5) Flash vapor at the LNG storage tank is held constant at approximately 5 MMscfd, about 1% of the feed gas.
6) Flash gas and boil-off gas compression are not considered.

Assumed feed gas conditions after acid gas removal are given below:

Pressure, bar a	45
Temperature, °C	40
Composition, mol %	
Nitrogen	0.12
Methane	86.05
Ethane	6.40
Propane	4.25
i-Butane	1.18
n-Butane	1.17
i-Pentane	0.31
n-Pentane	0.30
Hexanes +	0.22
Total	100.00

RESULTS

A series of simulations was prepared based on the assumptions given above. The data gathered from these simulations was then examined from several perspectives to better understand the implications of the trends within the information. First, the placement of the NGL recovery unit within the chilling section was considered. Next, the impact of varying levels of ethane recovery at each point in the chilling process was calculated. Finally, alternate NGL recovery configurations were applied at each point for comparison purposes. The resulting data demonstrated design optimums that pointed in different directions depending on whether the desired objective was process efficiency or plant capacity.

Both LNG and total liquid product rates were studied when determining the impact on production values. The total liquid product rate includes both LNG and NGL, which gives a different picture when LNG volume is reduced by NGL recovery. In many plants portions of the recovered NGL are reinjected into the LNG, though that option is not pursued here.

Three temperature levels, 40°C, -15°C, and -40°C, were used as the draw-off points for the feed to the NGL recovery unit using an expander process. NGL recoveries were varied by adjusting the demethanizer column pressure and recompressor power. Ethane (C_2) recovery was used as the metric for setting comparable performance levels for the NGL recovery section. The NGL recovery levels were adjusted to approximately 20%, 35%, and 80% C_2 recovery.

Draw points of -15°C, -40°C, and -59°C were chosen for the NGL recovery approach using a high pressure (HP) refluxed column. The refluxed column overhead is chilled to the next lower refrigeration temperature level, which dictated the NGL recovery.

Process Efficiency

Process efficiency is one of the primary performance metrics applied as a measurement of the benefit offered by a design approach. In many circumstances a higher efficiency will translate into a higher production rate. Process efficiency for this effort is measured in terms of Specific Power, the sum of the refrigeration and recompression power divided by the product rate. A series of figures was prepared illustrating the relationship between specific power and NGL recovery. These include:

- Specific power vs. NGL recovery feed temperature for an expander based NGL recovery process with recompression (Figure 1)
- Specific power vs. C_2 recovery for an expander based NGL recovery process with recompression (Figure 2)
- Specific power vs. C_2 recovery for a NGL recovery process using a high pressure refluxed column (Figure 3)
- Specific power vs. C_2 recovery for an expander based NGL recovery process without recompression (Figure 4)

In each case the results are plotted in two ways, considering the product as LNG only or as the sum of the LNG and NGL product rates.

The figure below presents the relationship between specific power and the feed temperature to an expander based NGL recovery unit. In this case each curve represents a series where C_2 recovery is held constant.

Figure 1
Specific Power vs. NGL Recovery Feed Temperature
For Expander Process with Recompression

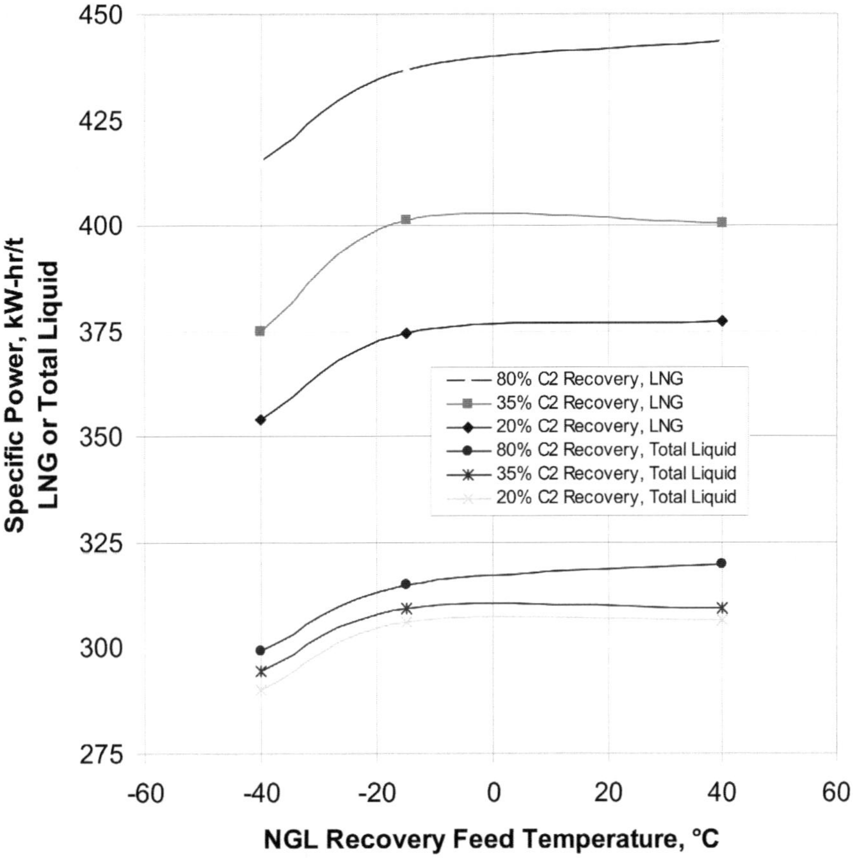

All curves, whether referenced against LNG only or total liquid, show a decided improvement with a -40°C feed point and a relatively flat range at higher temperatures. The placement of the curves reflects the higher specific power associated with increasing C_2 recovery. The separation is much more pronounced in the upper series of curves, which are referenced against LNG only.

Figure 2 below presents the relationship between specific power and the level of C_2 recovery. Each curve represents a series where the feed temperature to the NGL recovery unit is held constant.

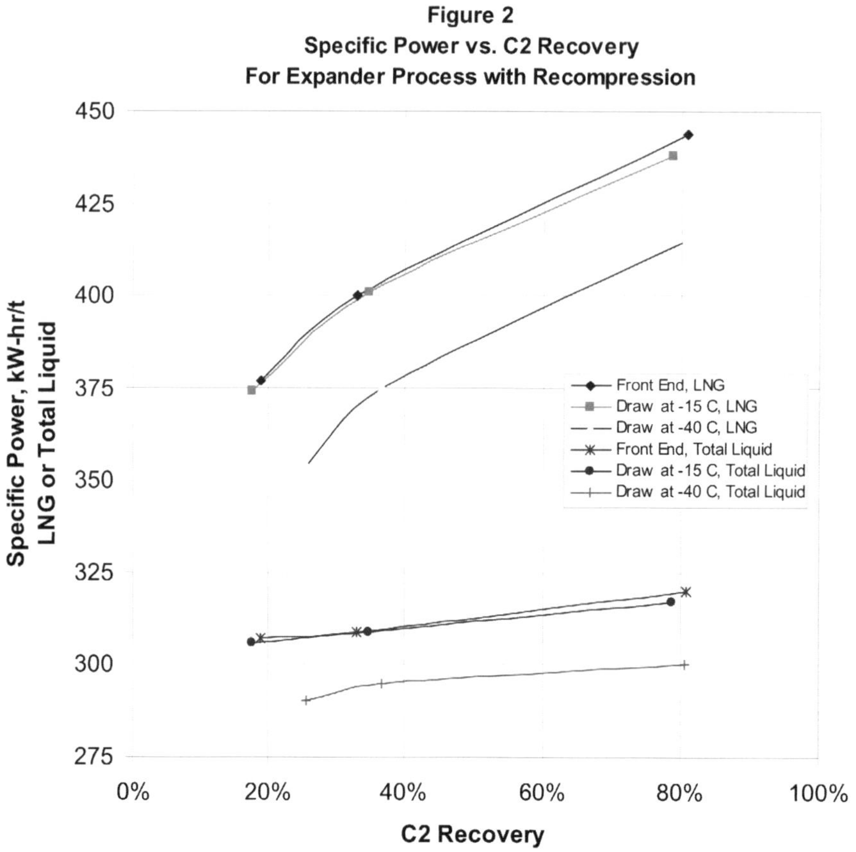

As expected, specific power improves at lower levels of NGL recovery. For the upper curves representing the specific power referenced against LNG only, higher levels of NGL recovery reduces the portion of the feed going to the LNG product. This virtually guarantees that these curves will be rising with higher levels of C_2 recovery.

Although the trend is in the same direction for specific power referenced against total liquid, the effect is not nearly as pronounced. Increasing C_2 recovery does not remove product from the specific power calculation, so the slope is based purely on the power required at varying recovery levels.

Note that the specific power curves for the cases using warmer feeds to the NGL recovery units are very similar. It is not until the feed draw point drops to -40°C that a significant change is seen, which in this case shows a notable improvement.

Figures 3 and 4 below show specific power for two process alternatives that do not use recompression. The former uses a high pressure refluxed demethanizer column that takes its feed from one refrigeration temperature level and its reflux duty from the next. The latter depicts an expander based NGL recovery unit without the normal recompression. This gets the two approaches on a comparable basis using only refrigeration compression power. Comparisons to expander based NGL recovery with associated recompression can be made by referring to the previous cases shown.

Figure 3
Specific Power vs. C2 Recovery
HP Refluxed Column

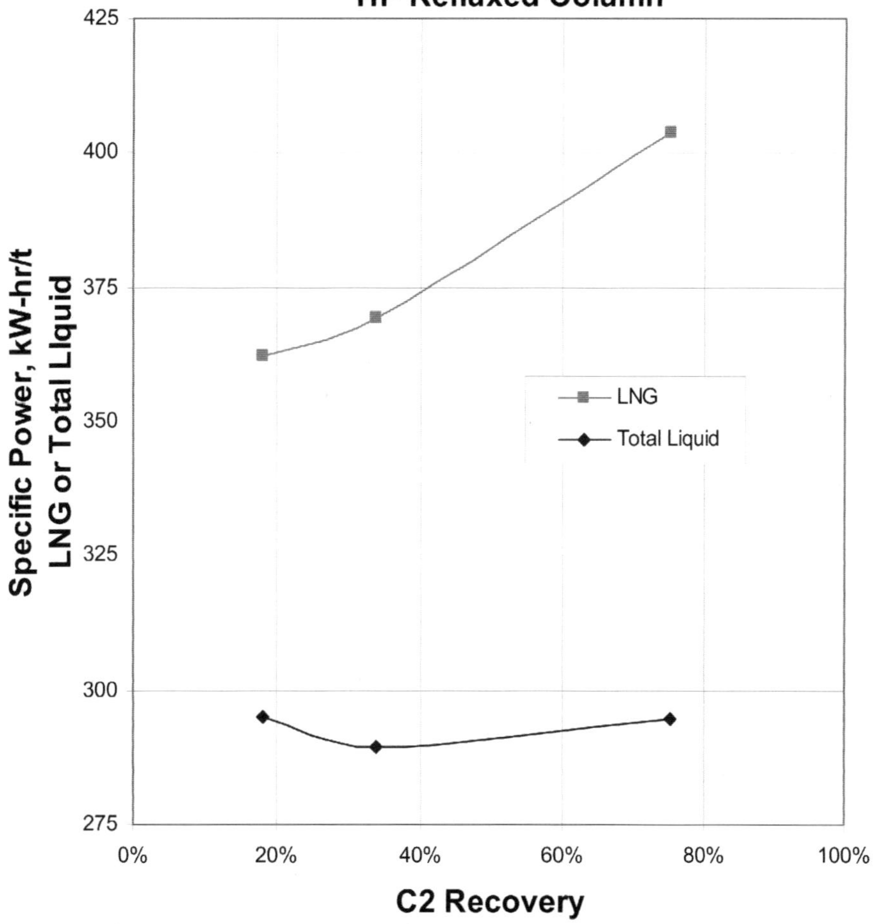

Figure 3 gives the specific power results using a high pressure demethanizer. The general trend of these curves shows a steadily increasing specific power at higher C_2 recovery levels when referenced against LNG only. Specific power for total liquid rate stays relatively flat as the benefits of recovering NGL and reducing the flow to the liquefaction section are offset by the effects of a leaner gas composition on liquefaction temperatures. The refluxed column approach has a similar specific power to the previously shown expander based NGL recovery case with a -40°C feed and using recompression.

Figure 4 below, which shows results from an expander based process without recompression, qualitatively depicts specific power trends similar to cases with recompression.

Figure 4
Specific Power vs. C2 Recovery
Expander Process Without Recompression

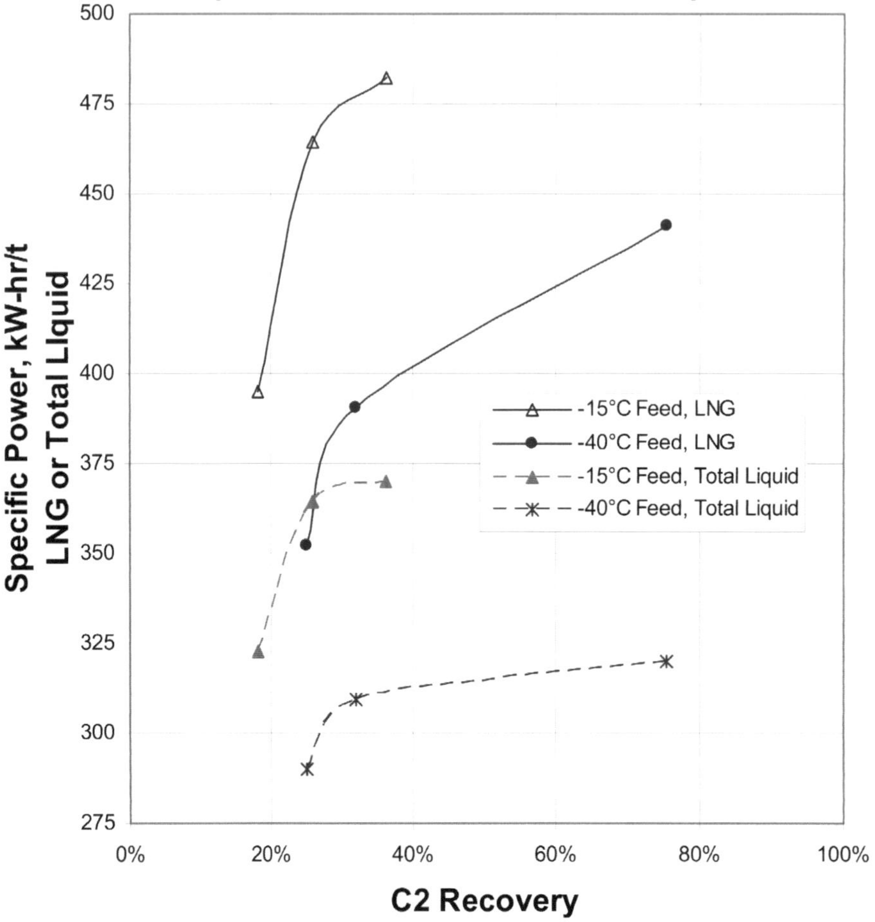

The lack of recompression both increases the specific power and increases the slope of the curve with higher C_2 recovery. The marginal effects are most prominent at low levels of C_2 recovery where mostly heavier components are removed. The impact on the refrigeration system caused by the reduced pressure to the liquefaction section more than offsets the savings in power from omitting recompression. The lower pressure liquefaction feed shifts the condensation curve to a lower temperature level, which increases the refrigeration power requirements.

Production Rate

Although the previous illustrations focused on the aspect of specific power, it is necessary to also examine the impact on production rate to get the total picture. When refrigeration power is fixed, but other sources of power such as recompression can vary, the production rate is not necessarily tied to the process efficiency.

Figure 5 below demonstrates exactly this point. Here, the product rates of the simulated cases are plotted against the feed temperature to an expander based NGL recovery unit. Again, curves representing LNG only and total product liquid are shown.

Figure 5
Product Rate vs. NGL Recovery Feed Temperature
For Expander Process with Recompression

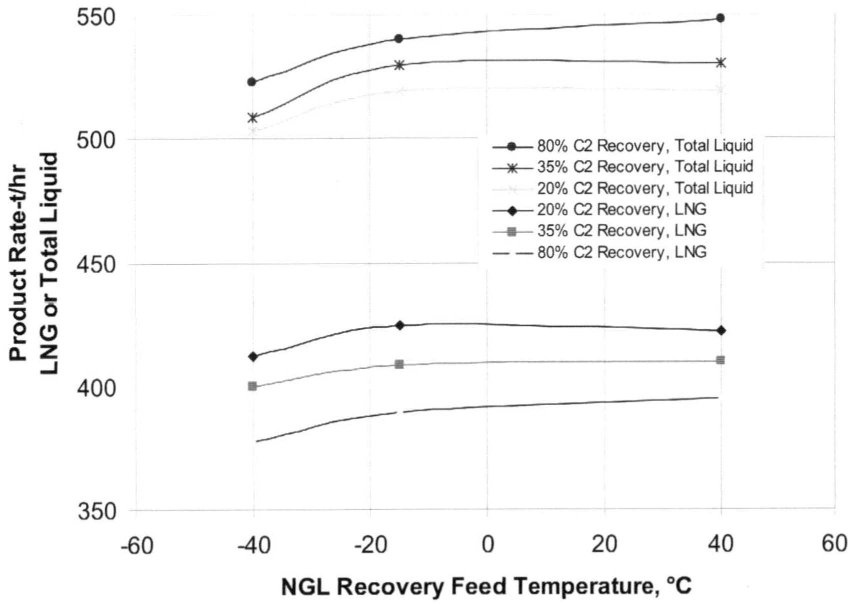

Note that the product rate drops off with the cold -40°C feed, while staying relatively flat in the warmer feed region. This is in contrast to the -40°C feed point previously showing the best specific power. Even though the case using a colder feed had a higher efficiency, the additional power input from the recompressor at warmer feed temperatures showed up in higher product rates.

Figure 6 below demonstrates another difference between the results based purely on process efficiency. Here, product rate is plotted relative to C_2 recovery for an expander based NGL recovery unit.

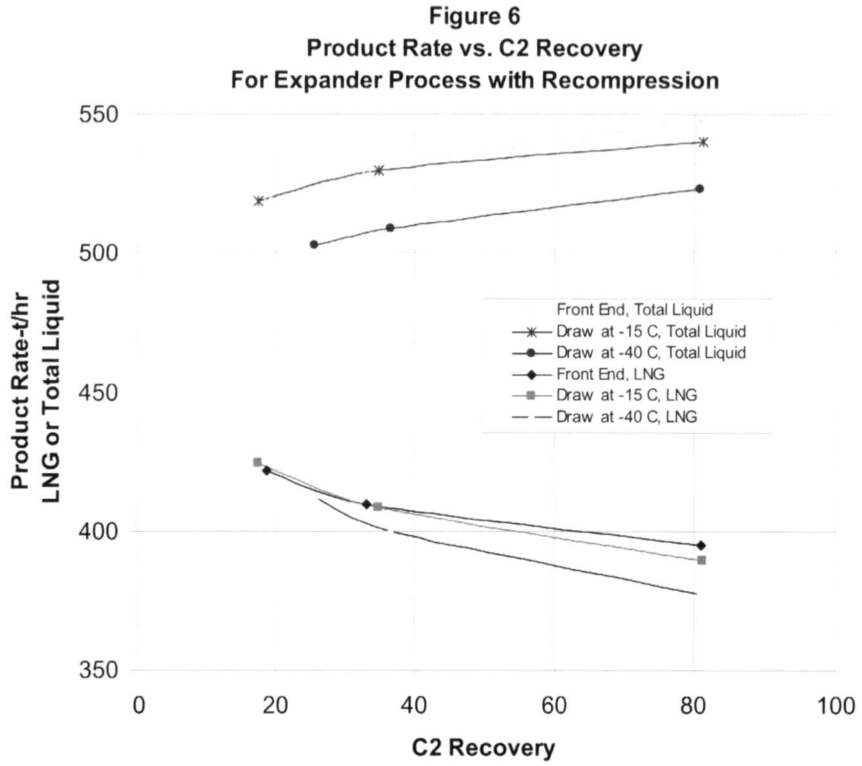

Again, the LNG rate declines as C_2 recovery increases; however the overall product rate increases at higher levels of C_2 recovery. The latter is in contrast to declining process efficiency with increasing C_2 recovery. Note that the highest total liquid product rate is with front end NGL recovery with the deepest level of C_2 extraction. This point is where the greatest amount of recompression power is required.

Figures 7 and 8 below shows product rate versus C_2 recovery for the cases where recompression power is not applied. This includes a high pressure refluxed demethanizer and an expander based NGL recovery process without recompression.

Figure 7
Product Rate vs. C2 Recovery
HP Refluxed Column

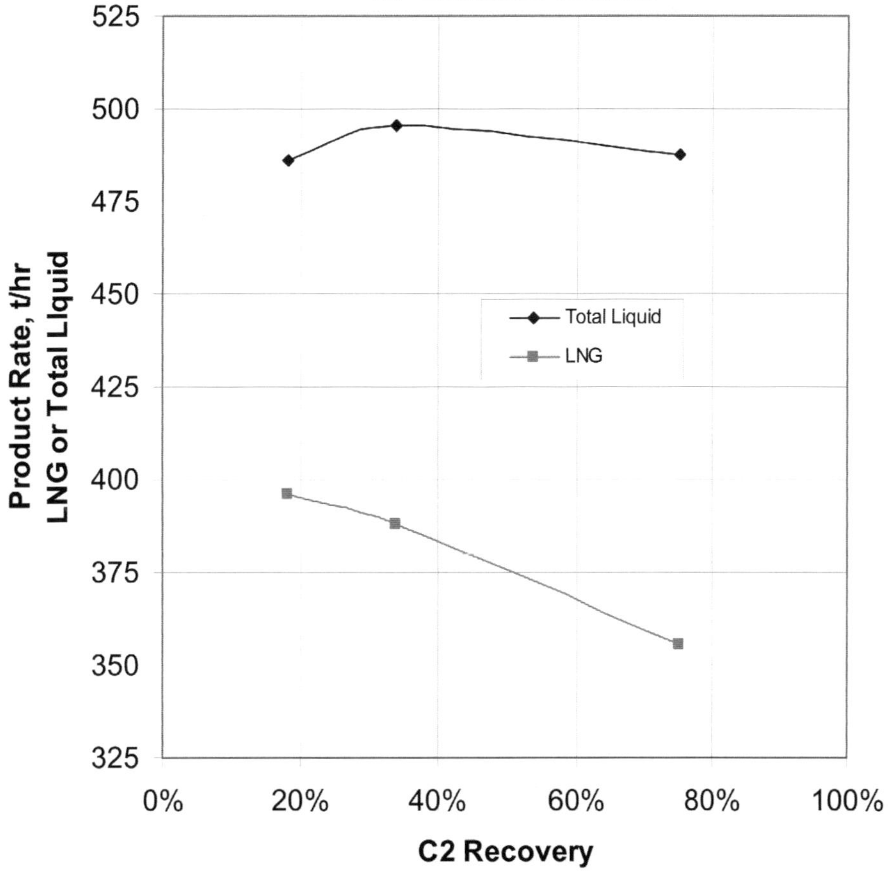

The total liquid production of the HP refluxed column approach stays fairly flat throughout the C_2 recovery spectrum with a slight optimum in the 30-40% C_2 recovery range. LNG production drops as expected with increasing recovery. The decline in LNG production is higher than for the expander process designs using recompression since the latter have an additional power input. Note that even though the refluxed column provides more product than the expander process when recompression is not included, the previous cases illustrating expander processes with recompression produced consistently higher product rates.

Figure 8 shows that when recompression is omitted from an expander based NGL recovery process, both LNG and total product liquid rates decline continuously with higher C_2 recovery.

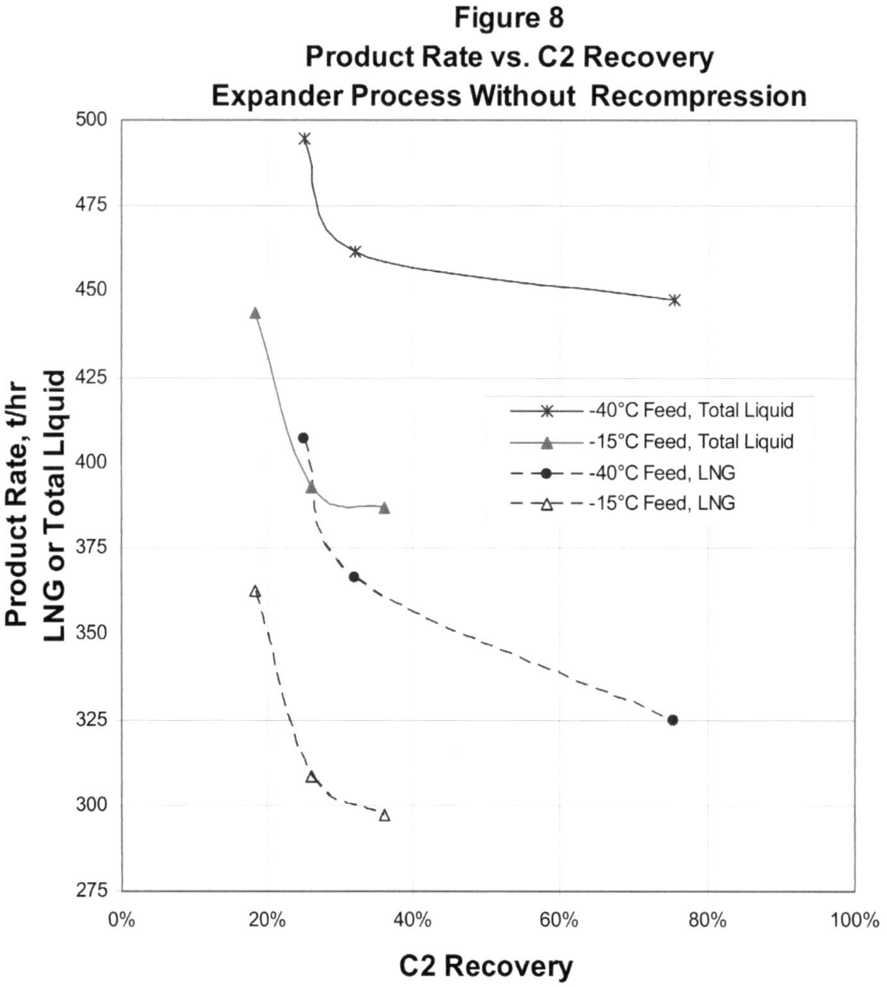

The decline in total product rate is in contrast with all other cases, reflecting the penalty of the lower pressure to the liquefaction section with increasing C_2 recovery.

Heating Value

In most cases significant NGL recovery will not be required beyond providing makeup for the refrigeration system. Occasionally, however, the LNG heating value may need adjustment to meet product specs. The figure below provides an indication of the impact of varying levels of NGL recovery on the higher and lower heating values (HHV and LHV) of the LNG product.

Figure 9
Heating Value vs. C2 Recovery

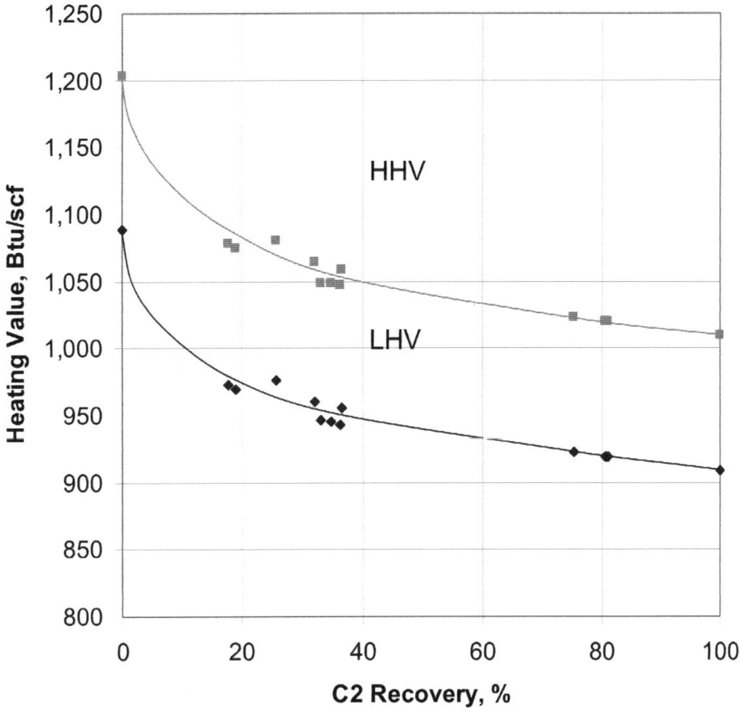

Some of the scatter in the plotted data results from the fact that the various NGL recovery configurations performed differently on recovery of heavier components. Systems using little or no recompression tended to have a higher demethanizer pressure, which decreased the separation effectiveness. Thus the NGL recovery arrangements using a -40°C feed had somewhat higher propane and butane losses than the corresponding processes with warmer feeds. This is illustrated by figures 10 and 11 below.

Figure 10
C3 Recovery vs. C2 Recovery

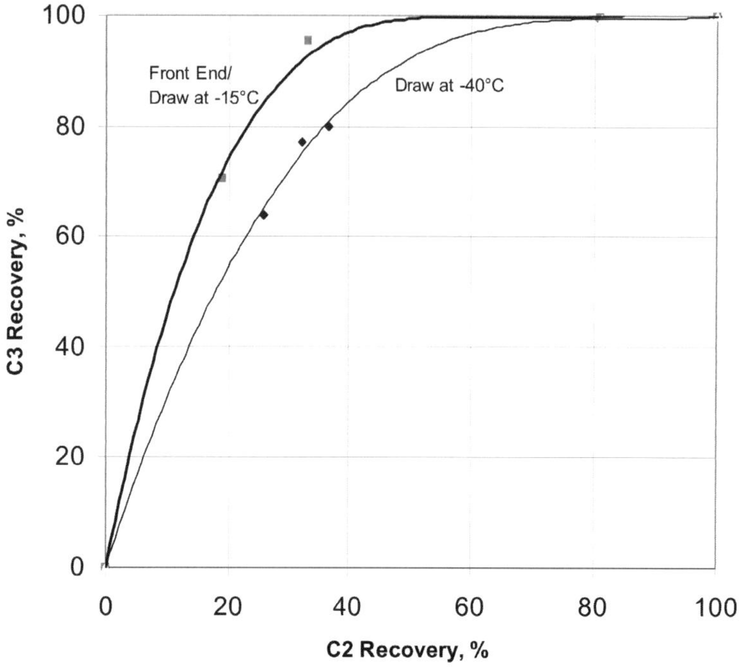

Figure 11
C4 Recovery vs. C2 Recovery

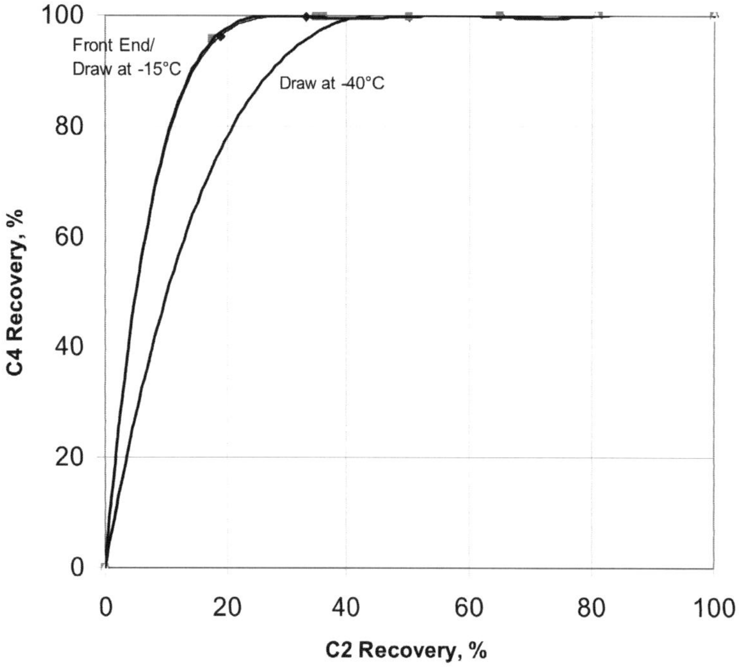

CONCLUSIONS

The optimum design of an NGL recovery system within an LNG production facility depends on the target objectives.

Low Specific Power/High Efficiency

From an efficiency standpoint there is a clear trend supporting an integrated NGL recovery system with a relatively cold feed. The data indicated a significant improvement in overall system performance with a -40°C feed to NGL recovery compared to warmer draw points. It is interesting to note that there was virtually no change in specific power from a front end design (40°C feed) to a -15°C draw.

Lower levels of C_2 recovery offer the highest efficiency, which is as expected. Higher levels of NGL recovery consume additional power and reduce the LNG product stream. When the specific power is measured relative to total liquid product, the overall trend is much less sensitive to the C_2 recovery level.

High Production Rate

Under the basis of design conditions laid out in this study, high efficiency is not necessarily synonymous with high production rates. While refrigeration power is restricted to a fixed value, recompression power associated with NGL recovery is unconstrained. This presents the opportunity for additional power to be applied to the overall system, which can have a positive impact on production rates.

This is best illustrated by noting that an NGL recovery feed temperature of -40°C, which corresponded to the best efficiency point, yielded the lowest production rate at all levels of C_2 recovery. Varying the feed temperature to NGL recovery from 40°C to -15°C had relatively little impact.

LNG production was highest at low C_2 recovery levels, while total liquid production was highest at high C_2 recovery levels. This demonstrated that although increasing NGL recovery pulled potential product away from the LNG stream, it allowed processing a higher feed flow, which resulted in more total liquid production.

For those processes where recompression power was not used, the highest product rate naturally corresponded to the highest efficiency point since the total power applied to the system remained constant.

Optimum System

As indicated in the sections above, this study did not indicate an optimum approach that was equally valid for all facilities. The constraints placed upon each design can shift the design philosophy in different directions. There does appear to be a best efficiency point that would support a colder feed to the NGL recovery system. There also appears to be an opportunity at warmer feed temperatures to apply additional power through recompression that can increase product yields in refrigeration constrained systems.

Although this study holds the LNG and NGL product streams as separate entities, most LNG production facilities fractionate and recombine a portion of the NGL with the LNG product. Exploration of various flow arrangements and recovery level combinations would add another dimension of complexity to the systems described above; however it is expected that the general trends of these approaches will be the same as those presented here.

REFERENCES

1. Campbell, R.E. and Wilkinson, J.D., “Hydrocarbon Gas Processing”, U.S. Patent No. 4,157,904 (1977).

2. Hagyard, P., Paradowski, H., and Gadelle, D., “Simultaneous Production of LNG and NGL”, paper presented at LNG 14 Conference, Doha, Qatar (2004).

3. Hudson, H.M., Wilkinson, J.D., Cuellar, K.G., and Pierce, M.C., “Integrated Liquids Recovery Technology Improves LNG Production Efficiency”, paper presented at GPA 2004 Spring National Meeting, New Orleans, LA (2004).

4. Kotzot, H., “Overview of the LNG Industry”, paper presented at GPA 2003 Spring National Meeting, San Antonio, TX (2003).

5. Sawchuk, J., Jones, R., Durr, C., and Davis, K., “BP’s Big Green Train: Benchmark Next Generation LNG Plant Designs”, paper presented at LNG 14 Conference, Doha, Qatar (2004).

6. Scurlock, R.G., editor, “History and Origins of Cryogenics”, Oxford University Press, Oxford, UK (1992).

Benefits of Integrating NGL Extraction and LNG Liquefaction Technology

Doug Elliot
IPSI LLC

Wesley R. Qualls
ConocoPhillips Company

Shawn Huang
ConocoPhillips Company

Jong Juh (Roger) Chen
IPSI LLC

R. J. Lee
(Formerly) IPSI LLC

Jame Yao
IPSI LLC

Ying (Irene) Zhang
IPSI LLC

AIChE Spring Meeting, April 2005
5th Topical Conference on Natural Gas Utilization
Atlanta April 10-14, 2005

Abstract

The combination of changing global markets for natural gas liquids (NGL) with the simultaneous increase in global demand for liquefied natural gas (LNG) has stimulated an interest in the integration of NGL recovery technology with LNG liquefaction technologies. Historically, the removal of "heavy" or high-freezing-point hydrocarbons from the feed to LNG plants has been characterized as "gas conditioning" and achieved using one or more distillation columns. While some attempts to provide reflux to the distillation columns marginally enhanced NGL recovery, little emphasis was placed on maximizing NGL recovery as a product from the LNG process. As such, the integration of the two processes was not a priority. Integrating state-of-the art NGL recovery technology within the *CoP LNG℠ Process*[1], formerly the Phillips Optimized Cascade LNG Process, results in a significant reduction in the specific power required to produce LNG, while maximizing NGL recovery. This corresponds to a production increase in both LNG and NGL for comparable compression schemes as compared to stand-alone LNG liquefaction and NGL extraction facilities. In addition, there are potential enhancements to the overall facility availability and project economics using the integrated concept. This integrated concept has been applied to three ongoing international NGL/LNG projects using the CoP LNG Process. In these cases, LNG production has increased by approximately 7%, while using the same process horsepower.

1.0 Introduction

Due to clean burning characteristics and the ability to meet stringent environmental requirements, the demand for natural gas has increased considerably over the past few years. Projections reflect a continued increase for the next several years. However, it is a clean burning methane rich gas that is in demand as opposed to the typical raw gas that exists in nature, which often includes additional components such as heavier hydrocarbons and other impurities. The heavier hydrocarbons, once separated from natural gas, are referred to as Liquefied Petroleum Gas (LPG) and Natural Gas Liquids (NGL). Impurities may include carbon dioxide, hydrogen sulfide, mercaptans, nitrogen, helium, water, and even trace contaminants such as mercury and trimethylarsine. Natural gas must be "conditioned" prior to liquefaction to remove undesired components. This "conditioning" normally takes place in separate or stand-alone facilities and typically includes the extraction of heavier hydrocarbons such as LPG and NGL. The "conditioned" gas is then typically fed to pipelines for distribution.

However, transportation to distant markets through gas pipelines is not always economically or technically feasible. As such, natural gas liquefaction has become a viable and widely accepted alternative. The economics of liquefying natural gas is feasible due mainly to the great volume reduction achieved upon liquefying, which creates the ability to store and transport large volumes. The demand for Liquefied Natural Gas (LNG) in North America has increased considerably as energy demands have increased at the same time wellhead gas supply to pipelines has decreased. LNG is a natural alternative to supplement gas pipelines as the infrastructure to process and burn the gas is largely in place. In addition, LNG is a highly reliable source of gas. For instance, ConocoPhillips' Kenai, Alaska facility, which utilizes the CoP LNG Process, has supplied LNG to Japan for over thirty-five years without missing a single cargo. Figure 1 is an aerial view of the recently constructed Atlantic LNG facility located in Point Fortin, Trinidad, comprised of three trains and a fourth under construction, all utilizing a modernized version of the CoP LNG Process.

Figure 1 – The Atlantic LNG Facility – Artist Rendition, Courtesy of Atlantic LNG Company of Trinidad & Tobago

Pentanes and heavier hydrocarbons, including aromatics having a high freezing point, must be substantially removed to an extremely low level in order to prevent freezing and subsequent plugging of process equipment in the course of liquefaction. In addition, heavy components must also be removed in order to meet BTU requirements of the LNG product. The heavy hydrocarbons separated from LNG, may then be utilized as petrochemical sales or for gasoline blending. In fact, NGL and/or LPG liquids may command a greater value than LNG. Many efforts have focused on recovery of these heavy hydrocarbons. However, most of the effort has been placed on removal of the heavy hydrocarbons in a separate NGL plant, located upstream of the LNG liquefaction facility.

Alternatively, since all components having a higher condensing temperature than methane will be liquefied in the liquefaction process, it becomes technically practical to integrate NGL recovery within LNG liquefaction. Duplication of processing equipment and refrigeration requirements are avoided with an integrated approach. In fact, a substantial cost savings may be achieved when NGL recovery is effectively integrated within the liquefaction process.

There have been attempts for NGL recovery within the LNG facility.(1) For example, lighter NGL components are recovered in conjunction with the removal of C_5+ hydrocarbons using a scrub column in a propane pre-cooled mixed refrigerant process, as disclosed in the literature.(2, 3) The NGL recovery column is often required to operate at a relatively high pressure (typically above 550 psig) in order to conserve refrigeration compression horsepower requirements. Although refrigeration horsepower is conserved utilizing a high-pressure column, separation efficiency is often significantly reduced due to less favorable operating conditions, *i.e.* lower relative volatility.

With careful integration of the NGL recovery and LNG processes, the overall efficiency of the integrated process can be significantly improved, thereby increasing NGL recovery and reducing specific power consumption.(4) This paper provides examples of NGL recovery technology integrated within the CoP LNG Process, while also presenting recovery and efficiency performance.

2.0 Traditional Stand-Alone Gas Plant Upstream of Liquefaction Plant

A number of NGL recovery processes have been developed for natural gas and other gas streams.(5, 6, 7). Among various NGL recovery processes, the cryogenic expansion process has become the preferred process for deep hydrocarbon liquid recovery from natural gas streams. Figure 2 depicts a typical cryogenic expansion process configuration. In the conventional turbo-expander process, feed gas at elevated pressure is pretreated for removal of acid gases, water, mercury and other contaminants to produce a purified gas suitable for cryogenic temperatures. The treated gas is typically partially condensed utilizing heat exchange with other process streams and/or external propane refrigeration, depending upon the gas composition. The resulting condensed liquid, containing the less volatile components, is then separated and fed to a medium or low-pressure fractionation column for recovery of the heavy hydrocarbon components. The remaining non-condensed vapor, containing the more volatile components, is expanded to the lower pressure of the column using a turbo-expander, resulting in further cooling and additional liquid condensation. With the expander discharge pressure essentially the same as the column pressure, the resulting two-phase stream is fed to the top section of the fractionation column. The cold liquid portion acts as reflux, enhancing recovery of heavier hydrocarbon components. The vapor portion combines with the gas in the overhead of the column. The combined gas exits the column overhead as a residue gas. After recovery of available refrigeration, the residue gas is then recompressed to a higher pressure, suitable for pipeline delivery or for LNG liquefaction.

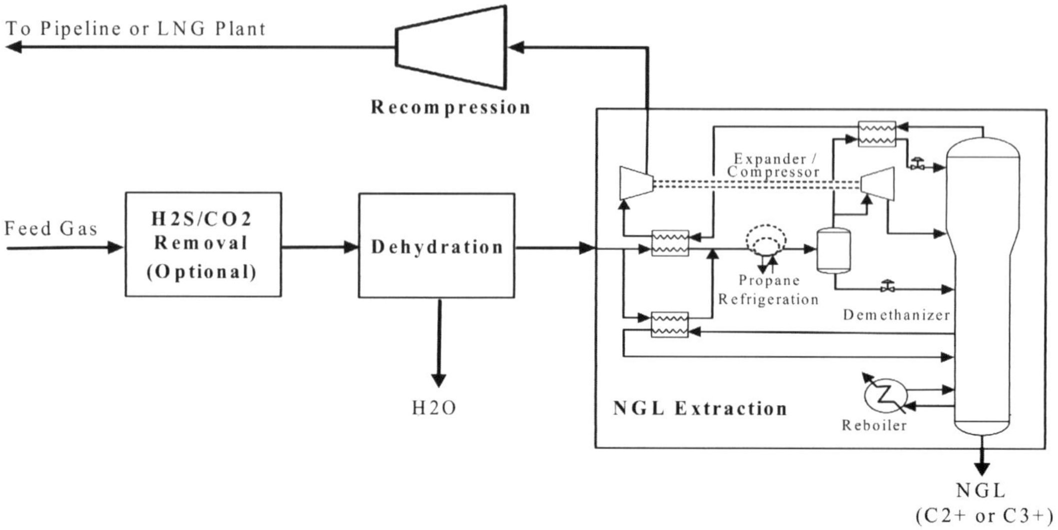

Figure 2 - Block diagram for typical NGL extraction plant

The fractionation column (as described) acts essentially as a stripping column since expander discharge vapors are not subject to rectification. As such, a significant quantity of heavy components remains in the gas stream. These components could be further recovered if subjected to rectification. In an attempt to achieve greater liquid recoveries, recent efforts have focused on the addition of a rectification section and methods to effectively generate an optimal rectification reflux stream, e.g. the gas subcooled reflux illustrated in Figure 2.

Figure 3 illustrates a typical block diagram of the LNG facility. Gas comprising predominantly methane enters the LNG facility at elevated pressures and is pretreated to produce a feedstock suitable for liquefaction at cryogenic temperatures. Pretreatment typically includes the removal of acid gases (hydrogen sulfide and carbon dioxide), mercaptans, water, mercury, and other contaminants. The treated gas is then subjected to a plurality of cooling stages by indirect heat exchange with one or more refrigerants, whereby the gas is progressively reduced in temperature until complete liquefaction. The pressurized LNG is further expanded and sub-cooled in one or more stages to facilitate storage at slightly above atmospheric pressure. Flashed vapors and boil off gas are recycled within the process.

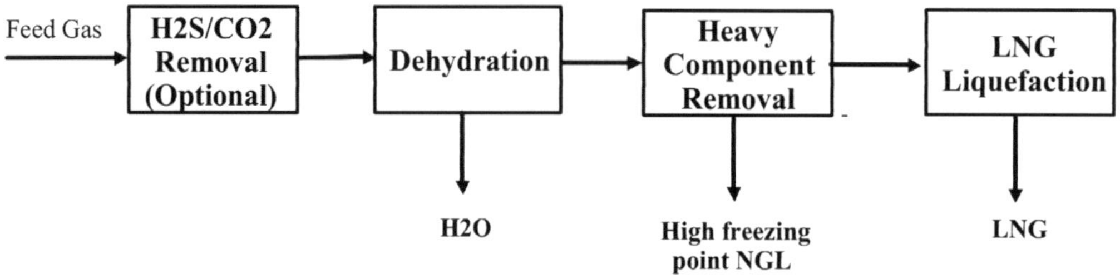

Figure 3 - Block diagram for typical LNG plant

Because LNG liquefaction requires a significant amount of refrigeration energy, the refrigeration system(s) represent a large portion of a LNG facility. A number of liquefaction processes have been developed with the differences mainly residing on the type of refrigeration cycles employed. The most commonly utilized LNG technologies are:

1) ***CoP LNG^{SM} Process*(8)** This process, formerly known as the Phillips Optimized Cascade LNG Process, utilizes essentially pure refrigerant components in an integrated cascade arrangement. The process offers high efficiency and reliability. Brazed aluminum exchangers are largely used for heat transfer area, providing for a robust facility that is easy to operate and maintain. Refrigerants typically employed are propane, ethylene and methane.

2) **Propane pre-cooled mixed refrigerant process(9, 10)** This mixed refrigerant process provides an efficient process utilizing a multi-component mixture of hydrocarbons typically comprising propane, ethane, methane, and optionally other light components in one cycle. A large spiral wound exchanger is utilized for the majority of heat transfer area. A separate propane refrigeration cycle is utilized to pre-cool the natural gas and mixed refrigerant streams to approximately –35 °F.

3) **Single mixed refrigerant process(11)** This process includes heavier hydrocarbons in the multi-component mixture, *e.g.* butanes and pentanes, and eliminates the pre-cooled propane refrigeration cycle. The process presents the simplicity of single compression, which is advantageous for small LNG plants.

Historically, the removal of heavy hydrocarbons from natural gas is considered part of feed conditioning. In most cases, the residue gas (comprising primarily of methane) from a NGL recovery plant is delivered to the LNG facility for liquefaction. It is common practice for NGL extraction to stand-alone as a separate plant from LNG liquefaction facilities for various commercial or geographical reasons. One such commercial reason is when NGL recovery and sales are desired well in advance of LNG. There may also be geographical reasons to take this approach such as cases where NGL liquids are required in a different location than LNG and where a long gas pipeline separates the two plants. As still another reason, the NGL recovery plant may already exist, prior to consideration of a LNG facility.

3.0 Process Description of an Integrated NGL and LNG plant

A block diagram for an integrated LNG and NGL process is presented in Figure 4. For the purposes of this paper, the CoP LNG Process was used for the LNG liquefaction technology. Treated natural gas is first cooled by utilizing refrigeration from within the liquefaction process in one or more stages and then introduced into a distillation column, or Heavies Removal Column. Figure 4 represents the simplest embodiment of NGL integration, where the Heavies Removal Column is not refluxed other than with condensed liquids contained within the column feed. Once the feed has entered the column, it is separated or in this case stripped. A bottoms stream, primarily comprised of NGL components, and a methane rich overhead stream are formed. The methane rich overhead stream is chilled, condensed, and in most cases subcooled within the liquefaction process. Once liquefied and subcooled, the stream is subsequently flashed to near atmospheric pressure in one or more steps in preparation for LNG storage. Flashed vapor is used as a methane recycle refrigerant with a portion heated and compressed for fuel. The liquid stream from the Heavies Removal Column is introduced to a second distillation column in one or more feed trays, depicted in Figure 4 as a Deethanizer. In the second column, the liquid stream is separated into a bottoms stream and a vapor overhead stream, primarily comprised of ethane and lighter components. The second column acts primarily as a deethanizer or partial deethanizer or in some cases a depropanizer or partial depropanizer, depending on the desired BTU level of the LNG product and desired level of propane and/or ethane recovery. The second column bottoms stream may be routed to further fractionation in order to separate the NGL and/or LPG liquids into the desired product slate.

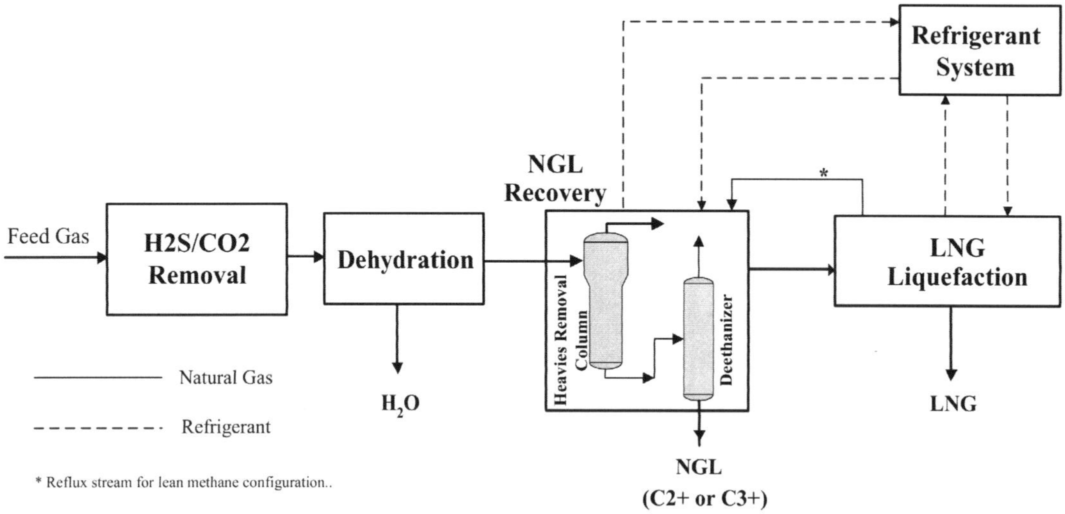

Figure 4 - Block diagram showing the integrated NGL and LNG process

NGL recovery integration not only reduces capital investment through re-utilizing essentially all equipment in the NGL facility for LNG production, but also improves overall thermodynamic efficiency. There are significant advantages in the following aspects:

- The overall integrated process reduces combined capital and operating costs.
- The integrated process reduces combined CO_2 and NO_X emissions by improving the thermodynamic efficiency of the overall process.
- Higher recovery of propane (and ethane) is achievable.
- Most NGL process equipment is already utilized in LNG liquefaction plants.

In the integrated process for cryogenically recovering ethane, propane, and heavier components, the Heavies Removal Column in the LNG facility replaces the NGL recovery column in the NGL plant, thereby reducing capital expenditure. Through adjusting operating conditions, the level of NGL recovery can be optimized in accordance with required specifications and the relative market prices of LNG, LPG and NGL. Only one common utility is required in the integrated plant, resulting in additional savings in capital expenditure. If well planned, the NGL recovery portion of the plant may be constructed at an earlier stage and later integrated into the LNG liquefaction plant. The opportunity for early NGL sales may significantly improve LNG project economics. The flexible design of an integrated plant allows for an easy transition between ethane recovery or ethane rejection, which is useful given relatively frequent changes in ethane demand.

Since integration of NGL recovery into the natural gas liquefaction process allows for higher recovery of heavier hydrocarbon components, the removal of liquefaction contaminants such as cyclohexane and benzene are also improved. This is important since these particular components, even at relatively low concentrations, may create freezing problems in the colder sections of the LNG process. Thus higher NGL recovery and less operational concerns are achieved at the same time that the front-end NGL plant is eliminated.

4.0 Variations of the Integrated Plants

Greater recovery of LPG and NGL products is possible by supplying a rectification section to the Heavies Removal Column and selecting an optimal method of reflux. Various configurations exist, depending on the component selected for recovery as well as the desired recovery level. Of course, multiple streams within the liquefaction section may be utilized for reflux to this column. The cases that follow are not an exhaustive list but rather the more common choices as integrated within the CoP LNG Process.

Case 1 No-reflux case

The simplest embodiment of integrated NGL recovery utilizes a Heavies Removal Column with the only reflux essentially from liquids contained within the column feed.

Case 2 Lean Methane Reflux Scheme

A lean methane stream is condensed within the liquefaction process and introduced as reflux to the Heavies Removal Column. In the CoP LNG Process, there are multiple sources that may be used for lean methane reflux, each containing extremely low concentrations of heavy components. Lean methane used as reflux enhances NGL recovery efficiency within the column, subsequently reducing NGL components in the overhead stream to a minimum. Thus, a higher NGL recovery is achieved even at relatively high operating pressures, in the range of 600 psig. Of course, lean methane reflux is more advantageous for ethane recovery operations.

Case 3 Deethanizer overhead reflux scheme

In this configuration, Heavies Removal Column reflux is generated from the deethanizer overhead. Refer to Figure 4. The deethanizer overhead is partially condensed with the liquids or a portion of the liquids introduced to the top of the deethanizer as normal reflux. The vapor or a portion of the vapor is compressed, partially condensed and introduced to the Heavies Removal Column as reflux. This reflux stream is rich in ethane, which provides an excellent choice for propane recovery. It is possible to operate the deethanizer as a demethanizer, providing a methane rich reflux to the Heavies Removal Column. Thus, operation may be easily switched between propane and ethane recovery simply by adjusting operating parameters.

5.0 Case Study Assumptions

The cases described above; no-reflux, lean methane reflux, and deethanizer overhead reflux were studied with the results presented in this paper. Each integration configuration was modeled at varying levels of propane recovery. For propane recovery specifications, a 2.0 mole% maximum C2:C3 ratio was used, consistent with commercial-quality propane. A case was modeled for ethane recovery as well, assuming an 85% ethane recovery target. For ethane recovery specifications, a 2.0 mole% maximum C1:C2 ratio was used, consistent with commercial-quality ethane.

The LNG production rate for each case examined assumes that gas turbine drivers are utilized to provide refrigeration requirements for the liquefaction process. The available power is based on 25°C (77°F) ambient air temperature. For consistency, gas-treating requirements are identical in all cases as well as the gas turbine horsepower. Table 1 summarizes the parameters for the case studies.

Table 1: Process Parameters for Case Studies

Feed Gas Composition			**Heat Sink for Cooling**	
Component	Mole Percent		Cooling Medium	Ambient Air
Nitrogen	0.10		Temperature (°F)	77
Carbon Dioxide	0.01			
Methane	85.99		**Inlet Feed Gas Pressure**	
Ethane	7.50		Pressure (psig)	1100
Propane	3.50			
i-Butane	1.00		**Liquefaction Cycle CoP LNG Process**	
n-Butane	1.00			

i-Pentane	0.30			
n-Pentane	0.20			
Hexane Plus	0.40			
Total	100.00			

The Net Present Value (NPV) calculation is based on a 20-year production life assuming a 10% discount rate. Installation costs for storage tanks, compressors, turbines, and other common equipment were assumed to be the same for all cases. The following premises are used for NPV comparisons.

- The no-reflux integration case requires the lowest capital expenditures. As such, it provides a convenient basis for comparison. Installation costs for the no-reflux integration or base case are assumed at $200 MM for 0% propane recovery.
- Installation costs for small changes in feed rates are estimated using the six-tenths rule.
- The stand-alone or non-integrated NGL recovery case required $5 MM for additional equipment and $5 MM for external propane refrigeration for a total of $10 MM over the base case.
- The lean methane reflux integration case required $4 MM for additional equipment over the base case.
- The deethanizer overhead reflux case includes $5 MM for additional equipment over the bases case.

6.0 Comparison of Non-Integrated Facilities With Integrated Facilities

Propane Recovery Cases

Table 2 compares the performance of the stand-alone or non-integrated case and three different integrated cases, no-reflux, lean methane reflux, and deethanizer overhead reflux. Incremental NPV calculations are included. For clarification of terms, the definition of specific power consumption is the required compressor power divided by feed gas flow rate (HP/MMSCFD Feed). As revealed in Table 2, the specific power consumption of the deethanizer overhead reflux case is about 4% lower than that of the stand-alone non-integrated case. The comparison also reveals that from the aspect of specific power consumption, integrated configurations utilizing no-reflux and lean methane reflux do not result in significant advantages. In fact, the results reveal that the integrated process with no-reflux requires more specific power than the non-integrated case. On the other hand, the integrated case utilizing deethanizer overhead as reflux resulted in higher LNG and NGL production capacities as compared to the other cases.

The study revealed that liquid recovery cases for integrated facilities utilizing no-reflux or lean methane reflux were not as efficient for propane recovery and did not allow for optimal heat integration. The integrated process utilizing deethanizer overhead as reflux overcame these limitations, allowing for higher efficiency and liquid recovery. In addition, since efficiency is improved and installation costs remain lower, a higher NPV is realized as compared to the non-integrated case, where NGL recovery is achieved in a stand-alone facility. This effectively demonstrates the economic advantage of integrating NGL recovery within LNG liquefaction. The cases presented efficiently integrated NGL recovery into LNG liquefaction technology while allowing for higher overall propane recovery as well as that of heavier components, in excess of

95% from the feed gas. The optimized integrated process allows for recovery of over 99% for propane and heavier hydrocarbon. Higher propane and NGL recovery is achieved while requiring less energy and also eliminating the stand-alone NGL plant, leading to significant savings in operating as well as capital costs.

Ethane Recovery

As revealed above, the integrated case using deethanizer overhead as reflux for the Heavies Removal Column is effective for high propane recovery. The operating parameters may also be modified quite easily for enhanced ethane recovery. Table 2 illustrated the results of the integration operation for ethane recovery conditions using the same feed conditions listed in Table 1. Ethane recovery above 85% was easily achievable for this integration case. Again, multiple sources of lean methane reflux may be utilized with comparable results.

The price of liquid ethane is historically cyclical, depending heavily on petrochemical feedstock demands. When liquid ethane demand is high, increasing ethane recovery may generate additional revenue. On the other hand, it is often desirable to leave ethane in the LNG product, while maintaining high propane recovery when the ethane market is depressed. Due to the cyclic nature of the liquid ethane, a facility providing flexibility to easily switch between propane and ethane recovery allows producers to quickly respond to market conditions.

Table 2: Performance of Non-Integrated and Integrated Cases

	Propane Recovery							Ethane Recovery
Case Description	Base	Integrated NGL and LNG Facilities						Integrated Facilities
		Integration With No-Reflux		Lean Methane Reflux		Deethanizer Overhead Reflux		Lean Methane Reflux
	Standalone		Deviation		Deviation		Deviation	
Overall Performance								
Feed Gas, MMSCFD	660	603	-8.7%	661	0.0%	690	4.5%	705
Ethane Recovery, %								85.8%
Propane Recovery, %	98.5%	95.0%	-3.6%	95.0%	-3.6%	95.0%	-3.6%	100.0%
Relative LNG Production @ 93% Availability, MTPA	100	93	-7.0%	102	2.4%	107	7.2%	90
Relative NGL Recovery, BPD	100	89	-11.2%	97	-2.6%	102	1.8%	
Relative Specific Power, HP/MMSCFD Feed	**100**	**110**	**9.5%**	**100**	**0.0%**	**96**	**-4.3%**	**94**
Incremental NPV (20 yrs, 10%), $MM		**-266**		**52**		**217**		

Economic Basis: LNG – $2.5/MMBTU; NGL - $17/BBL; Feed - $0.5/MMBTU

Comparison of Reflux Schemes in Integrated Processes

The relative specific power consumption for the three integrated configurations are compared and presented in Figure 5. For the no-reflux case, relative specific power first decreases but then increases as propane recovery increases. Once propane recovery is higher than about 75%, relative specific power becomes a strong function of the desired propane recovery level. Essentially, this is due to the fact that the liquid recovery section of the no-reflux scheme is not very efficient, requiring additional horsepower to achieve high propane recovery.

By comparison, the relative specific power of the deethanizer overhead reflux case continues to decreases as propane recovery increases. Given the same conditions, the deethanizer overhead reflux case requires the lowest specific power consumption with the no-reflux case requiring the highest. The trend is more evident at higher propane recoveries. At 95% propane recovery, the specific power of the deethanizer overhead reflux case is 4% less than the lean methane reflux case and 12% less than the no-reflux case. Of course, as propane recovery increases, the NGL product rate also increases. Generating natural gas liquids (NGL) requires less refrigeration than liquefying methane-rich natural gas (LNG). Therefore, the deethanizer overhead reflux configuration requires less specific power (HP/MMSCFD) at high propane recoveries and also improves overall facility thermal efficiency.

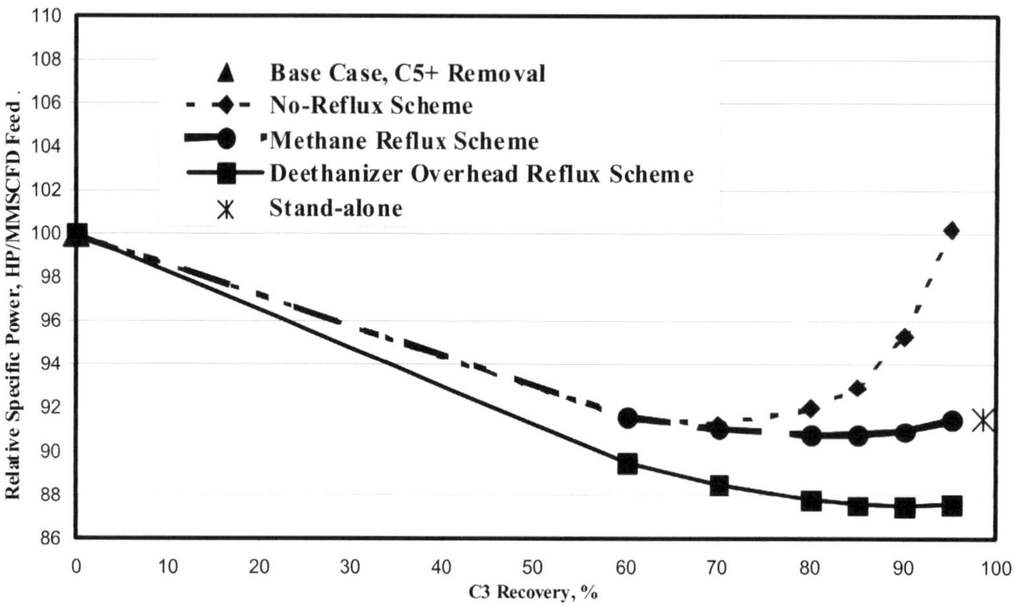

Figure 5 - Comparison of Relative Specific Power Requirements at Varying Propane Recovery for Different NGL Integration Cases

Figure 6 illustrates the dependence of incremental NPV on propane recovery level for the three different NGL recovery configurations. The basis for comparison of incremental NPV is the no-reflux case at 0% propane recovery. For the no-reflux base case, incremental NPV first increases but then decreases as propane recovery is increased. As one would expect from the specific power requirements results revealed in Figure 5, the incremental NPV of the deethanizer reflux case continues to increase as propane recovery increases. While, NPV for the lean methane reflux case is higher than the no-reflux case, the deethanizer overhead reflux case is clearly a better choice for high propane recovery requirements.

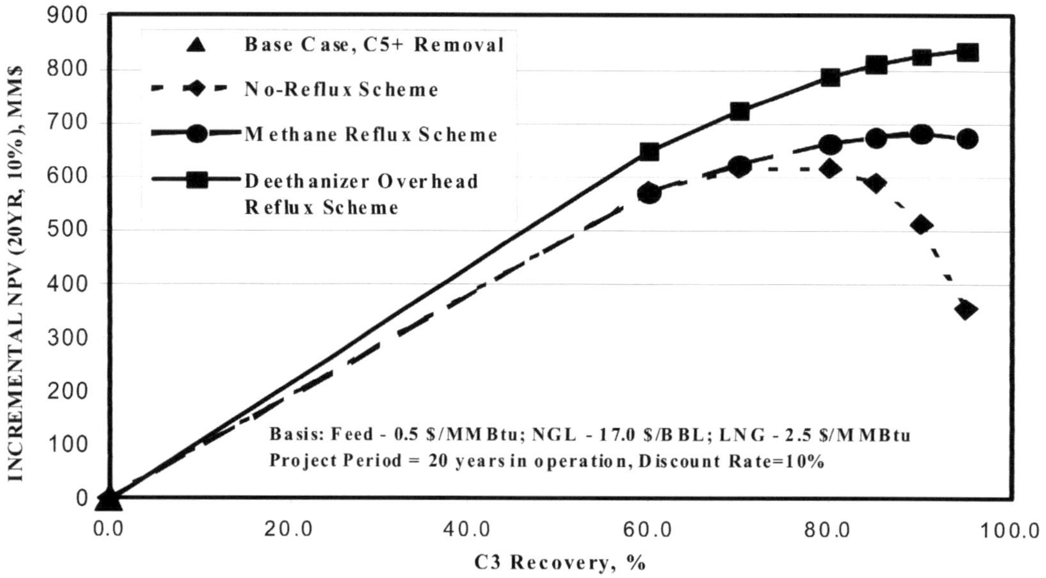

Figure 6 - Incremental NPV at Varying Propane Recovery for Differing Integrated Cases

Table 3: Comparison of Lean Methane Reflux and Deethanizer Overhead Reflux

Case Description	Lean Methane Reflux	Deethanizer Overhead Reflux
C3 Recovery, %	95	95
Heavies Removal Column Reflux		
Relative Flow Rate, M^3/Hr	226	100
C2 Content, mol%	4.6	56.3
Critical Pressure, Psia	769	978
Critical Temperature, ^oF	-100	35
Heavies Removal Column Top Tray		
C2 Content, mol%	5.4	10.1
Critical Pressure, Psia	793	865
Critical Temperature, F	-94	-78
Relative Specific Power, HP/MMSCFD Feed	**104.3**	**100**

Operating pressure of the Heavies Removal Column is ultimately limited by the critical pressure of the hydrocarbon mixture in the column. As the column pressure approaches the critical pressure, relative volatility decreases significantly, resulting in a decreased capacity for fractionation. However, the deethanizer overhead reflux configuration is capable of providing an ethane-rich reflux stream to the top of the Heavies Removal Column, driving the column further away from the critical pressure, thereby enhancing relative volatility. Therefore, overall separation efficiency is improved. Therefore, in order to achieve the same propane recovery levels as the lean methane reflux case, much less reflux is required, further resulting capital and operating cost savings. Table 3 provides a more detailed comparison of the deethanizer overhead reflux and the lean methane reflux configurations.

The deethanizer overhead reflux case allows higher-pressure operation of the Heavies Removal Column as compared to the lean methane reflux case. This is due to the fact that the deethanizer overhead reflux configuration results in a higher critical pressure for the Heavies Removal Column. This allows for separation at higher column pressures, which accordingly allows for more efficient use of refrigeration within the liquefaction (LNG) process. The end result is lower operating costs. Of course, the lower operating costs combined with lower installation costs results in overall improved project economics. It is clear that proper integration of deethanizer overhead reflux to the Heavies Removal Column provides for efficient propane recovery within LNG liquefaction technology.

7.0 Conclusions

Proper integration of NGL recovery technology within LNG liquefaction technology results in significant advantages by lowering overall capital cost requirements and improving both LNG and NGL production. Through careful process selection and heat integration, the integrated LNG/NGL facility results in lower specific power consumption and increased NPV as compared to non-integrated facilities. As demonstrated in this paper, given the same horsepower, an approximate 7% LNG production gain is realized through careful integration.

Utilization of deethanizer overhead as reflux to the Heavies Removal Column improves separation efficiency while maintaining higher column pressures for efficient and economical utilization of mechanical refrigeration. For high propane recovery, the deethanizer overhead reflux configuration requires less specific power and results in improved project economics when compared to configurations using lean methane reflux or no reflux to the Heavies Removal Column.

8.0 References

1. Chiu, C.-H., *LPG-Recovery Processes for Baseload LNG Plants Examined*. Oil & Gas Journal, 1997: p. 59-63.

2. Hudson, H.M., Wilkinson, J.D., Cuellar, K.T. and Pierce, M.C. *Integrated Liquids Recovery Technology Improves LNG Production Efficiency*. 82nd Annual Convention of the Gas Processors Association. 2003. San Antonio, Texas.

3. Wilkinson, J., Hudson, H., Cuellar, K. and Pitman, R., *Efficiency through Integration*. Hydrocarbon Engineering, 2002, 7(12): p. 23-26.

4. Lee, R.-J., Yao, J., Chen, J.J. and Elliot, D.G., *Enhanced NGL Recovery Utilizing Refrigeration and Reflux from LNG Plants*, U.S. Patent 6,401,486, (June 11, 2002).

5. Lee, R.J., Yao, J., and Elliot, D.G., *Flexibility, Efficiency to Characterize Gas Processing Technologies*. Oil & Gas Journal, 1999, 97(50): p. 90-94.

6. Yao, J., Chen, J.J. and Elliot, D.G., *Enhanced NGL Recovery Processes*, U.S. Patent 5,992,175, (November 30, 1997).

7. Yao, J., Chen, J.J., Lee, R.-J. and Elliot, D.G., *Propane Recovery Methods*, U.S. Patent 6,116,050, (September 12, 2000).

8. Houser, C.G., Yao, J., Andress, D.L. and Low, W.R., *Efficiency Improvement of Open-Cycle Cascaded Refrigeration Process*, U.S. Patent 5,669,234, (September 23, 1997).

9. Rentler, R.J., Macungie, P. and Sproul, D.D., *Combined Cascade and Multicomponent Refrigeration Method with Refrigerant Intercooling*, U.S. Patent 4,404,008, (September 13, 1983).

10. Newton, C.L., *Process for Liquefying Methane*, U.S. Patent 4,445,916, (May 1, 1984).

11. Swenson, L.K., *Single Mixed Refrigerant Closed Loop Process for Liquefying Natural Gas*, U.S. Patent 4,033,735, (July 5, 1977).

9.0 Endnotes

[1] CoP LNG services are provided by ConocoPhillips Company, Phillips LNG Technology Services Company and Bechtel Corporation through a collaborative

relationship with ConocoPhillips Company. CoP LNG, ConocoPhillips and its logo are trademarks of ConocoPhillips Company. Bechtel and its logos are trademarks of Bechtel Group Inc.

Author Biographies

Douglas G. Elliot, President, Chief Operating Officer and co-founder of IPSI LLC, an affiliate of Bechtel Corporation, in Houston, Texas, has helped develop much of the technology currently used in the gas processing industry and is the author of more than 50 technical publications, plus several patents. Doug has been a member of the Gas Processor's Association Research Steering Committee since 1972. He holds a B.S. degree from Oregon State University and M.S. and Ph.D. degrees from the University of Houston, all in chemical engineering. Doug is a Fellow of the American Institute of Chemical Engineers, and a Bechtel Fellow, the highest honor Bechtel offers for technical achievement.

Wes Qualls is currently a Director of LNG Technology Licensing in the ConocoPhillips' Global Gas Group. He has over 16 years experience in refining, NGL processing, petrochemicals, chemicals & plastics, and upstream technologies. His background includes process engineering and design, commissioning & startup, and facility operations. He currently manages licensing activities for both technical and commercial efforts for multiple LNG projects. He holds Bachelor of Science degrees in Chemical and Mechanical Engineering from Texas A&M University.

Shawn Huang joined ConocoPhillips in April 2001 as a Chief Engineer. He came to the U.S. in August 1987, after working as a research scientist at the Institute of Atomic Energy. He joined Exxon in Houston in October 1990 and Halliburton in February 2000 as a Principal. He received his Ph.D. degree in Chemical Engineering from the University of Idaho in December 1990, and his MBA degree from the University of Texas in May 2004.

R.J. Lee, Vice President of IPSI LLC, an affiliate of Bechtel Corporation, is a senior process engineer specializing in oil and gas processing, cryogenic industrial gas separation, and liquefied natural gas and enhanced NGL recovery technology. He graduated from National Taiwan University in 1981 with a B.S. degree in Chemical Engineering, and holds M.S. and Ph.D. degrees in Chemical Engineering from Purdue University, Indiana.

Roger Chen is Senior Vice President of IPSI LLC, an affiliate of Bechtel Corporation, and has more than twenty-eight years experience in research, process design and development in the areas of gas processing and oil and gas production. Roger holds a B.S. degree from National Taiwan University; M.S. and Ph.D. degrees from Rice University, all in Chemical Engineering, and is a member of AIChE, ACS and the Gas Processors Association Research Steering Committee.

Jame Yao, Vice President of IPSI LLC, an affiliate of Bechtel Corporation, in Houston, Texas, has 23 years experience in the development of gas processing and LNG technologies and is the author of more than 15 technical publications, and more than 15 patents. Jame holds a B.S. degree from National Taiwan University and M.S. and

Ph.D. degrees from Purdue University, all in Chemical Engineering. He is a member of the American Institute of Chemical Engineers.

Irene Zhang is a process engineer with IPSI LLC, an affiliate of Bechtel Corporation. She joined IPSI LLC in May 2002. Irene holds B.S. and M.S. degrees from Tsinghua University; and a Ph.D. degree from Rice University, all in Chemical Engineering.

Monitoring LNG Expanders using No- Load and Orifice Characteristics

Hans Kimmel
Ebara International Corporation
Sparks, NV 89434

AIChE Spring Meeting, April 2005
5th Topical Conference on Natural Gas Utilization
Atlanta April 10-14, 2005

Abstract:

In Liquefied Natural Gas plants the process efficiency and equipment reliability are of paramont importance. LNG expanders have been introduced in all recently constructed plants to improve the liquefaction process. The performance of LNG expanders is determined by two particular characteristics, the "No-Load Charateristics" and "The Orifice Characteristics". These typical characteristics of all liquid expanders provide a unique manner of reliably measuring the performance of the expander particularly for two-phase liquid-vapor LNG expanders. No sophistical instruments are required to accurately monitor the condition of the expanders: its mechanical parameters and hydraulic performance.

The Current Status of LNG Facility Standards and Regulations

Myron L. Casada
ABS Consulting, Risk Consulting Division
Donald C. Nordin
ABS Consulting

AIChE Spring Meeting, April 2005
5th Topical Conference on Natural Gas Utilization
Atlanta April 10-14, 2005

Abstract

With the large number of proposed liquefied natural gas (LNG) facilities in the United States (and worldwide), many new technical personnel are becoming involved in applying and interpreting LNG standards and regulations. Also, opponents of the new LNG developments are questioning whether existing safety and security requirements are adequate. In this paper, we will begin with an overview of existing standards and regulations pertinent to LNG facilities (for both onshore and offshore applications). That discussion will include documents from the National Fire Protection Association, the European Committee for Standardization, the Department of Transportation, and the United States Coast Guard. For offshore facilities, it will discuss the guidelines from some of the international classification societies. The paper will outline what is addressed (and what is not addressed) by these standards. The paper will then focus on and discuss the issues with those requirements and the changes that are currently under consideration by the standards organizations and federal regulators. The paper finishes with some specific recommendations for consideration by LNG facility developers.

Introduction

A key contributor to the safety of LNG facilities is the fact that they have, over the last 40 years, been built to some of the highest standards imposed in the petrochemical industry. These standards have matured and have been improved, even as they have been proven by the test of time. However, the industry has two major challenges facing it:

- Public opposition to traditional facilities, like onshore LNG import terminals
- Uncertainty about appropriate standards and regulations for nontraditional facilities, like LNG floating or fixed offshore import terminals

For traditional facilities, industry personnel need to have detailed knowledge of the appropriate standards, not only to implement them properly, but also to decide where a specific design in a specific location might need to exceed minimum standards. Also, if the public is going to have confidence in our ability to design, construct, and operate LNG facilities in its municipality, we need to be able to speak knowledgably about codes and standards when we interact with members of the public.

For offshore LNG terminals, regardless of their design, there are no industry standards or regulations that are as prescriptive or specific as those applicable to onshore facilities. Currently, the best sources of guidance for offshore terminal developers are the guidance documents developed by the marine classification societies (e.g., the American Bureau of Shipping, Det Norske Veritas, and Lloyd's Register of Shipping). Even with those sources, offshore terminal developers face significant uncertainty as they prepare and submit applications to the regulators for approval. This regulatory uncertainty presents significant economic risk to the developers, even though they try to make what they believe are the right decisions for the long-term safety of the facility.

In this paper, we will examine each of the major standards and regulations pertinent to LNG facilities. The examination of each document will cover:

- Technical/design areas included
- Management systems requirements (if any)
- Risk-based features (if any)
- Expected or proposed changes (if any)

We will also make recommendations regarding steps that we believe facility operators should take that go beyond the requirements of current codes, standards, and regulations.

Onshore LNG Facilities

The primary standards and regulations that apply to onshore LNG facilities in the United States are:

- NFPA 59A – *Standard for the Production, Storage, and Handling of Liquefied Natural Gas*
- EN 1473 – *Installation and Equipment for Liquefied Natural Gas - Design of Onshore Installations*
- 49 CFR Part 193 – *Liquefied Natural Gas Facilities: Federal Safety Standards*
- 33 CFR Part 127 – *Waterfront Facilities Handling Liquefied Natural Gas and Liquefied Hazardous Gas*

Each of these standards and regulations is discussed below.

NFPA 59A – *Standard for the Production, Storage, and Handling of Liquefied Natural Gas.* The National Fire Protection Association (NFPA) in the United States publishes NFPA 59A. The current version was issued in 2001, and an update is currently being prepared for release this year.

Technical/design areas included
NFPA 59A includes a wide variety of LNG facility considerations, including facility layout and spacing, process equipment requirements (e.g., for storage containers, vaporization facilities, piping systems, instrumentation and electrical equipment, LNG transfer equipment, fire protection equipment). The standard includes seismic design requirements for LNG facilities.

The standard also incorporates, by reference, codes and standards developed by a number of other organizations (e.g., American Society of Mechanical Engineers, the

American Society of Civil Engineers, the American Petroleum Institute, the American Concrete Institute, and the American Society for Testing and Materials).

Management systems requirements
NFPA 59A does not require any overall safety management system; however, there are requirements in the standard for safety, security, operations, maintenance, and training activities. There are also requirements regarding the competence of designers and fabricators. These requirements should lead a developer to implement typical safety management systems like (1) written safety programs, (2) operating procedures review and approval systems, (3) inspection, test, and preventive maintenance programs, (4) formal training programs, and (5) contractor evaluation and safety management.

Risk-based features
NFPA 59A does not require risk assessment nor does it suggest that risk techniques be used to define characteristics of the facility. However, it does require that consequence analyses be used to judge the acceptability of the facility and equipment siting. For example, the standard specifies maximum radiant energy levels and flammable vapor concentrations for specific scenarios (e.g., design spills and impoundment fires). Also, in several cases, it indicates that certain requirements of the standard can be altered if deemed acceptable to the "authority having jurisdiction." Preparation of any such submittals to the authority, using a risk-based decision-making approach, could increase the likelihood that such a submittal might be approved.

Expected or proposed changes
NFPA standards must be updated or revalidated at least every 5 years. The NFPA 59A committee has already met to prepare a new version to be released in 2005. No information regarding the nature of any proposed changes has been published yet.

EN 1473 – *Installation and Equipment for Liquefied Natural Gas - Design of Onshore Installations.* The European Norm standard EN 1473 was approved in 1997, based on an earlier British standard. Like NFPA 59A, it applies to onshore LNG facilities. However, it is a much more performance-based standard than NFPA 59A, allowing many of the design decisions to be made or justified based on the risk analyses that are required by the standard.

Technical/design areas included
The standard provides requirements for all of the types of LNG-related equipment typically found in a liquefaction plant, export terminal, import terminal, and other LNG storage locations. These requirements are presented in the body of the text and in normative (i.e., mandatory) appendices (which are provided for LNG tanks, pumps, vaporizers, and pipelines). A very useful "informative" (i.e., nonmandatory) Appendix H provides a description and illustrations of the various types of LNG storage tank designs (e.g., single, double, and full containment alternatives).

Management systems requirements
Unlike NFPA 59A, this standard explicitly requires implementation of a quality management system (QMS) and a safety management system (SMS). The QMS, which is required to meet ISO 9001, must apply to the design, procurement,

construction, and operation phases of the facility. The SMS must address the information generated in the hazard assessment for the facility.

Risk-based features

EN 1473 requires that a hazard assessment be performed to evaluate potential accident events. The standard allows either a probabilistic or a deterministic approach for the hazard assessment. In the probabilistic approach, events are evaluated based on their expected frequency and consequence. Events that do not meet established acceptance criteria (which are suggested by example frequency categories, consequence categories, and a risk matrix provided in informative Appendices J, K, and L, respectively) require the development of risk mitigation measures.

In a deterministic approach, the standard requires identification of "credible hazards," determination of the consequences of those hazards, and justification of the measures necessary to control the risks of each hazard.

Analysis techniques that are specifically referenced by EN 1473 for use in a hazard assessment include:

- Hazard and operability (HAZOP) analysis
- Failure modes and effects analysis
- Event tree methods
- Fault tree methods

The Center for Chemical Process Safety (Ref. 1) and the American Bureau of Shipping (Ref. 2) both provide documents that describe these techniques in some detail.

Regardless of what method is used to perform the hazard assessment, Section 4.6 of the standard also mandates that the design be subjected first to a preliminary process flow sheet review and then to a HAZOP analysis (when approved piping and instrumentation diagrams are available). The standard requires updates of the HAZOP analysis when major changes occur and that a management of change process be implemented for lesser changes.

Expected or proposed changes

This standard has not been revised since 1997; however, because of its performance-based approach, it is not likely to require updating. The authors are not aware of any impending changes.

49 CFR Part 193 – *Liquefied Natural Gas Facilities: Federal Safety Standards.* The Part 193 regulation is based in large part on NFPA 59A (the 1996 version) and as such has requirements that are very similar to that document. It was issued and is maintained by the Office of Pipeline Safety (OPS) in the Research and Special Programs Administration (RSPA) in the Department of Transportation (DOT). The regulation is the responsibility of OPS because LNG facilities in the United States are closely associated with, and often operated by, interstate pipeline companies that are already regulated by OPS.

Technical/design areas included

The regulation covers siting requirements, design, construction, equipment, operations, maintenance, personnel qualifications and training, fire protection, and security.

Management systems requirements
Like NFPA 59A, the regulation does not explicitly require a safety management system; however, it requires control of activities that are typically controlled by a safety management system. In the United States, a process facility, like an LNG facility that handles large quantities of flammable liquids or gases, would typically be covered under the Occupational Safety and Health Administration (OSHA) process safety management (PSM) standard (29 CFR 1910.119) and the Environmental Protection Agency (EPA) risk management program (RMP) rule (40 CFR 68). However, both of those regulations specifically exempt facilities that are regulated under the DOT regulations in 49 CFR Parts 193 and 195. The result of this coordination of regulatory approaches is discussed in the conclusions to this paper.

Risk-based features
Like NFPA 59A, this regulation defines basis events that require consequence analyses, but does not require more comprehensive risk-based analysis approaches.

Expected or proposed changes
DOT has recently issued a rulemaking disestablishing the RSPA (Ref. 3). The new Pipeline and Hazardous Materials Safety Administration will manage regulations for hazardous materials transportation. Various other functions of RSPA will be handled by the new Research and Innovative Technology Administration. The changes came into effect on February 20, 2005.

One of the first things on the new organization's agenda will be the petition by the City of Fall River, Massachusetts, for DOT to revise its regulations to require the establishment of new "minimum safety standards" for the location of an LNG facility. Specifically, the petition requests that DOT add requirements, including:

- A 2,500-foot distance from an LNG facility to population centers of 1,000 people (or 250 people if the people are elderly or children)
- One mile distance from areas of 5,000 population
- A ban on closure of bridges required for hospital access
- A ban on facilities in locations where safety and security zone implementation is not practical

At the time this paper was prepared there was not formal response from DOT on how it was going to address this petition. The City of Fall River also submitted a similar petition to the United States Coast Guard (Coast Guard). That petition is described later in this paper.

33 CFR Part 127 – *Waterfront Facilities Handling Liquefied Natural Gas and Liquefied Hazardous Gas.* 33 CFR Part 127 is the Coast Guard regulation that governs waterfront import and export LNG facilities or other waterfront facilities handling LNG. Its jurisdiction runs from the carrier or barge unloading arms to the last valve before the LNG tank.

Technical/design areas included
Part 127 provides much of its design requirements by referencing NFPA 59A (in particular for siting and layout, LNG storage containers, piping systems, instrumentation

and electrical, and LNG transfer equipment). Part 127 also includes other additional requirements for piping systems, hoses, piers and wharves, layout and spacing, electrical power systems, lighting systems, communications systems, warning systems, and sensing/alarm systems. Part 127 also includes a security section that only applies to waterfront LNG and liquefied hazardous gas facilities. (Since July 2004, waterfront LNG facilities must also meet the requirements of 33 CFR Subchapter H discussed below.)

Management systems requirements
Part 127 does not mandate a comprehensive safety management system; however, it specifies that the facility operator provide a number of procedures and documents similar to those that are common to most safety management systems, including:

- Operations Manual
- Emergency Manual
- Pre-transfer Inspections
- Maintenance
- Training
- Firefighting and Emergency Response

Also, there are requirements in other Coast Guard regulations that require additional safety management measures, like the requirement to report and investigate incidents resulting in release of LNG.

Risk-based features
Part 127 does not include any specific risk-based design or operational approaches; however, Part 127.017 allows an operator to provide a written request to have alternative procedures, methods, or equipment standards considered by the Coast Guard. An effective way to provide written support for such a request would be to use one of the analysis techniques defined in the Coast Guard's *Guidelines for Risk-based Decision Making*. That document (which is available on the Coast Guard's Web site) describes procedures for use of techniques such as change analysis, relative ranking, or what-if analysis for consideration of safety equivalency.

Expected or proposed changes
The City of Fall River also submitted a petition to the Coast Guard regarding suggested changes in Coast Guard regulations related to LNG facilities. The petition suggests that the Coast Guard establish thermal exclusion zones and vapor dispersion exclusion zones for LNG vessels (moored or in transit) comparable to LNG facility zones. In the *Federal Register* of November 3, 2004, the Coast Guard requested public comment on the need for such rulemaking. Public input was due to the Coast Guard by February 1, 2005. At the time this paper was written there was no feedback from the Coast Guard regarding how it was going to respond to the petition. It is not clear that the Coast Guard will find rulemaking warranted. The Coast Guard has indicated (Ref. 4) that it is considering the information available from the ABS Consulting LNG consequence methodology study for the Federal Energy Regulatory Commission (Refs. 5 and 6) and the Sandia National Laboratory report (Ref. 7) on releases from LNG carriers as it is looking for ways to improve and help ensure the safety of LNG transportation.

Offshore LNG Facilities

The ability and responsibility to regulate offshore terminals arises from the Deepwater Port Act of 1974 (DWPA) and is being jointly managed by the Coast Guard and the Maritime Administration (MARAD). Much of the information in this paper regarding regulation of offshore terminals came from recent testimony to Congress by Captain David Scott of the Coast Guard (Ref. 4).

Deepwater Port Act

The Coast Guard's authority to regulate deepwater ports (DWPs) is defined in the DWPA, and the regulations pertaining to the licensing, design, equipment, and operation of DWPs are provided in 33 CFR Parts 148, 149, and 150. Originally pertaining only to oil, the Maritime Transportation Security Act of 2002 (MTSA) added natural gas to the DWPA and shifted responsibility for licensing DWPs for natural gas (including LNG) from the Minerals Management Service to the Secretary of Transportation. The Coast Guard and MARAD have been formally designated by DOT to process license applications submitted under the DWPA.

Coast Guard Regulations

There are no current regulations that provide LNG-specific design requirements for offshore LNG terminals. Because offshore LNG terminals represent new concepts, the Coast Guard is applying a "design basis" approach rather than mandating a series of prescriptive requirements. Under a design basis approach, each concept is evaluated on its own technical merits, using relevant engineering standards and concepts that have been approved by recognized vessel classification societies and other competent industrial and technical bodies. Many of the codes and standards listed above for land-based installations are being used by Deepwater Port Applicants. When this is done, it is very important that the different risks and issues surrounding the offshore installation be identified. The use of a structured risk-based decision-making process will facilitate the identification of these issues and risks. One aspect of the DWPA is the solicitation for comments by other agencies with an interest in the facility. It is here that many of the land-based codes and standards will be invoked, due in part because of the familiarity that other agencies have with them. Being able to identify the offshore or project-specific issues related to the installation may help streamline this review process.

Classification Society Rules

The development of offshore LNG terminals is an example of technical innovation by industry. Associated with such innovation is the need to define design approaches that ensure that an acceptable level of safety and environmental protection is achieved. Classification societies have responded to the lack of existing standards for offshore terminals by developing new rules. The American Bureau of Shipping (ABS), Det Norske Veritas, and Lloyd’s Register of Shipping have all developed preliminary rules, which are available to industry and the government as they define an approval process for these terminals. This paper uses the *ABS Guide for Building and Classing Offshore LNG Terminals (ABS Guide)* as an example of what these new documents provide (Refs. 8 and 9). The other societies’ rules differ in some ways, but for sake of brevity, only the *ABS Guide* will be described.

ABS Guide for Building and Classing Offshore LNG Terminals. ABS published the first version of this guide in 2003 and has prepared several updates based on industry and government agency input.

Technical/design areas included

ABS's Guide provides criteria that cover both steel and concrete gravity-based structures, as well as floating systems. It includes design requirements for:

- Structures
- Containment Systems
- Position Mooring Systems
- Process Facilities
- Layout and Arrangements
- Hazardous Area Classification
- Process Support and Service Systems
- Electrical Systems and Installations
- Instrumentation and Control Systems
- Safety Systems

The *ABS Guide* provides both requirements for specific design features and the analytical approaches (e.g., dynamic analyses, strength analyses) that will be used to evaluate the system designs.

Management systems requirements

The *ABS Guide* does not include specific requirements for management systems.

Risk-based features

The *ABS Guide* requires that a risk assessment be performed to (1) identify significant hazards and accident scenarios that may affect the terminal and (2) consider the benefit of existing or potential risk control options. The objective of the risk assessment is to identify areas of the design that may require the implementation of risk control measures to reduce identified risk(s) to an acceptable level. To accomplish that objective, the risk assessment is required to be a systematic process that can identify situations where a combination or sequence of events could lead to undesirable consequences, such as property damage, threats to personnel safety, and environmental damage.

The risk assessment must consider, as a minimum, the following events:

- Damage to the primary structure due to extreme weather, impact/collision, dropped objects, helicopter collision, exposure to unsuitably cold temperatures, exposure to high radiant heat
- Fire and explosion
- Loss of primary liquid containment (for a duration to be determined based on an approved contingency plan)
- LNG leakage
- Release of flammable or toxic gas to the atmosphere or inside an enclosed space

- LNG containment roll over (i.e., thermodynamic instability due to LNG stratification)
- Loss of ability to offload LNG or discharge gas ashore
- Loss of any one critical component in the process system
- Loss of any single component in the station keeping/mooring system (for floating systems)
- Loss of electrical power

The risk control options (prevention and mitigation measures) deemed necessary by the risk assessment are considered part of the design basis for the terminal.

The *ABS Guide* recommends that early in the project a risk assessment plan be developed, documented, and submitted for review prior to conducting the risk assessment. During review of the plan, an agreement will be reached on the extent of the classification society's participation and/or monitoring of project-related risk studies. The society's participation in and/or monitoring of key tasks (e.g., hazard identification meetings) is considered necessary in order to establish a minimum level of confidence in the risk assessment results.

Expected or proposed changes
ABS has been active in soliciting industry and government comments on the *ABS Guide* and as offshore terminal designs are submitted for review, it is likely that the *ABS Guide* will continue to be revised.

Conclusions

From the review of the current LNG codes, standards, and regulations it is clear there are gaps in the documents we currently have available for designing and regulating LNG facilities, both onshore and offshore, in the United States. We believe these gaps exist in the following areas:

- Lack of requirements for broad safety management systems
- Limited application of risk-based tools and risk assessment
- Significant uncertainty regarding the regulatory approval process for offshore terminals

We would like to offer recommendations for consideration by organizations developing new LNG facilities that address each of these gaps.

Recommendation 1: Consider voluntary implementation of broad safety management safety systems.

EN 1473 explicitly requires the implementation of an SMS. However, the U.S. DOT regulations and NFPA 59A do not have similar requirements. Also, because of the exemption for facilities regulated under 49 CFR Part 193, the OSHA PSM and EPA RMP regulations do not apply to U.S. LNG facilities. However, we believe that effective implementation of such systems (or similar SMSs) can contribute to long-term safety of LNG facilities and should be considered by terminal developers, even though it is not required.

Recommendation 2: Consider using risk assessment methodologies for LNG facilities as a tool to help evaluate facility design, operations, and maintenance.

This is also a requirement of EN 1473 and is included in the *ABS Guide* as an important design process step. In addition, it is recommended by the Society for International Gas Tanker and Terminal Operators (SIGTTO) in its document *LNG Operations in Port Areas – Recommendations for the Management of Operational Risk Attaching to Liquefied Gas Tanker and Terminal Operations in Port Areas* (Ref. 10).

Recommendation 3: Consider working closely with the Coast Guard and MARAD as they continue to define the technical requirements for and the process under which additional offshore terminals will be approved.

At the time this paper was written, two offshore LNG terminals have been approved. However, because the technology is changing rapidly, additional terminal designs are likely to present different problem in the licensing process. Whether the process uses the classification society guidance documents or depends solely on a Coast Guard "design basis approach," input by all stakeholders is going to be an important part of moving the regulatory development and approval process along.

Biographies

Myron L. Casada, P.E.

Mr. Casada is Vice President of Marine, Offshore, Ports and Terminals for ABS Consulting and has over 30 years of engineering and risk assessment experience. Mr. Casada has provided extensive consulting in the risk-based decision-making and risk assessment areas for government agencies, including the Coast Guard, the Defense Department, and the Nuclear Regulatory Commission. He has been involved in ABS Consulting's support to the Coast Guard in the areas of risk-based decision making, facilitating projects involving the safety of liquefied natural gas transportation in U.S. ports, security threats to hazardous materials shipping, and safety in drill rig movements.

He also has coordinated much of ABS Consulting's support to the maritime industry in meeting the IMO International Ship and Port Facility Security Code, having helped write both ABS's *Ship Security Guide* and ABS Consulting's *Port Security Guide*.

Mr. Casada holds a B.S. degree in engineering and an M.S. degree in engineering administration (with a concentration in system reliability and risk assessment), both from the University of Tennessee.

Donald C. Nordin

Mr. Nordin has 18 years of experience in the marine, offshore and fabrication arenas. As a Senior Consultant with ABS Consulting, he acts as a primary liaison with clients to develop solutions to complicated engineering problems.

In the LNG arena Mr. Nordin has performed studies to identify the risks associated with the development of offshore LNG facilities. He has been involved with the development of offshore LNG facilities from project conception through the regulatory filing process and remains involved with these projects as they progress. He has performed the studies necessary to select an appropriate site for the development of an offshore LNG facility based on a set of project performance goals. He has performed risk analysis and design reviews of ports and marine and offshore structures associated with oil and gas production and LNG. Mr. Nordin has conducted risk assessments for a variety of marine applications, including serving as a subject matter expert and marine systems coordinator for the development of ABS's integrated tanker risk model.

Mr. Nordin holds a bachelor of science degree in marine engineering and is a Coast Guard Licensed Third Assistant Engineer for Steam and Motor vessels, unlimited horsepower.

References

1. Center for Chemical Process Safety of the American Institute of Chemical Engineers, Guidelines for Hazard Evaluation Procedures, Second Edition with Worked Examples, Center for Chemical Process Safety, New York, NY, 1992.

2. American Bureau of Shipping, Guidance Notes on Risk Assessment Application for the Marine and Offshore Oil and Gas Industries, American Bureau of Shipping, June 2000.

3. "Pipeline and Hazardous Materials Safety Administration; Establishment and Delegation of Powers and Duties," Federal Register, Vol. 70, No. 33, Friday, February 18, 2005.

4. "Statement of Captain David L. Scott on the Coast Guard's Role in LNG Safety and Security Before the Committee on Energy and Natural Resources Subcommittee on Energy," U. S. Senate, February 15, 2005.

5. Consequence Assessment Methods for Incidents Involving Releases from Liquefied Natural Gas Carriers, Report 131-04, ABSG Consulting Inc., May 4, 2004.

6. "Notice of Availability of Staff's Responses to Comments on the Consequence Assessment Methods for Incidents Involving Releases from Liquefied Natural Gas Carriers," United States of America Federal Energy Regulatory Commission, Docket No. AD04-6-000, June 18, 2004.

7. Guidance on Risk Analysis and Safety Implications of a Large Liquefied Natural Gas (LNG) Spill Over Water, SAND2004-6258, Sandia National Laboratories, December 2004.

8. Guide for Building and Classing Offshore LNG Terminals, American Bureau of Shipping, April 2004.

9. "The Changing Regulatory Scheme Applicable to Offshore LNG Terminals," W.H. Daughdrill and C.A. Karpowics, Ecology and Environment, Inc., and T.W. Grove and R. Gallagher, American Bureau of Shipping, OTC 15298, Offshore Technology Conference, May 5-8, 2003.

10. LNG Operations in Port Areas — Recommendations for the Management of Operational Risk Attaching to Liquefied Gas Tanker and Terminal Operations in Port Areas, First Edition, Society of International Gas Tanker and Terminal Operators, Witherby & Company Limited, 2003.

Biography of Speaker

Mr. Casada (pronounces name like "Canada") is currently Vice President for Marine, Offshore, Ports, and Terminals at ABSG Consulting Inc. in Houston, a risk management and engineering consulting firm affiliated with the America Bureau of Shipping.

In that role, he is responsible for structuring projects to support clients in the LNG and related industry sectors, with current efforts underway that include risk assessment, consequence analysis, marine engineering, site evaluation, seismic qualification, and regulatory compliance assistance. Mr. Casada has 30 years experience in the risk assessment and safety consulting industry. For the last 10 years he has been involved in ABS Consulting's support to the United States Coast Guard's risk-based decision-making program, including Coast Guard projects involving LNG operational plans, consequence analyses, and security assessments.

Operating Experiences with an Integrated Selective Catalytic Reduction System (SCR) Operating with Submerged Combustion Vaporizers (SCV) at a North American Base Load LNG Vaporization Facility

David Hawkins
BD Heat Recovery Division, Inc.

AIChE Spring Meeting, April 2005
5th Topical Conference on Natural Gas Utilization
Atlanta April 10-14, 2005

Background

Since 2002 – and continuously since early 2003 a base load LNG vaporization facility has been operating with an integrated system comprised of four submerged combustion vaporizers (SCV) and two selective catalytic reduction systems (SCR). The system is configured with two SCV's operating with one SCR. This paper presents the system layout and the design considerations that resulted in the finalized design. In addition, the paper discusses some of the operating issues and remedial actions taken to resolve these issues.

System Configuration

One of the challenges of installing a SCR system to operate with a SCV is the low temperature and saturated condition of the flue gases leaving the vaporizer. SCR systems inject ammonia into the dirty gas stream; this mixture is then converted by catalytic action into nitrogen (N_2) and water (H_2O). In order to carry out the conversion the flue gases must be at a certain reactive temperature. This temperature can vary depending on a number of factors such as, catalyst type, selected volume, pressure drop and formation of ammonia salts. Temperatures are usually greater the 350°F and frequently in excess of 600°F. SCV's are a highly efficient form of vaporization and the outlet flue gas temperature is normally between 60°F to 120°F with the normal operating temperature around the 60°F level.

Additional heat is therefore required to elevate the flue gases to the reactive temperature. In the case of the facility discussed in this paper, the external heat was taken from the vaporizer prior to the exhaust gases passing through the water bath. The integrated system is shown in diagram 1.

Diagram 1

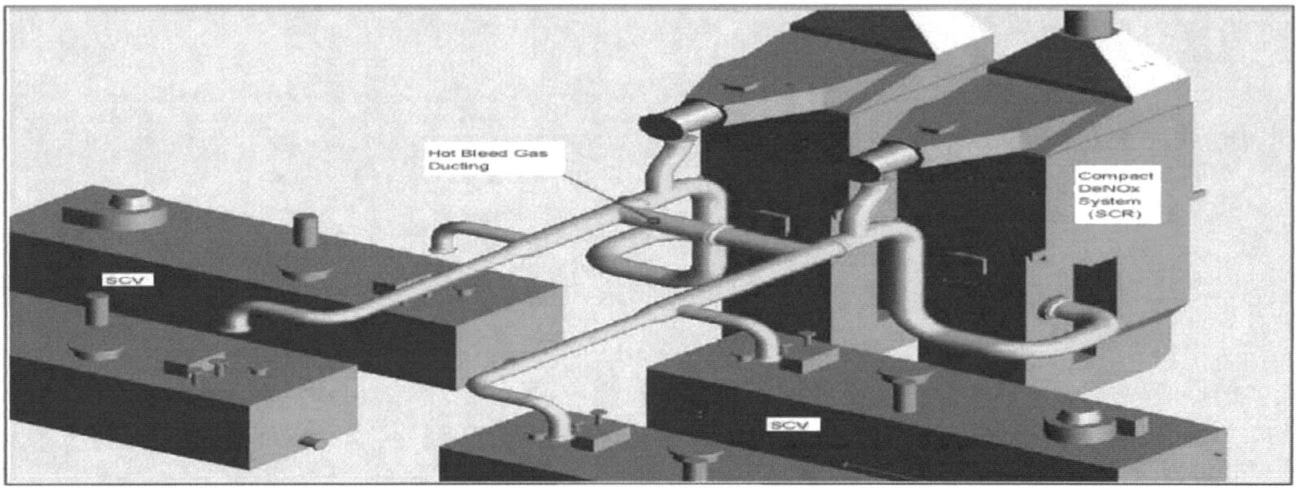

The cool saturated flue gases exit from two vaporizers and are individually ducted to a common plenum prior to entering the single SCR. Hot bleed gases taken from beneath the water bath at approximately 1,700°F pass through a water-cooled control valve (located in the SCV water bath) prior to being ducted to the SCR system. The SCR system can operate with both SCV's or just a single SCV, even at reduced loads.

Prior to entering the SCR, the hot bleed gases are separated supplying heat to two locations. These locations are identified as the preheater and post-heater sections. A detailed schematic (diagram 2) shows the layout of the SCR system, which is termed by our company as "The Compact DeNOx System."

Diagram 2

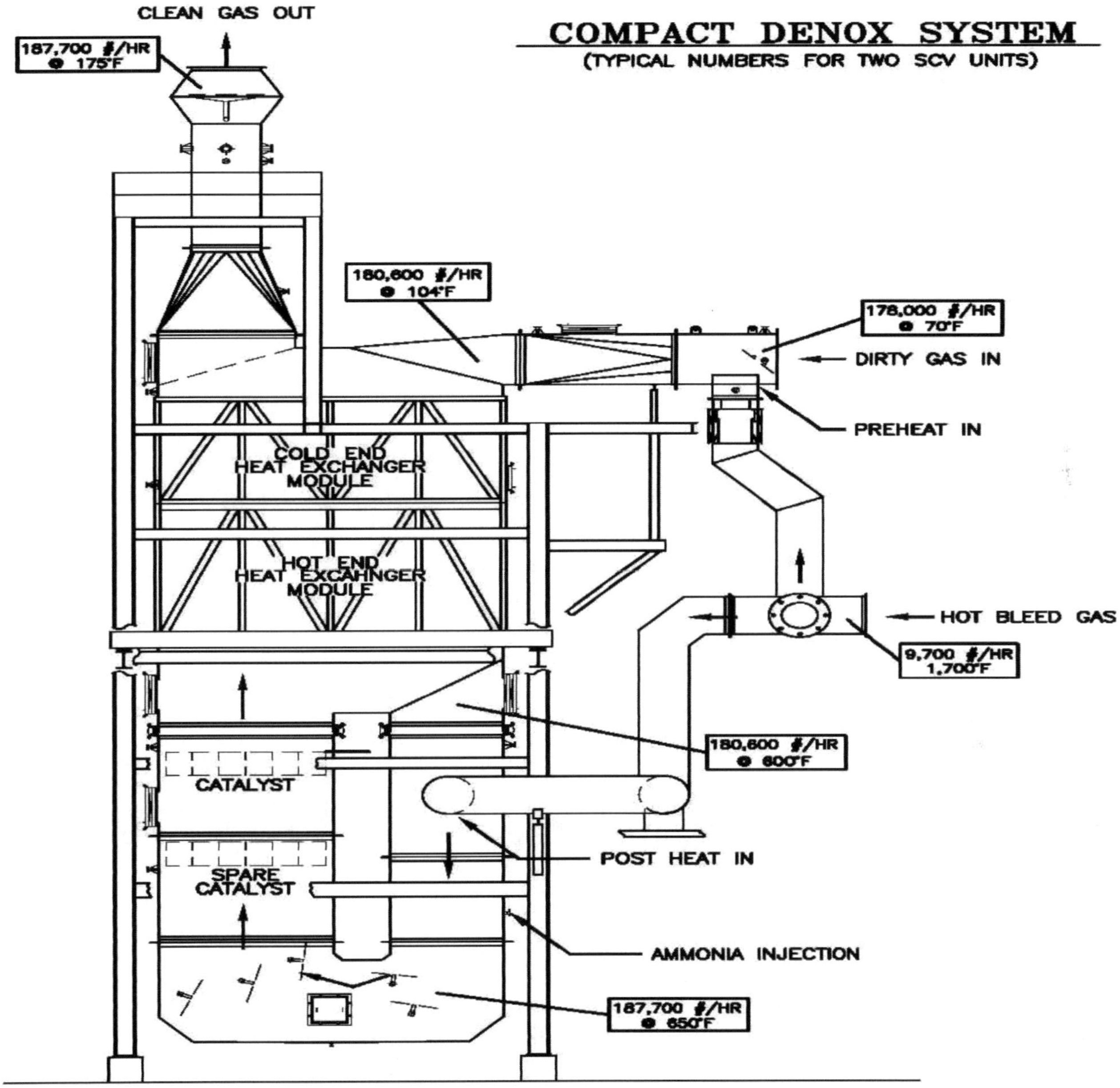

To prevent the acidic saturated flue gases (ph ~ 6) from entering the cold stage of the heat exchanger, hot bleed gas is mixed (utilizing a static gas mixer) with the exhaust gases adding approximately 30°F to the cold gases.

Even though the cold section of the heat exchanger is manufactured from stainless steel 304, avoiding water droplets increases the life expectancy. The flue gases are heated in the cold stage of the heat exchanger to approximately 225°F prior to entering the hot stage (manufactured from carbon steel) where the gases are raised to 600°F. The flue gases are further heated to 650°F with the addition of 1,700°F hot bleed gas. Aqueous ammonia (19%) is then directly injected into the dirty flue gas stream. A static gas mixer incorporated into the duct produces a vortex, which provides thorough mixing. Additional static gas mixers are used throughout the system to distribute the ammonia, NOx, and temperature evenly across the catalyst face.

The hot gases are then ducted 180 degrees before passing up through a single layer of catalyst where the NOx is converted into Nitrogen (N_2) and water (H_2O). A spare catalyst layer is designed into the system so that additional catalyst can be added in the future as the reactivity of the original catalyst declines.

After passing through the Catalyst, the hot gases provide the heat source (via the heat exchanger) for heating the cold gases that initially entered the system. Finally the clean gases exit the system and are dispersed into the atmosphere with a maximum of 5-ppm NOx and 5-ppm ammonia slip.

Design Considerations

Heating System

Five methods of supplying the required heat input to the SCR system were considered

- Steam Coils
- Direct Gas Fired Burner Installed in the Pre and Post Heating Locations
- Indirect Gas Fired Air Heater
- Direct Gas Fired Air heater
- Hot Bleed Gases taken from the Submerged Combustion Vaporizer

Steam Coils

Due to the location of the steam plant plus pressure and temperature restrictions this option was not viable.

Gas Fired Burner

This option is employed on about 90% of our Compact DeNOx Systems installed in the USA and Europe. However, this option was considered unsuitable for the LNG application because the potential exists of a gas leak occurring in the SCV. This firing method would result in direct flame contact to the leaking methane.

Indirect Gas Fired Air Heater

With this option a separate gas fired air heater would be installed. A burner would supply heat to a refractory lined furnace. A separate air blower would supply air to tubes installed in the furnace; the hot air would then be piped to the pre and post heating locations in the system. This option was rejected because of cost and the addition of a second emission source.

Direct Fired Air Heater

This option uses a separate but direct gas fired Air Heater. The combustion point would be located at a significant distance from the pre and post heating locations. A blower would supply air over the gas-fired burner. The resulting hot air would pass to the pre and post heating locations at a higher pressure than present in the SCR system. Consequently, the potential for any direct flame contact with leaking gas was considered negligible. In addition, the combustion gases pass through the DeNOx system eliminating a secondary emission source. This option was built into the design as a back up to the chosen hot bleed gas system.

Hot Bleed Gas

The SCV supplier first suggested this option, which was chosen as the preferred heating system. A small portion of the gases generated by the large single burner (100 mm Btu/hr) would be extracted prior to the water bath. Based on the extraction point it was anticipated, that the flue gases would be approximately 1,700°F. At close to 99%, the SCV is an inherently efficient method of vaporization. Therefore, extracting flue gases prior to the water bath has an effect on the overall system efficiency. However, it was determined both from an operational and capital cost basis, this was the most cost effective choice. Unfortunately, this type of system had never previously been installed; therefore a backup system was part of the original design.

Once the bleed gas option was selected, the main considerations became the inter-connecting ducting between the SCV and the SCR and the use of high temperature control valves. To reduce the maintenance and improve reliability, water-cooled control valves were selected and located in the water bath of the SCV.

For the hot bleed gas ducting, the following insulation options were considered.

Refractory Lined Ducts – There were concerns that if the refractory degraded plugging of the heat exchanger and catalyst could occur.

Externally Insulated Ducts – The metallurgy required at 1,700°F was expensive and expansion issues could have been problematic.

Internally insulated Ceramic Fiber Lined Stainless Steel Ducts – This was the preferred option. Stainless steel 304 was used to avoid corrosion problems, should condensation occur between the duct wall and the internal insulation. Ceramic fiber insulation has been used for a number of years in steam reformers and was considered to be reliable. A ceramic fiber hard core was installed to prevent the soft fibers from being eroded.

Heat Exchanger

The counter-flow heat exchanger is a welded plate type manufactured to provide zero leakage and consequently no crossover contamination from the dirty to the clean side. The counter-flow design is essential to reduce the exchanger size and permit small gas-to-gas approach temperatures of 50°F or less which, results in a more efficient system. The installed heat exchanger was designed as a two-stage unit. To provide protection against corrosion caused by the saturated exhaust gases, the cold end plates are manufactured from stainless steel 304. The hot section operates above the water dew point and is manufactured from carbon steel.

Static Gas Mixers (SGM)

The static mixer utilizes the effects of a vortex (picture 3), produced when a disc or delta wing is positioned at a specific angle to a gaseous stream. There are several locations in the SCR system that use the beneficial effects of this devise.

Picture 3 (Vortex Formation)

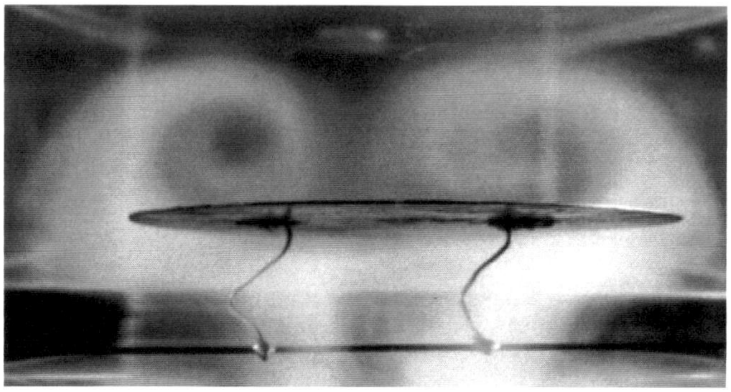

Pre & Post Bleed Gas Locations – The static gas mixer is used to mix the 1700°F bleed gases with the dirty exhaust gases.

Ammonia Injection – The SGM is used to mix the atomized 19% aqueous ammonia with the hot dirty exhaust gases. The SGM technology permits the direct injection of the aqueous ammonia into the gas stream without the use of external vaporizers. There are three discs at this location and each one has an ammonia injection nozzle located at the center. Therefore only three ammonia nozzles are required for the complete SCR process.

Homogenizers – After the ammonia has been injected into the exhaust stream, additional SGM's and turning vanes are provided to ensure the homogeneous distribution of the gas, NOx, ammonia and temperature. Good distribution of these elements is required to maximize the catalyst function and reduce the ammonia slip to a minimum.

Injection Nozzles

Three Lechler nozzles are installed to atomize the 19% aqueous ammonia (using small amounts of 100 psig, dry, compressed air). The ammonia is supplied at a pressure of 60 – 70 psig.

Catalyst

The NOx catalyst is a Haldor Topsoe DNX- 930, which is a low pressure drop honeycomb design comprised of Titanium Dioxide, Vanadium Pentoxide and Tungsten Trioxide.

Picture 4 (Typical Catalyst Module)

.
A spare catalyst layer is designed into the system so that additional catalyst can be added in the future, as the reactivity of the original catalyst declines.

Flow Modeling

A plexi-glass 35:1scale model is produced incorporating the static gas mixers. Adjustments are made to produce the required flow patterns for the exhaust gases, ammonia injection, temperature and NOx distribution.

System Pressure Drop

The Compact DeNOx system is designed for an overall pressure drop of 12-inches WC. Higher pressure drops can be used to reduce system size however the SCV blowers will require additional horsepower resulting in an increase in operational costs. The reverse is true if lower pressure drops are selected.

Operating Issues & Remedies

There have been two areas where design decisions have resulted in operational issues, and although two parts of the system have been affected, the root cause can be traced to the hot bleed gas heating system.

Hot Bleed Gas System

Shortly after start-up, water entered the hot bleed gas piping; it is still unclear if this was a result of water droplet entrainment from the water bath or a leak from the SCV distribution system. The water saturated the hard core ceramic internal liner leaving the soft fibrous insulation unprotected from the gas stream velocities. This resulted in insulation releasing from the duct walls plugging the catalyst with ceramic fiber. With the internal insulating material dislodged, the hot bleed gas system could not be used and therefore the SCR system operated with the saturated exhaust gases for an extended period of time.

A different type of internal insulation was installed, but again water affected the system. The problem was finally remedied by changing the bleed gas duct to an externally insulated system. The expansion problems that were an initial concern have not materialized and the connecting duct issues appear to have been solved. A droplet separator has also been installed upstream of the SCR system.

The seals on the high temperature control valves have produced maintenance issues; this in turn has led to water incursion into the bleed gas system. These valves have been rebuilt and the temperature of the bleed gas reduced to prolong the life expectancy of these seals.

Catalyst

Due to the low temperatures present in the water bath of the SCV and the moisture content of the LNG gas being fired in the burner, condensation takes place in the water bath resulting in a net increase of water. Combustion gases passing through the water bath produce several dilute acids such as carbonic and nitric acid. These acids reduce the ph to the 3 – 4 level. This acidic water is dosed with Sodium Carbonate to balance the ph, prior to being returned to the Charles River; unfortunately Sodium is a poison to the NOx catalyst.

When the SCR system was operated without bleed gas heating, the saturated gases passed over the catalyst, before its removal. Once the dried out catalyst was reinstalled and heat restored to the system. Sodium deposited on the catalyst during the saturated operation remained; this significantly reduced the reactivity of the catalyst, which had to be replaced.

Initially the complete water bath was neutralized with sodium carbonate, now an overflow system has been installed and only the water going to discharge is treated. However, a small amount of water is injected into the SCV burner to limit NOx formation, and this water contains the sodium carbonate. Provided the SCR system is operated at design conditions and temperatures, this sodium should pass through the system without creating problems. Further operating experience will determine if this is correct. An alternative to sodium carbonate would be to use ammonia as the dosing chemical; this would alleviate the potential for sodium poisoning, although increasing operating costs.

Conclusions

As a number of companies consider building and permitting additional vaporization facilities in various parts of the United States, the integrated SCV and SCR System is a viable option, particularly in circumstance where low NOx burners cannot meet emission levels. The system installed in downtown Boston is meeting and exceeding all the environmental requirements of 5-ppm NOx and 5-ppm ammonia slip. The majority of the operational issues can be traced back to the hot bleed gas heating system and while this is the most efficient system, the reliability has to be questioned.

To reduce the amount of entrained water droplets entering the SCR system, installation of a moisture separator in the upstream ductwork is recommended.

If high pressure and temperature steam is available this is the simplest form of supplying the necessary heat. In most cases however, this option is not available. Installation of a direct-fired remote air heater can duplicate the hot bleed gas system although at a lower efficiency. This option reduces the SCR's reliance on the SCV and will permit independent operation.

Reducing the gas temperature from 1,700°F to approximately 1,100°F will permit the use of stainless steel 304 (less expensive than the Inconel used with1,700°F) with external insulation, thus avoiding the potential fouling of the catalyst and heat exchanger.

If a bypass system cannot be installed to isolate the SCR and permit SCV operation, then during extended periods of cold operation, the catalyst should be removed to prevent degradation. In the case presented the catalyst can be removed in approximately 6 – 8 hours.

More operating experience needs to be acquired to see if dosing with Sodium Carbonate will have a long-term effect on catalyst life expectancy. However, an alternative is available with the use of ammonia albeit at a higher operating cost.

Acknowledgements

Balcke-Dürr AG - the original developer of the Compact DeNOx System
GEA Ecoflex GmbH – the current supplier of the Compact DeNOx Technology
Professor Hans Ruscheweyh – the developer of the Static Gas Mixer
Aker-Kvaerner – the initial design engineering company for Cabot- Distrigas
Distrigas – The management and operations personnel for all their assistance and candidness in discussing the plant operations.

Disclaimer

The views expressed in this paper are solely the author's and are based on site visits, discussions and knowledge of the design decisions that resulted in the system presented. They may not reflect the views of other participants in the project and should not be construed to represent the views of the current owner of the facility under discussion, Tractebel-Distrigas.

ITM Syngas: Ceramic Membrane Technology for Lower Cost Conversion of Natural Gas

Douglas L. Bennett[1], Michael F. Carolan[1], Christopher M. Chen[1], Dale M. Taylor[2], Merrill A. Wilson[2], Theodore R. Ohrn[3]*

[1.] *Air Products and Chemicals, Inc., 7201 Hamilton Blvd, Allentown, PA, 18195 USA*
[2.] *Ceramatec, Inc., 2425 South 900 West, Salt Lake City, UT, 84119 USA*
[3.] *SOFCo-EFS Holdings LLC, 1562 Beeson Street, Alliance, OH 44601 USA*
* *Correspondence should be addressed to C. M. Chen, email: chencm@airproducts.com, tel: +1 (610) 481-3315*

Prepared for Presentation at the AIChE Spring National Meeting
Atlanta, GA April 12, 2004

Introduction

The ITM Syngas Team, led by Air Products and including ChevronTexaco, Ceramatec, SOFCo-EFS, and other partners, in collaboration with the U.S. Department of Energy, is developing Ion Transport Membrane (ITM) technology for the production of synthesis gas. ITM ceramic membranes are fabricated from non-porous, multi-component metallic oxides and operate at high temperatures (typically over 700 °C) with exceptionally high oxygen flux and selectivity. A conceptualization of the ITM Syngas technology is shown in Figure 1. Oxygen from low-pressure air permeates, as oxygen ions, through the ceramic membrane and is consumed through chemical reactions, thus creating a chemical driving force that pulls oxygen ions across the membrane at high rates. The oxygen reacts with natural gas in a partial oxidation process to produce a hydrogen and carbon monoxide mixture (synthesis gas).

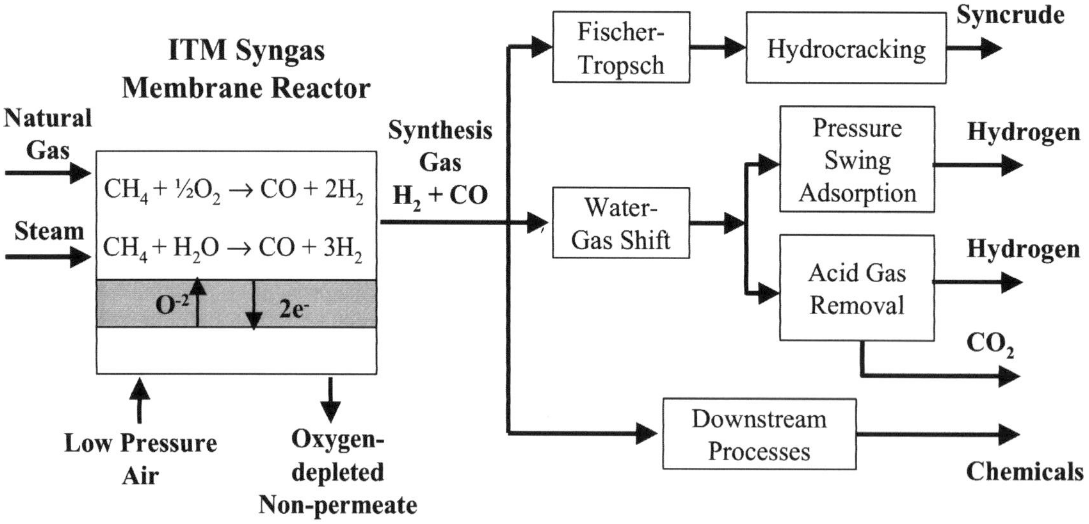

Figure 1. Conceptual ITM Syngas process showing multiple synthesis gas applications

Technology Applications

The ITM Syngas process is a breakthrough technology that combines air separation and high-temperature synthesis gas generation processes into a single ceramic membrane reactor, with significant savings in the capital cost of synthesis gas production. Because synthesis gas is a feedstock for a range of different processes, ITM Syngas represents a technology platform that has numerous applications, such as gas-to-liquids; hydrogen; ultra-clean fuels, including liquid transportation fuels; and chemicals such as methanol (Figure 1).

Process designs and cost estimates were developed for single-train Fischer-Tropsch Gas-to-Liquids plants producing 6,000 to 18,000 barrels/day. The capital cost of the synthesis gas process was projected to be 30% lower for ITM Syngas compared to conventional technology employing oxygen-blown Autothermal Reforming (ATR) with a cryogenic Air Separation Unit (ASU). The ITM Syngas process used a more conservative steam/carbon (S/C) ratio of 1.5, compared to a S/C ratio of 0.6 used for the ATR. Future advances in the development of the ITM Syngas technology could lead to lower S/C ratios and further increase in capital and operating cost savings. The 6,000 barrel/day GTL plant was evaluated for a Floating Production Storage and Offloading (FPSO) application, and footprint reductions of up to 40% were identified compared to ATR/ASU.

Applications to co-produce hydrogen and carbon dioxide, such as use of the hydrogen as a decarbonized fuel for centralized or distributed power applications, were also evaluated. ITM Syngas processes producing 150 to 760 MMSCFD (million standard cubic feet per day) of fuel-grade hydrogen were compared to conventional processes based on ATR/ASU. Economic evaluation of the ITM Syngas process showed the potential for 30% capital cost savings in the syngas generation process area and over 20% capital cost savings for the overall synthesis gas/H_2/CO_2 production process. The ITM Syngas process also had a higher thermal efficiency of 74% compared to 71% for the oxygen-blown ATR process. The capital cost and efficiency advantages for ITM Syngas are due to the combination of oxygen separation and high temperature synthesis gas production into a single unit operation and the ability to use low-pressure air as the oxidant source.

Membrane Design

Within the ITM Syngas development program, a planar ceramic membrane was designed and fabricated that is able to achieve high oxygen fluxes while withstanding the thermo-mechanical stresses encountered during operation.[1] The membrane design incorporates micro-channel features to reduce the thermo-mechanical stresses that occur within the membrane assembly because of the pressure loads and the exothermic oxidation reactions. The micro-channel features also provide for efficient distribution of the gases (high pressure natural gas/synthesis gas and low pressure air) to the membrane surfaces. Figure 2 illustrates the ITM Syngas micro-channeled planar membrane.

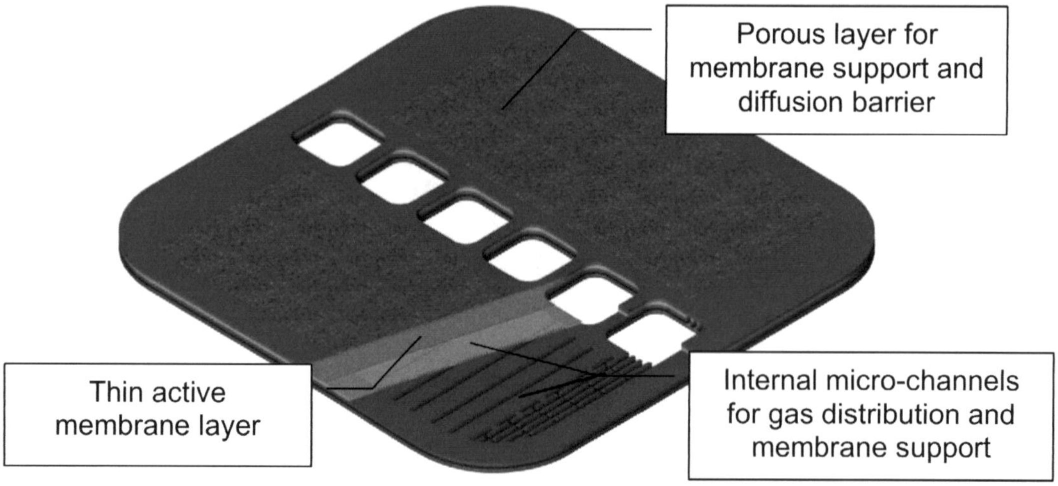

Figure 2. Cut-away view of the ITM Syngas ceramic membrane design showing the multi-layer structure.

The planar membrane design enables very efficient packing of membrane surfaces within the reactor volume. An ITM Syngas module is assembled by stacking membranes vertically. Figure 3 shows a submodule simulant built from 4 full-size ceramic membranes. Ceramic spacers are placed between the individual membranes to create channels for the flow of natural gas and synthesis gas between wafers. Various sizes of modules consisting of multiple pilot-scale membranes or of single full-size membranes have been joined and sealed using a ceramic sealant suitable for operating temperatures up to 1100°C [2].

Figure 3. Full-size ITM Syngas ceramic membranes shown as part of a 4-membrane submodule simulant

The design of the ITM Syngas planar membranes is multi-faceted where thermal and fluids design criteria must be integrated into a thermo-mechanically durable design that can be manufactured with low-cost commercial processes. These design facets are discussed below.

Thermo-Mechanical Design

Global Pressure Loads -- The economics of the ITM Syngas process are most attractive when a high-pressure natural gas feed and a low-pressure air feed are used to produce a high-pressure synthesis gas product to match downstream process pressures, thus avoiding the expenses associated with compressing the air or the synthesis gas. The large pressure difference across the membrane puts the ceramic membranes in hydrostatic compression since the high-pressure natural gas/synthesis gas flows outside the membrane and the low-pressure air flows inside the membrane. The compressive forces are counter-balanced by a continuous network of internal support ribs, shown in Figures 2 and 4, and by taking advantage of the fact that for ceramic materials the compressive strength exceeds the tensile strength by a factor of about five.

Local Support of Thin Membrane Layer – Locally within a membrane, the thin active membrane layer bridges the internal flow channels and is supported by the channel walls or ribs. Additional stiffening of the membrane is achieved by laminating a relatively thick porous layer onto the thin dense active layer. This multi-layer dense membrane and porous support structure re-distributes the localized bending stresses and provides for controlled gas-phase diffusion – discussed in more detail below.

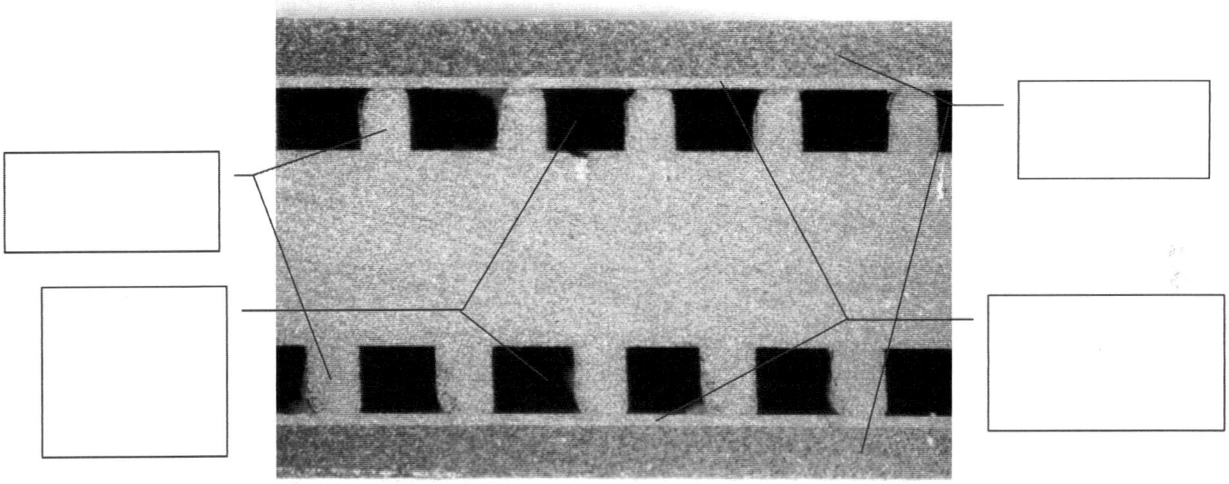

Figure 4. Internal features of the ITM Syngas membrane showing multi-layer structure of uniform ceramic composition

Thermal/Fluids Design

Interconnected Gas Channels – Just as the mechanical support of the membrane requires contiguous support ribs, the internal gas flow requires contiguous channels. This is accomplished by laminating multiple featured layers together such that parallel channels communicate through neighboring layers and allow the gas to wind between layers as it flows from the membrane inlet to outlet. By properly designing the dimensions of the membrane manifolds and the flow channels, good pressure and flow distributions are maintained.

Global Heating of Natural Gas/Synthesis Gas – The overall conversion of natural occurs by an exothermic partial oxidation reaction. As this gas stream flows over and is reacted on or near the surfaces of the membrane, the ceramic membrane develops thermal gradients that induce thermo-mechanical stresses. These stresses are minimized by operating the air and natural gas/synthesis gas streams in a cocurrent flow configuration that leads to reduced temperature differentials and more favorable thermal profiles.

Localized Heating – Solid-state diffusion of oxygen ions through the ceramic membrane is a thermally activated process. Thus as exothermic oxidation occurs, the oxygen permeation increases, allowing for more oxidation. During membrane operation, this local phenomenon can become unstable unless a secondary diffusion limit is implemented. By adding a porous layer adjacent to the thin dense membrane layer, gas phase mass transfer can be controlled by the porous layer structure, including pore size, porosity, tortuosity and layer thickness. In this manner, the oxygen flux will be controlled by the characteristics of the thin dense membrane layer and constrained by the diffusion-limiting porous layer.

Membrane Fabrication Using Laminated Object Manufacturing (LOM)

The primary challenge in commercializing micro-channel devices has been the ability to mass produce the micro-channel components in the preferred materials at a commercially viable cost. The ITM Syngas planar membranes were designed with a multi-layer structure that could be manufactured using commercial-scale production and automation equipment.

In order to attain dimensional tolerances, a Laminated Object Manufacturing (LOM) method was developed. This LOM process begins with ceramic powders within a specified process window for surface area and particle size. These powders are blended with a mix of organic binders, plasticizers and solvents to form a flowable ceramic slip. This slip is then cast onto a polyester film that dries and cures into a flexible, ceramic-filled plastic tape. This process is called tape casting. Once cured, these tapes can be featured by numerous machining technologies, such as CNC laser machining and punching. The featured, plastic layers are pressure-laminated and bonded together. Once these layers are laminated, they are heated to remove the organics and to densify the ceramic powders. The densification step is called sintering. The key steps in this process are shown in Figure 5.

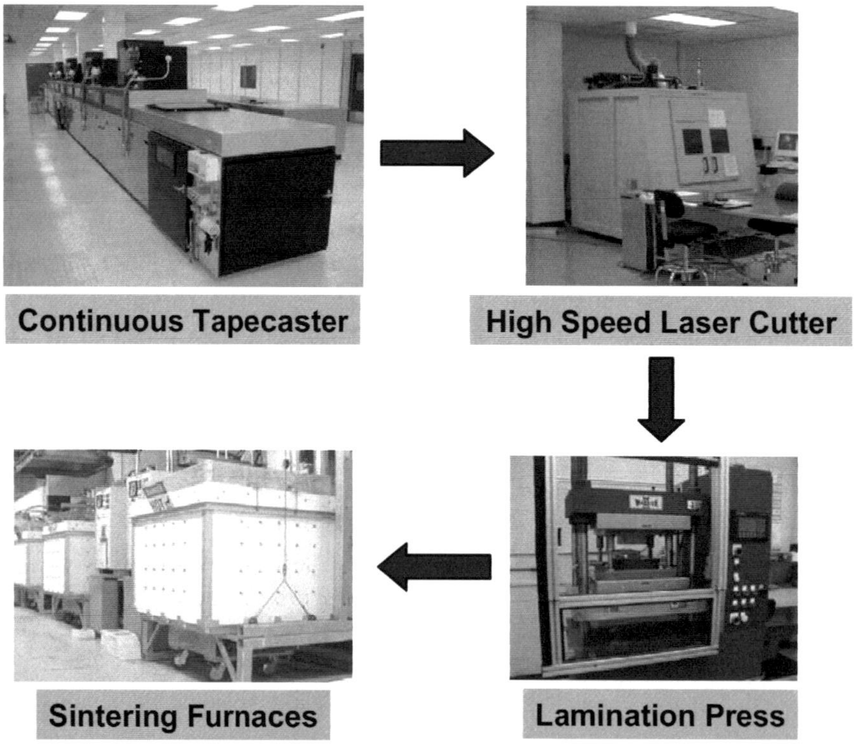

Figure 5. LOM (Laminated Object Manufacturing) method for ceramic planar membranes

Test Results

A laboratory Process Development Unit (PDU) was constructed at Air Products and is used to test the pilot scale planar ceramic membranes. The PDU operates at nominally 24 MSCFD (thousand standard cubic feet per day) of synthesis gas capacity, and is used to evaluate membrane performance under full process operating conditions.

Recent PDU tests with pilot-size membranes included operation under synthesis gas simulating the conditions of a commercial ITM Syngas reactor including temperatures between 825 and 900°C and syngas pressures between 215 and 425 psig. Membranes demonstrated the ability to operate through a range of temperature, pressure, and composition conditions. Figure 6 shows that the ITM Syngas program achieved fluxes at the commercial flux target under simulated commercial conditions.

Recent tests also included the successful demonstration of ceramic to ceramic seals [2,3] and ceramic to metal seals [4] at full process conditions. The ceramic to ceramic seals are used to join the sintered ceramic components together to form a single module. The robustness of the ceramic to ceramic seals was demonstrated by the modules remaining leak tight as they were cycled through the range of temperature, pressure and gas composition. The ceramic to metal seals are used to join the ceramic modules to metal tubes supplying the air feed and withdrawing the spent air from the modules. These seals also remained leak tight against the full pressure differential of up to 410 psid at temperatures up to 900°C.

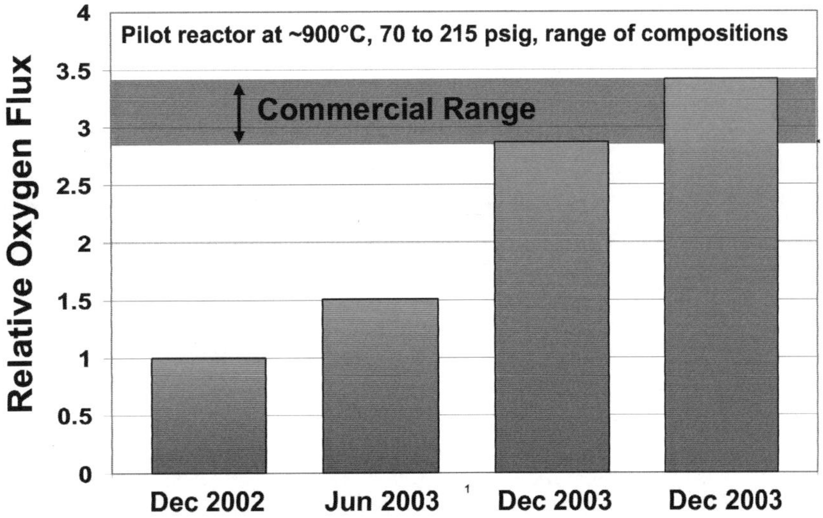

Figure 6. Achievement of oxygen fluxes at the commercial target was demonstrated first in December 2003. Later tests demonstrated fluxes above the commercial target.

Conclusions

Significant progress has been made in designing and manufacturing planar ceramic membranes, and in demonstrating targeted membrane oxygen fluxes, which are necessary steps toward commercializing the ITM Syngas technology. The potential economic benefits of the technology continue to be very promising.

Acknowledgement

The authors gratefully acknowledge the work of ITM Syngas team members at Air Products and Chemicals Inc., Ceramatec Inc., ChevronTexaco Energy Research and Technology Co., Norsk Hydro, Eltron Research Inc., SOFCo-EFS Holdings LLC, Penn State University, University of Pennsylvania, and Pacific Northwest National Lab. The co-funding by the U.S. DOE (DE-FC26-97FT96052) is also gratefully acknowledged.

References

[1] US patent application US 2004/0186018 A1, M. Carolan et al., "Planar ceramic membrane assembly and oxidation reactor system"
[2] US Patent application US 2004/0182306 A1, D. Butt et al., "Method of forming a joint"
[3] US Patent application US 2004/0185236 A1, D. Butt et al., "Method of joining ITM materials using a partially or fully-transient liquid phase"
[4] US Patent 6302402 B1, S. Rynders et al., "Compliant high temperature seals for dissimilar materials"

104b C1 Catalytic Partial Oxidation Over Short-Contact-Time Reactors: Methane to Synthesis Gas

Anthony H McDaniel, James E Miller, Pauline Ho, Steven F Rice, Vasilis Papavassiliou, William A Steen

Current methods of synthesis gas production have a number of deficiencies including high energy usage, emission of unwanted CO_2 and NO_x, and the requirement for additional processing to adjust the H_2:CO ratio. We have applied design, experiment and modeling towards the development of a revolutionary reactor system, named Catalytic Hot Oxygen Reactor or CHOR, that directly reacts methane with substantially pure O_2 to produce synthesis gas with the desired 2:1 H_2:CO ratio. Recent efforts have been aimed at elucidating the relative roles of the catalyst and the gas phase in the overall reaction and improving the catalyst stability.

Sandia is a multiprogram laboratory operated by Sandia Corporation, a Lockheed Martin Company, for the United States Department of Energy's National Nuclear Security Administration under contract DE-AC04-94AL85000.

Choosing the Proper Technology to Monetize Large, Remote Natural Gas Reserves

Dan H. Brannon – ConocoPhillips
Robert M. Lesnick – ConocoPhillips

Abstract: Recent positive developments in LNG and GTL technologies have created new opportunities to commercialize large discovered natural gas reserves in remote regions of the world. These technologies are capital intensive and require long project lives to be economic. Choosing the proper technology can be difficult given long-term market and technological development uncertainties. This report identifies the key commercial drivers and differences between the two technologies and compares and contrasts the benefits and concerns of each.

ConocoPhillips Positions in LNG and GTL

ConocoPhillips is uniquely qualified to develop both LNG and GTL projects worldwide. As the third largest U.S. oil company, with assets in excess of $US 80 billion, it is an integrated major energy company with upstream and downstream operations in over 40 countries with superior project management and project financing skills.

The merger of Conoco and Phillips Petroleum in 2002 combines a long history of successful LNG development. Conoco pioneered the design and construction of the first commercial LNG carrier, the *Methane Princess* in 1959. Phillips has reliably operated one of the first LNG plants (Kenai, Alaska) since 1969, never missing a LNG delivery in its 35 years of operation. Currently the company has five projects under development in Asia/Pacific, the Middle East and Atlantic Basin areas for the production of LNG. ConocoPhillips is also pursuing development of several LNG import facilities in the US and Europe. The company has recently licensed its proprietary ConocoPhillips LNG Process (formerly known in the LNG industry as the Phillips Optimized Cascade LNG Process) in Australia and to several Atlantic Basin LNG projects through its strategic collaboration with the Bechtel Corporation, bringing the total capacity utilizing this technology to over 30 mtpa worldwide.

Kenai LNG Plant

GTL Demonstration Plant – Ponca City, OK

ConocoPhillips has developed its own proprietary GTL technology that results in higher conversion efficiency than other GTL processes. The company operates the fourth largest GTL plant in the world at its refinery location in Ponca City, Oklahoma. In operation since 2003, the Ponca City plant is capable of manufacturing up to 400 BPD of GTL products. The plant was designed to provide engineering scale-up data for a world-scale 80,000 BPD GTL plant.

In addition to the successful completion of this plant, the company has executed an agreement with Qatar Petroleum Corporation to investigate the feasibility of utilizing North Field gas to support a two-plant, 160,000 bpd GTL operation.

Global Gas Environment

There is no shortage of natural gas in our world. Gas reserves many times over annual consumption have been identified. Historically, a large percentage of these reserves could not be economically developed and delivered to market. Increasing demand for natural gas, improvements in project scope and scale, and employment of new technologies have made it possible to consider development of large remote gas fields that otherwise would have been considered stranded.

Natural gas markets continue to grow, supported by new demand created by clean and efficient gas-based electric power generation technologies. Recently, countries with deeply developed gas markets have experienced periodic supply shortfalls. In some areas the tightening of supply has caused gas values to rise and in some cases to reach price parity with oil. This trend has created significantly higher margins to support more expensive remote developments. Refer to Figure 1 to observe this trend in the United States.

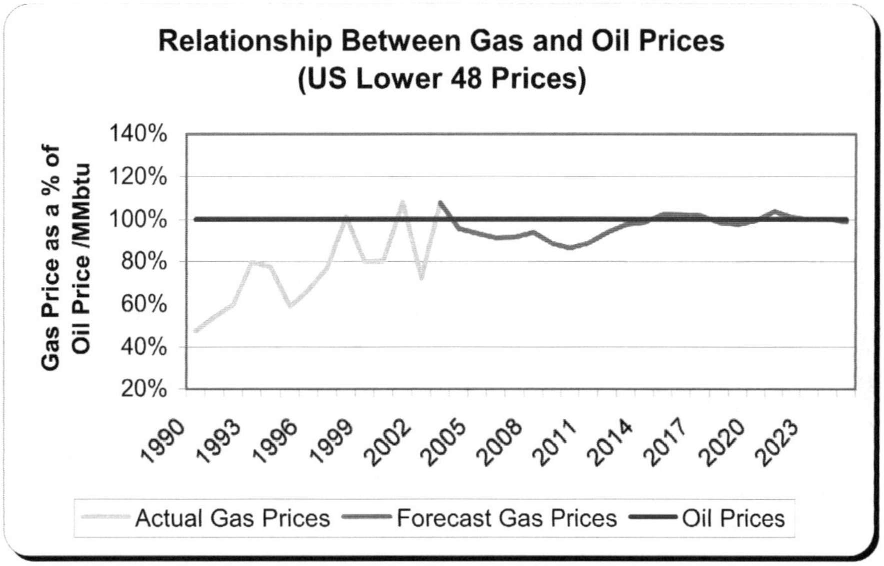

Figure 1 Source: US Energy Information Agency

The current options for the development of natural gas reserves fall into two general categories: Physical Change and Chemical Change technologies.

Physical Change Options

Physical change technologies move gas to market by compression, or change of state (liquefaction). These technologies are simple and efficient. Gas transported by pipeline and Liquefied Natural Gas (LNG) are the most common physical change applications. Gas transport by pipeline is the predominant access to market. When too distant, LNG can be an economic alternative to access traditional markets. Under special circumstances gas may also be transported as compressed natural gas (CNG), although no world scale project employing this technology has been developed to date. Both LNG and CNG are highly capital intensive, require specialized waterborne carriers and complex commercial structuring that often limit the application of these technologies.

Chemical Change Options

Traditional chemical change technologies transform natural gas (primarily methane) into more complex liquid compounds such as methanol, di-methyl ether (DME) and ammonia. These processes are also highly capital intensive, and are less thermally efficient than physical change alternatives. Although the market value for these products exceeds that of methane as a fuel, the market size of each individual product is too small to support the development of more than a few large natural gas developments.

"Gas to Liquids" (GTL) technologies use the syngas (CO and H_2) produced from methane and oxygen to feed a Fischer Tropsch catalytic process that converts these gases into naphtha and diesel. However, since no economic world-scale GTL developments have been completed to date, there exists a higher degree of uncertainty regarding GTL development costs and operational efficiencies than with other proven technologies.

An attractive aspect of GTL developments is that the products are clean (as is the case with LNG) and are sold into markets that are large and easily accessed by existing transport systems. It is likely that a production source will be closer to a GTL market than a LNG market because diesel markets are regional while gas markets of size are limited and often distant. Figure 2 shows the relative effect transportation distance has on LNG and GTL projects. It can be seen that it is more cost effective to transport a liquid fuel like GTL diesel long distances than it is to transport gas in the liquid state (LNG).

Note on figures in this report

The costs and product values for LNG and GTL projects vary greatly and are highly dependent on the location of project, product markets, quality of gas, project timing and fiscal regimes. For comparison purposes, unless otherwise noted, the figures in this report are based on Middle-East project developments with US product destinations.

Although these assumptions may not fully optimize a specific project, they do form a basis for legitimate comparisons between LNG and GTL alternatives.

Price forecasts were taken from the US Energy Information Agency (EIA) (2002 $US):

World Crude Oil Price: $24.53/bbl

US Gas Import Price: $4.24/MMBtu

Other Key Assumptions:

LNG Vessel Size: 200,000 cbm

LNG/GTL Plant Inlet: 1.1 Bcfd

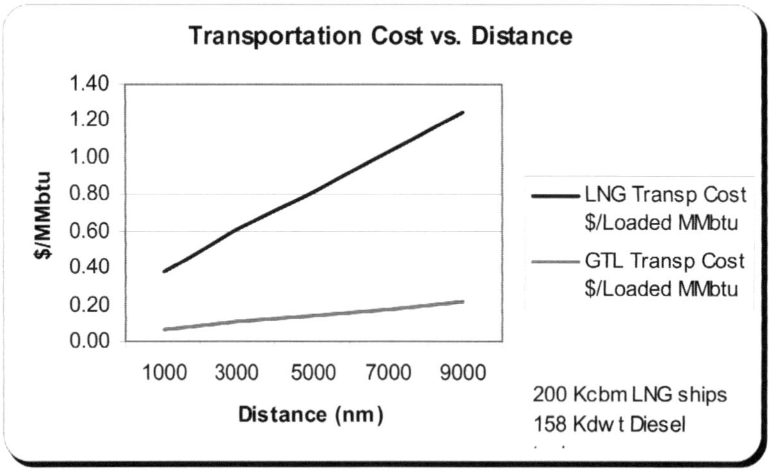

Figure 2 Source: ConocoPhillips

Recent Trends

Recent reductions in project development costs (both physical and chemical change based) combined with incentivized gas development energy regulations and strong product values has sparked interest in a number of remote natural gas development projects worldwide.

Construction and unit production costs have been driven downward for several reasons. The overall plant size and corresponding equipment size has increased dramatically, resulting in economies of scale being realized with the larger plant sizes. In addition, more reliable equipment and new control technology allow less redundancy in design and improved process efficiency. Competition among engineering and construction firms has also lowered EPC margins. The benefits of competition is clearly evident regarding the construction of cryogenic LNG ships where the introduction of additional ship building capacity has cut the cost of a new ship by one-third. Figure 3 shows recent and anticipated trends in investment costs for new LNG and GTL plants.

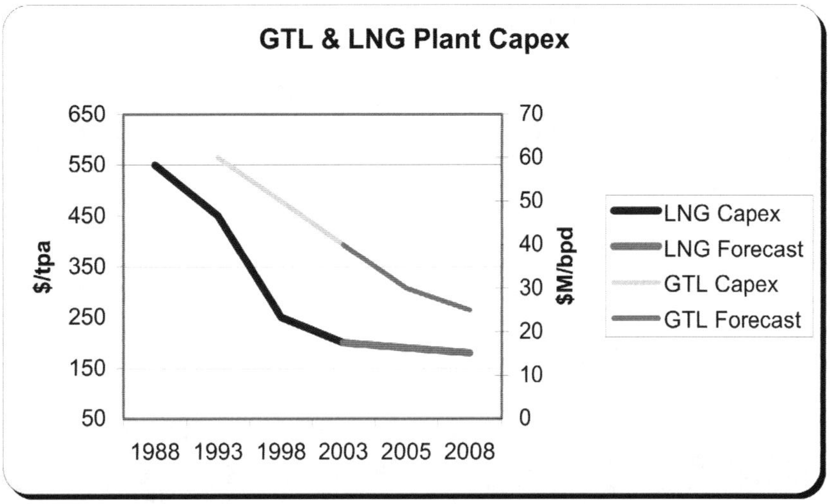

Figure 3 Source: General Industry Publications

Additionally, the regulatory and business environment for natural gas developments has improved. Demand for cleaner fuels continues to grow worldwide. Host governments in gas-rich countries are aggressively facilitating development of their reserves by offering improved fiscal terms, lower tax rates and other incentives to simulate gas exploration and production. Continued strong energy prices hold the promise of robust project returns. As a result, investment capital from both the public and private sector is being made available to fund these large gas export and conversion projects.

LNG vs GTL Project Driver Comparison

Although the above trends support the development of additional remote gas reserves, selecting the best gas resource and technical option from a risked economic perspective often remains a challenge for the oil and gas industry. Although there are many common critical success factors associated with the development of large natural gas reserves, there are some important differences between LNG and GTL applications.

Common Critical Success Factors

ConocoPhillips has identified seven critical success factors for new large remote natural gas development projects. These factors determine the relative attractiveness of a potential investment and are critical in qualifying for project funding.

1. Stakeholder alignment

Gaining rights to upstream development and sharing interests throughout the elements of the value chain help to create seamless operations and can reduce project development cycle time that may add value to the project.

2. Gas reserve quantity and quality

Recoverable gas reserves should be at least 4 TCF and be able to have sustained field flow rate of 500-1000 MMcfd for a period of 25 years to be suitable for large LNG or GTL projects. In addition, low cost, high rate wells with large ultimate recoveries enhance the project.

Generally project economic metrics are enhanced when: a) there are high volumes of condensate and recoverable natural gas liquids, and b) the content of inert gas (N_2) and acid gasses (CO_2 and H_2S) is low.

3. Advantaged site

Locations with lower construction costs, available low cost utilities and process water, and an adequately skilled labor force are favored over isolated sites. "Location Factor" is a comparative measure to an international standard of additional or reduced construction and operating costs based on a location's remoteness and productive labor capacity. Location factor generally has a greater effect on GTL projects. This is due to the concentration of investment required at the GTL plant location, whereas the majority of a LNG's project capital resides offshore (LNG ships and regasification terminal). Location Factor can marginally improve or significantly impair a project's net present value (NPV).

Available export infrastructure and gas field distance to that infrastructure are critical in determining the overall cost structure of the project. Many remote gas fields will require additional investments to develop supporting greenfield export facilities, thus making those sites cost disadvantaged to brownfield projects that have some infrastructure in place.

4. Expansion potential

Projects with adequate reserves and market potential can capture expansion synergies, improve incremental production costs and add long-term value to the development.

5. Market accessibility

Project economics are improved where: a) shipping distances and the associated costs are reasonable, b) there is affordable access to distribution and marketing infrastructure, and c) there is a sufficient market of long-term, quality customers.

6. Strong local partners

Experienced and financially secure local partners can provide critical assistance in project development, construction management and stakeholder relations.

7. Supportive host government

The highly capital intensive and long-lived nature of complex LNG and GTL projects require fiscally favorable terms, and fiscal transparency and consistency. Realistic expectations of resource owners, governments, and contractors must prevail when negotiating reserve ownership, gas transfer price, profit splits, or cost and market projections. It is important for interests to be aligned to increase the overall project value and reduce project execution risk. Governments that demonstrate a willingness to expedite a project can capture a window of opportunity that might otherwise cause a project to be postponed or canceled.

LNG Project Drivers

LNG projects are capital intensive, often requiring several billion dollars of investment. However, these costs are spread throughout its value chain including: upstream development, liquefaction plant, shipping, regasification and distribution infrastructure. A successful LNG development requires the cooperation and often the participation of all of its value chain members. Thus LNG projects tend to be very commercially complex and underpinned by long-term contracts to ship and market production.

The lack of excess LNG ships and spare terminal capacity has limited the development of a LNG spot market and has contributed to the delay of many LNG development and expansion projects. Market access requires a world-scale investment ($US 500 million or greater) of its own in the form of a regasification terminal. Although many new terminals have been proposed in Asia, Europe and the United States, it is improbable that at such costs, any new terminal will be built at risk. A new terminal project will likely require substantial capacity commitment before it is taken past the permit stage.

LNG ships are highly specialized and are generally purpose-built for a specific development. These carriers are costly and the shipping element can be the most expensive component of the project. Distance to market therefore can dramatically affect project returns since longer distances require more ships and increase product losses and operating expenses. Recently created ship-building competition and the potential to introduce very large LNG carriers will continue to reduce the impact of distance on project returns. However, these gains are likely to begin eroding over the next decade as demand for these ships grows and steel prices rise. This trend is shown in Figure 4.

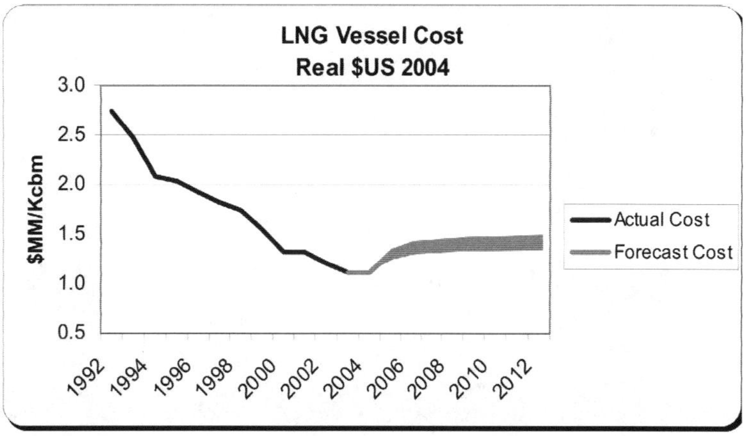

Figure 4 Source: Maritime Business Strategies LLC

LNG projects also require large sunk investment costs in assets that are generally underutilized at project commencement (LNG storage, dock and harbor facilities). Thus, LNG projects see significant improvement when sales ramp-up periods are completed and when projects are expanded to more fully utilize all facilities.

GTL Project Drivers

GTL projects are also capital intensive, but this investment is concentrated in the manufacturing plant. GTL products are easily shipped and sold into large existing markets. Thus, project and commercial structure complexity is much simpler than a similar sized LNG development.

GTL plants have higher unit operating costs and lower process efficiencies than LNG plants. These aspects require careful design consideration. A well-designed GTL plant with good conversion efficiency can have a thermal efficiency in the range of 65% and a carbon efficiency (carbon retained as salable product) in excess of 80%. These efficiencies, coupled with plant uptime in the range of 92%, are required to achieve strong economic metrics.

Large-scale plants are also an important contributor to project viability. Projects today are proposed with plant sizes of 70,000 bpd and greater. A plant of this size will generally consist of four individual trains. Although the individual train sizes have increased by about 1/3, the overall plant capacity has increased significantly when compared to proposals of only a few years ago. This economy of scale is an important driver of the project economics and recent GTL proposals show that the major GTL technology players are technically and commercially comfortable with this larger scale.

The amount of pre-investment for underutilized assets is less than that required for LNG. Thus, GTL does not experience the same magnitude of expansion benefits as an LNG plant.

Figure 5 shows a relative comparison of LNG and GTL technologies for a hypothetical development in the Middle East. Operating margins (green) are used to pay taxes and royalties and are otherwise shared throughout the development's value chain to improve rate of return beyond the cost of capital.

Several important technology comparisons can be made using Figure 5. Shipping, regasification and marketing costs (orange) are substantially more costly for LNG vs GTL. However, GTL's higher capital costs (yellow) and lower process efficiencies (shown as fuel and losses in black) causes its overall cost of service to exceed that of LNG. Although GTL's margin on a unit basis is higher than that of LNG, it should be noted that GTL's lower thermal efficiency results in less product manufacture (less units) and must be used to pay back higher capital costs.

Both technologies offer adequate project margins when oil and gas equivalent values are above $20.00/bbl.

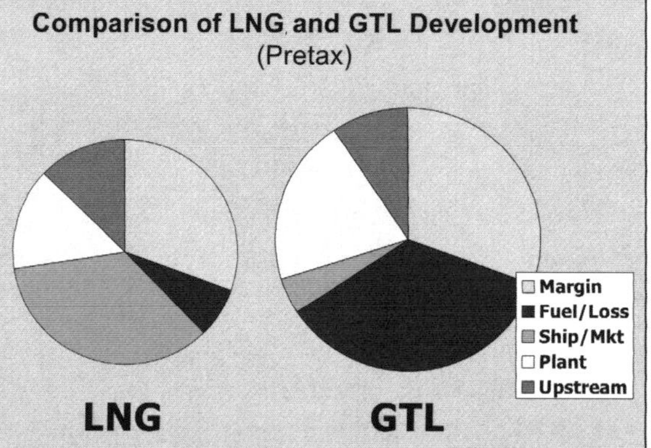

Figure 5

Until recently, LNG has generally been the only large gas export technology available, and consequently has been the preferred development option. GTL, however, is quickly emerging as a commercially viable gas monetization alternative. GTL becomes particularly attractive when LNG markets or market access is limited or slow to develop and causes a delay in development. A one-year delay of a world scale project awaiting gas market disposition may reduce the net present value of a project by as much as $US 100 million. GTL may also be favored when there are long shipping distances to market, or when a natural gas-rich country wishes to diversify its development portfolio.

Summary

Abundant natural gas resources exist in the world today. However, several important criteria must be met for successful development to occur. Key to any successful LNG or GTL project is high quality gas resources, strong partnerships and supportive fiscal terms and regulations. The choice between LNG and GTL options can be difficult and often is determined by the availability and strength of market. ConocoPhillips is uniquely qualified to support the evaluation of technical options. ConocoPhillips provides superior project development and execution results by supplying experienced resources in project management, project structuring, and marketing management along with its proprietary technological solutions and infrastructure for market access.

104d Enabling Offshore Production with Microchannel Process Technology Systems

Francis P Daly, Tad Dritz, Terry Mazanec

Velocys Technology will enable monetizing natural gas by the production of methanol or middle distillates on a floating production, storage, and offloading (FPSO) facility. The challenges of offshore chemical production are two-fold: (1) size, weight, and motion sensitivity of facilities capable of converting the gas to liquid product; and (2) high capital cost of such facilities. The technology presented herein addresses both of these key hurdles by greatly reducing the size of the process equipment, and accommodating co production of oil and natural gas derived, liquid products on a single FPSO.

Dealing with offshore associated gas has long been an oil and gas industry challenge. Historically, offshore associated gas was either flared or vented. As environmental restrictions began to limit these alternatives, re-injecting the gas into the reservior grew to be the solution of choice. Today, the economic evaluation of remote (without access to a pipeline) fields must include the cost of re-injecting associated gas. For this reason, many discoveries are deemed uneconomical to produce and are left untapped.

Monetizing associated gas fulfills the need to limit flaring and offers an attractive alternative to re-injection. Four possible gas monetization strategies are available: (1) compress and pipeline the gas to shore, (2) generate electricity offshore and transmit it to on-shore markets via subsea cable, (3) produce and transport compressed (CNG) or liquified (LNG) natural gas using specialized tankers, or (4) convert the gas to a saleable liquid product, either methanol or Fischer Tropsch (FT) liquids. Sending natural gas to shore via a pipeline is generally the preferred choice, but this option is typically not available for deepwater reserves. Each of the remaining options has unique advantages. However, converting gas into liquid products is the most attractive since liquid chemicals are valuable commodities that can be easily and safely transported.

Velocys Technology systems greatly intensify the processes required to convert natural gas to liquid products, thereby making the entire production process compact, and more suitable to the space-constrained offshore environment. The Velocys Technology system for converting gas to liquid products couples syntheis gas production using a novel steam methane reformer with a highly efficient synthesis reactor. Velocys is developing synthesis reactors for production of both methanol production and FT liquids.

The key requirements of an offshore facility include:

- Insensitivity to motion

- No need for on-board oxygen generation

- Compactness

- Low freshwater requirement.

The Velocys Technology systems excel at all these requirements and remain economical at lower volume production scales. Processing chemicals in microchannels results in high velocities which provide freedom from gravitational and other external forces, which is critical on floating processing operations that can experience high intensity wave moments during inclement weather. The increased velocity in microchannel reactors equates to low residence time, and therefore low reactor volume and size. However, the small footprint of the reactor is only part of the space savings. Due to the abililty to tightly control temperature profiles in the synthesis reactors, higher conversion rates can be achieved; thereby, reducing recycle rates and compression requirements.

In addition, Velocys systems are expected to have better energy efficiency than competing technologies because microchannel steam methane reformers can operate effectively with approximately one-third less water than conventional systems. Velocys' reformer system has been demonstrated with a steam-to-carbon ratio of 2:1, compared to a 3:1 steam-to-carbon ratio which is required in conventional reformers. Process water efficiency is imperative for offshore processing where every gallon of freshwater must be desalinated using energy intensive water-making equipment.

Velocys is actively collaborating with companies and governmental agenccies to design, develop and demonstrate breakthrough production systems. For instance, Velocys is currently working with the U.S. Department of Defense to demonstrate the feasibility of a transportable FT liquids production system.

This paper will provide the current status, demonstration test results and ecocomic analyses for both FT and methanol systems. Results will indicate that the use of the Velocys Technology systems could enhance economic returns and in some cases could transform marginal fields, including certain dry gas offshore discoveries, into attractive prospects.

104e Fischer-Tropsch Synthesis: Impact of Water on Iron and Cobalt Catalysts During Synthesis

Burtron H Davis, Gary Jacobs, Karuna Chaudhari, Mingsheng Luo, Tapan K Das

Most studies of the role of water on the activity and/or selectivity of iron or cobalt catalysts are derived directly from conversion data or by adding additional water, thereby changing reaction variables. We have defined the impact of water during synthesis keeping other variables constant while adding water. This is accomplished by starting the synthesis with a feed that includes an inert gas and then replacing all or some of the inert gas with water. With the iron catalyst it appears that both water-gas-shift and FT synthesis reactions occur on a common site and that high water content removes these sites through irreversible oxidation. The situation is more complex with the cobalt catalyst. In some special cases water has a positive impact on the conversion while in others it has a negative impact. With higher loadings of cobalt, water has a negligible impact. However, at high levels of water addition (25-30 volume %) water causes an irreversible decline in catalytic activity and at least part of this is due to oxidation of the cobalt metal.

104f High Conversion Efficiency Reactor for Offshore Production of Methanol

Francis P Daly, Tad Dritz, Terry Mazanec

Methanol production is shifting away from the United States and other high-cost natural gas areas to locations with low-cost stranded, natural gas fields. Many of these fields are associated with deepwater oil reservoirs. Currently, stranded offshore associated gas is either flared or reinjected into the reservoir. A more economical approach would be to convert this underutilized energy source into methanol or other transportable liquid products. To realize this goal, the methanol production process must be adapted to operate on a floating production, storage, and offloading (FPSO) vessel.

Velocys has begun developing a novel reactor that will greatly intensify the methanol synthesis reaction, thereby making the entire production process more compact, and more suitable to the space constrained offshore environment. The concept is a microchannel process technology reactor. When combined with a novel steam methane reformer (SMR) that Velocys is also developing, the resulting methanol process will meet the requirements of an offshore system, including extreme compactness and low freshwater consumption.

The first step in methanol production is a synthesis gas (syngas) production process. Since Velocys' SMR system generates syngas, the current focus is on the second step - methanol synthesis - where syngas is catalytically converted to methanol by the following reactions:

$CO + 2H_2 ==> CH_3OH + 90.7$ kJ/mole

$CO_2 + 3H_2 ==> CH_3OH + H_2O + 49.5$ kJ/mole

Because both reactions are exothermic and result in a volume reduction, conversion is favored by both low temperature and high pressure. Today's methanol plants are based on a pelletized, copper catalyst that must be operated at 50 atm and temperatures exceeding 210°C to ensure sufficient catalyst activity.

The rate of heat removal in conventional reactors is much slower than the rate of heat release from these exothermic reactions. Additionally, conventional reactors must use pelletized catalysts which have low mass transfer effectiveness factors. Powdered catalysts enable much improved mass transfer rates, but cause excessive pressure drop. Because of these inefficiencies, a low fraction (less than 30 percent) of synthesis gas is converted per pass. Though operating at higher pressure would increase the conversion per pass, the trade-off between capital costs and energy costs has resulted in plant designs with large recycle loops to achieve acceptable yields. Hence, we believe that considerable opportunity exists for developing an improved methanol synthesis process by improving the heat and mass transfer efficiency of the reactor.

Microchannel reactors can greatly increase the per pass conversion for methanol synthesis due to much higher heat transfer rates, catalyst forms with high effectiveness factors, and procise temperature control. Interleaving microreaction channels with heat exchange channels allow for rapid removal of the exothermic heat of reaction generated during synthesis. Additionally, the local temperature can be reduced along the length of the channels. This design enables fast kinetics at the beginning of the reactor to convert the bulk of the syngas into methanol, while favoring equilibrium in the latter stages to achieve superior per pas conversion. These advantages lead to significantly lower capital costs and energy consumption.

For the strongly equilibrium-limited methanol synthesis reaction, the reaction driving force potential is set by local temperature. The Velocys reactor will be hottest at the reactor entrance, and will approach the catalyst highest operating temperature of 270 to 300°C to assist kinetics. The coolest point will be at the reactor exit, and may well be below 200°C or even 180°C to maximize equilibrium driving force potential. A series of isothermal zones control the temperature along the reaction channel. The coolant channel zones have embedded microchannel features which enable partial boiling of water at variable pressures to sequentially decrease temperature.

These systems will have much higher heat and mass transfer performance. Mathematical modeling shows greatly improved per pass conversion with either the commercial available catalysts or improved formulations. By leveraging these advantages, this system could achieve single-pass conversion levels above 80 percent.

Benzene Crystallization Risks in the LIQUEFIN LNG process

J-C de Hemptinne
Institut Français du Pétrole

AIChE Spring Meeting, April 2005
5th Topical Conference on Natural Gas Utilization
Atlanta April 10-14, 2005

Abstract

In natural gas liquefaction processes , the gas feed is cooled to very low temperatures (-160°C). The presence of traces of heavy hydrocarbons components may therefore result in crystallization, and hence plugging. The present work investigates the Solid - Liquid - Vapor equilibria and proposes a way to predict the risks of crystallization.

In a first section, the phase diagrams that may be encountered with natural gas mixtures at these very low temperatures are presented. A literature review shows that few data are available for operating conditions of LNG processes .

In a second section, the crystallisation model based on a cubic equation of state, coupled with a Huron-Vidal mixing rule and combined with a solid phase model is presented. This model requires the crystallization heat capacity of pure components, which must be determined using the few data available. Nevertheless, it appears that the resulting model is rather predictive when compared to data that were not used in the regression.

Finally, the model is applied in different zones of the LIQUEFIN process commercialized by Axens. The results have been taken into account for a specific design.

Introduction

Due to the increase of the world natural gas consumption, the distance between production and consumption sites is ever increasing. Pipeline transportation is not economical for these long distances and only liquid phase transportation is feasible. As a consequence, many high capacity gas liquefaction plants are presently in project. A typical natural gas composition is provided in Table 1.

Table 1: Typical composition of the natural gas (with a high benzene content) to be liquefied (mole percent)

N2	5%
CH4	86%
C2	5%
C3	2%
iC4	0.5%
nC4	0.8%
iC5	0.1%
nC5	0.3%
nC6	0.1%
nC7	0.06%
nC8	0.02%
nC9	0.01%
CO2	0.01%
benzene	0.05%
cyclohexane	0.05%

Table 1 also indicates that trace amounts of heavy components, and in particular benzene, may remain present in the feed. These components may crystallize in the process conditions, with the risk of plugging. It is essential to control the risk of crystallization.

The purpose of this work is to investigate these risks. The paper first presents the phase diagram that is encountered in natural gas treatment. In a second section, a model is proposed for calculating the risk of crystallization. Finally, the phase diagrams of some fluids whose composition is representative of process conditions, are presented.

Natural gas phase diagrams

Pressure - temperature phase diagrams of multicomponent mixtures may become complex if solid phases are considered (Tiffin et al. 1979,). Our purpose is to investigate the risk of a first solid phase appearing. For this purpose, the study of binary mixtures containing methane and a heavy component is sufficient. The choice of the heavy component depends on the crystallization temperature when it is pure. Table 2 shows that the components presenting the largest risk are, in that order, p-Xylene, cyclohexane, benzene, o-xylene and normal decane. The components heavier than C6, however, are almost absent. A comparison of the crystallization temperature of benzene and cyclohexane in a mixture with methane is provided in Figure 1. It shows us that in a mixture, benzene crystallizes well before cyclohexane in cooling. Hence, we will focus on the benzene crystallization behavior.

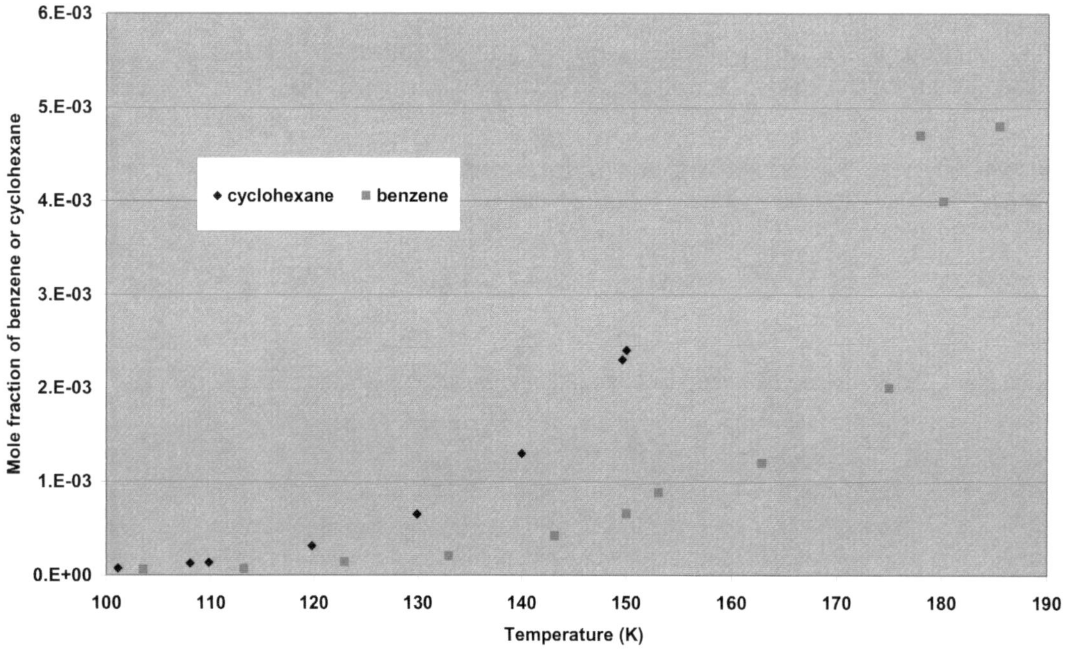

Figure 1 Comparative crystallization temperature from SLE data of Neuman *et al.* (1972).

Table 2: Pure component crystallization data (DIPPR , 2003)

	Triple point Temperature	Triple point Pressure	Enthalpy of fusion	Volume of fusion
	K	Pa	J/kmol	m3/kmol
p-Xylene	286.4	5.76E+02	1.71E+07	1.86E-02
cyclohexane	279.7	5.36E+03	2.74E+06	8.30E-03
Benzene	278.7	4.76E+03	9.87E+06	1.04E-02
o-Xylene	248	2.18E+01	1.36E+07	1.17E-02
nC10	243.5	1.39E+00	2.87E+07	2.71E-02
m-Xylene	225.3	3.18E+00	1.16E+07	1.56E-02
nC9	219.7	4.31E-01	1.55E+07	2.17E-02
nC8	216.4	2.11E+00	2.07E+07	1.97E-02
nC7	182.6	1.83E-01	1.41E+07	1.39E-02
cyclopentane	179.3	9.07E+00	6.09E+05	1.20E-02
Toluene	178.2	4.75E-02	6.64E+06	5.90E-03
nC6	177.8	9.02E-01	1.31E+07	1.26E-02
nC5	143.4	6.86E-02	8.40E+06	1.65E-02
nC4	134.9	6.74E-01	4.66E+06	9.17E-03
methylcyclopentane	130.7	2.25E-04	6.93E+06	1.40E-02

The study of the phase diagrams of binary mixtures will help us understand the existing experimental data, that will be needed for the model development and validation.

Kohn (1981) has classified the solid - liquid - gas phase diagrams into four types. Only the first type is of interest here, as it is the one encountered for the methane - benzene system. Its pressure - temperature projection is sketched in Figure 2. For the light component, one recognizes the vapor pressure line, the solidus line and the sublimation line. For the heavy component, only the vapor pressure line is drawn. The locus of critical points is visible as a line that connects the critical point of each of the pure components. In this diagram, this line is cut by two dashed lines that represent three-phase solid-liquid - vapor equilibria (SLVE).

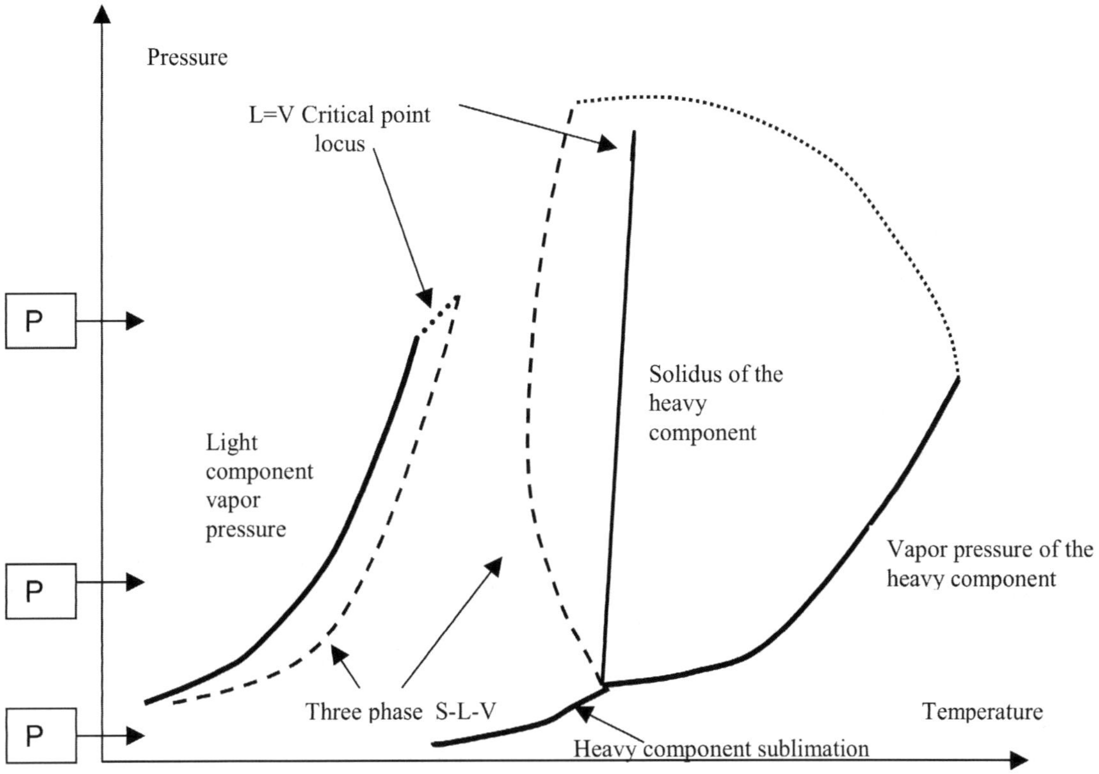

Figure 2 Sketch of a the benzene – methane phase diagram

The meaning of these curves is best understood when investigating a cut at constant pressure. We investigate here an intermediate pressure, that is higher than the three phase point of either components, but lower than the critical point of both components (i.e., for the benzene - methane mixture, that yields that type of diagram, between 0.0047 and 46 bar). In these conditions, Figure 3 is found.

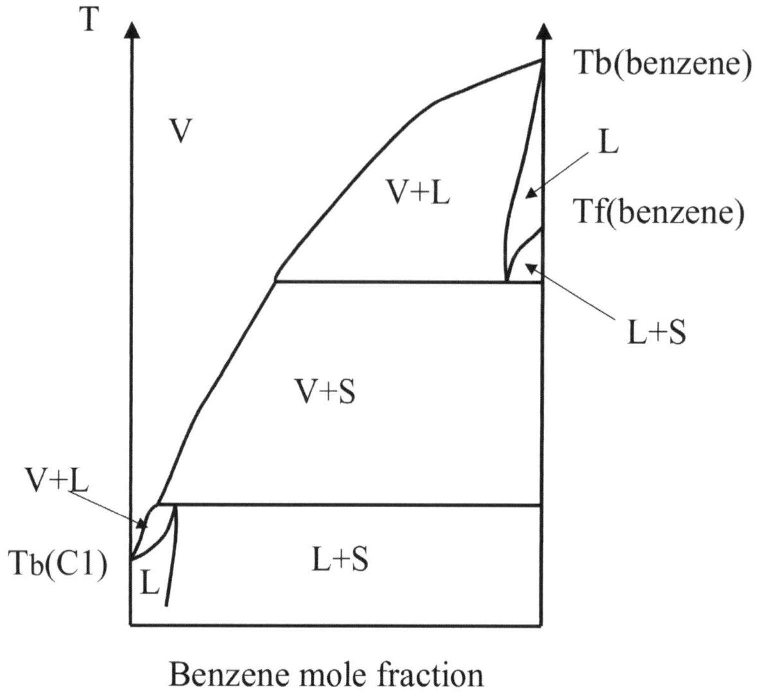

Figure 3 Sketch of the methane - benzene binary diagram at fixed pressure.

At high temperature, the mixture is always in the vapor phase. In a mixture that is rich in benzene, the liquid-vapor equilibrium (VLE) region is first encountered upon cooling. A liquid phase appears, that is richer in benzene than the vapor phase. If the initial mixture was sufficiently rich in benzene, the mixture may become entirely liquid before reaching the limit of solid formation. This high temperature solid - liquid equilibrium line (HT-SLE) forms the limit below which benzene crystallizes. In this work, it is assumed that the heavy component always crystallizes as a pure component. This may not always be true, but it represents the highest risk of crystallizing at a given temperature.

If the temperature is further lowered, a first (high temperature) three phase line is reached. At this temperature, the liquid of the VLE becomes unstable and starts crystallizing. Hence, the liquid will disappear upon cooling, until only a vapor and a solid phase remain. Below this temperature, and if the system contains a large amount of methane (which is our case in natural gas liquefaction processes), no VLE is encountered, but benzene will crystallize directly out of the vapor phase, forming a vapor - solid equilibrium (VSE). Note that, since the natural gas contains more than two components, the three phases will coexist. Yet, it is important to notice that a solid phase may deposit from the vapor in the absence of a liquid phase.

When the temperature reaches very low values, a new VLE region is encountered, where the liquid phase is rich in methane. It is so rich that the solid benzene can dissolve in this liquid phase. Hence, a second (low temperature) three-phase line is found. If the system is rich in methane, and in the VSE region, an unexpected phenomenon will occur, that is that upon cooling, the solid benzene will melt and become liquid. At still lower temperatures, a regular VLE is observed.

The Figure 3 shows that when temperature is lowered from the VLE, a single liquid phase is reached. Finally, at extremely low temperatures, the solid phase appears again (LT-SLE).

Figure 4 shows the benzene - methane three - phase lines as published by Luks *et al.* (1981). They provide as a function of temperature and pressure, the composition of the liquid phase. The graph clearly shows how the two three phase lines change with pressure. The lower three-phase line ends at the methane critical point. As a matter of fact, from that point onward, it becomes impossible to make the difference between the liquid and the vapor phase. The high temperature three-phase line continues until much higher pressures because the critical points locus of the benzene - methane mixture reaches high pressures.

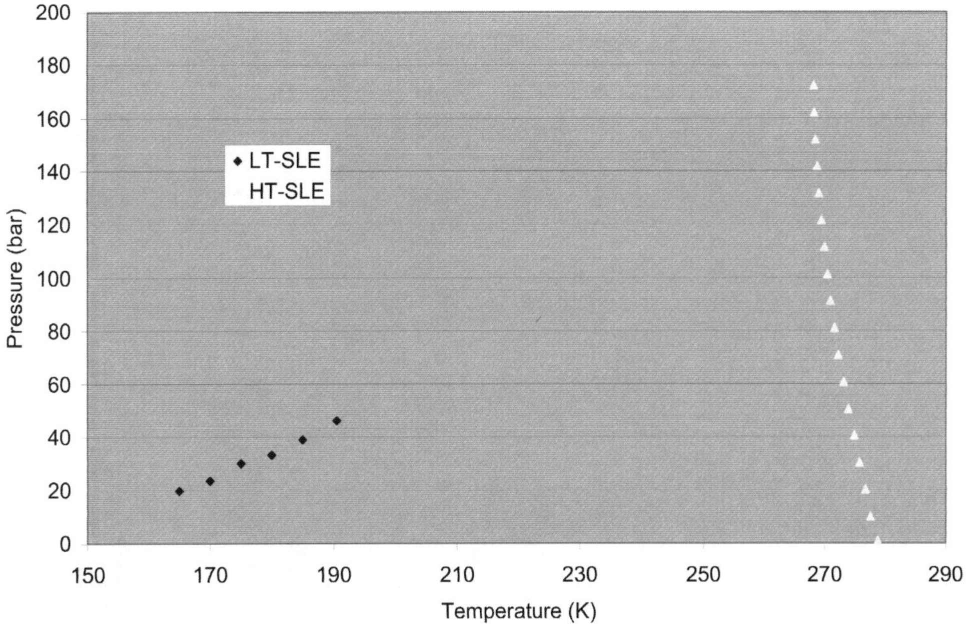

Figure 4 Pressure-Temperature projection of the methane - benzene three-phase lines, according to the data of Luks et al. (1981)

We note that the conditions that may be of interest to NGL processes are those where vapor - solid equilibrium (VSE) is encountered. Unfortunately, very few such data have been found. Only Rijkers et al. (1992) publishes such data, but for rather high benzene concentrations.

$$b = \Omega_b \frac{RT_c}{P_c} \quad (3)$$

The temperature dependence of the *a* parameter is calculated using:

$$\alpha(T_R) = T_R^{C_3(C_2-1)} \exp[C_1(1 - T_R^{C_3 C_2})] \quad (4)$$

The properties of the components used are provided in table 3.

Table 3 : Component properties used in this work.

Component	Critical Temperature (K)	Critical Pressure (bar)	C1	C2	C3
Nitrogen	126.25	33.944	0.5764	0.9093	0.6765
H2S	373.55	90.08	0.1596	0.8708	2
Methane	190.55	46	0.1195	0.904	2
CO2	304.19	73.82	0.2047	0.8197	2
Ethane	305.43	48.84	0.2091	0.8681	1.6875
Propane	369.82	42.5	0.2153	0.8498	1.8108
Butane (n)	425.15	38	0.1626	0.8568	2.4165
Butane (iso)	408.13	36.48	0.6712	0.8859	0.8903
Pentane (n)	469.65	33.69	0.8201	0.9228	0.8755
Pentane (i)	460.39	33.81	0.1691	0.8535	2.4876
Cyclopentane	511.7	45.1	0.1415	0.8565	2.5391
nC6	507.35	29.69	0.4252	0.8118	1.4723
cyclohexane	553.45	40.73	0.0494	0.9273	6.5471
Benzene	562.6	49.24	0.0842	0.8732	3.4496
nC7	540.15	27.36	0.2493	0.8256	2.3311
Toluene	591.72	41.09	0.2516	0.8351	2.0436
nC8	568.75	24.82	0.2159	0.8271	2.7241
o-Xylene	630.25	37.328	0.2575	0.8367	2.2195
p-Xylene	616.15	35.4109	0.2172	0.8466	2.5718
m-Xylene	616.97	35.413	0.2387	0.8408	2.4116
nC9	594.55	22.879	0.1724	0.8426	3.4558
nC10	617.45	20.974	0.1523	0.846	3.9327

The parameters of the mixture are obtained using a linear rule for *b*:

$$b = \sum_i b_i x_i \quad (5)$$

and the Huron-Vidal (1979) mixing rule for *a*:

$$\frac{a}{b} = \sum_i x_i \frac{a_i}{b_i} - \frac{g_\infty^E}{\ln 2} \quad (6)$$

The excess Gibbs energy is expressed using a modified NRTL:

$$g_\infty^E = \sum_{i=1}^{N} z_i \frac{\sum_{j=1}^{N} b_j z_j \exp\left(-\alpha_{ij} \frac{A_{ji}}{RT}\right) \cdot A_{ji}}{\sum_{k=1}^{N} b_k z_k \exp\left(-\alpha_{ij} \frac{A_{ki}}{RT}\right)} \quad (7)$$

As a result, each set of two components is associated with 3 interaction parameters at any given temperature: $\alpha_{ij}=\alpha_{ji}$, A_{ij} and A_{ji}. In addition, these parameters are function of temperature:

$$\alpha_{ij} = \alpha_{ji} = \alpha'_{ij} + \beta_{ij} T \text{ (often taken = 0.2, except in case of need)} \quad (8)$$

$$\frac{A_{ij}}{RT} = a_{ji} + \frac{b_{ij}}{T} + \frac{c_{ij}}{T^2} \quad (9)$$

Binary interaction parameters

Most of these parameters have been fitted on binary VLE data, usually at room or above room temperature. Even though temperature effect is taken into consideration through equations (8) and (9), this represents a drawback considering the very low temperature of the process. As an example, we present the deviation on the VLE plots, obtained for the binary methane - benzene system (Figure 5). This binary system is the most important one, since benzene is the component that first crystallizes out of the methane-rich natural gas.

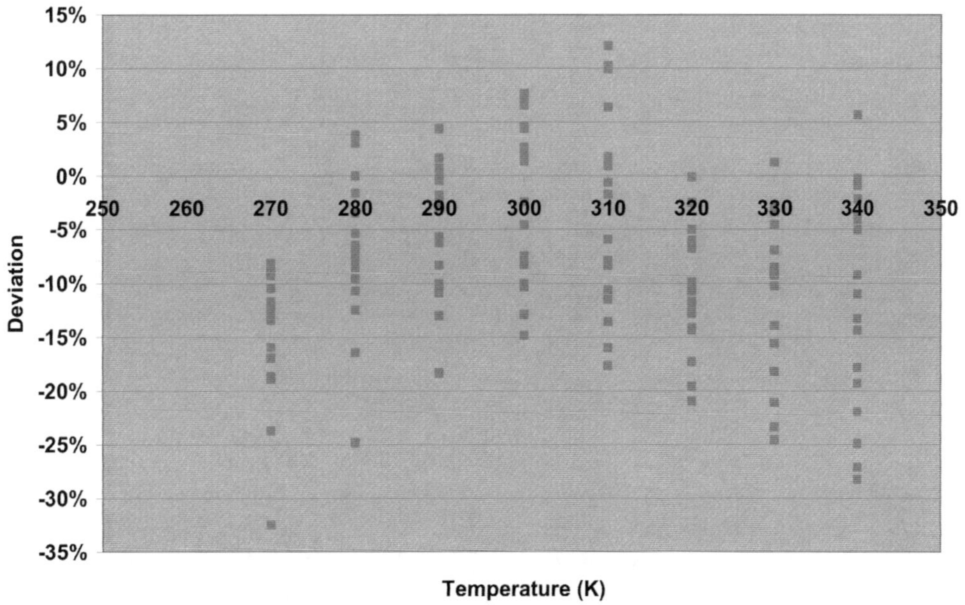

Figure 5 : Deviations between experimental and calculated bubble pressures for the methane benzene binary system, as a function of temperature.

Deviations are of the order of 30% on bubble pressures, in accordance with the uncertainties on the available data.

Heat capacity of fusion

The heat capacity of fusion is obtained by plotting the function $\ln\left(\frac{f_{i_i}}{f_{L_i}^0}\right)$ as a function of temperature, for each SLE data point. Indeed, when crystallization is observed, we may state:

$f_{i_i} = f_{Si_i}^o$, where f_i is the fugacity of component i in the stable phase. Hence, the logarithm of this ratio is equal to the right hand side of equation (1). As mentioned earlier, all quantities but the heat capacity of fusion are known. It is therefore possible to fit this parameter to such a plot, as shown in Figure 6.

Figure 6 shows the logarithm of the fugacity ratio for two types of solid equilibrium data. The sublimation data are obtained from low temperature pure benzene vapor-solid equilibrium; the lowest temperature data, however, come from benzene crystallization out of liquid methane. It is clearly visible that the two curves are not colinear. The reason for this is most probably due to the imperfections of the Huron-Vidal mixing rule that is used. When sublimation data are used, no mixing rule is needed, as all data concern the pure component. On the opposite, the low temperature SLE data concern very small benzene concentrations in almost pure methane. Hence, in the latter case, the infinite dilution fugacity is used, which is extrapolated from the equation that has been validated on high temperature VLE data.

Yet, in the present study, we focus our attention on the low temperature and high dilution behavior. Doing so, we find Δc_{pi} = -200,000 J/kmol/K for benzene.

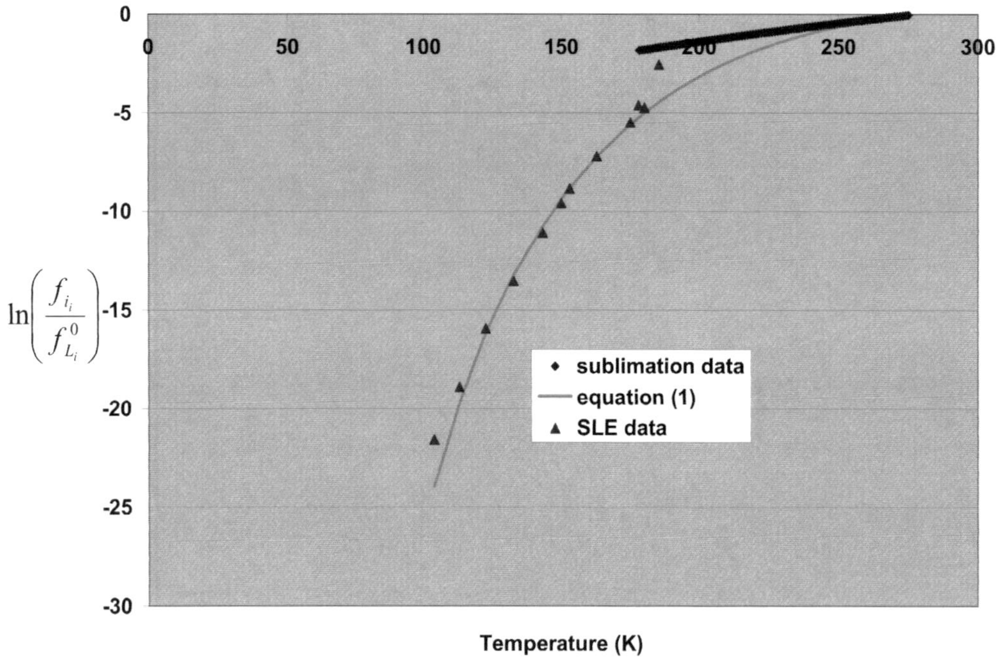

Figure 6: Correction term (equation 1) as a function of temperature for benzene crystallization. The diamonds refer to sublimation data; the triangles to benzene crystallization in methane.

Model validation

The model that is thus developed has first been validated by extrapolation of pressure and temperature conditions, using the data of Rijkers *et al.* (1992) as shown in Figure 7. This plot shows two distinct types of behavior. At higher temperatures, and high concentrations of benzene, SLE lines are seen. The data points and the model (line) agree within 5°C. The model overpredicts crystallization temperatures which is acceptable for a safe design. At lower temperature, a vapor-solid equilibrium line is shown. Again, the model qualitatively agrees with the data. The crystallization temperature increases with decreasing pressure. At lower pressures (process conditions), we expect to overpredict the crystallization temperature, again a safe result. Unfortunately, no data could be found showing crystallization from the vapor phase at benzene concentrations that are closer to the process conditions (less than 1%).

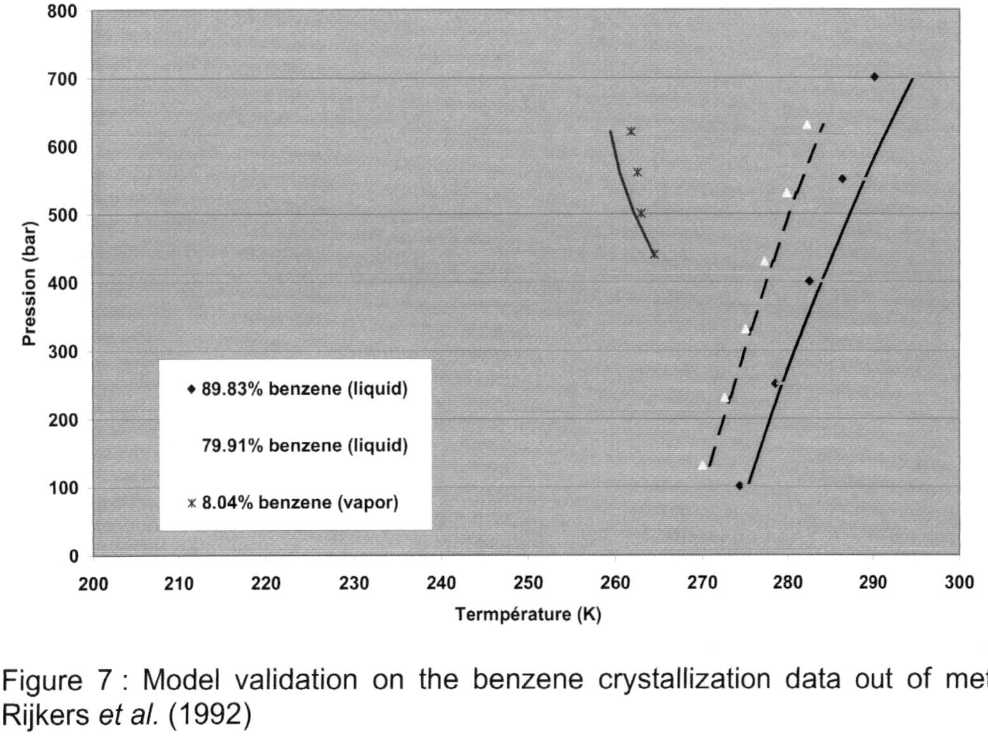

Figure 7 : Model validation on the benzene crystallization data out of methane of Rijkers *et al.* (1992)

Process analysis

A typical LIQUEFIN Process is sketched in Figure 8. The feed gas whose composition is shown in Table 1 goes through a pre-cooling unit where the temperature drops to -60°C. It then goes through a demethanizer before being further cooled to its final temperature of -160°C. Crystallization must be avoided at any place in the process, because of the risk of plugging the heat exchangers. We investigate first the phase diagram of a fluid with the feed composition, and in a second stage that of a fluid whose composition is that of the final fluid.

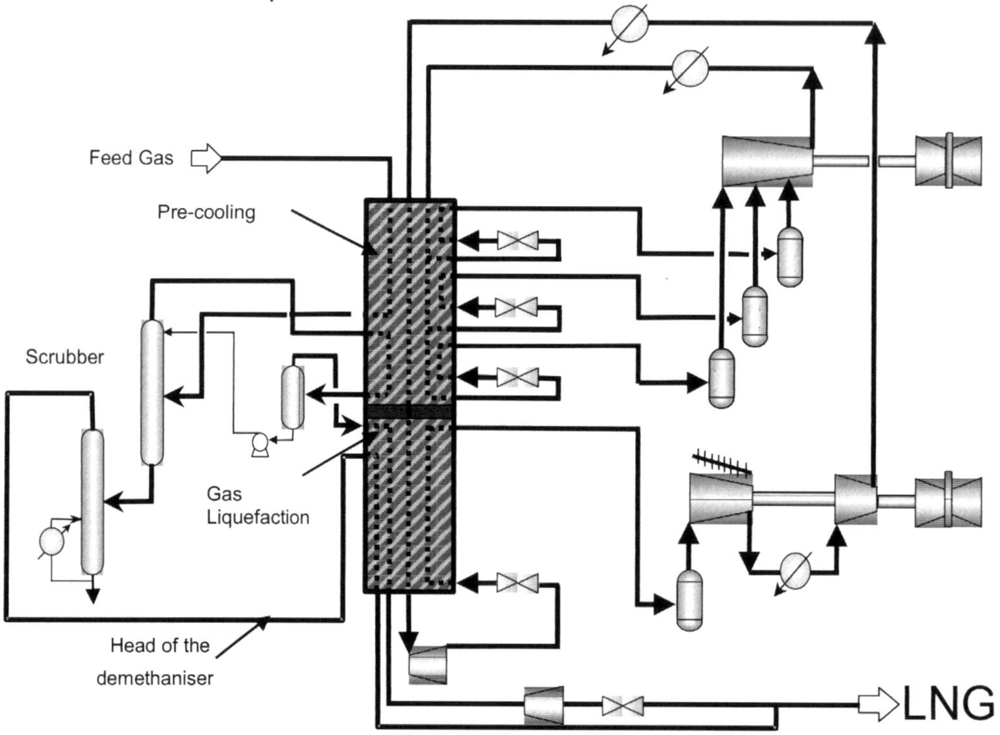

Figure 8 Sketch of the LIQUEFIN process

Feed gas phase diagram

Using the feed composition given in Table 1, Figure 9 is obtained. The Liquid - Vapor (L-V) phase envelope is predicted. Where 'S' is found, benzene crystallizes. There is a clear risk below -90°C, but pressure has an effect as well:

- When pressure is very low (atmospheric), the solid appears much earlier (at higher temperature). This can be explained rather easily by observing that in these conditions, the liquid phase is rather concentrated in benzene.
- In the liquid region, it is seen that increasing pressure results in a higher solubility of the solid. This is because this liquid phase remains somewhat compressible as it is essentially composed of methane, and therefore the temperature is close to the critical temperature. Hence, at higher pressure, the liquid is denser which means it is a better solvent for the benzene crystals. Note that the phenomenon had been opposite if the liquid had been less compressible, as is the case with

pure benzene: the higher the pressure, the larger the tendency to form a solid phase.

P (bara)\T (°C)	-100	-90	-80	-70	-60	-50	-40
100	F+S	F	F	F	F	F	F
95	F+S	F	F	F	F	F	L-V
90	F+S	F	F	F	F	F	L-V
85	F+S	F	F	F	F	F	L-V
80	F+S	F	F	F	F	L-V	L-V
75	F+S	F+S	F	F	F	L-V	L-V
70	F+S	F+S	F	F	F	L-V	L-V
65	F+S	F+S	F	F	L-V	L-V	L-V
60	F+S	F+S	F	F	L-V	L-V	L-V
55	F+S	F+S	F	L-V	L-V	L-V	L-V
50	F+S	F+S	F	L-V	L-V	L-V	L-V
45	F+S	F+S	L-V	L-V	L-V	L-V	L-V
40	F+S	F+S	L-V	L-V	L-V	L-V	L-V
35	F+S	F+S	L-V	L-V	L-V	L-V	L-V
30	F+S	L-V+S	L-V	L-V	L-V	L-V	L-V
25	L-V+S	L-V	L-V	L-V	L-V	L-V	L-V
20	L-V+S	L-V	L-V	L-V	L-V	L-V	L-V
15	L-V+S	L-V	L-V	L-V	L-V	L-V	L-V
10	L-V+S	L-V+S	L-V	L-V	L-V	L-V	L-V
5	L-V+S	L-V+S	L-V	L-V	L-V	L-V	L-V
1	L-V+S	L-V+S	L-V+S	L-V	L-V	L-V	L-V

Figure 9: Phase diagram of the feed, given in table. L-V means liquid - vapor; F means fluid phase. When 'S' is included, benzene is crystallized.

LNG phase diagram

A typical composition of the head of the demethanizer is shown in Table 4. For this exercise, Benzene specification in the demethanizer overhead has been relaxed to 10ppm instead of the generally considered 1ppm. The corresponding phase diagram is presented in Figure 10.

Table 4: Typical molar composition of the head of the demethanizer.

N2	1.5%
CH4	91.5%
C2	6.0%
C3	1.0%
benzene	10 ppm

P (bara)\T (°C)	-170	-160	-150	-140	-130	-120	-110	-100	-90	-80	-70	-60
40	F+S	F	F	F	F	F	F	F	F	L-V	L-V	F
38	F+S	F	F	F	F	F	F	F	F	L-V	L-V	F
36	F+S	F	F	F	F	F	F	F	F	L-V	L-V	F
34	F+S	F	F	F	F	F	F	F	L-V	L-V	L-V	F
32	F+S	F	F	F	F	F	F	F	L-V	L-V	L-V	F
30	F+S	F	F	F	F	F	F	F	L-V	L-V	L-V	F
28	F+S	F	F	F	F	F	F	F	L-V	L-V	L-V	F
26	F+S	F	F	F	F	F	F	F	L-V	L-V	L-V	F
24	F+S	F	F	F	F	F	F	L-V	L-V	L-V	L-V	F
22	F+S	F	F	F	F	F	F	L-V	L-V	L-V	L-V	F
20	F+S	F	F	F	F	F	F	L-V	L-V	L-V	L-V	F
18	F+S	F	F	F	F	F	L-V	L-V	L-V	L-V	F	F
16	F+S	F	F	F	F	F	L-V	L-V	L-V	L-V	F+S	F
14	F+S	F	F	F	F	F	L-V	L-V	L-V	L-V	F+S	F
12	F+S	F	F	F	F	L-V	?	?	L-V	L-V	F	F
10	F+S	F	F	F	F	L-V	L-V	L-V	L-V	L-V	F	F
8	F+S	F	F	F	L-V	L-V	L-V	L-V	L-V	F+S	F	F
6	F+S	F	F	F	L-V	L-V	L-V+S	L-V	L-V	F+S	F	F
4	F+S	F	F	L-V	L-V	L-V	L-V+S	L-V	L-V+S	F+S	F	F
2	F+S	F	L-V	L-V	L-V	L-V+S	L-V+S	L-V+S	F+S	F+S	F	F
1	F+S	L-V	L-V	L-V	L-V	L-V+S	L-V+S	L-V+S	F+S	F+S	F	F

Figure 10: Phase diagram of the head of the demethanizer with 10 ppm of benzene (as shown in Table 4.

This figure again shows the liquid - vapor equilibrium zone as 'L-V', and the single phase region as 'F'. Three different crystallization zones can be identified:

- Below -170°C benzene crystallizes from the liquid phase.
- In the same way as in Figure 9, we observe benzene crystallisation at low pressure within the liquid - vapor zone.
- In opposition with what is seen in Figure 9, the fluid is very rich in methane, and as a result the vapor zone, below the liquid- vapor equilibrium, is visible. In this zone, where the fluid is single phase vapor, crystallization also occurs, at rather high temperatures (up to -70°C), even though the benzene concentration is as small as 10ppm molar.

This last crystallization zone can be explained when considering the Txy sketch of Figure 3. It was explained how, at a fixed pressure, a first crystallization zone from the vapour phase could be observed at moderate temperature. When temperature decreases, the crystals dissolve in the liquid that appears. The crystals reappear at much lower temperature, out of the liquid methane.

Unfortunately, no experimental data illustrating this behavior could be found.

Conclusion

This work was executed in order to evaluate the risk of benzene crystallization in the LIQUEFIN gas liquefaction process. The work is presented in three stages:

In order to understand the physical phenomenon of crystallization, the phase diagram of the methane - benzene binary system is first investigated in detail. It is shown that two liquid - solid and one solid - vapor equilibrium zones exist, in addition to the well-known liquid - vapor equilibrium.

In a second stage, a thermodynamic model is developed, that uses the Peng-Robinson equation of state with the Huron - Vidal Mixing rule. The parameters for this model have been determined based on high temperature VLE data. It is obvious that the model is used at temperatures that are well below the temperature of the data used. Hence, care has been taken that the extrapolation to lower temperatures is as acceptable as possible. An additional parameter (fusion heat capacity) is determined using low temperature SLE data.

In a final stage, the model is used to calculate the phase diagram of two typical compositions encountered in the LIQUEFIN process. The first composition is rich in heavy end components. As a result, a liquid phase is always present. This liquid may be concentrated in benzene, which increases the crystallization risk at higher temperatures and low pressures. The second composition is very pure methane. As a result, a single vapor phase may exist at high temperature and low pressure. Even though the benzene concentration is very small in this context (10ppm molar), a crystallization risk from the vapor phase exists which justify the choice for a lower benzene specification in the demethanizer overhead. This phenomenon is of importance for the design of the LNG process, but no experimental data validating this behavior has been found.

References

DIPPR, Brigham Young U. edition, Release: January, 2003

Huron, M.J & Vidal, J.: "New mixing rules in simple equations of state for representing vapour-liquid equilibria of strongly non-ideal mixtures" *Fluid Phase Eq*, 1979, 3, 255-271

Kohn, J.P.& Luks, K.D. : « L-L-V Equilibria in Cryogenic Mixtures », *GPA RR-49,* 1981

Luks,K.D.;Hottovy,J.D.;Kohn,J.P., *J. Chem. Eng. Data,* 1981, 26, 4, 402-403

Neumann,A.; Mann,R.; von Szalghary,W.D., *Kaeltetech. Klim.,* 1972, 24, 5, 145-149

Peng, D-Y & D.B. Robinson-A new two-constant equation of state-*Ind. Eng. Chem. Fundam.* 1976-15---59-64

Rijkers,M.P.W.; Malais,M.; Peters,C.J.; De Swaan Arons, J., *Fluid Phase Equilib.,* 1992, 77, 327-342

Vidal, J.-Thermodynamics: Applications in Chemical Engineering and the Petroleum Industry , Editions Technip, (2003)

The Thermodynamic Model

Model description

Models for the calculation of phase equilibrium are based on the fugacities. Phase equilibrium is observed when the fugacities of each component in each phase are equal. A cubic, Peng-Robinson (1976), equation of state is used here for calculating the liquid and vapor fugacities. The mixing rule of Huron-Vidal (1979) is used so that a better representation of the non-idealities is obtained.

A solid phase appears if for any of the components the solid fugacity is lower than the fugacity of the equilibrium phase. This solid fugacity $f_{S_i}^o$, is calculated as follows (Vidal, 2003):

$$\ln\left(\frac{f_{S_i}^o}{f_{L_i}^o}\right) = \frac{\Delta H_i^f}{RT}\left(1-\frac{T}{T_i^f}\right)+\frac{\Delta c_{pi}}{R}\left[1-\frac{T}{T_i^f}-\ln\left(\frac{T}{T_i^f}\right)\right]+\Delta v_i(P-P_a) \quad (1)$$

This equation expresses the change in Gibbs energy between the liquid and the solid phase, at the system pressure and temperature. On the right hand side, the two first terms express the deviation due to the temperature difference between the fusion temperature and the system temperature. Usually, only the first term, that is proportional to the fusion enthalpy (ΔH_i^f) is used. However, as we consider very large temperature deviations (more than 100K), it is important to include the second term, that tells us that this fusion enthalpy varies with temperature. The proportionality factor is here Δc_{pi}, the fusion heat capacity. Finally, the third term of equation (1) is required when the system pressure differs from that at which the fusion properties are provided (which is usually atmospheric pressure). This last term is proportional to the volume of fusion, Δv_i. Both enthalpy of fusion and volume of fusion are available in databases (table 2). The heat capacity of fusion will be fitted on SLE data.

In equation (1), $f_{L_i}^o$ is the fugacity of the pure component in its liquid phase at the system pressure and temperature. Obviously, this is an unstable state, as the pure component is usually solid in reality. However, using an equation of state, it is possible to extrapolate and obtain this value. In this work, the Peng-Robinson (1976) cubic equation of state is used:

$$P=\frac{RT}{v-b}-\frac{a(T)}{v^2+2bv-b^2} \quad (2)$$

where the *a* and *b* parameters are calculated from the critical coordinates, using the critical constraints:

$$a(T)=\Omega_a\frac{R^2T_c^2}{P_c}\alpha(T_R) \quad (2)$$

Predicting Environmental Impacts on Multi-Train LNG Facility using Computational Fluid Dynamics (CFD)

Wing K. Yee, Dan Lin, Vibhor Mehrotra, and Philip Diwakar
Bechtel Corporation
Houston, TX 77056

AIChE Spring Meeting, April 2005
5th Topical Conference on Natural Gas Utilization
Atlanta April 10-14, 2005

Abstract

Production rate at Liquefied Natural Gas (LNG) plant can be very sensitive to the inlet temperatures of compressor turbine drivers and air-cooling equipments. In order to achieve optimal production rate, the surrounding environmental impact on the multi-train LNG facility has to be studied. Computational Fluid Dynamics (CFD) is a tool that can recreate the multi-train LNG facility in virtual reality and predict the interaction of the LNG facility with the surrounding environment using numerical analysis. It can also explain several phenomena that may be left unaccounted during the design stage.

This paper demonstrates the use of CFD modeling at Bechtel, in designing the LNG plant layouts, by studying the interaction of the multi-train LNG facility with the environment. The simulations of hot air self-recirculation, recirculation between equipments and knockout affects of multiple trains in a LNG facility will be presented. Results of proposed mitigation and the improvement in LNG plant production will also be presented.

In addition, CFD simulation on existing LNG facilities will be presented and compared with field measurements. It was accomplished by collecting weather data from the site at different inlets to compressor turbine drivers for a period of time. These results were carefully analyzed to construct a theory for the observed phenomenon. Finally CFD modeling was utilized to validate the proposed theory on how weather can impact the LNG Plant production rate.

1. Introduction

The worldwide demand for natural gas has been growing quite rapidly. In order to meet the demand larger-capacity Liquid Natural Gas (LNG) plants are needed[1]. With large pieces of equipment allocated within a confined plot area[2], the local environment around the plant (air temperature, wind direction and wind speed) will have a higher impact on the equipment and plant performance. For example, equipment such as air-cooled condensers requires air to cool off the refrigerants. If the inlet temperature is higher, the heat transfer rate decreases resulting in less refrigerant condensation. In order to offset this effect, more power is needed for the gas compression and refrigeration. If this phenomenon happens during a period when additional power is not available, it results in loss of LNG production.

Computational Fluid Dynamics (CFD) simulation can be used to study and mitigate the impact of surrounding environment on LNG Plant Production[3]. Often during the design stage of a LNG plant, equipments are located in a limited space. The impacts of the exhaust from other equipment or the air circulation between buildings are hard to predict without the use of fluid dynamic analysis (CFD). If a problem can be foreseen before the design is finalized, it may save time and money. CFD can also be used to study the layout of an existing plant. Sometimes there are phenomena that are caused by the natural environment that were not accounted for during the design stage. CFD can be used to evaluate the cause of such situations and suggest quick fixes based on the simulation model.

2. Typical CFD Study of Air Re-Circulations

When large numbers of heat generation equipment are located in a limited plot space, interactions between those equipments are unavoidable. Many types of equipment at a LNG plant involve drawing in and exhausting air. When the equipments are close together, limited amount of fresh air is available and some equipment may start to draw in exhaust air from others or themselves. Such contamination will result in loss of cooling surface area and/or higher inlet temperature above ambient. The temperature contamination can be classified into two categories and examples are shown below.

2.1 Self-recirculation

Due to the site layout or wind condition, sometimes the exhaust air can recirculate back to the inlets of the same unit. As mentioned before, the efficiency of the equipment will be reduced when the inlet air temperature rises. CFD simulations can predict such phenomenon under various site-specific conditions. Using an air-cooled refrigerant condenser unit as an example. A CFD model for a double banking of refrigerant condensers was constructed. Different wind speed and direction were tested and the effects on the refrigerant condensers were studied. Figure 1 shows the temperature profile just underneath the fans of the refrigerant condensers. The simulations predicted that when the wind comes perpendicular to the length of the refrigerant condensers, it creates the largest heated zone. Figure 2 shows the velocity vectors of the airflow and self-recirculation can be seen. Self-recirculating air usually moves along the leading edge of the refrigerant condensers, thus the longer the edge; more self-recirculating air will merge together to form a bigger heated zone. The side edges also have some recirculation due to the corners of the structure. When the airflow passes the corners, it creates trailing waves behind the corner and recirculation can occur within these zones. .

When the wind is blowing along the length of the refrigerant condensers, the self-recirculation is much smaller because the edge facing the wind is shorter. However, under strong wind conditions, the turbulence created from the corners can swirl back into the inlets of the refrigerant condensers. From these simulations, the plant can be re-oriented or mitigations can be studied to minimize the self-recirculation during cross winds.

In addition to predict unforeseen problems or validate the designs, CFD can also be used for mitigation studies. Continuing with the refrigerant condensers example, there are several ways to minimize the self-recirculation of the exhaust gas. One of the mitigation is to install horizontal walls to the side to limit the amount of heated air reentering the unit. In Figure 3, the effect of horizontal wall is compared with one without. It is clear that there is less self-recirculation when the horizontal wall is installed. This effect was also demonstrated in a previous paper published by Bechtel[1]. Another alternative is to extend the hoods of the fans, however the cost of construction is much higher. CFD can model several mitigation methods and predict the efficiency. This makes the decision making easier without the penalty of constructing different solutions in real life for experiments.

2.2 Temperature Contamination from Other Exhausts

When multiple large heat releasing equipments are located within a limited plot area, contamination from other sources is also possible in addition to self-recirculation. Downstream equipments can draw heated exhaust from upstream equipments. This is a more common scenario in multi-train LNG plants and harder to fix once a plant is built. Unlike self-recirculation, there is no typical worst-case scenario. Demonstrating with a multi-train LNG plant simulation, the CFD simulation shows the airflow path and the temperature increase. In Figure 4, the flow pattern shows how some exhaust air is drawn into the downstream equipments. Since wind can be coming from any directions, the upstream heat source varies in degrees and directions.

Large vessels, towers and buildings can also block air feeding into air-cooled equipment. Turbulence generated behind these structures can also create local recirculation zones. If air-cooler equipment is located within these zones, the equipment can be starved for fresh air. Using CFD in the design phase can minimize some of the recirculation problems. In a simulation, the spaces and orientation of the buildings can be adjusted, and different kind of weather conditions can be prescribed. Different mitigations can also be examined, such as changing the orientation of the plant or relocating some of the equipments. Although there are no complete solutions for these problems, analyzing different scenarios can minimize the impacts on production. Manipulating the design of a virtual plant is a lot less expensive than making changes in a real facility. And CFD often takes less time to get the results than testing with a physical model. Combining the cost and timing benefits, CFD will gradually play a more important role in the design process.

3. Analyzing Existing Plants

CFD can be very useful in analyzing existing plant performance since the model can be validated against certain data measured at specific locations of the plant. It can also assist in understanding how equipment performs when weather condition changes. An example would be how the wind speed can affect the inlet temperature of a certain piece of equipment. Weather and various equipment inlet temperature data at a LNG plant was collected and statistically analyzed. The fluctuation in the inlet temperature is due to the change in weather pattern. CFD was utilized to determine the cause of the inlet temperature fluctuation. The maximum and minimum of the wind speed under the predominant wind direction were used in the simulations. In Figure 6, the air path to the compressors inlets is shown and it appears that it is not approaching in a direct path. This suggests some of the air is drawn into other equipments upstream and the equipment has to draw air from an indirect path. If there is heated air in this path, the compressor inlet temperature will increase. In this case, the air path is bent and the air enters the compressors building before being drawn to the inlets. The compressor building has heat sources such as exhaust stacks and piping and it can heat up the air. At lower wind speed, the air has a longer resident time to be heated inside of the compressors building. At higher wind speed, the resident time is shorter and the air is less heated before reaching the inlets. This phenomenon is location dependent. Each piece of equipment can have a different airflow path and the affects from the wind speed can be different.

In this demonstration, CFD is able to show how wind speed can affects the performance of downstream equipments. This study also demonstrates that CFD modeling is a useful tool to improve efficiency of existing LNG plants.

4. Results and Conclusion

This paper has demonstrated the use of CFD in the pre- and post-design of LNG plants. As multi-train LNG plants become more common, the impacts from the surrounding environment are vital to the production. CFD can be a faster and cheaper way to analyze the fluid dynamics around the plants. It can produce insights to the plant designers or engineers and an informed decision can be made. As a tool for mitigation, various parameters can be changed in the virtual space and the most effective solution can be found. Computing technology continues to advance and more complicated problems can be solved in a shorter time using CFD simulations. If utilized correctly, CFD simulation can play a very important role in design and construction of very large-scale LNG plants.

5. References

1. Avidan, A., Messersmith, D. and Martinez, B., “LNG liquefaction technologies move towards greater efficiencies, lower emissions”, Oil & Gas Journal, August 19, 2002

2. Mehrotra, V., Berkoe, J., and Messersmith, D., “Evaluating Performance of Air cooled Heat Exchangers in LNG Plants”, AIChE Spring National Meeting, New Orleans, La, March 31-April 3, 2003

3. Berkoe, J., “Fluid dynamics visualization solves LNG plant recirculation problem”, Oil & Gas Journal, March 29, 1999

6. Biography of Speaker

Dr. Wing K. Yee Received his Ph.D from University of Utah in Chemical Engineering. Specialized in large scale CFD (Computational Fluid Dynamics) simulation in parallel computing. Currently he is working for the Bechtel Corporation in Houston. He continues to do CFD work in the Advanced Simulation and Analysis Group for the Oil, Gas and Chemical Industry. These studies include air flow on large buildings, and industrial burners.

7. Figures

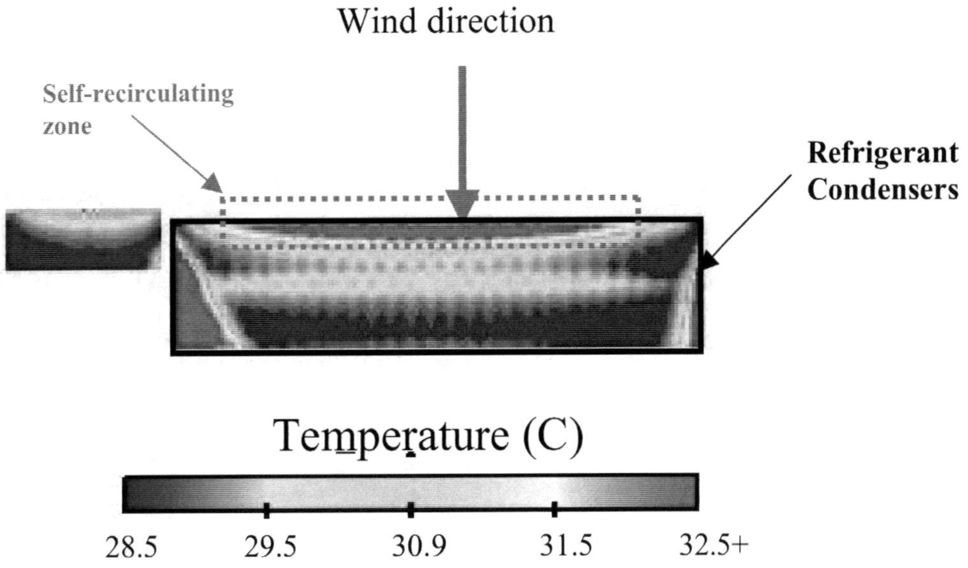

Figure 1. Temperature profile underneath double-banking refrigerant condensers. 6 m/s wind comes from the indicated direction.

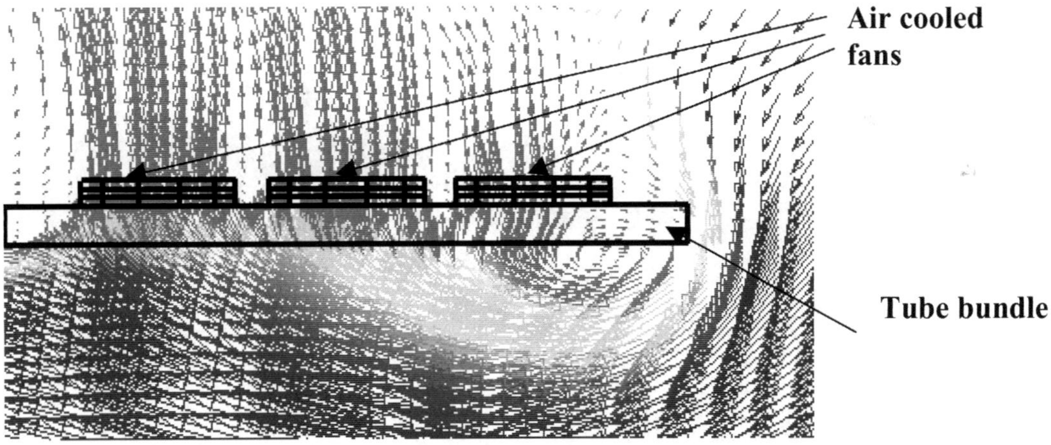

Figure 2. Velocity vectors showing self-recirculation of the heated exhaust air from a refrigerant condensers.

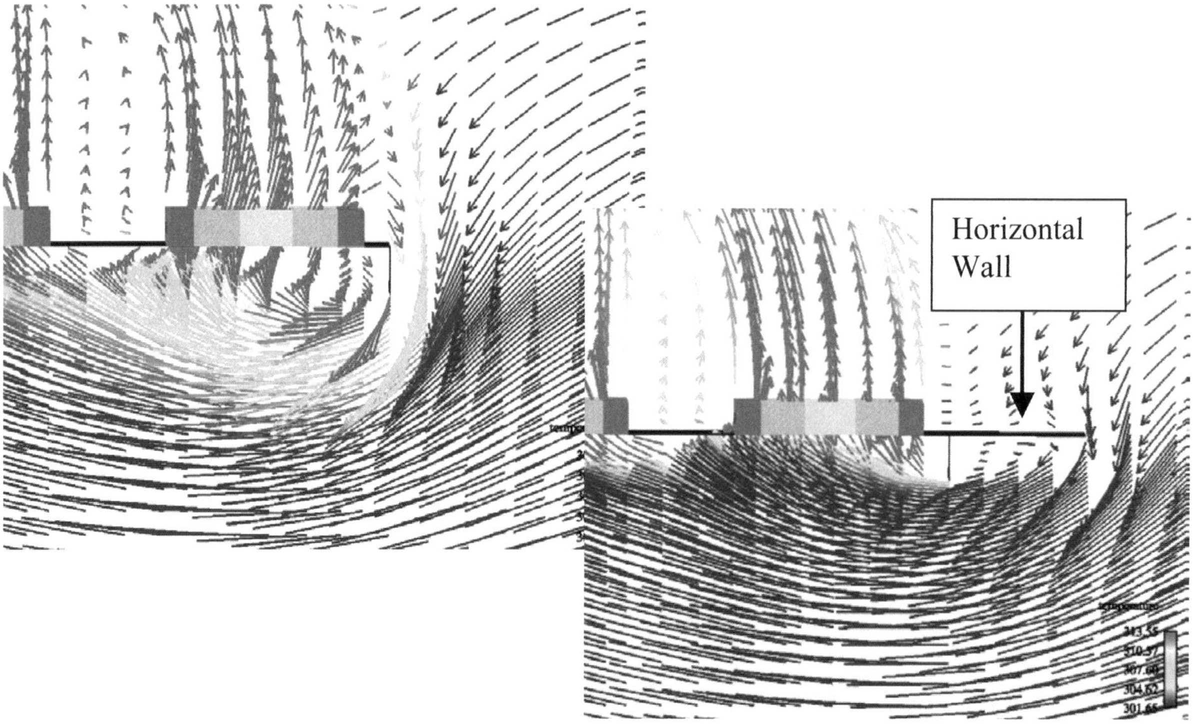

Figure 3. Air self-recirculation comparison with the installation of horizontal wall.

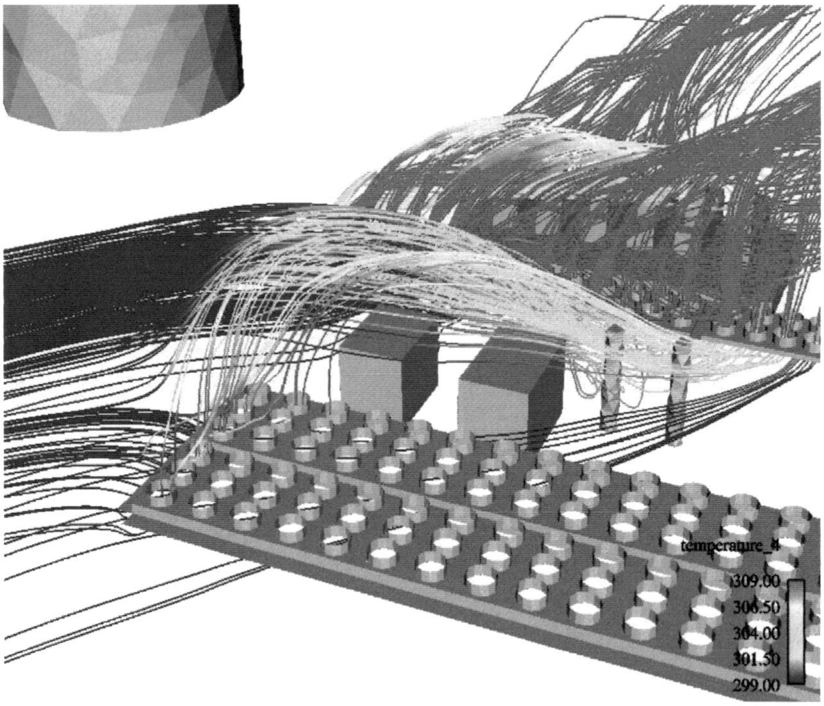

Figure 4. CFD simulation shows exhaust air from one equipment recirculation to another equipment's inlet.

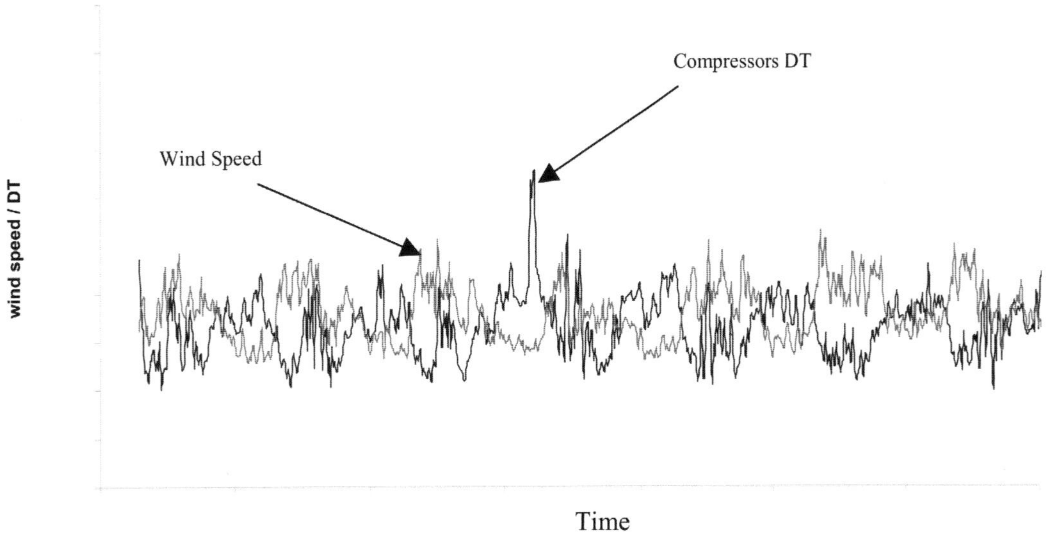

Figure 5. Correlation between the Compressors inlets temperature and the wind speed data. The compressors data are represented as a change of temperature from the ambient.

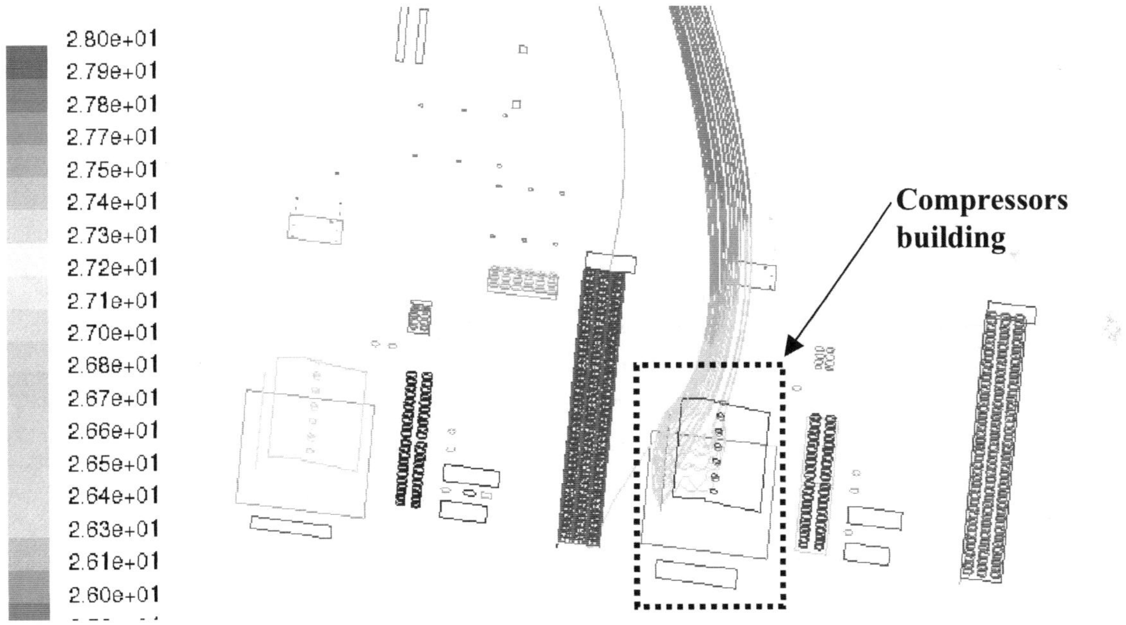

Figure 6. Path lines show the source of the compressors inlets. The air is heated before entering the compressors and the amount of fresh air drawn in is less.

Biography of Speaker

Dr. Wing K. Yee received his Ph.D. from the University of Utah in Chemical Engineering. He specialized in large scale CFD (Computational Fluid Dynamics) simulation in parallel computing. He is currently working for the Bechtel Corporation in Houston. Dr. Yee continues to do CFD work in the Advanced Simulation and Analysis Group for the Oil, Gas and Chemical Industry. These studies include air flow on large buildings, and industrial burners.

2005 AIChE Spring National Meeting: LNG Sessions

Authors: Lyn Himmelberger and Patrick Billman

Organization: Cryoquip, Inc.

Position/Title: n/a

Mailing Address: 25720 Jefferson Avenue
Murrieta, CA 92562 USA

Phone: 951-677-2060

Fax: 951-677-2066

E-mail: lhimmie@aol.com

Title of the Paper: LNG Vaporization Technology: The Environment vs. Energy Sourcing

ABSTRACT

Increasing constraints dominate the selection of best available technologies. The optimal decision for LNG vaporization varies depending upon an ever-changing landscape of environmental concerns and proven technologies. Major sources of heat remain the same, but how they are applied individually and in combination may well determine the immediate and long-term viability of a project. Picking a single technology is no longer sufficient to assure both regulatory approval and commercial success. Increasing scrutiny is being given to life cycle cost and environmental impact while being balanced against initial capital cost.

This paper reviews the predominant sources of vaporization energy, the over-used and the under-utilized, with emphasis on the latter. The technical and economical aspects of viable vaporization processes are examined in light of constraining environmental concerns. Certain characteristics of LNG send-out terminals are fixed, but many are subject to change and improvement.

By optimizing the relatively fixed characteristics and combining those with best available technologies, the overall project can be efficient and effective, reliable and flexible, while satisfactorily addressing local and global environmental issues. The technical, commercial and environmental benefits and shortcomings of several technologies will be discussed alone and in combination with others. Life cycle costs will be reviewed for several combination.

Introduction

LNG as a fuel source is regaining momentum in the USA. Interest peaked but quickly dissipated again over 25 years ago. Companies involved in delivering the product to market today face a substantially different set of parameters and constraints that have developed since the prior installations were done. The new restrictions and environmental concerns dictate a review and possible reconsideration of the technologies applied during the first wave of LNG receiving and send out facilities.

Optimal design of the LNG vaporization process is crucial to assure an LNG terminal's long range operability in light of the ever changing regulatory standards, on stream reliability requirements, peak demand response, capital cost, future expandability, simplicity, maintenance cost control, and operating expense. All of these aspects combine to create a reliable system while minimizing the life cycle cost of the plant

We will examine the major potential sources of energy for vaporization, and the sources of energy inherently available at LNG send out terminals. Review both past LNG vaporization techniques, and the technologies already successfully being applied in the technically aligned cryogenic air separation industry. Characteristics of each, showing relative energy requirements, costs, emissions and plot plan are discussed. These existing techniques will be combined into various hybrids that optimize the favorable characteristics while minimizing some of the less desirable effects. Although not exhaustive, the effort will focus on proven concepts combined in creative ways to maximize efficiency without sacrificing sound engineering practice.

For as large as the LNG industry may be currently, some future projections indicate growth that will quadruple capacity by 2020. Imports climbed by more than 100 % from 2002 to 2003 and we have recently seen the reactivation of existing facilities. At the time of this writing, in excess of 50 receiving / send out projects and upgrades are in the proposal stages. Most are large projects.

With the advent of larger projects comes the need to survey recent developments in technology that enhance productivity and reduce Life Cycle Costs for the new installations. Many improvements are necessary to support the projected growth throughout the supply chain. Recent developments in recovery and transportation technology have increased access to the supply and decreased the cost of transport, while increasing the safety. Similar improvements at receiving and send out terminals are required to lower Life Cycle Costs while reducing Environmental Impact.

Seawater for vaporization has become a major issue from an environmental standpoint. With natural gas prices hovering around three times the traditional levels, uncertainty and higher costs associated with the current methods of LNG vaporization are coming under increased scrutiny. In addition, the costs for just handling the large quantities of seawater to effect LNG vaporization now runs into tens of millions of dollars.

New technologies have been developed in other aligned cryogenic markets, while the LNG supply industry has remained relatively the same in the US market. Other techniques which have been developed in parallel cryogenic markets including He, H2, Ar, Ethylene, CO2, N2 and O2 have already been successfully applied to LNG for numerous applications. Some of these include LNG to CNG refueling stations, LNG boil off to flare vaporizers, peak shaving, mobile LNG supply modules, and other developing applications. Also, numerous new cycles have been proposed for vaporizing LNG.

Fully optimized receiving and send out terminals may incorporate integrated energy loops combined with hybrid vaporization systems. This paper will touch on some of the potential technology enhancements and combinations that help facilitate further optimization of these terminals. Our purpose is to develop alternatives that are easy to operate, economical, practical and environmentally friendly.

Study

In its simplest form, heat energy is available from air, fuel, refrigeration process cycles and water in sufficient quantities to vaporize large flows of LNG. Historically, the majority of vaporization of LNG has been done by using water and fuel. Specifically, Open Rack Vaporizers [ORV's] and Submerged Combustion Vaporizers [SCV's] have predominated for receiving and send out terminals. Environmental issues as well as operating costs, have recently tempered the original interest in these previous stalwarts.

In other aligned cryogenic applications, the focus has concentrated on other technologies. Ambient air predominates in the air separation industry for vaporizing all types of cryogens. Perhaps this is because of the extensive familiarity the industry has with air, not only as a feed stock for its major products, but also as a process medium. The preferred device is a Natural Draft Ambient Air Vaporizer [AAV's] because of its simplicity and ease of operation. The AAV has no moving parts, requires no power or fuel and is sure to function as designed as long as there is air. Fan Assisted Ambient Air Vaporizers [FAV's] are also used when space constraints and ambient conditions dictate. Other popular techniques include Shell and Tube [S&T] vaporizers utilizing waste heat recovery [WHR] streams from compressor cooling water, Water Circulating heaters [WCV's], and compact, indirect contact, submerged combustion heaters called Vaporizing Fired Thermal Units [VFTU's].

Similarly, as the air separation industry is fond of using air to vaporize their products, so the petroleum industry is fond of using petroleum distillates. This has spawned the use of numerous single and multi-component heating / refrigeration cycles.

With all the available methods of vaporizing a cryogen, just investigating all the possible routes to warming LNG can be overwhelming. In addition, most companies focus on supplying just one or sometimes two approaches and typically see their technology as a panacea for any problem statement. The responsibility to sort through the real (vs imagined) advantages and disadvantages falls to the user. Difficult as it may seem at first, combining technologies into hybrid systems offers significant advantages, both commercially and environmentally.

Analysis

A review of some of the more critical design parameters and their impacts is needed. Assuming that all of the technologies under review actually are able to deliver LNG at the design pressure and temperature required, we will concentrate on the costs, benefits and drawbacks of the various technologies. Specifically, we will look at the relative Capital Cost [CC], Installation Cost [IC], Operating Cost [OC] including power, maintenance and fuel, Space Required [SR], Simplicity vs Complexity [SC], and Life Cycle Cost [LCC] for various time periods.

Our design basis is one billion cubic feet per day [1BCF/D] at a pressure of 1200 psig to the pipeline. The technology listings will be in alphabetical order.

Technology		[CC]	Ranking	[IC]	Ranking
Amb Air Vap [Nat Draft]	[AAV]	2.6	[9]	3.5	[10]
Air w/ Intermediate Fluid	[AIF]	3.0	[11]	4.7	[14]
Direct Seawater S&T*	[SS&T]	3.5	[12]	2.4	[8]
Direct Seawater ORV*	[ORV]	4.6	[15]	3.7	[12]
Fan Assisted Amb Air	[FAV]	2.9	[10]	3.9	[13]
Heated Water	[WCV]	2.0	[6]	1.8	[4]
Reverse Refrig Cycle	[RRC]	4.5	[14]	9.3	[15]
Seawater Intermediate Fluid*	[SIF]	4.1	[13]	3.5	[11]
Sub Combust Direct Contact	[SCV]	1.1	[2]	3	[2]
Submerged Combust Indirect	[VFTU]	1.0	[1]	1.0	[1]
Waste Heat Recovery	[WHR]	0.4		0.3	
Hybrid I [AAV + WHR]	[H I]	1.6	[4]	2.0	[5]
Hybrid II [FAV + WHR]	[H II]	1.6	[5]	2.9	[6]
Hybrid III [AAV+ORV+WHR]	[H III]	1.4	[3]	1.7	[3]
Hybrid IV [FAV + VFTU]	[H IV]	2.3	[8]	2.5	[9]
Hybrid V [AAV + VFTU]	[H V]	2.1	[7]	2.3	[7]

- Includes cost of seawater handling system and pumps

What is the energy requirement associated with each? As a base case, the horsepower for the LNG send out pumps and the basic facility is assumed to be the same for each case and is treated as zero. For 1 BCF/d we assumed one GE LM 2500 gas turbine for the LNG pumps and running the facility. The fuel to generate additional power to run the equipment below is added assuming a power turbine efficiency of 30%.

[OC] = Operating Costs

Technology		[OC] HP	Rank	[OC] Fuel	Rank	[OC] Maint
Amb Air Vap [Nat Draft]	[AAV]	0	[1]	0	[1]	V Low
Air w/ Intermediate Fluid	[AIF]	High	[15]	Low	[10]	High
Direct Seawater S&T	[SS&T]	High	[12]	0	[1]	Moderate
Direct Seawater ORV	[ORV]	High	[11]	0	[1]	High
Fan Assisted Amb Air	[FAV]	Mod	[10]	0	[1]	Low
Heated Water	[WCV]	Low	[7]	V High	[15]	Mod
Reverse Refrig Cycle	[RRC]	0 ?	[2]	0	[1]	V High
Seawater Intermediate Fluid	[SIF]	High	[14]	0	[1]	High
Sub Combust Direct Contact	[SCV]	High	[13]	V High	[14]	Mod
Sub Combust Indirect	[VFTU]	Low	[8]	V High	[13]	Mod
Waste Heat Recovery	[WHR]	Low	[1]	0	[1]	Low
Hybrid I [AAV + WHR]	[H I]	0	[3]	0*	[1]	V Low
Hybrid II [FAV + WHR]	[H II]	Low	[5]	0*	[1]	Low
Hybrid III [AAV+ORV+WHR]	[H III]	Low	[6]	0*	[1]	Mod
Hybrid IV [FAV + VFTU]	[H IV]	Mod	[9]	High	[12]	Mod
Hybrid V [AAV + VFTU]	[H V]	Low	[4]	High	[11]	Mod

*No additional fuel required

What is the environmental impact of these technologies? The base case is the power turbine [1 model GE LM 2500] firing enough to generate electrical power to run the LNG send out pumps. Low NOx assumes all fired equipment at 30 ppm NOx. Ultra low NOx assumes an SCR producing 5 ppm, including the power turbine.

		Environmental Impact		Percent of base case	
Technology		Relative [CO] [CO2] [NOx]	Rank	Low Nox [30 ppm]	Ultra Low NOx [5 ppm]
Amb Air Vap [Nat Draft]	[AAV]	1.0	[1]	18	3
Air w/ Intermediate Fluid	[AIF]	2.0	[12]	23	4
Direct Seawater S&T	[SS&T]	1.7	[8]	21	4
Direct Seawater ORV	[ORV]	1.7	[9]	21	4
Fan Assisted Amb Air	[FAV]	1.5	[5]	20	3
Heated Water	[WCV]	2.5	[13]	83	14
Reverse Refrig Cycle	[RRC]	1.0 ?	?	?	?
Seawater Intermediate Fluid	[SIF]	2.0	[11]	23	4
Sub Combust Direct Contact	[SCV]	2.8	[14]	72	12
Sub Combust Indirect	[VFTU]	1.9	[10]	70	11
Waste Heat Recovery	[WHR]	Included			
Hybrid I [AAV + WHR]	[H I]	1.0	[2]	18	3
Hybrid II [FAV + WHR]	[H II]	1.2	[3]	19	3
Hybrid III [AAV+ORV+WHR]	[H III]	1.2	[4]	19	3
Hybrid IV [FAV + VFTU]	[H IV]	1.6	[7]	45	8
Hybrid V [AAV + VFTU]	[H V]	1.6	[6]	44	7

How would the space required and relative complexity of each system compare?

[SR] = Space Required [SC] = Simplicity [lower #] vs Complexity [higher #]

Technology		[SR]	Rank	[SC] Rank
		Land based/ Platform		
Amb Air Vap [Nat Draft]	[AAV]	High / Mod	[15]	[1]
Air w/ Intermediate Fluid	[AIF]	Moderate	[4]	[13]
Direct Seawater S&T	[SS&T]	Low	[1]	[5]
Direct Seawater ORV	[ORV]	Moderate	[2]	[6]
Fan Assisted Amb Air	[FAV]	Moderate	[8]	[2]
Heated Water	[WCV]	Moderate	[7]	[12]
Reverse Refrig Cycle	[RRC]	High	[14]	[15]
Seawater Intermediate Fluid]	[SIF]	Moderate	[3]	[14]
Sub Combust Direct Contact	[SCV]	Moderate	[6]	[3]
Sub Combust Indirect	[VFTU]	Moderate	[5]	[4]
Waste Heat Recovery	[WHR]	Low		
Hybrid I [AAV + WHR]	[H I]	Moderate	[10]	[7]
Hybrid II [FAV + WHR]	[H II]	Moderate / Low	[9]	[8]
Hybrid III [AAV+ORV+WHR]	[H III]	Moderate	[12]	[11]
Hybrid IV [FAV + VFTU]	[H IV]	Moderate	[11]	[10]
Hybrid V [AAV + VFTU]	[H V]	Moderate	[13]	[9]

This page shows the Net Present Value [NPV] at 10, 14 and 18 %. Life Cycle Costs are shown at present costs in $ millions including capital expenditure, power, fuel and maintenance over the 25 year life of the plant.

Technology		[LCC] for 25 years 10%	14%	18%	Ranking
Amb Air Vap [Nat Draft]	[AAV]	50	49	48	[4]
Air w/ Intermediate Fluid	[AIF]	105	94	86	[9]
Direct Seawater S&T	[SS&T]	67	58	52	[5]
Direct Seawater ORV	[ORV]	90	78	71	[8]
Fan Assisted Amb Air	[FAV]	78	71	66	[6]
Heated Water	[WCV]	316	239	191	[15]
Reverse Refrig Cycle	[RRC]	160	149	141	[11]
Seawater Intermediate Fluid	[SIF]	85	73	64	[7]
Sub Combust Direct Contact	[SCV]	285	215	170	[14]
Sub Combust Indirect	[VFTU]	254	191	151	[13]
Waste Heat Recovery	[WHR]				
Hybrid I [AAV + WHR]	[H I]	33	31	30	[1]
Hybrid II [FAV + WHR]	[H II]	40	37	34	[3]
Hybrid III [AAV+ORV+WHR]	[H III]	38	34	31	[2]
Hybrid IV [FAV + VFTU]	[H IV]	172	136	112	[12]
Hybrid V [AAV + VFTU]	[H V]	157	124	103	[10]

This page shows the Net Present Value [NPV] at 10, 14 and 18 %. Life Cycle Costs are shown at present costs in $ millions including capital expenditure, power, fuel and maintenance over the 40 year life of the plant.

Technology		[LCC] For 40 years 10%	14%	18%	Ranking
Amb Air Vap [Nat Draft]	[AAV]	50	49	48	[4]
Air w/ Intermediate Fluid	[AIF]	109	95	86	[9]
Direct Seawater S&T	[SS&T]	71	59	53	[5]
Direct Seawater ORV	[ORV]	94	80	71	[8]
Fan Assisted Amb Air	[FAV]	81	72	66	[6]
Heated Water	[WCV]	343	249	194	[15]
Reverse Refrig Cycle	[RRC]	162	150	141	[11]
Seawater Intermediate Fluid	[SIF]	90	74	64	[7]
Sub Combust Direct Contact	[SCV]	311	224	173	[14]
Sub Combust Indirect	[VFTU]	277	199	153	[13]
Waste Heat Recovery	[WHR]				
Hybrid I [AAV + WHR]	[H I]	34	31	30	[1]
Hybrid II [FAV + WHR]	[H II]	41	37	34	[3]
Hybrid III [AAV+ORV+WHR]	[H III]	39	34	31	[2]
Hybrid IV [FAV + VFTU]	[H IV]	185	140	114	[12]
Hybrid V [AAV + VFTU]	[H V]	169	128	104	[10]

It should be noted that the relative ranking of the equipment does not adequately reflect the absolute amounts of power, fuel, maintenance or total costs. The charts below attempt to show the relative magnitude of the differences of the various technologies in some of the categories. NOx, for example, is shown for base systems, typical low NOx technology and addition of Selective Catalytic Reduction [SCR's]. The [AAV's] have no CO, NOx or CO2 emissions. These serve as the base case = 1.0. Fired equipment systems have the highest emissions levels followed by those having high power consumption, whether supplied by power turbines on site or conventional plants. Either way the emissions occur, either on site or at the power plant.

Also, simplicity and complexity are relative and are shown on an arbitrary scale in an attempt to reflect the relative magnitude as opposed to any absolute value. Some systems are inherently complex by their very design and number of components, while those with no moving parts are inherently simple. No attempt was made to put a value on simplicity or complexity, just a relative ranking.

Costs, power requirements, fuel consumption, space required, emission levels and all other data were gathered from an assortment of catalogs, equipment specifications and numerous conversations with users of the equipment as well as engineers who specify and install the equipment. All values are approximate and will actually occupy a range determined by the specifics of a particular site. Therefore, actual rankings may vary somewhat for each specific site and design conditions.

No attempt was made to represent any one supplier's particular equipment, but a serious effort was made to homogenize numbers from various sources to be broadly representative of the various discrete technologies discussed. It is very likely that particular vendors can do better than the averages shown for any given technology, and likely some will do worse. Within the scope of an overview such as this, it is impossible to adequately represent the best of the best. Consider all the charts, graphs, and representations with a suitable amount of skepticism and speak with the various technology suppliers directly for your specific requirements.

Following is a summation of the various rankings for each technology. The subtotal adds the rankings for Capital Cost + Installation Cost + Operating Costs for Power and Fuel + Emission rankings + Space Required + Simplicity vs Complexity. That subtotal is weighted into the final ranking at 50% (subtotal / 2) with the ranking for the Life Cycle Cost making up the other 50% . The resultant number is shown at the right.. This was an attempt to display an overall ranking for each technology taking into account a sum of all the rankings. The results are shown in the order of the result, lowest combined total to highest.

[LCC] For 25 years at 14% + [CC + IC + OC + NOx + SR + SC] / 2

Technology		Ranking
Hybrid I [AAV + WHR]	[H I]	[1]
Hybrid II [FAV + WHR]	[H II]	[2]
Hybrid III [AAV+ORV+WHR]	[H III]	[3]
Ambient Air Vaporizer [Nat Draft]	[AAV]	[4]
Direct Seawater Shell & Tube	[SS&T]	[5]
Fan Assisted Ambient Air Vaporizer	[FAV]	[6]
Submerged Combustion [Indirect]	[VFTU]	[7]
Direct Seawater ORV	[ORV]	[8]
Hybrid V [AAV + VFTU]	[H V]	[9]
Seawater using Intermediate Fluid	[SIF]	[10]
Submerged Combustion [Direct Flame Contact in Water]	[SCV]	[11]
Hybrid IV [FAV + VFTU]	[H IV]	[12]
Reverse Refrigeration Cycles	[RRC]	[12]
Heated Water Circulating Vaporizer	[WCV]	[14]
Air using Intermediate Fluid	[AIF]	[15]

Conclusions

Capital Cost is only one measure of a product's value in a large complex system. As we can see, the Installed Costs can vary considerably in both total dollar value and percentage of original Capital Cost. Life Cycle Cost for large LNG receiving and send out terminal vaporization is dominated by the Fuel usage, and Power requirements.

Since Capital Costs are relatively competitive for most of the systems reviewed, they tend to be overshadowed by other considerations including Environmental Impact. Not surprisingly, systems which require low / no Power and low / no Fuel also offer the lowest Environmental Impact as well. Hybrid systems are appealing from a Life Cycle Cost standpoint and in addition offer some advantages in Environmental Impact.

A review of the Life Cycle Costs for 25 and 40 years reveals that as hurdle rates increase, the gaps between systems narrow on cost. Also, little difference exists at the various hurdle rates except where there are large ongoing costs for Fuel and to a lesser extent, Power.

In absolute terms, the no Fuel, no Power hybrid with Waste Heat Recovery has the lowest Life Cycle Cost while other hybrids with little power and no additional Fuel are close behind.

The best non-hybrid system is Natural Draft Ambient Air, with Shell & Tube Direct Seawater, Fan Assisted Ambient Air, Seawater Intermediate Fluid, and ORV's rounding out the field below $ 100,000,000 Life Cycle Cost.

Since the two overriding issues for a new plant are Life Cycle Cost and Environmental Impact, below is a more precise ranking in each individual category for all the technologies. In these two important facets, overall Environmental Impact from NOx, CO and CO2 contribution, as well as Life Cycle Cost figured at 14% over a 25 year life, we see the essence of the decision driving criteria in today's market place.

In the next ranking for the combined categories of Life Cycle Cost + Environmental Impact, it is interesting to see the subtle shifts, in the best and worst rankings depending upon a specific focus. Also of interest is the grouping of various technologies together at the top or bottom of the list in spite of the particular focus.

Below is a list of the alternatives presented in the order of Environmental Impact from least to most:

Technology		Ranking
Ambient Air Vaporizer [Natural Draft]	[AAV]	[1]
Hybrid I [AAV + WHR]	[H I]	[2]
Hybrid II [FAV + WHR]	[H II]	[3]
Hybrid III [AAV+ORV+WHR]	[H III]	[4]
Fan Assisted Ambient Air Vaporizer	[FAV]	[5]
Hybrid V [AAV + VFTU]	[H V]	[6]
Hybrid IV [FAV + VFTU]	[H IV]	[7]
Direct Seawater Shell & Tube	[SS&T]	[8]
Direct Seawater ORV	[ORV]	[9]
Submerged Combustion [Indirect]	[VFTU]	[10]
Seawater using Intermediate Fluid	[SIF]	[11]
Air using Intermediate Fluid	[AIF]	[12]
Heated Water Circulating Vaporizer	[WCV]	[13]
Submerged Combustion [Direct Flame Contact in Water]	[SCV]	[14]
Reverse Refrigeration Cycles	[RRC]	[15]

Note: Some of these technologies were unduly burdened by the high NOx levels of the power turbine used to develop the horsepower to run the equipment vs the base NOx levels of the competing fired equipment.

Below is the list of the alternatives presented in the order of Lowest Life Cycle Cost from lowest to highest:

Life Cycle Cost alone:

Technology		Ranking
Hybrid I [AAV + WHR]	[H I]	[1]
Hybrid III [AAV+ORV+WHR]	[H III]	[2]
Hybrid II [FAV + WHR]	[H II]	[3]
Ambient Air Vaporizer [Natural Draft]	[AAV]	[4]
Direct Seawater Shell & Tube	[SS&T]	[5]
Fan Assisted Ambient Air Vaporizer	[FAV]	[6]
Seawater using Intermediate Fluid	[SIF]	[7]
Direct Seawater ORV	[ORV]	[8]
Air using Intermediate Fluid	[AIF]	[9]
Hybrid V [AAV + VFTU]	[H V]	[10]
Reverse Refrigeration Cycles	[RRC]	[11]
Hybrid IV [FAV + VFTU]	[H IV]	[12]
Submerged Combustion [Indirect]	[VFTU]	[13]
Submerged Combustion [Direct Flame Contact in Water]	[SCV]	[14]
Heated Water Circulating Vaporizer	[WCV]	[15]

Below is a list of the alternatives reviewed in the order of lowest Environmental Impact + lowest Life Cycle Cost from lowest to highest:

Technology		Ranking
Hybrid I [AAV + WHR]	[H I]	[1]
Ambient Air Vaporizer [Natural Draft]	[AAV]	[2]
Hybrid II [FAV + WHR]	[H II]	[3]
Hybrid III [AAV+ORV+WHR]	[H III]	[4]
Fan Assisted Ambient Air Vaporizer	[FAV]	[5]
Direct Seawater Shell & Tube	[SS&T]	[6]
Hybrid V [AAV + VFTU]	[H V]	[7]
Direct Seawater ORV	[ORV]	[8]
Seawater using Intermediate Fluid	[SIF]	[9]
Hybrid IV [FAV + VFTU]	[H IV]	[10]
Reverse Refrigeration Cycles	[RRC]	[11]
Air using Intermediate Fluid	[AIF]	[12]
Submerged Combustion [Indirect]	[VFTU]	[13]
Submerged Combustion [Direct Flame Contact in Water]	[SCV]	[14]
Heated Water Circulating Vaporizer	[WCV]	[15]

Air Emission Control Technologies for Submerged Combustion Vaporizers

Kamal Shah, Judy Wong and Gary Reeves
Aker Kvaerner Inc.

AIChE Spring Meeting, April 2005
5th Topical Conference on Natural Gas Utilization
Atlanta April 10-14, 2005

ABSTRACT

There is a growing demand for natural gas in the United States (U.S.) and it is anticipated that the demand will exceed domestic production capacity. Predictions by the National Petroleum Council (NPC) on "Balancing natural gas policy – Fueling the demands of a growing economy" in September 2003 forecast that the traditional North American producing area will provide only 75% of the long term U.S. gas needs. At present 2% of the total U.S. natural gas demand is met by imported liquefied natural gas (LNG). Over the next 20 years, supply from the imported LNG is expected to make up over 15% of the total U.S. gas demand. This will require construction of new LNG import terminals.

One of the main components of the LNG import terminal facility is the LNG vaporizer, which vaporizes the LNG from liquid to gaseous form. The natural gas is then transported to the users in a pipeline. Various LNG vaporizer technologies have been utilized in the world today. Each technology has its own advantages and disadvantages depending on the terminal locations, economic benefits and environmental impact. Most of the existing land based LNG terminals in the U.S. use Submerged Combustion Vaporizer (SCV) technology as a primary means for vaporizing LNG due to its high thermal efficiency and ease of start up and operation. Also the SCV technology has an excellent safety record.

Although SCV technology has high thermal efficiency, excellent operability and good safety record, the conventional SCV generates air emissions due to natural gas combustion producing nitrogen oxides (NOx) and carbon monoxide (CO) among others. Many areas of the U.S, where the LNG import terminals are being considered, require air emission controls to mitigate the environmental impacts. This paper discusses various technologies that have been recently developed to reduce NOx and CO emissions in the exhaust gases emitted from the SCV and provides a relative economic comparison.

1. INTRODUCTION

Due to the growing demand for natural gas in the United States and expected decline in the domestic natural gas production capacity, gas prices have increased over the last two years. Last year natural gas prices reached above $5 per million BTU. Recently in January 2005, members of the United State Senate Energy Committee heard from the industry leaders, environmental groups and government regulators regarding how best to increase the domestic supply for the nations energy needs. Because of the inadequacy of domestic production, imported LNG is expected to make up over 15 % of the total U.S. demand for natural gas over the next 20 years. Europe is likely to follow the U.S. in moving towards importing natural gas as well. This will require construction of new LNG import terminals.

One of the main components of the LNG import terminal facility is the LNG vaporizer, which vaporizes the LNG from liquid to gaseous form. Various LNG vaporizer technologies have been utilized in the world today. Each technology has its own advantages and disadvantages depending on the terminal locations, economic benefits and environmental impact. The submerged combustion vaporizer (SCV) technology has been utilized as a primary means for vaporizing LNG due to its high thermal efficiency and ease of start-up and operation. Also the SCV technology has an excellent safety record.

Although SCV technology has high thermal efficiency, excellent operability and a good safety record, natural gas combustion involved in the process leads to air emissions of nitrogen oxides (NOx) and carbon monoxide (CO). LNG terminals are being considered in localities where air emission controls are required to mitigate the environmental impacts. This paper discusses various technologies that have been recently developed to reduce NOx and CO emissions in the exhaust gases emitted from the SCV. Technologies considered are low NOx burner, ultra-low NOx/CO, selective catalytic reduction (SCR) and NOx oxidation with scrubber. Emissions and relative costs associated with each technology are presented.

2. SUBMERGED COMBUSTION VAPORIZER (SCV)

Figure 2.1 below depicts a schematic diagram that shows the main features of a SCV. It is comprised of a stainless steel tube bundle immersed in a warm water bath. LNG flowing through the bundle is vaporized by the heat transferred from the water bath operating at a temperature range between 60°F and 120°F. The lower operating temperature range is desired to achieve higher thermal efficiencies. Water bath temperature is maintained by hot fuel gas combustion products from the SCV burner that utilize combustion air supplied by a blower. The burner is mounted at one end of the water bath partially submerged and combustion products discharge into the water bath via the sparger tubes. Combustion products form a frothing two-phase mixture which rises in the water bath because of its lower density. This rising flow is confined within the weir assembly which surrounds the tube bundle. The froth flows up through the tube bundle and over the weir, where disengagement of the combustion product gases occurs. The water falls back over the weir and then re-circulates through the heat exchanger tube bundle. Because of the direct heat transfer from the hot flue gases to the water and the high agitation caused by the frothing two phase flow, heat transfer efficiency for the SCV is very high, typically in the 96 to 100% range. The high efficiency allows a compact design with small plot area requirements.

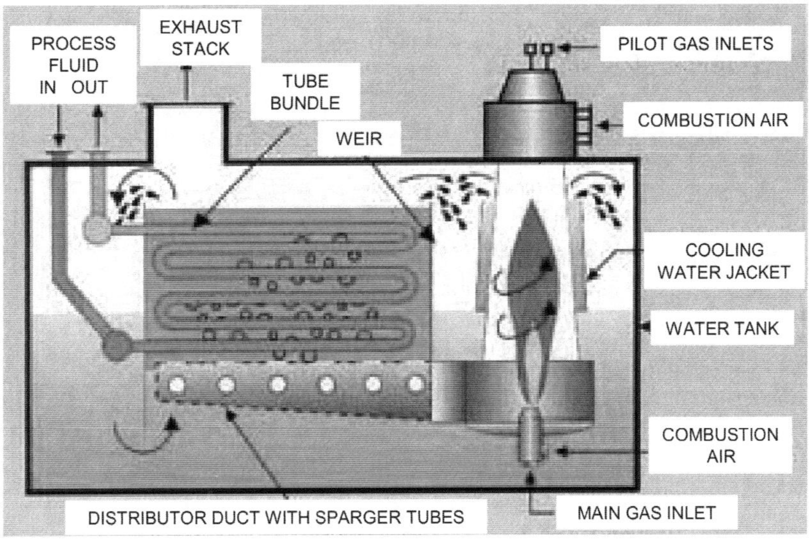

Figure 2.1 Submerged Combustion Vaporizer (Kaldair TX-LNG SCV)

3. MITIGATION OF AIR EMISSIONS FROM SCV

The main disadvantage of SCV is the atmospheric discharge of nitrogen oxides (NOx) and carbon monoxide (CO) from combustion of natural gas involved in the process. Various technologies have been developed to reduce the amount of these undesirable pollutants from SCV. There are two major approaches to mitigate emissions. The first approach is to improve the burner or combustion control technologies to reduce generation of these pollutants. The second approach involves post combustion treatment of the flue gases from the SCV combustion to reduce these pollutants.

The technologies associated with burner design or combustion control to minimize formation of hazardous pollutants include the following:

- Conventional SCV with water injection
- Low NOx burner / Flue Gas Recirculation (FGR)
- Flameless Thermal Oxidizer (FTO) "Ultra-Low NOx / CO" (1)

The technologies associated with post combustion requiring treatment of combustion products to reduce undesirable pollutants are as follows:

- Selective Catalytic Reduction (SCR)
- LoTOx™ oxidizer with EDV® scrubber (2)

Relative emissions, installed costs and operating costs for each of these options are presented below in Table 3.1 for an SCV with a duty of 100 MMBtu/hr. Fuel gas high heating value (HHV) of 1127 BTU/SCF with SCV thermal efficiency of 99% (HHV basis) and water bath temperature of 68°F are used as the basis for comparison.

A detailed description for each option is provided in the following sections. Results shown in Table 3.1 below were developed based on the information available from suppliers of the technologies and equipment. Where the information was not available, relevant literature and in-house cost resources were utilized.

Table 3.1 Comparison of relative emissions and costs for various options

OPTION	NOx (ppmv)	CO (ppmv)	Relative Capital Cost	Relative Operating Cost
SCV with Conventional Burner with Water Injection	38-40	50-80	100	100
SCV with Low NOx Burner / Flue Gas Recirculation (FGR)	18-20	18-22	130	102
SCV with FTO "Ultra-Low NOx/CO"	<5	<10	~305	103
SCV with Selective Catalytic Reduction (SCR)	<5	<10	305	105
Conventional SCV with LoTOx™ oxidizer and EDV® scrubber	10	50-80	210	103

All emissions shown above are corrected to 5% O_2 (dry basis). Operating costs include costs for power, fuel, caustic for pH control, and ammonia injection associated with SCR operation. Prices for fuel and power are assumed at $4/MMBtu and $0.08/kW-hr respectively.

3.1 Conventional SCV with water injection

As described in Section 2 above, SCVs normally use vaporized LNG as a fuel and they are integrally-heated vaporizer units utilizing fuel combustion process. Emissions from SCV fuel combustion include nitrogen oxides (NOx), carbon monoxide (CO), carbon dioxide (CO_2), volatile organic compounds (VOCs), and particulate matter (PM_{10}). Since LNG is free of sulphur, sulphur dioxide is normally not a concern associated with SCV emissions.

Natural gas combustion produces three types of NOx: thermal NOx, prompt NOx and fuel NOx. Most of the NOx are thermal NOx formed through thermal dissociation and subsequent reaction of nitrogen (N_2) and oxygen (O_2) molecules in the combustion air. The formation of thermal NOx is affected by three furnace-zone factors: (1) oxygen concentration, (2) peak temperature, and (3) time of exposure at peak temperature. Prompt NOx is formed through early reactions of nitrogen molecules in the combustion air and hydrocarbon radicals from the fuel. Prompt NOx reactions occur within the flame and are usually negligible when compared to the amount of thermal NOx. However, prompt NOx levels may become significant with ultra-low NOx burners. Fuel NOx results from the evolution and reaction of fuel-bound nitrogen compounds with oxygen. Because of the low fuel nitrogen content of LNG gas, NOx formation through the fuel NOx mechanism is insignificant. With conventional burners and water injection into the flame through the back of the burner to lower the flame temperature, total NOx emission from SCV combustion is normally in the range of 38-40 ppmv. The rate of CO emissions depends on the efficiency of natural gas combustion. CO emission increases when the burner is improperly tuned and/or operating at off-design levels. With conventional burners, total CO emission from SCV is normally in the range of 50-80 ppm. The SCV water bath is usually kept slightly basic by injecting caustic to counteract the effect of carbon dioxide dissolved in the water bath. This removes some NOx produced by the SCV burner. The advantage of this removal is accounted for in the NOx levels presented in this paper.

3.2 SCV with Low NOx Burners / Flue Gas Recirculation (FGR)

Low NOx burners are designed to "stage" combustion. Staging partially delays the combustion process, resulting in cooler flame temperature which suppresses thermal NOx formation. NOx emission reduction of about 50% can be observed with low NOx burners. NOx emission level is expected to be about 18-20 ppmv and CO emission level is about 18-22 ppmv

An alternate method is to use a flue gas recirculation (FGR) system. A portion of the cold flue gas is recycled from the stack and combined with combustion air prior to being fed to the burner. The FGR system reduces NOx emissions primarily because the recirculated gas is colder and acts as diluents to reduce the combustion temperature, thus suppressing the thermal NOx formation. To a lesser extent, FGR also reduces NOx emission by lowering the oxygen concentration in the primary flame zone. NOx emission from FGR is compatible to the reduction obtained from a low NOx burner but usually produces a higher CO emission than the staged low NOx burner.

Operating cost associated with SCV with low NOx burner system includes costs for fuel, power and caustic. Power cost is slightly higher than the conventional burner because a larger blower is required to supply higher excess air required for a low NOx burner.

3.3 SCV with FTO "Ultra-Low NOx/CO"

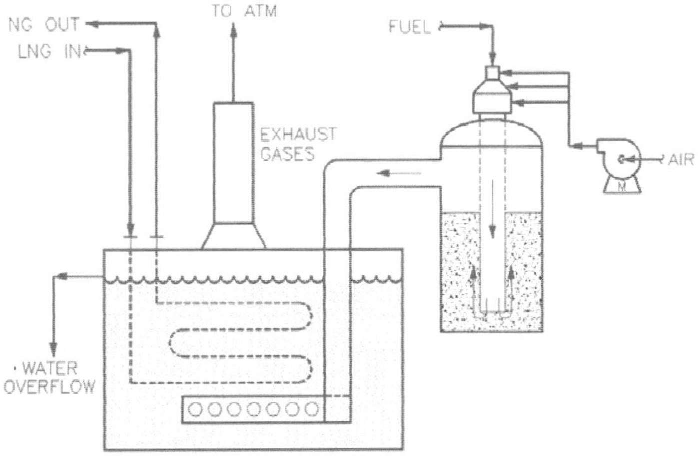

Figure 3.1 SCV with FTO Ultra-Low NOx/CO

FTO "Ultra-Low NOx/CO" is a recently developed technology developed by Linde / Selas Fluid Processing Corporation (SFPC)/ T-Thermal that employs a flameless thermal oxidizer (FTO) to reduce emissions. A typical system utilizing FTO is depicted above in Figure 3.1. The system consists of a refractory-lined oxidation vessel with a large centered dip-tube. About half of the oxidizer is filled with randomly packed inert ceramic pieces forming a uniform matrix. Hot fume from the burner, air and supplemental trim fuel enter the system at the top of the dip-tube and are mixed thoroughly before entering the ceramic matrix. Air is added to provide the oxygen required for oxidation and to ensure that the mixture does not exceed the lower flammability limit (LFL). Oxidation occurs within the ceramic matrix where the temperature is kept at approximately 1800°F or less by adjusting the fuel/air ratio. This low operating temperature keeps thermal NOx from forming and the long residence time provides complete combustion producing almost no residue hydrocarbons and low CO level.

This technology is effective in lowering the NOx and CO levels. It can lower the emission levels to less than 5 ppm for NOx and 10 ppm for CO. The capital cost is comparable with a selective catalytic reduction units. However the initial cost is higher than the low NOx burner. The technology is newly developed and additional work is ongoing by the supplier to commercialize it. Operating cost for a SCV with FTO "ultra low NOx/CO" system includes costs for fuel and power and is expected to be slightly higher than the low NOx burner system.

3.4 SCV with Selective Catalytic Reduction (SCR)

The SCR system involves injecting NH_3 into the flue gas in the presence of a catalyst to reduce NOx emissions by converting NOx product to nitrogen and water. SCR is based on the following reactions between ammonia and NOx at high temperature:

$$6NO + 4NH_3 \rightarrow 5N_2 + 6H_2O$$
$$2NO + 4NH_3 + 2O_2 \rightarrow 3N_2 + 6H_2O$$
$$6NO_2 + 8NH_3 \rightarrow 7N_2 + 12H_2O$$

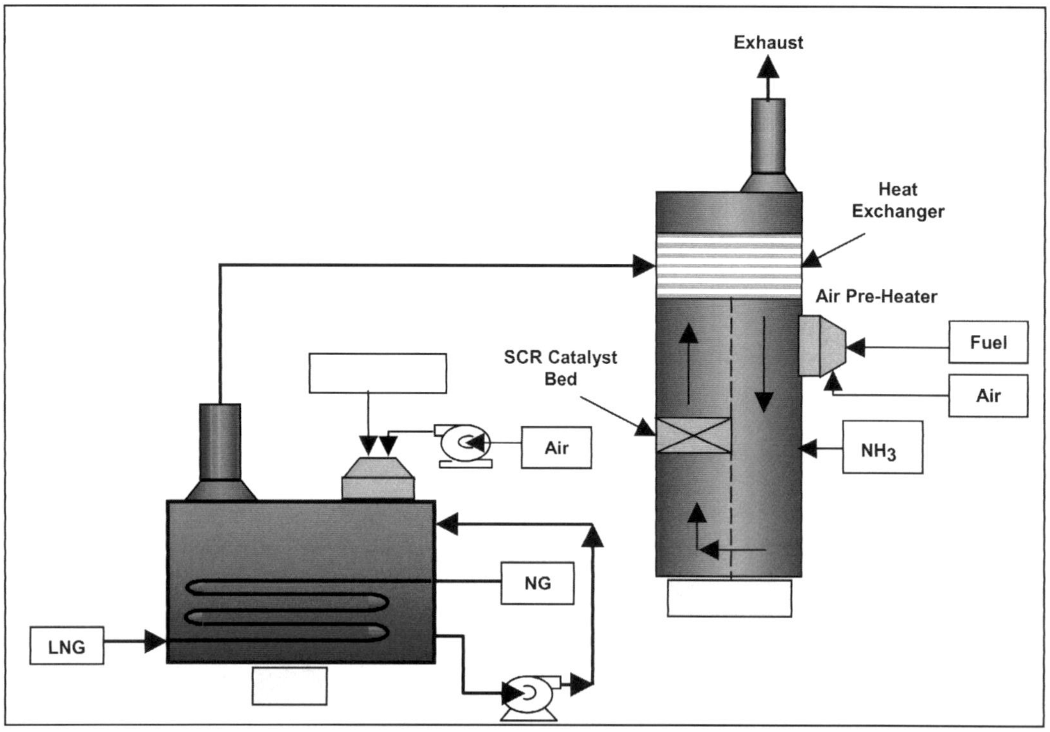

Figure 3.2 SCV with SCR

Figure 3.2 represents a typical system utilizing Selective Catalytic Reduction (SCR) for SCV emission reductions. The exhaust from SCV flows to a heat exchanger where it is pre-heated by the hot gas from the SCR catalyst bed. The warm gas exits the heat exchanger and is heated further by an integral burner to reach a temperature around 650°F required for catalytic reduction reaction. Ammonia is then injected into the gas stream before it reaches the SCR catalyst bed. The NOx and NH_3 combine on the NOx catalyst's surface, forming an ammonia salt intermediate that subsequently decomposes to produce elemental nitrogen and water. Clean gas from the catalyst bed exchanges heat with the cool incoming SCV exhaust and then exits the stack at around 140°F.

The system can be designed to remove 90-95% of the NOx with 3-5 ppm NH_3 slip in the exhaust gas. With the addition of an oxidizing CO catalyst, about 90% of the CO can be removed in the high temperature zone. The operating cost associated with SCV with SCR system includes costs for additional fuel for reheating the flue gas and ammonia.

3.5 SCV with LoTOx™ oxidizer system and EDV® scrubber

Wet scrubbers are commonly used to remove SO_2 and particulates but are not very effective in removing NOx because of the relative insolubility of NOx in water. NOx in combustion gas typically consists of approximately 95% NO and 5% NO_2 with low solubility of 0.063 g/l and 1.26 g/l respectively at 77°F and 1 ATM pressure.

NO and NO_2 can be oxidized with ozone to form N_2O_5 which is highly soluble in water. N_2O_5 is easily converted to HNO_3 with wet scrubbing based on the following reactions:

$$NO + O_3 \rightarrow NO_2 + O_2$$
$$2NO_2 + O_3 \rightarrow N_2O_5 + O_2$$
$$N_2O_5 + H_2O \rightarrow 2HNO_3$$

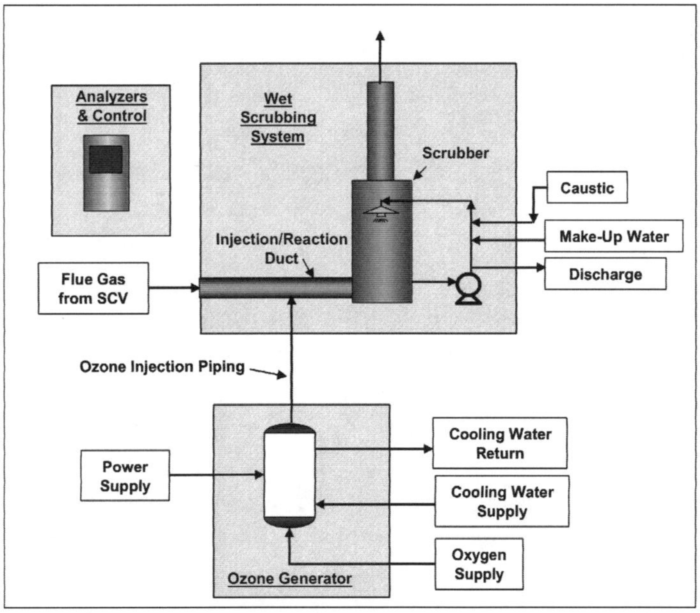

Figure 3.3 Low Temperature Oxidizer Scrubber System

Figure 3.3 depicts a typical system utilizing low temperature oxidation technology. Ozone is injected into Injection/Reaction duct where it is mixed with saturated SCV flue gas and reacts quickly with NOx to form highly soluble N_2O_5. N_2O_5 then reacts rapidly with moisture in the SCV flue gas to form nitric acid (HNO_3). N_2O_5 and HNO_3 are absorbed into the liquid phase in the downstream scrubber which is kept slightly basic with caustic to improve absorption of nitric acid. Scrubber systems are highly efficient in scrubbing N_2O_5 and the reaction into the liquid phase is irreversible and NOx emissions can be reduced to 10 ppmv or lower. Any un-reacted or excess ozone in the oxidized flue gas readily absorbs into the aqueous medium. The dilute acidic scrubber liquid can be neutralized with a caustic before release to a sanitary sewer or wastewater treatment facility. The spent liquid is high in nitrates making it undesirable for a normal biological wastewater treatment system and that can be a concern if the local regulatory requirements do not allow the discharge of a high nitrate salt containing water. This problem can be mitigated by using potassium hydroxide as the caustic source and selling the resulting spent scrubber liquid as a fertilizer.

The removal of NOx to about less than 10 ppmv is set by the need for avoiding potential for ozone slip. Careful design and operation would minimize any excess ozone going to atmosphere that does not react with NOx or dissolve in the scrubbing liquid. The CO reduction catalyst can not be utilized due to the low temperature in the system.

Operating cost associated with oxidation system includes costs for power, oxygen and cooling water requirements for the ozone generator. For small installations, the oxygen needed to generate ozone, is most economically provided by a liquid oxygen tank and vaporizer. The system, in the U.S. gulf coast, is typically supplied and maintained by an oxygen supplier at a cost of approximately $70 per short ton of oxygen. For a large SCV, typical oxygen consumption is about 2 short tons per day. For large terminal with many SCVs, oxygen is most economically produced on site as a gas. Oxygen suppliers will typically build, own and operate the oxygen generation plant for a larger capacity. The cost of oxygen for this type of arrangement is typically $50 per short ton.

4. OTHER EMERGING TECHNOLOGIES

Various ongoing work at different research facilities[(3)(4)] may lead to future development of suitable technology in reducing pollutants from SCV exhaust gases. One such development is concentrated on low temperature NOx reduction process using a combined system of pulsed corona discharge and catalysts. The process utilizes plasma enhanced selective catalytic reduction (PE-SCR) system to convert NOx into elemental nitrogen in the presence of a catalyst. The reaction using ammonia takes place at about 300°F; heating of the SCV exhaust gas is therefore required. This process is still in a developmental stage and much work needs to be done to determine its applicability for SCV emission reduction.

Also, there are precious metal catalysts available today, which can work with ammonia injection effectively at about 400°F in ways similar to the higher temperature catalyst used in SCR described above. Again the flue gas has to be heated as for the SCR system, which uses a lower cost metal catalyst.

5. CONCLUSION

SCVs are an extensively used vaporizer technology in the LNG import terminals. The major challenge for use of SCVs is the emissions associated with fuel gas combustion. Environmentally sound technologies have been developed to control SCV emissions. This paper presented various options for mitigating emissions from SCVs as well as their relative capital and operating costs. The selection of optimum emission control technology is site specific and depends on the local regulatory requirements. In addition one must be aware that the science of emission control is being applied by the industry at a rapid rate, and that a strategy selected when a project is sent for permitting may not be the best option by the time the facility is designed and constructed.

6. REFERENCES

1. TECHNOLOGY UPDATE from Linde / Selas Fluid Processing Corporation(SFPC) / T-Thermal, introduction of FTO "Ultra Low NOx" submerged combustion vaporizer.

2. Jeffery Sexton et al, "LoTOx technology demonstration at Marathon Ashland Petroleum LLC's refinery in Texas City, Texas", NPRA annual meeting, March 23-25, 2004, San Antonio, Texas.

3. H.H.Kim, et al, "Low temperature NOx reduction processes using combined systems of pulsed corona discharge and catalysts, Journal of physics D: Applied physics, 34 604-613

4. T. Czech, et al,"Pulsed and DC streamer corona induced plasmas for NOx removal from exhaust gases" 2 nd Nat. symp. PLASMA'95, 26-28 June, 1995, Warsaw, 205-208

Limitations in LNG Vaporization Process Selection

Joseph H. Cho, Ph.D., P.E.
Principal Technical Professional
Gopal Mathur
Technical Professional Leader
Heinz Kotzot
Section Leader, LNG & Gas Processing
Charles Durr
Director LNG, GTL and Gas Processing

Kellogg Brown & Root, Inc., Houston, Texas 77002

AIChE Spring Meeting, April 2005
5th Topical Conference on Natural Gas Utilization
Atlanta April 10-14, 2005

ABSTRACT

The optimum choice of an LNG vaporization system for an LNG terminal facility is primarily affected by the terminal's site selection, the environmental conditions, regulatory limitations and operability considerations.

The selection of the vaporization process is influenced by the permit approval process. In order to comply with the LNG industry's demands for minimizing life cycle costs, many vaporization processes have been proposed and developed

This paper reviews the environmental issues imposed on LNG vaporization processes and discusses the technical and economical features of viable vaporization processes. The processes discussed in this paper include the following:

- Open Rack Vaporizers with seawater;
- Submerged Combustion Vaporizers with Selective Catalytic Reduction units;
- Shell and Tube Vaporizers with closed loop heating medium circulation and Fired Heaters with flue gas Condensing Heat Exchangers;
- Shell and Tube Vaporizers with closed loop heating medium circulation and Reverse Cooling Towers;
- Ambient Air Vaporizers with supplemental Fired Heaters.

A life cycle cost comparison of each process configuration is included to assist in the process selection.

INTRODUCTION

The choice of vaporization process affects not only the terminal's capital expenditure, but can have a significant impact on its operating cost. Studies have shown that for a Submerged Combustion Vaporizer (SCV) system the fuel cost can be about 40% of the terminal operating cost with a 1,000 MMScfd send-out rate. The vaporization system determines the terminal's emission rate.

The baseload Liquefied Natural Gas (LNG) regasification terminals have primarily used two types of vaporizers: Open Rack Vaporizer (ORV) and SCV. In addition to these vaporizers, several other types of vaporizers have been used. These include the Shell and Tube Exchange Vaporizer (STV) with or without intermediate fluid. Currently, many different vaporizers and vaporization schemes have been developed that use either one type of vaporizer or a combination of different types of vaporizers.

Selection of the vaporization process is important for the fit-for-purpose design of the terminal. The selected process should be economical and reliable, using proven technologies, meet the regulatory requirement and minimize the environmental impact.

This paper discusses the applicability and limitations of several vaporization processes currently available, and provides basic guidelines for the design of vaporization processes of planned LNG terminals.

ENVIRONMENTAL CONSIDERATION

The selection of a vaporization process is site and project specific, based on the design conditions of each LNG terminal. The Federal Energy Regulatory Commission (FERC), has jurisdiction for the approval of a terminal in the U.S., and they review environmental and engineering design for the proposed project. Historically, the environmental phase is concerned with the typical air and water emissions, noise, etc. The environmental report must also address the detailed technical design features of emission sources. Additional concerns raised recently are: endangered plant and animal species, cultural protection, wetlands, and of course the general public safety.

Air Emission

Air quality is determined primarily by the type and amount of contaminants emitted into the atmosphere, the size and topography of the air basin, and meteorological conditions. The emission estimation should include discharges from the vaporizer, emergency generator, gas turbines, fire water pump engines, vent/flare system, and any fugitive reactive organic gas sources.

An LNG terminal is classified as a natural gas distribution source as per Standard Industrial Classification (CIS) code 4924. Emissions exceeding those limits require an emission permit under the Code of Federal Regulation. The regulations governing emissions vary from state to state and sometimes even from county to county. Western states and highly populated Eastern states impose relatively strict environmental requirements.

The total concentrations represent the combined air quality impact of emissions, dispersion, atmospheric reactions and transport that has occurred for all existing sources inside and outside of the project site. Since the vaporization system could be the dominant source of terminal emissions, the NO_2, SO_2 and CO discharge levels have to be evaluated in the selection of vaporization process. The expected total emissions for the selected vaporization system have to be lower than those of the Federal and State ambient air quality standards.

Liquid Discharge

The potential liquid effluent sources from terminals can include:

- Process wastewaters such as water blowdown from SCV's, leakage from heat transfer fluid, area wash down waters, etc.

- Cold seawater from ORV's
- Potentially contaminated storm water
- Sanitary wastewater or treated effluent

Parameters to be included in the wastewater discharge permit are invariably site specific, taking into consideration wastewater characterization and intake water quality. The design wastewater effluent outfall is critical due to the potentially stringent limitations for effluent wastewater discharge and the concern for recirculation of colder seawater into the seawater intake pumps.

The combustion of the gasified LNG produces water in the SCV bath through condensation at the low exhaust temperature. Figure 1 shows the typical amount of water blowdown for a 100 MMBtu/h capacity SCV. which has to be neutralized.

The seawater supply requires chlorination to protect the system, especially the heat transfer surface, against biological fouling. Chlorination is generally provided by means of injecting a sodium hypochlorite solution (commercial bleach) into the suction of the seawater pumps on a continuous basis.

When using seawater for cooling purposes, the World Bank Guidelines state that: "The effluent should result in a temperature increase of no more than 3°C at the edge of the zone where initial mixing and dilution take place. Where the zone is not defined, use 100 meters from the point of discharge [1]".

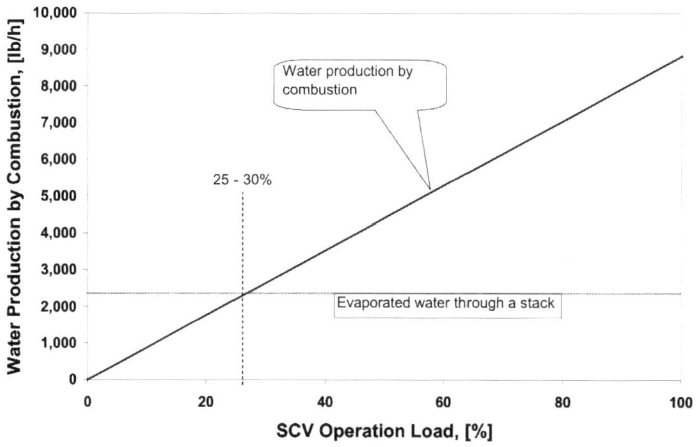

Fig. 1 – Water Blowdown from SCV Operation

When providing heat for LNG vaporization the discharge seawater temperature decreases. Although the regulations for allowable seawater temperature change were initially developed for heating seawater, they are also applied to cold return water, since currently no regulations exist for discharging cooled water. The seawater discharge from an ORV is typically around 10°F cooler and can be blended into a large body of water, in order to keep average temperatures within the 3°C temperature difference at the boundary to satisfy the World Bank Guidelines.

Other Environmental Issues on Seawater

The available heat sources for LNG vaporization include using a portion of vaporized LNG as fuel, ambient air or seawater and a combination of both. The ideal heat source would be an integrated heat cycle with an adjacent power plant or other plant that needs to release process heat. This heat could then be used for LNG vaporization. The integration with a power plant can also chill the gas turbine inlet air by using the LNG's cold energy. The cooled air stream increases the air mass to be compressed and hence increases the gas turbine efficiency and power output [2].

The seawater intake system is to be designed and operated in accordance with the guidelines of the Clean Water Act. The objective of the regulation is to minimize mortality of all types of sea-life due. The rule states that new facilities, with design intake flow equal to or greater than 10 million gallons per day, must meet certain requirements of:

- Water intake level
- Water intake velocity ;
- Intake flow limitations relative to fresh water stream flow or tidal excursion volume;
- Employment of technologies that will reduce the level of adverse environmental impact.

The environmental assessment on seawater usage is a sensitive issue for domestic terminals in the U.S. [3]. The project which uses seawater as a heat source must consider the effects of cold water on Ichthyoplankton, which consists of fish larvae and eggs. These would be free floating and readily pulled into the seawater circulating system [4]. The impact on the ecosystem changes has to be properly investigated and methodology needs to be developed to mitigate the loss of fish species.

Micro-organisms that are present in the seawater system can also be affected and destroyed by the temperature drop and strong turbulent water circulation. Conservation and restoration of the ecosystem shall be included in the Environmental Impact Assessment (EIA) study. Regulators demand more information about the selection of inlet screens in the seawater intakes. The seawater organisms are also affected by the residual sodium hypochlorite as an anti-fouling chemical, used for seawater system protection from bio-fouling. Providing an alternative method to control bio-fouling of the seawater supply system will be effective in persuading the permitting authority as well as the environmentalist.

Many seawater intake designs have been developed to minimize the environmental impact. One design includes a dual barrier system to protect marine life from entering the seawater intake. Downstream of the coarse trash screen and upstream of the seawater pumps, a secondary barrier is installed. This aquatic filter barrier is constructed of a passive marine life exclusion system, consisting of a series of fine mesh fabric filter or wedge-wire screen panel. The size of the screen can be selected to reduce impact on marine life.

APPLICATIONS AND LIMITATIONS

Open Rack Vaporizer (ORV) Option

ORV's use seawater as their sole source of heat. Seawater supplied from an overhead distributor flows over the outer surface of long finned tube panels and vaporizes the LNG inside the tubes, while cooling the seawater. ORV's are widely used in Asia and Europe, and are well proven in baseload LNG regasification service.

Great care must be taken to assure a continuous and uniform flow of seawater to the panels to avoid formation of ice and the subsequent loss of performance. Some other issues that must be addressed are:

- Water Availability: Volume of water available at the project site must be evaluated.
- Return water discharge temperature: A site specific environmental study to determine the impact to aquatic life must be carried out. Depending on send-out temperature and seawater temperature during winter months, supplementary heat sources may be required.
- Return water recirculation: Modeling will be required to insure that cold discharge water does not re-circulate back to the intake side.

- Return water de-chlorination and aeration: This will require an environmental assessment.
- Back-up vaporization system: Temporarily shutdown the water intake pumps may be necessary due to the aquatic life considerations.
- Peak shaving facility: Even though the seawater supply system is typically equipped with a spare seawater pump, a spare ORV cannot be utilized for peak shaving without extra capacity from the seawater piping. Common practice is to install a spare vaporizer which utilizes a different heat source (e.g. submerged combustion vaporizers) This can be used as a spare as well as for peak shaving. However, the emission from the spare unit must be taken into consideration. The expected operating hours and emission levels from the spare vaporizer must be included in the EIA.
- Seawater quality: Contaminated seawater can affect the ORVs' heat exchange surface coating. Suspended solids, mainly silt, should be limited, since it also contributes to the erosion of the coated surface.

Submerged Combustion Vaporizer (SCV) Option

The SCV system requires approximately 1.5 % of the total vaporized LNG as fuel. Increasing natural gas prices have a negative impact on the economics.

The SCV's are designed to utilize the low pressure fuel gas from the boil-off gas, augmented by send-out gas. Depending on the vaporizer capacity, single or multiple burners may be used. The combusted gas is sparged into a water bath to utilize the heat of condensing water from the flue gas. The LNG passes through tubes that are submerged in the water bath. The water acts as an intermediate fluid for transferring the heat from the combustion process to the LNG. This also requires electric power to run the combustion air blower.

Flue Gas Recirculation (FGR) technology, originally developed and applied to the incineration industry, has been adopted in the design of low emission SCV. Since the flue gas is water saturated and at low temperature, a portion of flue gas is recycled to the blower suction which helps lower the combustion temperature, resulting in 25% lower NOx emission levels compared to conventional SCV [5]. Recycling of flue gas reduces combustion capacity by about 6% [5]. This can be compensated by an increase in the blower power.

The SCV's have a proven track record, especially as backup for ORV. Their turn-down ratio is 10:1.

Limitations of SCV application:

- Emissions of NOx and CO: The available technology can reduce the NOx and CO emissions from the SCV's below 40 ppmv of each component. However, these emission levels do not satisfy the emission restrictions in some environmentally sensitive areas. A Selective Catalytic Reduction (SCR) system, which has been widely used in other industries, may be combined with SCV to reduce NOx levels in the flue gas.
- Efficiency Decrease: SCV's can be designed for an efficiency as high as 99%, based on lower heating value of fuel. Addition of selective catalytic reduction unit (SCR) for NO_X control reduces the efficiency by 2-3% as some exhaust gas reheating is required.

- Water quality: High chloride contents (over 30 ppm) in water makeup will cause pitting corrosion on the heat exchanger tube surface. Therefore the makeup water quality needs to be carefully measured and controlled.
- Plot space of SCV: Generally, a SCV with SCR requires larger plot space than an ORV for the same thermal capacity.
- Limited operating experience: Two integrated SCV and SCR units have been operating at Distrigas in Everett, MA. since 2003. Figure 2 shows the schematic diagram of low NOx system used in Distrigas.
- Complexity: The addition of SCR units to SCV's increases capital and operating costs, and operating complexity.
- Blowdown of water from the SCV to control pH and treatment.

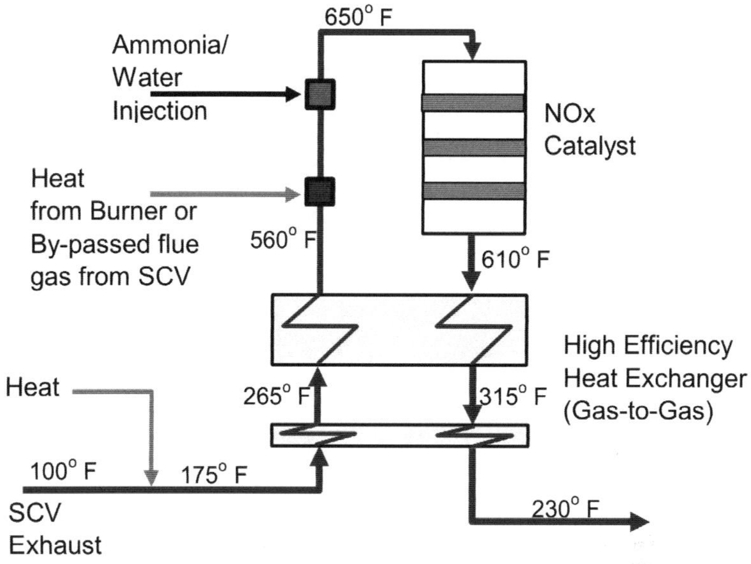

Fig. 2 - Schematic Diagram of deNOx System
(Courtesy of BDT Engineering)

Fired Heater Option

Fired Heaters (FH) have been widely used in process plants. The FH burner is sealed to eliminate the possibility of flash back with complete combustion inside the burner. The controlled flame inside the heater eliminates the possibility of flame impingement on the tube surface. In a receiving terminal the FH will indirectly vaporize the LNG by heating a Heat Transfer Fluid (HTF) that is in contact with the LNG through a Shell and Tube Vaporizer (STV).

The high heat transfer coefficients result in a compact design, reducing space requirements. Various kinds of HTFs are available: water, water solutions with ethylene glycol, polyethylene glycol, or methanol. The selection of the type of HTF depends on its physical-chemical properties, operating costs, proven track records, and environmental and safety considerations.

STV's have not been widely used for the baseload LNG vaporizer, compared to ORV's and SCV's. The STV's contribute less than 5% of total vaporization capacity in the world's LNG terminals [6].These vaporizers can be classified into two categories:

- direct heat exchange type and
- indirect type with intermediate fluid.

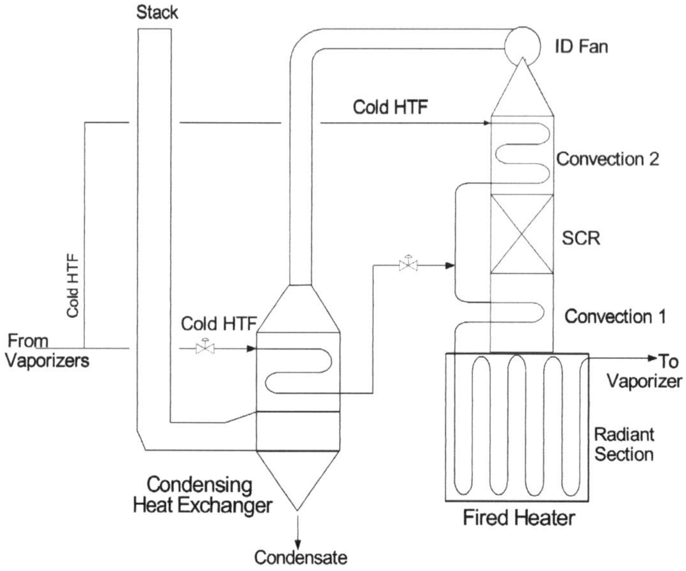

Fig. 3 – FH Equipped with Condensing Heat Exchanger and Selective Catalytic Reduction Unit

Since this study focuses on heat supply to the vaporizers, the details of STV are not addressed.

A SCR system can easily be fitted into the FH. The SCR can reduce the NOx and CO emission to comply with environmental requirements. A FH equipped with SCR eliminates over 99% of NOx and CO emission. The SCR can be installed between convection coils where the flue gas temperature is still high enough for the catalytic reaction.

However, the conventional FH has a lower thermal efficiency compared to the SCV because of the high exhaust temperature. The conventional FH with 89% of thermal efficiency (LHV basis) is designed for an exhaust temperature of 300 – 350°F. One of the reasons for designing it with a high exhaust temperature is to avoid water condensation in the flue gas. Acid gas contained in the flue gas dissolves in the condensate, resulting in acidic condensate and requiring special corrosion resistant material for the convection tubes. However, the vaporized LNG contains only a trace amount of sulfur, but CO_2 is present in substantial quantities after combustion.

The concept of the condensing heat exchanger (CHX) is based on removing latent heat from the flue gas. The flue gas enters the Teflon coated exchangers through a carbon steel inlet plenum at the top and flows downward across the horizontal banks of exchanger tubes. The flue gas exits the heat exchanger through a Fiberglass Reinforced Plastic (FRP) outlet plenum on the bottom of the exchanger as shown in Fig. 4. In a typical application water is heated in the Teflon CHX, flowing countercurrent to the flue gas [7].

Fig. 4 – Flue Gas Condensing Heat Exchanger

As the flue gas temperature approaches the water vapor dew point, condensation occurs on the tubes. Droplets of condensate form and fall as constant rain over the tube bundle. This enhances the latent heat transfer and at the same time cleans the tube surface [7]. The condensate is collected and removed at the bottom of the heat exchanger.

CHX optimization studies indicate that the flue gas temperature gas can be reduced below the water bubble point, which improves the thermal efficiency to a level almost equivalent to SCV's. Increasing the thermal efficiency of the condensing heat exchanger results in lower operating costs and improves the economics. The additional cost for the installation of the exchanger can be recovered by savings of fuel costs. There will be some increase in the air blower power consumption due to the added heat transfer surface of the CHX. Depending on the fuel gas price, the payback period can be just a few years..

The following limitations apply to the FH option:

- Condensate discharge from the condensing heat exchanger has to meet environmental regulation;
- The largest size today handles about 300,000 lb/h of flue gas, which is equivalent to the heat recovery from a FH capacity of 300 MMBtu/h [7]. Fired heaters of this capacity have not been used in LNG vaporization service;
- On-site ammonia storage is required;
- Catalysts have to be regularly replaced after their useful life time.

Reverse Cooling Tower (RCT) Option

Heat can be extracted from ambient air in a Reverse Cooling Tower (RCT) as shown in Fig. 5. The recovered heat is integrated with STV's for LNG vaporization. During the winter, when not enough heat can be extracted from the atmosphere to vaporize the desired LNG quantities, partial or full supplemental heating is required. Supplemental heating can be achieved by circulating a portion of the HTF through a fired heater, heating to about 180 ^{o}F and mixing with cold HTF to yield a 55°F mixed stream temperature. The overall economics are improved by reduced fuel consumption as heat is extracted from the atmosphere.

The cold HTF is sent to the RCT, where HTF is warmed by heat exchange with the water as a heat absorbing fluid. The cold water from the tower basin is circulated to the top of the tower. The cold water cascades down through the tower and is warmed by the air pulled through the tower through induced draft fans. The incoming air is cooled as it travels up the tower and heats the cold water cascading down. Any moisture present in the incoming air is also condensed in the tower. The warm water in the tower basin heats up the HTF flowing through the coil in the basin. A simple plate and frame type exchanger, located outside the tower basin, can also be used to exchange heat between RCT water and HTF.

When the RCT is utilized as the source of heat for vaporization, the vaporizer surface area increases about six fold, because of the low temperature differential and poor heat transfer caused by reduced viscosity of the heat transfer fluid (likely to be limited to propylene glycol). Lower operating and differential temperatures also result in larger heat transfer fluid circulation pumps which consume more electric power. Unfortunately, water cannot be used as the heat transfer fluid because of freeze concerns at the operating temperature of 55°F in and 35°F outlet temperature. For a supplimental heat source, low NOx burners with SCR units will not be necessary becasue of the limited duration of FH operation.

Some other issues that must be addressed are:

- The water condensate from the tower must be continuously discharged and safely disposed. The possibility of contamination of the circulating water in the RCT by heat transfer fluid through tube leakage must be taken into consideration;
- The circulating water stream is likely to require treatment chemicals such as biocides which in turn may require some treatment of the effluent water;
- This RCT concept has not yet been proven in a commercial LNG application;
- A full capacity supplementary fired heater is likely to be required because of the low heat recovery from the RCT during the winter season;

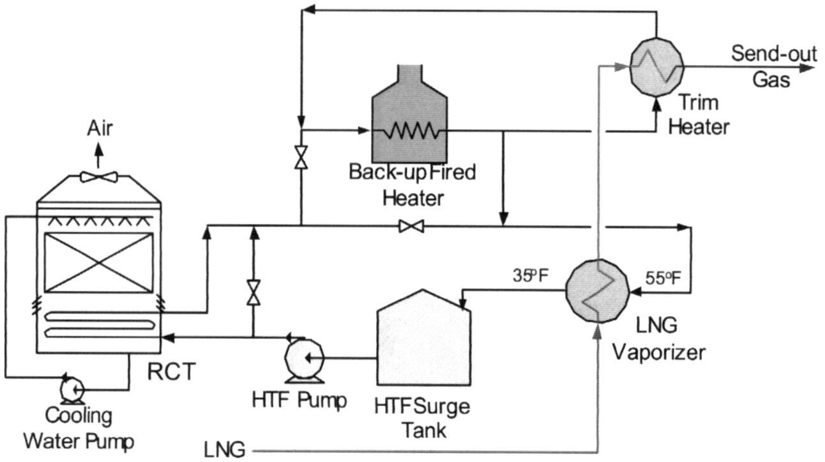

Fig. 5 – Configuration of Reverse Cooling Tower Option

- Large plot area is required;
- The choice of HTF fluid is limited.

Ambient Air Vaporizer (AAV) Option

Two types of ambient air vaporizers can be considered for LNG application.

- The direct air-to-LNG contact vaporizer, using air in either a natural or a forced draft vertical arrangement. Such vaporizers are in service for peak shaving plants at locations where the LNG vaporization duty and flow rate is relatively small (Fig 6). At this smaller scale plot space is not a major consideration. These systems are based on using single or multiple units in banks with common interconnecting pipes. Traditionally, these vaporizers have been designed with half of the parallel units on-line while the other half are off-line for defrosting.
- The indirect air-to-intermediate fluid-to-LNG vaporizer. This scheme uses an intermediate fluid between a shell-and-tube LNG vaporizer and conventional aircoolers to reheat the fluid by ambient air.

Fig. 6 – Ambient Air Vaporizer (Courtesy of Cryoquip)

Exclusive use of ambient air can totally eliminate emission and noise, if no supplemental heating is required. The water formed from condensation and melting ice must be collected. However, this water is environmentally neutral and can be easily disposed of. These vaporizers will also require construction of an LNG containment sump to be built under the vaporizers, resulting in additional cost and potentially raising some other environmental issues.

Following are summarized technical features of this vaporization scheme:

- The direct contact long vertical heat exchange tubes facilitate a downward air draft, because the air density entering at the top is lighter than the cold air at the bottom (reverse chimney effects);
- Fog and no-wind conditions control the design number of Vaporizer Trains. A trim heater is needed to supplement heating during these conditions and low ambient temperature;
- Spacing requirements must be evaulated for site-specific conditions.
- The air heater arrangement with intermediate fluid increase the operating cost by pumping the intermediate fluid besides the fans of the air heater.

Great care must be taken to assure continuous vaporization during low ambient temperature conditions and foggy days with no wind. In order to avoid the dense icing on the surface of the heat exchanger tubes, frequent defrosting with a 4-8 hour switching cycle is required. The intervals will vary depending on temperature and humidity. Long cycle times lead to dense ice on the surface of the heat exchange tubes, requiring long defrosting time. The optimum cycle time has not yet been established.

The airheater with intermediate fluid option is in operation since early 2004 at Dahey, India, and has been working well since startup. However, a secondary heating system, for periods of low ambient temperature is also installed (Fig 7).

Some other issues that must be addressed are:

- A large number of units in operation will have much larger fog generation than a couple of units, therefore dense fog dispersion must be studied. When the air is calm, the fog will tend to remain in the vaporization area;
- The large plot space demand has to be included in the life cycle cost;
- Experience and risk have to be included in the evaluation because this equipment does not have operational experience as a base load vaporizer;

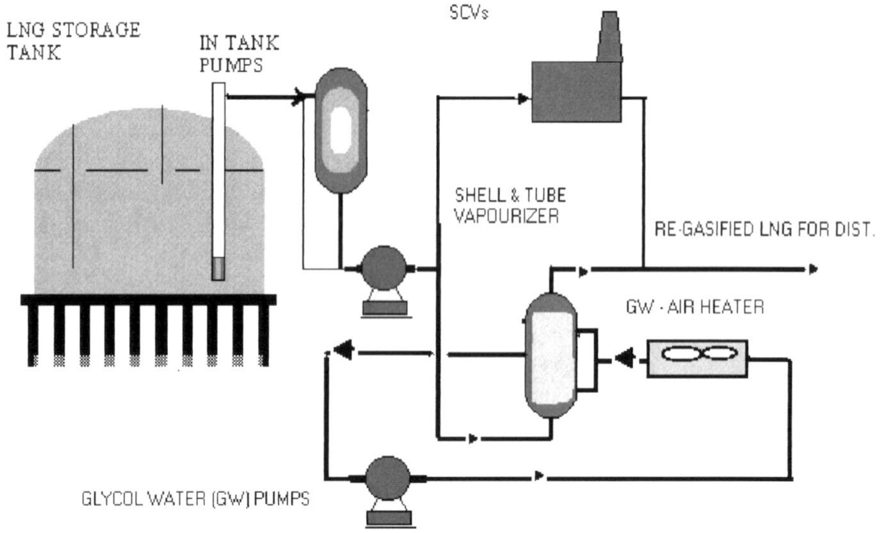

Fig. 7 – Dahej Terminal Vaporization Scheme

- No public information is available for heat capacity vs. ambient temperature to determine vaporization performance;
- Expected defrosting performance needs to be confirmed in various operating conditions such as low temperature, high humidity, long operation, fog, no wind, etc;
- Back-up facilities need to be included based on the terminal design availability and specific meteorological conditions.

ECONOMIC COMPARISON

Economic evaluations for the vaporization schemes discussed in the previous sections were evaluated with a pre-tax life cycle cost model developed by KBR. The expenses (capital, maintenance, and operating) of each scheme are calculated and used to determine an expense based Net Present Value (NPV). No tax credit was considered in this model because of revenue exclusion.

It was assumed that the terminal is located on the Gulf of Mexico. Following are the basic vaporization conditions:

- Vaporization Rate: 1,500 MMScfd
- Send-out pressure: 1,300 psig
- Send-out temperature: 34°F (at the battery limit)
- Vaporization heat duty: 900 MMBtu/hr

Schemes for the economic comparison are tabulated in Table 1. The details of these vaporization schemes can be found in the previous sections.

Table 1 – Vaporization Schemes

Vaporization	ORV Opt.	SCV Opt.	FH Opt.	RCT Opt.	AAV Opt.
Vaporizer	Open Rack Vaporizer	Submerged Combustion Vaporizer	Shell and Tube Vaporizer	Shell and Tube Vaporizer	Ambient Air Vaporizer
Heat Sources	Seawater	Fuel Burning at Submerged Combustion	Fuel Burning at Fired Heater	Ambient Air	Ambient air
Description	Required seawater intake and outfall facility	Integrated with SCR	Integrated with CHX	HTF is heated by ambient air	Ambient air draft vaporizer
Operating Vaporizer	7	8	7	7	22
Spare Vaporizer	1	1	1	1	2
Thermal Efficiency (LHV basis)	-	97*[1]	97	89*[2]	89*[2]

Notes
1. Thermal efficiency penalty has been considered for heating up the flue gas available for the SCR operation.
2. For backup facility, installation of simple fired heaters was assumed without condensing heat exchanger.

In addition to the capital expenditure of each vaporization scheme, there are other important parameters affecting economic evaluation, such as fuel cost, electric power cost, and efficiency of the thermal equipment.

The AAV option requires the highest capital expenditure, while the FH option shows the lowest capital cost because of a relatively simple configuration. However, the SCV option and the FH option have higher operating costs than any other options since the main heat source comes from burning a portion of the vaporized LNG.

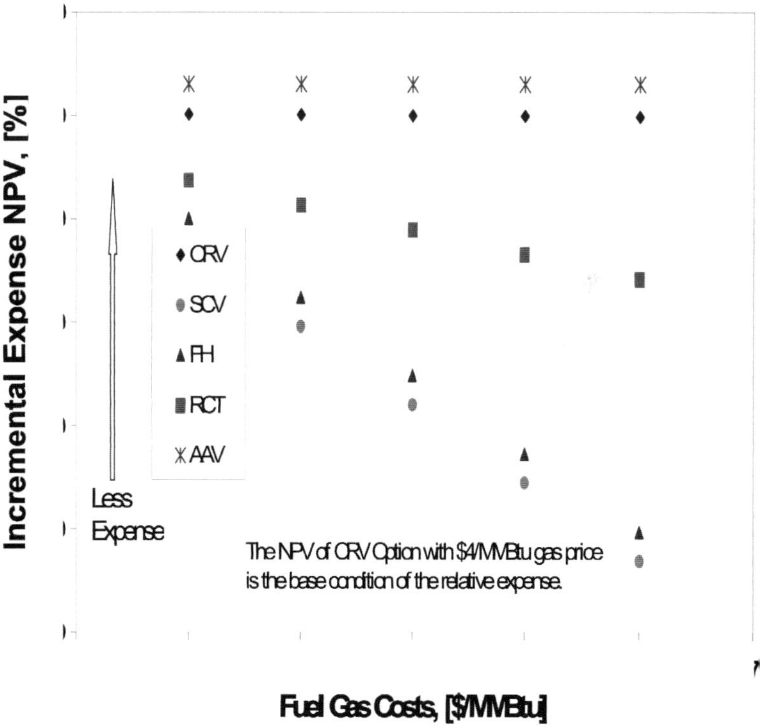

Fig. 8 – Sensitivity of Incremental Expense NPV

The best option will have the lowest relative NPV. Only the AAV Option shows a lower NPV than the ORV Option (-30%), all other options have a higher NPV than the ORV option. The relative vaporization costs illustrate the same trend as the relative NPV. These vaporization costs were estimated based on the yearly operating costs (fuel, electricity and maintenance costs) and yearly allocated capital costs that are directly related to the vaporization system (vaporizers and heat supply system only). Depreciation period and discount rate were considered in estimating the yearly allocated capital expenditure.

The overall economic evaluation is primarily affected by fuel gas cost and capital cost. In order to evaluate the economic effect of the fuel gas and electricity generating costs, sensitivity studies with these costs were performed. The sensitivity studies for options with fuel gas costs are presented in Fig. 8. This sensitivity study was carried out with a fixed electricity cost (8 ¢/kWh) and a discount rate of 12%.

The SCV option and the FH option generate a higher NPV than the other options and are naturaly quite sensitive to the fuel gas cost, while the ORV and AAV options are not affected by the fuel gas cost. As fuel gas cost increases, the NPV of all option increases. The fuel gas price will have to fall below $1/MMBtu, for the SCV option and the FH option to have some economic advantage over the ORV option, which is highly unlikely.

The sensitivity analysis with varying annual gas costs shows that the vaporization costs of the SCV and FH options increases directly with the fuel gas costs. The lowest vaporization cost is achieved by the ORV and the AAV because of the much lower fuel energy required for vaporization.

CONCLUSION

Site specific environmental issues which affect vaporization and the permitting process have to be carefully reviewed in the selection of the appropriate vaporization process.

The environmental impact on seawater usage is a sensitive issue for the U.S. domestic terminals. The seawater design must ensure that there are no adverse effects on the marine life. In order to use seawater as a vaporization heat source, conservation and restoration of the ecosystem should be well addressed in the seawater system design.

Newly developed vaporization schemes have been discussed for their practicability and limitations. Two fuel firing vaporization schemes, SCV+SCR and FH+CHX, can meet strict environmental regulations. These vaporization schemes have adopted relatively proven technology; however, these schemes result in significantly higher fuel costs than reusable energy schemes.

Two reusable energy schemes, RCT and AAV, show many advantages but require high capital costs in return for lower operating costs. These cost savings have to be evaluated against the operational risk of unproven schemes. Many issues discussed in this paper still need to be studied further for site and project specific applications prior to selecting the most appropriate vaporization schemes for a specific baseload LNG import terminals.

REFERENCES

1. World Bank Group: "Table 4, Liquid Effluent from Onshore Oil and Gas Production, Oil and Gas Development," Pollution Prevention and Abatement Handbook (1998).
2. Cho, J.H., Patel, H., Coyle, D.A., and Durr, C.A. : "A Case Study for Integrated LNG-to-Power : Technical Comparison of Available Cost Reduction Methods," Proceedings, presented at the AIChE 2004 National Meeting, New Orleans, 25-29 April 2004.

3. “Coast Guard Suspends Offshore 2004 Terminal Applications,”LNG in World Markets, October 2004, Vol. 16, No.9, p.14-16,
4. “Sinking Feeling : Offshore U.S. Terminal Developers Run into Cost, Environmental Challenges,” LNG Express, Vol. XIV, No. 10, p. 2-3, 2004.
5. Personal communication with Ed Vogel at Selasfluid.
6. World LNG Source Book 2001, Gas Technology Institute, Illinois, Sec. 4, p. 14-128, 2001.
7. Web page, http://www.chxheat.com/perform.html.

BIOGRAPHY OF SPEAKER

Dr. Joseph Cho is a Principal Technical Professional in the LNG, GTL and Gas Processing Division at KBR. He is a registered professional engineer and has been in the LNG industry for fifteen years. Prior to KBR, he was involved in designing LNG receiving terminals in Korea for 12 years. He graduated from the University of Oklahoma with a Ph.D. degree in Petroleum Engineering. He has authored and co-authored more than 30 papers in SPE, JPT, JCPT, and AIChE.

Soot and Coke Formations in MLNG TIGA Regeneration Gas Heater and Impacts on Downstream Operations

S. R. Moslim, S. A. Manaf

Operations Department, Malaysia LNG Sdn. Bhd
Bintulu, Sarawak, Malaysia

AIChE Spring Meeting, April 2005
5th Topical Conference on Natural Gas Utilization
Atlanta April 10-14, 2005

ABSTRACT

Liquefied Natural Gas (LNG) business has grown to be exceptionally competitive in the recent years and new challenges are imminent. To remain competitive, much effort particularly in the aspect of increasing LNG plant thermal efficiency is pursued. This is achieved through effective and optimum heat and energy integration. Malaysia's PETRONAS LNG complex in Bintulu, Sarawak, comprises of MLNG Satu (1983) and MLNG Dua plants (1995). Each of these MLNG plant facilities consists of three independent process trains with total LNG deliverable capacity of 7.6 mtpa and 7.8 mtpa for MLNG Satu and MLNG Dua respectively. In 2003, the complex was expanded with two more trains for MLNG Tiga Sdn Bhd, bringing the total capacity to about 23 mtpa. The three plants is being operated as one integrated site, all being part of the Gas Sector of PETRONAS. MLNG Tiga started operations in March 2003 with a Waste Heat Recovery Unit (WHRU) that is located at the downstream of Propane Gas Turbine exhaust gas. The WHRU consists of two parallel convection banks, which is installed with Regeneration Gas Heater coils and Heat Transfer Fluid coils per convection bank. The Regeneration Gas and Heat Transfer Fluid banks are installed to recover and utilize the heat available in the exhaust gas from the Gas Turbine Drivers. Since initial operations, soot and coke formations in the Regeneration Gas Heater coils and the resultant downstream equipments fouling particularly the Mercury (Hg) Adsorber and integrity issue with the heater coils and have posed major operational challenges to MLNG Tiga. This paper shares the investigation that was carried out to understand the mechanism of soot and coke formations in the Regeneration Gas Heater, the immediate measures that were taken to reduce soot and coke formation in order to achieve longer Hg Adsorber run length and hence ensure sustainable LNG productions; and the long-term study that is on going to eliminate further soot and coke formations.

Keywords: Coking, Fouling, Liquefied Natural Gas (LNG), Mercury Adsorber, Waste Heat Recovery.

1 INTRODUCTION

1.1 PETRONAS LNG Complex

PETRONAS LNG Complex; nestled in a small town of Bintulu, in the coastline of Sarawak, Malaysia, consists of 3 LNG Plants namely Malaysia LNG (MLNG), Malaysia LNG Dua (MLNG Dua) and Malaysia LNG Tiga (MLNG Tiga). The first LNG plant comprises of 3 independent process modules came on stream in 1983. Vast gas discoveries in Central Luconia area coupled with escalating demand for LNG led to the major expansion of the existing LNG facilities under MLNG Dua plants, which came on stream in 1995 and later followed by MLNG Tiga plants in 2003. The construction of 3 Liquefied Natural Gas (LNG) plants over a period of 25 years has anchors Malaysia's position as one of the world's largest and most reliable LNG supplier; with Bintulu being the world largest producer of LNG at a single location with total capacity of 23 million tonnes per annum.

1.2 LNG Process

An LNG plant can be separated into 4 main sections, namely the upstream facilities, gas treating, liquefaction and storage/terminal sections. The first section of the plant is the upstream facilities that receive raw feed supply via 125km pipeline from the Central Luconia basin, which then separates the condensate from the gas phase. The raw natural gas is then metered and sent to the individual module gas treating section. The raw feed gas consists of acid gasses namely carbon dioxide (CO_2) and hydrogen sulphide (H_2S), water and mercury (Hg), which needs to be removed to meet product specification and to prevent freezing in the liquefaction section. Hg also needs to be removed to prevent corrosion to the aluminum alloys in the Main Cryogenic Heat Exchanger (MCHE). The acid gases are removed using an amine-based solvent called Sulfinol and later incinerated before being discharged safely to the environment. After the removal of the acid gasses, the gas then enters a molecular sieve bed and sulphur impregnated activated carbon bed where the water content is reduced to less than 1ppm and the Hg content is reduced to less than $10ng/Sm^3$ respectively. The clean and dry natural gas is then liquefied using Air Products and Chemicals Inc. (APCI) license process. The APCI process uses 2 types of refrigerants, which are propane and mixed component refrigerant MR (mixture of nitrogen, methane, ethane, propane and butane). Natural gas is first pre-cooled by propane before it is finally chilled to –151 deg C in the MCHE by MR. The final cooling down to –161 degC is by Joule-Thompson expansion through a control valve. After the completion of the liquefaction process, the LNG is then piped to the storage area where it is kept at atmospheric pressure. With the arrival of an LNG tanker, LNG is then pumped to the ship via loading arms at the jetty head.

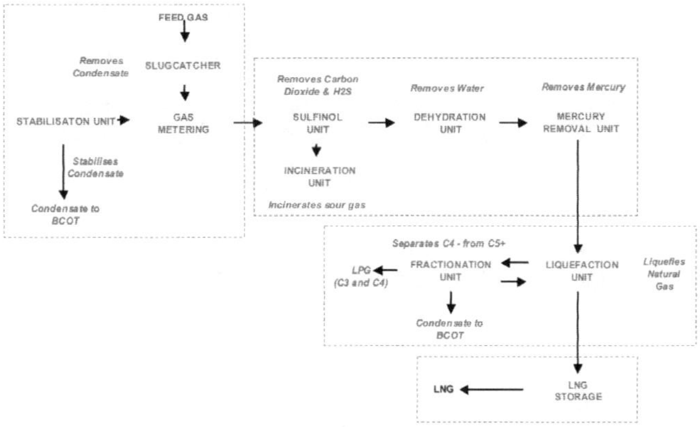

Figure 1. Simplified block diagram of LNG process

1.3 Dehydration and Mercury Removal Unit in MLNG Tiga LNG Process

Each of the two MLNG Tiga LNG Modules has a Dehydration Unit (U-91300) and Mercury Removal Unit (U-91350). The feed to U-91300 is wet natural gas from the Sulfinol Unit (U-91100), which has been pre-cooled in a Feed Gas/High High Pressure (HHP) Propane Cooler (E-91401) to condense as much of the water as possible, in order to reduce the load on the molecular sieve dryer beds (C-91301A/B/C). The natural gas from the Natural Gas Separator (V-91301) then flows downward through two driers beds, while the third drier bed is regenerated. The driers are regenerated with reverse flow, entering at the bottom and exiting at the top, as the regeneration gas is driest and hottest when it enters the bed. The regeneration gas is fed to Regeneration Gas Heater (RGH) (E-91303) where the gas is heated by the flue gas from the Gas Turbine driving the Propane Compressor (GT-91440). RGH is located at the Waste Heat Recovery Unit (WHRU). The WHRU consists of two parallel convection banks, which is installed with RGH coils and Heat Transfer Fluid coils (E-95101) per convection bank. RGH and E-95101 banks are installed to recover and utilize the heat available in the exhaust gas from GT-91440.

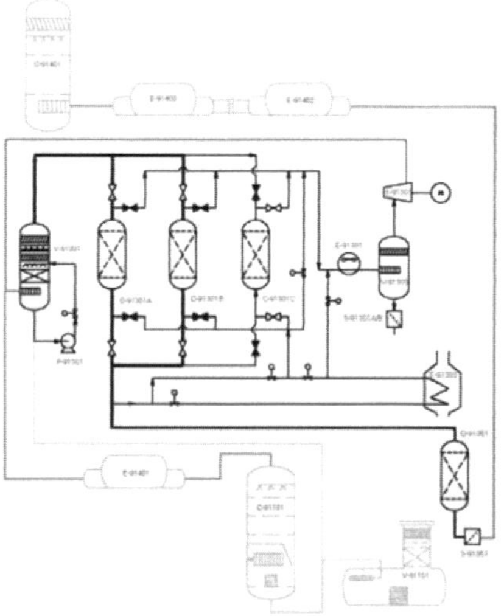

Figure 2. Simplified flow diagram of Dehydration and Mercury Removal Unit

Since initial operations, soot and coke formations in RGH coils and the resultant downstream equipments fouling particularly the Hg adsorber and integrity issues with the RGH coils have become major operational challenges to MLNG Tiga, as it results with unscheduled LNG Modules shutdown and high maintenance and operating costs. The first part of this paper provides an insight on the understanding of soot and coke formation mechanisms in the RGH. Next, this paper highlights the impacts of soot and coke mechanisms to MLNG Tiga. The final part of this paper discusses the immediate measures that were taken to reduce coke and soot formation in order to achieve longer Hg adsorber run length and hence ensure sustainable LNG productions; and the long-term study that is on going to eliminate further soot and coke formations.

2 UNDERSTANDING OF SOOT AND COKE FORMATIONS

2.1 Soot and Coke

Thermal degradation, decomposition or cracking of hydrocarbon results with the formation of soot and coke. The former is the results of cracking of gaseous hydrocarbon, which is typically black in colour, fine and fluffy in nature. The latter is the results of cracking of liquid hydrocarbon, which is typically black in colour and hard in nature. Different type of hydrocarbon cracks at different temperature levels. The heavier the hydrocarbon, the lower the temperature that cracking starts to occur, regardless whether a catalyst is present or not. Refer to Table 1 for the typical catalytic and non-catalytic decomposition/dehydrogenation/cracking temperature for various types of hydrocarbon.

Table 1. Typical catalytic and non-catalytic decomposition/dehydrogenation/cracking temperature for various types of hydrocarbon (Oortwijn, 2004)

Hydrocarbon	Catalytic and Non-Catalytic Decomposition / Dehydrogenation / Cracking Temperature
C1	800 degC
C2	540 degC
Gaseous paraffin (C3+)	<540degC
Long chained paraffin (C15+)	>340degC when exposed for infinite time (10years) >420 when exposed for short term (minutes)

2.2 Formation Factors

The formation of soot and coke occurred since the initial start-up of MLNG Tiga modules in 2003. Immediately after this was realised, a series of actions were taken to understand the problem. Through these actions, it was found that continuous formation of soot and coke in the RGH are attributed by the combination of 3 factors, namely feed composition in the regeneration flow into the RGH, temperature in the RGH and the presence of catalyst.

2.2.1 Feed Composition

The feed supply to the RGH consists of hydrocarbon in the range of methane (C_1) to decane (C_{10}), which enters the heater coils predominantly in gaseous phase and a small amount in liquid phase. The formation of liquid hydrocarbon in the regeneration gas is believed to occur in the flow control valve at the inlet to the RGH, as this valve has a significant pressure drop and the hydrocarbon gas is at or near dew point. It is also believed that liquid hydrocarbon is formed at the coils inlet header because during the purge cycle, the RGH coils header is exposed to purge gas that comes from the scrub column overhead, which is at -50degC. This results with cooling of the RGH inlet header. During the heating cycle, the regeneration gas with a hydrocarbon dew point of approximately 20degC will sees the cold metal, which would results with condensation of liquid hydrocarbon.

2.2.2 Temperature

Extensive checks of actual operating conditions and design criteria for MLNG Tiga RGH coils revealed that there are major differences between them. It was found that during operation at design regeneration flow of 980t/d, 670t/d of the flow went into the RGH coils while 310t/d of the flow by-passed the RGH coils. This resulted with RGH outlet bulk temperature during heating steady of approximately 500degC, which is 180degC higher compared to the design figure of 320degC. During heating ramp-up and ramp-down, the actual RGH outlet film temperature increased and peaked to a higher temperature of approximately 523degC. Comparison with MLNG Dua RGH was also made and it was found that the peaked RGH bulk temperature during heating ramp-up and ramp-down steps is only 440degC. The flue gas temperature for MLNG Tiga propane gas turbine design turndown case (60%) was found to be specified as 515degC, which was considered as very low. This implies that the RGH was over designed with respect to the other operating modes of the gas turbine and consequently the process gas that exits the RGH will exceed the design temperature of 320 degC. Actual propane gas turbine flue gas temperature is in the range of 550degC to 575degC (Kodde, 2004). The exposure of gaseous and liquid hydrocarbon to the actual high temperature levels had resulted with the formation of soot and coke in the RGH.

2.2.3 Catalyst

In addition to that, the RGH material is Incoloy 800, which consists of Nickel (Ni), Chromium (Cr) and Iron (Fe) that acts as catalyst for hydrocarbon cracking. Analysis of Module 8 RGH soot and coke samples confirmed the presence of all three metals with actual component ratio of Ni = 22.7wt%, Cr = 54.3wt% and Fe = 23wt%, which is close to that of Incoloy 800, which has a component ratio of Ni = 32.5wt%, Cr = 45wt%, and Fe = 21wt% (Hosoya, 2004).

2.3 Formation Mechanism

To further understand the formation mechanisms that occur in the RGH coils, a micro structural investigation was carried out based on MLNG Tiga soot and coke samples. Fresh and spent sulphur impregnated activated carbon samples were also used for comparison with soot and coke samples to ensure representative results were achieved. The analysis was conducted by Shell Global Solutions (SGSI), Technical Advisor for MLNG Tiga. The intention of the investigation was addressed through transmission electron microscopy (TEM) and scanning transmission electron microscopy (STEM) with associated elemental analysis by energy dispersive X-ray analysis (EDX). Some additional evidence was provided by conventional scanning electron microscopy (SEM) and bulk elemental analysis by inductively coupled atomic emission spectroscopy (ICP-AES). There are 2 main formation mechanisms that were identified through the investigation, namely catalytic mechanism and condensation mechanism.

2.3.1 Catalytic Mechanism

Catalytic mechanism involves the formation of soot at the active Incoloy 800 metals through chemisorptions of hydrocarbon species from the gas phase that subsequently dehydrogenate to form carbonaceous deposits. Soon after that, the active metal underneath the carbonaceous deposits diffuses through the deposit, to present a fresh active site at which further deposition occurs. The action by the active metal sustains catalytic soot production. This process is catalysed by Cr, Ni and Fe. Soot formed through this mechanism is characterised by the presence of long filamentous fibers of carbonaceous material, which often have a small metallic cluster at their head. The filaments may be hollow and of a lacy appearance, and are described as nanotubes. This is shown in Figure 3. The amount of soot formed through this mechanism is estimated to be approximately 20kg/month. This process occurs preliminary during dryer stand-by and secondary during dryer heating ramp-up (Oortwijn, 2004). Catalytic soot formation is considered more dominant compared to the second mechanism, which is condensation mechanism.

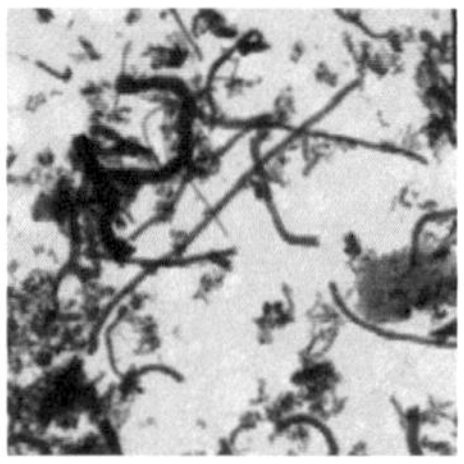

Figure 3. Soot formed by catalytic mechanism 5000x magnification view (Titchener and Smith, 2004).

2.3.2 Condensation Mechanism

Condensation mechanism involves the formation of coke from heavy hydrocarbons mists (C_{10} to C_{12}) that condense and forms droplets on the RGH metal surfaces, on the surface of a soot layer already present, or within the gas phase. Subsequent dehydrogenation results in the formation of coke deposits in the RGH coils. Coke formed through this mechanism has globular form, hard, brittle texture (Titchener and Smith, 2004). Coke formed through this mechanism is also known as shot coke. This is shown in Figure 4.

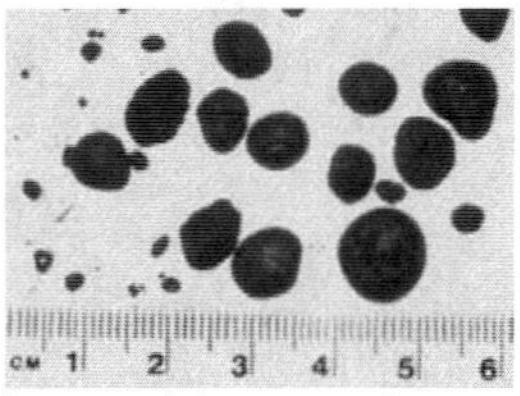

Figure 4. Coke formed by condensation mechanism

3 IMPACTS OF SOOT AND COKE FORMATIONS

3.1 Soot and Coke Movements

During dryer bed regeneration, it is believed that soot and coke from the RGH are carried together with the regeneration gas, which flows upwards through the dryer bed. The soot and coke will then be deposited at the bottom of the dryer bed. When the dryer bed is switched from regeneration to adsorption mode, the flow direction through the bed is reversed. Soot and coke are consequently pushed to the downstream Hg adsorber bed together with the main natural gas flow. This movement is consistent with Module 7 and 8 process data observations, whereby the normalised pressure drop (dP) across the Hg adsorber increased in a step wise fashion immediately after each dryer bed has completed the regeneration step. Refer to Figure 5 for an example for process data in Module 8.

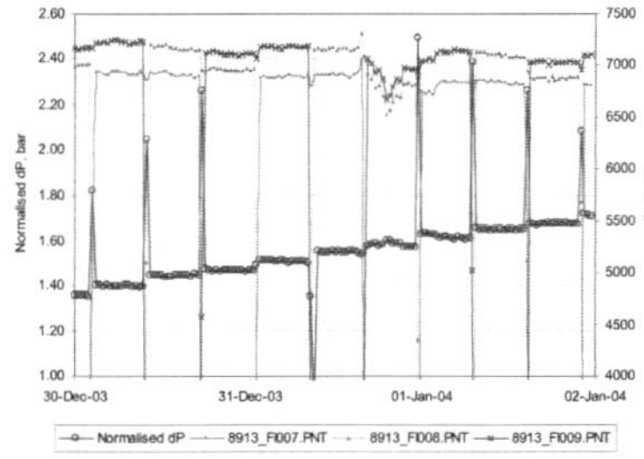

Figure 5. Step wise fashion increase of normalised pressure drop (dP) over Module 8 Hg adsorber immediately after a dryer bed (shown by flow through the bed 8913_FI007/8/9) have completed regeneration

3.2 Hg Adsorber Fouling

Deposition of soot and coke to Hg adsorber increases the bed pressure drop (dP) in an exponential manner. This reduces the life span of the bed from typically between 3 to 5 years to less than 6 months. An example of the normalised dP profile in Module 8 Hg adsorber is shown in Figure 6. After the maximum allowable pressure drop across the bed is reached, the LNG module has to be shutdown for duration of 7 days for the removal and replacement of the fouled bed that consists of ceramic balls and sulphur impregnated activated carbon. The loss of LNG production during each shutdown for Hg bed change out is estimated to be RM81M. The cost of bed internals replacement is estimated to be RM300k. Refer to Figure 7 for conditions of the fouled Hg bed.

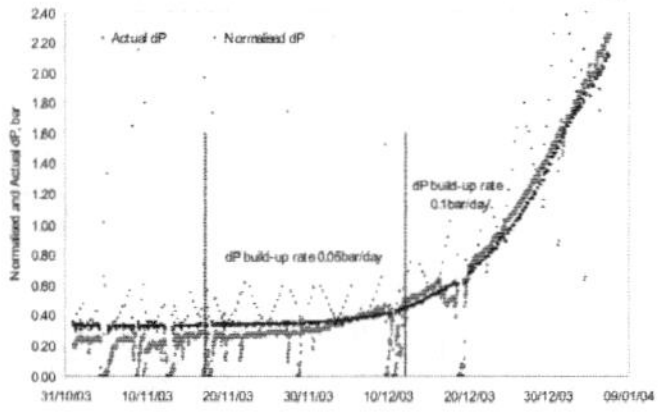

Figure 6. Pressure drop over Module 8 Hg adsorber

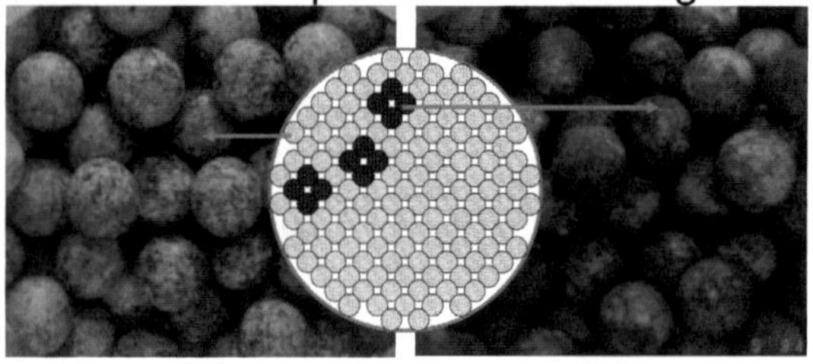

(a) (b)

FIGURE 7. (a) Top view of 1" ceramic balls that was fully covered with fine black dust, (b) Top view of 1" ceramic balls with 3 significant black spots

3.3 Hg Adsorber Dust Filter Fouling

Most of the soot and coke are captured and builds-up in the Hg adsorber bed but some would pass through the bed and ends up in the downstream Hg adsorber dust filter. This results with very frequent filter element change-out, which is approximately 3 times/year as compared to typically between every 3 to 5 years. Each change-out results with high maintenance cost that is estimated to be RM70k.

3.4 RGH Coils Tube Thinning

The deposition of soot and coke in the RGH had caused some blockages to the coils tubes. However, the blockage did not occur evenly whereby some tubes were fully blocked, some partially blocked while some were not blocked at all. This resulted with uneven and restrained thermal expansion in the tubes, which in turn resulted with irregular movements in the coils that caused coils deformation and tube rubbing or thinning failure. This failure was first observed in Module 8 DOSH shutdown in September 2004. Refer to Figure 8 for pictures of the damages. Investigation revealed that severe rubbing of the RGH tubes with the coils top bundle guides had caused significant loss of tube metal. Ultrasonic (UT) examination of Module 8 RGH showed that 2 tubes had lost 1.35mm to 1.7mm from its original installed tube thickness and considered to have less than the minimum thickness allowed by design code requirements, while 2 other tubes had lost its fins and still have wall thickness within design allowable limit. Subsequent investigation showed that the locations of the blocked tubes were consistent with the location of the coils deformation. This lead to the conclusion that that the root cause of the rubbing failure was due to blockage of RGH coils tubes (Takahashi, 2004). However, mechanical installation was also considered as a potential cause and could not be ruled out as a potential cause of the rubbing problem (Moslim, 2004). The seriously damaged 2 tubes were dismantled and the connections with the inlet and outlet headers were capped-off. The bundle guides were extended to guide the second tubes from the wall for the dismantled tube portions. Extended vertical baffles were also installed to avoid bypass of flue gas flow by covering the empty spaces of the removed tubes. In order to avoid similar rubbing damages in the future, the tubes in contact with the top tube guides were protected with additional stainless steel sacrificial sleeves integrated with several layers of fins. Since the coking and the mechanical installation in both Module 8 and Module 7 could be considered as rather similar, a decision was made to shutdown Module 7 for the inspection of the RGH (Moslim, 2004). During the shutdown, somewhat similar deformations and rubbing/thinning failure were observed in Module 7 RGH coils. However, the tube thinning was not as severe as Module 8. Nevertheless, similar design improvements were installed in Module 7 to ensure the integrity of the coils is not jeopardised.

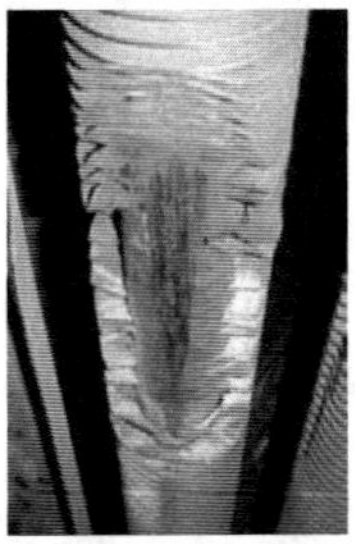

Figure 8. Rubbing damages at RGH

4. SHORT TERM MEASURES TO SUSTAIN LNG PRODUCTION AND LONG TERM DESIGN IMPROVEMENTS

4.1 Short Term Measures

A series of short term measures were taken to reduce soot and coke formation in order to achieve longer Hg adsorber run length and hence ensure sustainable LNG productions, namely process condition change, re-rating of the Hg bed, installation of restriction orifice at the downstream of the RGH, installation of coke catcher materials in the Hg adsorber, and usage of alternative sulphur impregnated activated carbon type. The details of the measures are described in the next sub-sections. Figure 9 depicts the overview of the impacts of the measures.

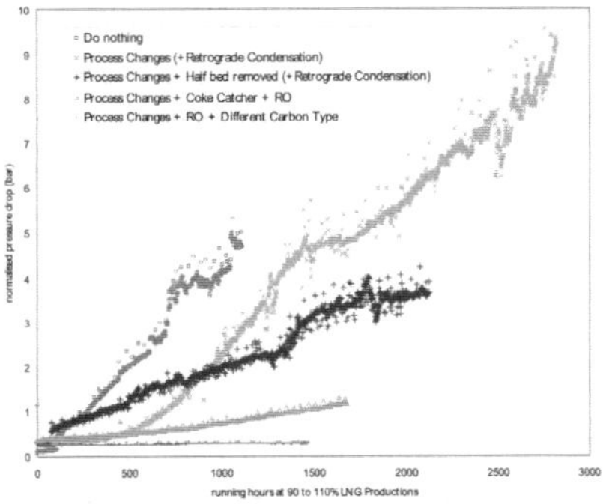

Figure 9. Overview of impacts of short term measures

4.1.1 Process Condition Change

Process condition change was the first measure that was taken because the change is simple and does not require any hardware modifications. The intention of the change was to reduce the film and bulk temperature in the RGH and hence minimise soot and coke formation. This was achieved by increasing the regeneration flow through the RGH from 980t/d to 1400t/d (by increasing the regeneration temperature set-point from 320degC to 330degC, and thus closing the by-pass valve across the RGH). This resulted with a reduction of the bulk and film temperature from 500degC to 340degC; and 523degC to less than 420degC respectively. The corresponding Hg adsorber dP build-up reduced from more than 0.1bar/day to 0.055bar/day (on average).

4.1.2 Re-Rating of Hg Adsorber

The weakest point in the Hg adsorber is the outlet distributor, which is designed for a maximum allowable dP of 4.05bar. This value defines the maximum allowable dP for the bed and dictates when the bed have to be shutdown for fouled bed replacement. A re-rating calculation was done with the intention of evaluating the true maximum allowable dP for the whole bed, which would prevent early unnecessary unscheduled shutdown. The outcome of the re-rated calculation confirmed that the true maximum allowable dP for the bed is 10.02bar, which allowed an additional 2 months bed run, before bed replacement.

4.1.3 Restriction Orifice Installation at RGH outlet

An attributing factor to the formation of coke is believed to be liquid formation in the flow control valve at the inlet to the RGH, as this valve has a significant dP and the hydrocarbon gas is at/near dew point. In order to prevent liquid from going into the coils, a restriction orifice was installed at the downstream of the heater coils. The restriction orifice was sized to have similar dP (3bar) as the inlet flow control valve. This modification was implemented in Module 7 and 8. The effectiveness of this measure is yet to be confirmed in the next Hg bed shutdown, where the bed will be inspected.

4.1.4 Coke Catcher Installation at Hg Adsorber

In August 2004, the configuration of Module 7 Hg adsorber loading scheme was modified with the intention of increasing the Hg adsorber run length, and avoids an unscheduled shutdown particularly during the peak LNG demand period. The selected loading scheme ensures that a minimum sulphur impregnated mass transfer zone (MTZ) height is available to guarantee that the bed is able to remove the designed feed gas Hg content of 1510ng/Sm^3 and reduced the value to less than 10ng/m^3; some margin is available on top of the MTZ to cater for the possibility of retrograde condensation due to the high dP; and sufficient bed height is available to ensure even flow distribution through the bed to prevent bed channeling. The selected loading scheme consists of coke catcher materials (Raschig rings and Halder Topsoe TK30 rings), which have higher void fractions than ceramic balls that is placed on top of the carbon bed to collect most of the soot in the open ring structures. In the first 3 months of the trial, the actual dP build-up in Module 7 bed was observed to be 10times slower compared to the build-up without the coke catcher materials.

4.1.5 Usage of Alternative Hg Bed Sulphur Impregnated Activated Carbon Type

From initial operations, Module 7 and 8 Hg bed were installed with naturally broken sulphur impregnated activated carbon adsorbent, which is not extrudates type. These activated carbons are typically not so uniform in sized, and have a wider particle size distribution, and smaller average particle size. In September 2004, Module 8 Hg bed was replaced with extrudates adsorbent as a trial. In the first 3 months of the trial, the Hg bed dP seemed to be very low and showed very minute build-up. However, the Hg bed dust filter dP seemed to build-up much faster, whereby the maximum allowable dP is reached after 2 months of operations. It is believed that the usage of extrudates activated carbon had allowed soot to pass though the Hg bed but ends up in the downstream dust filter.

4.2 Long Term Solution to Eliminate Soot and Coke Formation

With the understanding of the cause of the soot and coke formations, a decision was made to redesign the RGH to eliminate soot and coke formation. The new design case has been identified as option K2, where a new RGH will be designed, fabricated and installed to a lower propane gas turbine flue gas path. The new RGH coils will be located at the existing location for the third heat transfer fluid coils bank. The first implementation is planned for Module 7 in July 2005, which will be followed by Module 8 in March 2006. After the installation of the new K2 RGH coils, the old RGH coils will be blank-off. The new K2 RGH coil is designed to meet the following process and mechanical criteria.

4.2.1 Process

The new K2 RGH will have an operating temperature below 400degC. The K2 RGH coils will avoid exposure of the regeneration gas to propane gas turbine flue gas temperature of above 400degC. The maximum film temperature in the new coils will be less than 400degC at all times. The new coils materials will not contain Nickel and Chromium in the tubular coil materials. The coils location will ensure that the Heat Transfer Fluid (HTF) and RGH design duties are not compromised. The new design will eliminate the purge cycle to mitigate the possibility of shot coke formation.

4.2.2 Mechanical

The new design will ensure that the mechanical support system is sufficient to prevent tube-thinning (rubbing) damage to coils. The design will ensure that only minimum changes are required at the existing WHRU structure and support.

5. CONCLUDING REMARKS

In conclusion, the experience with soot and coke formations in MLNG Tiga RGH is considered as a unique challenge as this is the first time that such problem had happened in the LNG industry. The challenges that were faced in MLNG Tiga had served as valuable lessons learnt, particularly in the areas of heat and energy optimisation in WHRU design. The design that started with great intention of optimizing heat and energy recovery by installing the RGH coils together with the HTF coils in the propane gas turbine flue gas path has resulted with significant expensive challenges to sustain LNG productions and RGH coils integrity risks due to the continuous formation of soot and coke in the RGH coils.

The experiences in Tiga RGH have proven the importance of combination of gaseous and liquid hydrocarbon, high temperature and coils material selection to ensure that it would not result with a conducive environment for soot and coke formations that eventually would cause severe fouling of the heater coils and downstream equipment.

The immediate and short term measures that were implemented in MLNG Tiga namely RGH process condition change, re-rating of the Hg bed, installation of restriction orifice at the downstream of the RGH, installation of coke catcher materials in the Hg bed, and usage of extrudates Hg bed adsorbent with the intention to combat fouling of the Hg bed and extend the run length of the adsorber as long as practically possible had proven to be effective. Significant additional run lengths were achieved and unscheduled shutdown had been delayed which in turn results with additional LNG production opportunity.

REFERENCES

Hosoya K, “Additional Investigation of Coke in Reactivation Gas Heater (RGH E-91303), December 2004.

Kodde, A.J., ‘Investigation Into Rising Pressure Drop Over Tr-8 C91351 at MLNG Tiga’, March 2004.

Moslim, S.R., ‘Risk Assessment for Module 7 Regeneration Gas Heater Coil Rubbing Failure’, October 2004.

Oortwijn, P., ‘MLNG3 E91303 (Coke and Redesign)’, July 2004.

Takahashi, Y, ‘Investigation Report on Damages of Train 8 PR WHRU Regeneration Has Heaters During Shutdown in September 2004’, October 2004.

Titchener K.J., Smith G.C., ‘Analysis Report; Micro structural Analysis of Activated Carbon and Soot Samples ex. Petronas/MLNG’, September 2004.

Biography of Speaker

Siti Rafidah Moslim graduated with First Class Honours in Bachelor of Chemical Engineering from The University of Adelaide, Australia in December 2000. In March 2001, she joined Malaysia Liquefied Natural Gas (MLNG) a subsidiary of Petroliam Nasional Berhad (PETRONAS) based in Bintulu, Sarawak.

Her first assignment in MLNG was as a Process Technologist responsible for the troubleshooting of plant problems and development of plant changes for Liquefied Petroleum Gas (LPG) Plant and Terminal (Storage and Loading Facilities). She was closely involved in interface discussions with Petronas Dagangan Berhad (PDB) LPG bottling plant and managed a number of technical (process) and HSE issues in the discussions with OGP, PDB with respect to LPG bottling plant.

In October 2002, she joined MLNG Tiga Start-up Team as a Process Engineer where she was closely involved in the Commissioning and Start Up activities; and Process troubleshooting for new MLNG Tiga plant facilities. This includes Gas Treating facilities (Sulfinol, Dehydration and Mercury Removal units), Liquefaction facilities (Natural Gas, Multi-Component Refrigerant (MCR) and Propane loops) and Utilities / Common facilities (Slug catcher, Gas Metering, Condensate Stabilization unit, Nitrogen Unit, Instrument Air Unit, Flare and Fire Fighting Systems). She also work together with MLNG Tiga Project Team and EPCC Contractor in demonstrating the performance of new Tiga facilities, which is achieved through LNG Module 7 and 8 Performance Test run (PTR) and Operational Tests for individual process facilities.

LNG Vapor Cloud Exclusion Zones for Spills into Impoundments

Jerry Havens and Tom Spicer
Chemical Hazards Research Center
Ralph E. Martin Department of Chemical Engineering
University of Arkansas

AIChE Spring Meeting, April 2005
5th Topical Conference on Natural Gas Utilization
Atlanta April 10-14, 2005

ABSTRACT

LNG spills on land are most likely into dike/impoundment areas, introducing the need to consider effects of dike/impoundment walls on vapor holdup and dispersion. The DEGADIS and FEM3A models are approved for determination of vapor cloud exclusion zones, but DEGADIS is limited to prediction of dispersion from area sources over flat, obstacle-free terrain, and hence is not directly applicable to this important dispersion scenario. FEM3A, which is directly applicable, has only recently been approved; consequently, ad hoc methods have been widely used to determine input conditions for DEGADIS to estimate dispersion downwind of a dike system for such releases. The phenomenology of vapor holdup and vapor mixing with air in a dike system has been studied in wind tunnel and field experiments, and the fallacy of using an overly simplistic estimation of the time and rate of vapor cloud overflow has been demonstrated. In some cases, the methods used have resulted in exclusion zone determinations that do not provide for public safety as intended by 49 CFR 193. This paper summarizes the current state of knowledge about this important vapor dispersion scenario, including key experimental data that address the question. Applications of FEM3A to determine exclusion zones for a hypothetical spill within a typical dike/tank configuration, as well as shortcut-use of FEM3A to describe vapor overflow from the same impoundment for use as input information for DEGADIS simulation of the dispersion downwind from the dike, are illustrated.

1. INTRODUCTION AND BACKGROUND

Vapor cloud dispersion exclusion zones for credible LNG releases from primary containment and transfer systems are required by 49 CFR 193 and NFPA 59A. Since such spills are inevitably into dike/impoundment areas, estimation of such vapor cloud exclusion zones requires consideration of the effects on gas holdup and dispersion of the dike/impoundment walls. The DEGADIS and FEM3A models are approved for determination of such vapor cloud exclusion zones, but DEGADIS is limited to prediction of dispersion from area sources over flat, obstacle-free terrain and is therefore not directly applicable because of the atmospheric flow diversion that is caused by the dike/tank system. Because FEM3A, which is directly applicable to this vapor dispersion scenario, has been approved in 49 CFR 193 only recently, various ad hoc methods continue to be used to determine the source conditions (rate and area of vapor release) for input to DEGADIS to predict dispersion downwind of a dike system. Such methods are commonly based on determination of the rate of evaporation from the spill, coupled with application of a liquid-to-vapor expansion factor, to estimate the time and rate that a vapor cloud overflows the impoundment wall. The overflow gas is typically assumed to be pure methane and the overflow rate, which can be time varying, is used as input to DEGADIS. Such methods can result, for time-limited spills (for which 49 CFR 193 provides), in the conclusion that the gas will not overflow the dike, or that if it does, its overflow will be substantially delayed and will occur at a diminishing rate as a result of various heat transfer effects that decrease the LNG evaporation rate with time.

The accompanying figure, adapted from the documentation of the SOURCE5 model published by the Gas Research Institute to determine input information for DEGADIS prediction of vapor cloud exclusion zones extending downwind of a dike system, indicates several dike system configurations in common use. LNG spill evaporation rates which diminish with time are expected with any of these systems because of dike wall/floor surface cooling, and such effects can be important in reducing the required vapor cloud exclusion zone. Indeed, the choice of dike system offers the potential for minimizing the vapor exclusion zone by judicious design of liquid surface contact areas as well as incorporation of insulating materials to reduce heat transfer from the dike system surfaces to the spilled LNG. However, there is no basis for the assumption that the overflow gas is pure, and consideration of a wind flow across the facility strongly suggests that mixing of the evolving LNG vapors from the LNG spill surface with air would occur.

A brief review of the history of this issue will clarify how we got to the present state. 49 CFR 193 was developed at least partly under the guidance of Mr. Walter Dennis, now deceased, of the United States Department of Transportation. DEGADIS was not yet born, and a

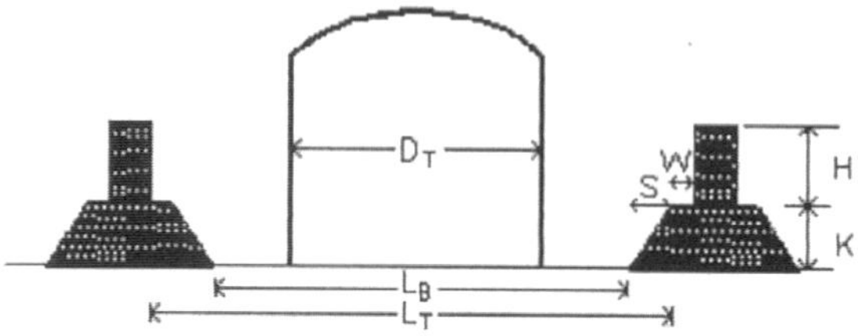

RECTANGULAR DIKE WITH SLOPING EMBANKMENT AND VERTICAL FENCE

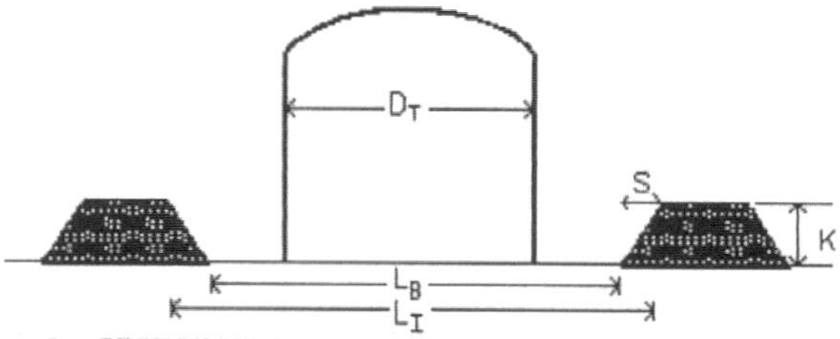

RECTANGULAR DIKE WITH SLOPING EMBANKMENT ONLY

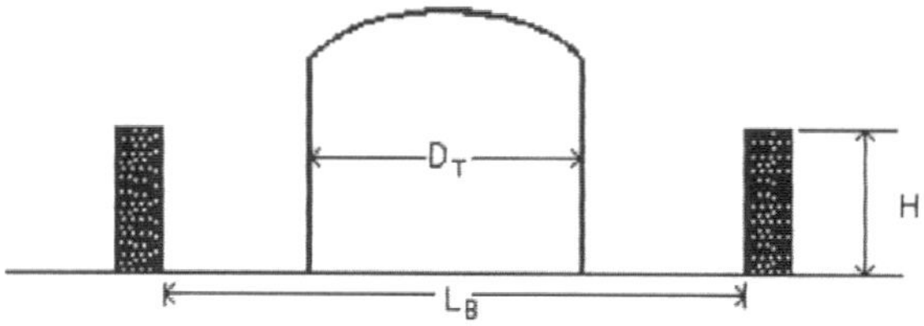

RECTANGULAR DIKE WITH VERTICAL WALLS

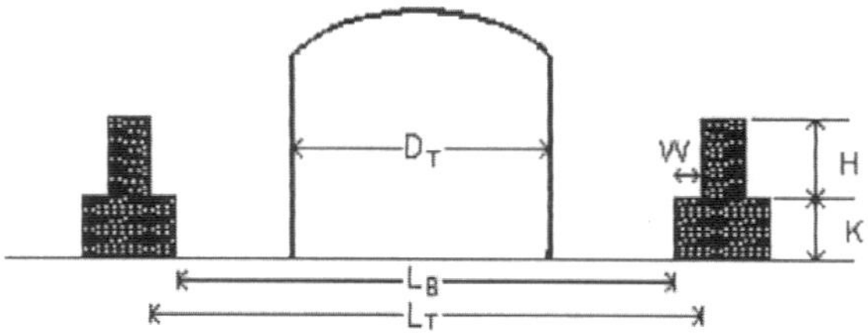

RECTANGULAR DIKE WITH VERTICAL THICK WALL AND FENCE

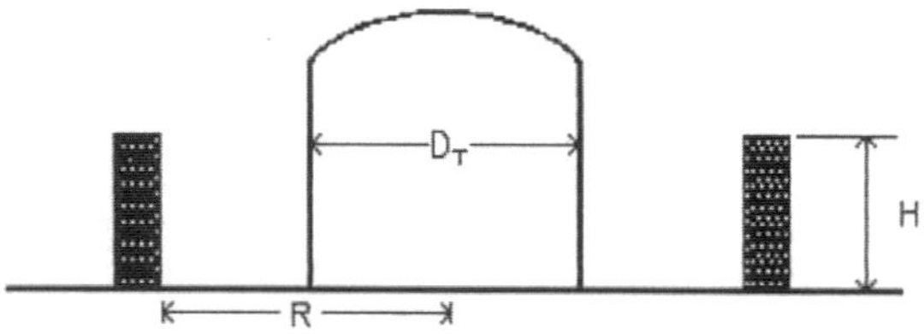

CIRCULAR DIKE WITH VERTICAL WALLS

vapor dispersion model, referred to as the "MTB" (for Materials Transportation Bureau) model, was specified initially in the regulation. The MTB model was not designed to provide for dense gas effects and was subsequently displaced by DEGADIS, which had been developed, as its name implies, for DEnse GAs DISpersion. However, the DEGADIS model was, and remains, directly applicable only to dispersion over flat, obstacle-free terrain. Mr. Dennis, with whom the principal author enjoyed a direct association, recognized that the "holdup" of vapor in a dike system could have important effects on the ultimate downwind dispersion of the vapor cloud. Consequently, at his recommendation, the Department of Transportation (DOT) and the Gas Research Institute (GRI) initiated in 1983 a research program to study this question.

Some of the key results of this research program, which continues today, will be discussed below, but at this point it is noted that DOT (Mr. Dennis) suggested, at the outset of the program, interim modeling methods for use until such time as the research program results were available. It is those interim measures, which included, at least implicitly, the assumption that no vapor air mixing occurs within the impoundment before the vapor cloud overflows, that have continued to be used and are the subject of this paper. It is a primary conclusion of this paper that the research program results provided clear, undisputable evidence that the interim measures so suggested were not justified.

2. EXPERIMENTAL DATA AND FEM3A SIMULATION

Vapor holdup and vapor/air mixing in an impoundment volume have been thoroughly studied in field and wind tunnel experiments, and the fallacy of using an overly simplistic estimation of the rate and concentration of the vapor cloud that overflows the dike has been demonstrated. All test data to date indicates that vapor/air mixing inside the impoundment is an important determinant of the downwind vapor cloud travel.

<u>The "Falcon" Field Tests</u>

In 1987, DOT and GRI conducted the "Falcon" tests at the Department of Energy's Liquefied Gaseous Fuels Spill Test Facility at the Nevada Test Site (see References). The Falcon tests (see photograph of a typical test, thought to be Falcon 1) were large scale field tests to determine the effectiveness of vapor fences to decrease downwind flammable cloud travel distances following LNG spills into a fenced (containment) area. The purpose of the tests was to provide large scale test data to answer this question, as well as to provide data for the ongoing evaluation and verification of the FEM3A model.

The Falcon test series involved five moderate-scale (20 to 66 cubic meter) releases of liquefied natural gas (LNG) into a fenced (vapor-containment) area with dimensions 44 m x 88 m x10 m (tall). The total (vapor containment) volume within the fence was 38,720 cubic meters. The table following shows the LNG volumes spilled in each test, along with the volume of LNG vapor (at approximately -260F) that would be evolved, assuming a liquid to gas volume expansion factor of 270. The last column shows the ratio of the aforementioned volume of LNG vapor formed to the containment volume (Vapor Fill Ratio), clearly indicating that the dike could have contained all of the LNG vapor if it was undiluted with air.

Falcon LNG Tests (1987)

Falcon	LNG Spilled (m^3)	LNG Vapor Formed (m^3)	Vapor Fill Ratio
1	66.4	17,928	0.463
2	20.6	5,562	0.144
3	50.7	13,689	0.353
4	44.9	12,123	0.313

However, flammable gas/air mixtures overflowed the fence (vapor containment) and flammable concentrations extended downwind of the fenced area in all four tests (the 2.5% gas concentration extended to approximately 240 meters (787 feet) in Falcon 1). Hence the Falcon Test results indicate clearly that neglect of vapor/air mixing within the impoundment cannot be justified, since if the vapor evolved were pure, and did not mix with air, the vapor clouds so formed would have never exceeded the volume of the impoundment, and consequently would have remained there within.

Chemical Hazards Research Center (CHRC) Wind Tunnel Tests

DOT and GRI jointly sponsored a wind tunnel test program at the University of Arkansas CHRC intended to develop definitive, verified methodologies and procedures for regulatory application of wind tunnel simulation, independently or conjunctively with a selected computational fluid dynamics model, to predict dispersion distance where diffusion is influenced by: (a) eddy entrainment from excess capacity LNG vapor detention systems, (b) wake turbulence from on site structures and natural obstacles, and (c) topographically induced diversion or meander. The first goal, which is the subject of this paper - prediction of "eddy entrainment from excess capacity LNG vapor detention systems" essentially means the prediction of the mixing with air of the LNG vapor which evolves from a spill and the entrainment (out of the impoundment volume) by the wind of that vapor air mixture.

The following discussion focuses on a specific wind tunnel experiment that has been carefully studied at the CHRC: the atmospheric dispersion of vapors from an LNG spill into the annular space between an LNG tank and its surrounding dike. These experiments were performed independently of, but parallel with, the Falcon field tests to study the effects of vapor holdup on dispersion downwind of LNG spills into impounded areas. The primary goal of the wind tunnel experiments and the Falcon tests was to determine the effect of a typical dike and tank structure on the flammable gas extent resulting from an LNG release into the impoundment area, and to provide data relating to the dispersion pattern which could be used to verify the FEM3A model for such application. After describing the results obtained from study of the steady-state dispersion pattern that results from a continuous steady release of LNG, the present discussion focuses on the aspects of the wind tunnel experiment that address the question of the transient development of the gas cloud in the impoundment and the time and rate of overflow of the vapor cloud.

The tank and dike configuration used for the experiment is illustrated here, in a photograph (taken during a flow visualization experiment) of a downwind-direction view of a section of the wind tunnel floor containing the model tank/dike system. The tank and dike configuration is a 1:150 scale model intended to represent a typical peakshaving facility with a nominal 1 BCF storage tank. The tank was located in the center of the dike, on a platform topped with a mesh screen (flush with the tunnel floor surface), that covered a gas source box mounted below the floor. A steady flow of 33.4 slpm carbon dioxide traced with 0.5 slpm (~1.5%) propane was introduced uniformly through the wind tunnel floor into the impoundment area. 150:1 scaling dictates that the equivalent field scale LNG release would be the steady evaporation, at the spill rate, of a continuous release of 0.6 m^3/s LNG, here considered to be pure methane. The experiment design information is summarized below.

	FIELD	TUNNEL
Release Rate	0.6 m^3 LNG/sec	33.9 l CO_2/min
Dike side length (inside)	95.4 m	0.636 m
Dike side length (outside)	169.5 m	1.13 m
Tank Wall Height	32.5 m	21.6 cm
Tank Top Height	42.5 m	28.3 cm
Tank Diameter	46.5 m	31 cm
Dike height	5.6	3.7 cm
Wind Speed	4.9 m/s at 10 m	40 cm/s at 6.7 cm
Atmospheric stability	neutral	neutral

The model experiments were releases of carbon dioxide at ambient temperature (typically 23 °C) and atmospheric pressure (typically 14.2 psia) for the following reasons:

- The density of carbon dioxide (at ambient temperature and pressure) is approximately equal to the density of LNG vapor (considered here to be pure methane) at its normal boiling point.

- It is not possible to simultaneously (scale) model the turbulent mixing processes and the surface-to-cloud heat transfer effects in the wind tunnel.

- The isothermal releases, uncomplicated by heat transfer effects, allow for more direct study of the validity of the Reynolds Number independence assumption upon which direct scaling of the lab scale experiments to field conditions would be based.

The scale model experimental conditions, as well as scaled results cited in this paper, assume Reynolds Number independence and consequently are based on Froude Number scaling, as follows:

$L_m = L_f/150$ $\qquad$ $U_m = U_f/150^{1/2}$ $\qquad$ $Q_m = Q_f/150^{5/2}$

where L_m = model length scale
L_f = field length scale
U_m = model velocity scale
U_f = field velocity scale
Q_m = model volumetric gas flow rate
Q_f = field volumetric gas flow rate

It follows that the model time scale (T_m) and the field scale (T_f) are related as $T_m = T_f/150^{1/2}$.

Steady-State Experiment Results

Complete description of the experimental results and their use in the ongoing verification effort which supports the FEM3A model (approved for use by 49 CFR 193 in 2000) has been published (see References). Key results are summarized here to set the stage for consideration of the use of the experimental data and FEM3A simulations to consider the central issue of this paper: the air/vapor mixing which occurs following a release of LNG into an impoundment system, and the rate and characteristics of the vapor/air mixture that overflows the impoundment.

The following figure shows the FEM3A predicted steady-state extents of the ufl, lfl, and lfl/2 gas concentrations for this vapor dispersion scenario. Wind tunnel and field scales are shown. The typically observed bifurcation of the cloud by the tank/dike structure is clearly evident, as are the presence of highly mixed, but spatially nonuniform vapor air regions within the impoundment (see near field view of the simulated concentration field). The concentration of the gas overflowing the dike walls for this particular configuration is slightly above the ufl concentration, suggesting a factor of five or more dilution of the vapor overflowing the dike. It is noted that this FEM3A simulation used a provisional (k-epsilon) turbulence closure scheme currently being evaluated at CHRC under a sub-contract with the Department of Energy.

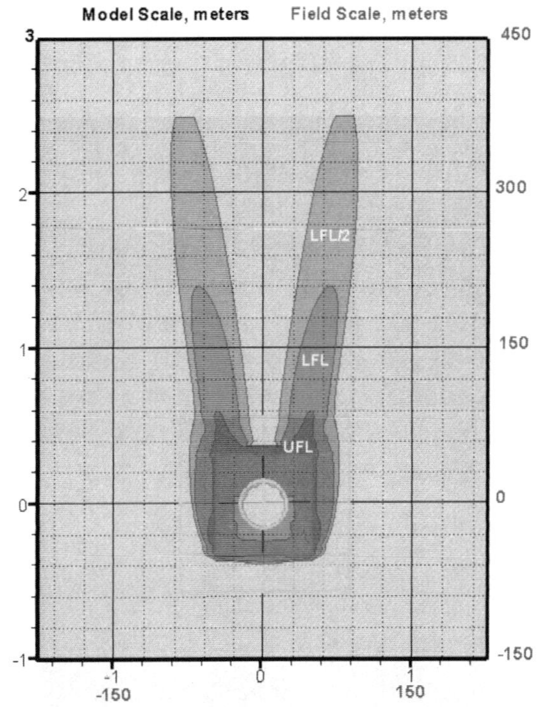

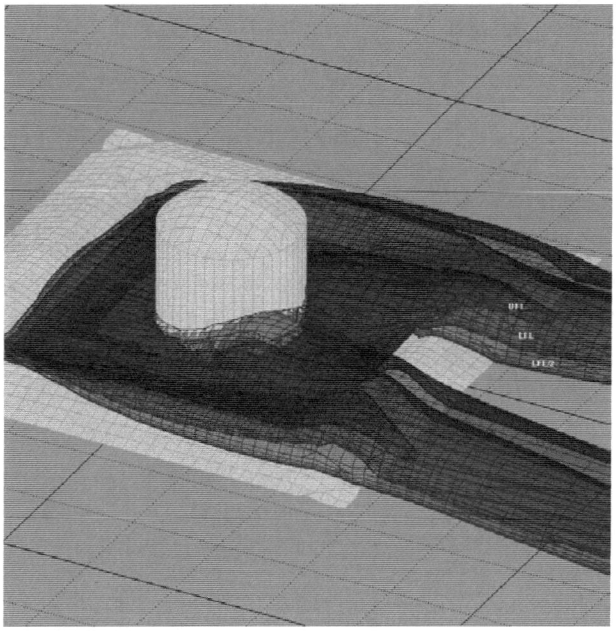

The downwind distance (wind-tunnel scale) is estimated to be 2.6 meters, and the approximate downwind distance observed in the wind tunnel for this experiment was 3.4 meters. This is an approximately 30% underprediction of the distance indicated by the wind tunnel results. Current efforts are directed to bringing this agreement closer by consideration of more appropriate description of the turbulence closure methods used as well as boundary condition specification issues that have been shown to be important.

We have estimated, using this simulation, the concentration and mass rate of methane vapor overflowing the (field-scale) impoundment and have input that information to the DEGADIS model to determine the maximum downwind extent of the lfl/2 gas contour. The result is approximately 480 meters beyond the dike, which is in good agreement with the scaled wind tunnel data.

Consideration of Impoundment Overflow Conditions

The results described above were keyed to consideration of the steady state results obtained from the FEM3A simulation as well as experimental data. We consider here additional information that is provided by the FEM3A simulation regarding the transient development of the vapor cloud within the impoundment, and the resulting overflow of the dike, all of which we have checked for consistency with observations of the experimental results.

The purpose here is to illustrate the use of the FEM3A model to determine the overflow that would (or would not) occur for a time-limited release of LNG within the impoundment. The case under consideration assumes that the methane gas source begins at time zero (in the simulation) and is represented as a step change from zero rate to an evaporation rate equal to the spill rate (0.6 m^3/s LNG). Consequently, the simulation is of a vapor input rate at approximately 162 m^3/s. This assumes a liquid-to-vapor expansion ratio (at the boiling point temperature) of 270.

The volume (field-scale) within the impoundment is approximately 60,000 m^3. If the gas remained pure with a volume expansion factor of 270, the dike would overflow in approximately 370 seconds (6 minutes). However, scaling the FEM3A simulation results (150:1) using the time scale factor indicated above indicates that the vapor cloud, reflecting an order of magnitude dilution, would overflow the dike in about 21 seconds. Furthermore, the FEM3A simulation results indicate that the maximum downwind extent for the (field) steady-state release is reached at approximately six minutes, thus predicting that if this spill were limited to six minutes duration, by utilizing automatic shut down features, the downwind extent of the vapor cloud would already be approximately 325 meters downwind of the dike edge when the LNG spill was stopped.

3. CONCLUSIONS

Extensive data from wind tunnel and field experiments indicate that the assumption that LNG vapor evolving from a spill within an impoundment area would not mix with air, thus lengthening the time to vapor cloud overflow, is untenable.

Since the lower flammability limit of LNG vapor/air mixtures is approximately 5% (reflecting a twenty-fold dilution with air) it follows that vapor/air mixtures overflowing the impoundment can pose hazards to the public and should be provided for in the determination of vapor cloud exclusion zones.

It is unlikely that a “one-size-fits-all” estimate of the amount of air dilution that would occur within an impoundment volume can be specified.

There appear to be only two demonstrated approaches to the determination of the effect upon vapor dispersion of the presence of dike/impoundment walls – physical or mathematical modeling.

Since lumped parameter models, such as DEGADIS, cannot be directly applied to this problem, the most cost effective alternative appears to be computational fluid dynamic modeling.

The FEM3A model, which is approved for such use in CFR 193 and NFPA 59A and is available to interested users, has been illustrated herein and is recommended for this application.

REFERENCES

Falcon Tests

Brown, T. C., R. T. Cederwall, S. T. Chan, D. L. Ermak, R. P. Koopman, K. C. Lamson, J. W. McClure, and L. K. Morris, "Falcon Series Data Report: 1987 LNG Vapor Barrier Verification Field Trials," GRI-89/0138, June 1990.

Chan, S.T., Journal of Hazardous Materials, 30 (1992) 195-224.

Chemical Hazards Research Center Tests and FEM3A Model Development Program

Havens, J. A. and T. O. Spicer, "Development of an Atmospheric Dispersion Model for Heavier-than-Air Gas Mixtures," Final Report to U.S. Coast Guard, Contract DT-CG-23-80-C-20029, May 1985, Volumes I, II, and III.

Havens, J. A. and T. O. Spicer, "Evaluation of 3-D Hydrodynamic Computer Models for Prediction of LNG Vapor Dispersion in the Atmosphere," Series of reports (including final report) to the Gas Research Institute, GRI Contract No. 5083-252-0788, 1983-88.

Havens, J. A. and T. O. Spicer, "Modification and Extension of LNG Vapor Dispersion Simulation Methods for Regulatory Application," series of reports to Gas Research Institute, GRI Contract No. 5086-252-1287, 1988-present.

Havens, J. A. and T. O. Spicer, "LNG Vapor Dispersion Prediction with the DEGADIS Dense Gas Dispersion Model," Gas Research Institute Report No. 89-0242, 1990.

Havens, J. A. and T. O. Spicer, "Feasibility Assessment of a Conjunctive Modeling Approach for LNG Vapor Dispersion Prediction," Gas Research Institute Report, 1991.

Havens, J. A., T. O. Spicer, and H. L. Walker, "Regulatory Application of Wind Tunnel Models and Complex Mathematical Models for Simulating Atmospheric Dispersion of LNG Vapor," Gas Research Institute Report No. 92-0257, August 1994.

Havens, J. A., T.O. Spicer, and H. L. Walker, "Evaluation of Mitigation Methods for Accidental LNG Releases: Volume 1/5--Wind Tunnel Experiments and Mathematical Model Simulations to Study Dispersion of a Vapor Cloud Formed following LNG Spillage into a Diked Area Surrounding a Storage Tank," Topical Report for Gas Research Institute, November 1996.

Havens, J. A., T. O. Spicer, and H. L. Walker, "Evaluation of Mitigation Methods for Accidental LNG Releases: Volume 2/5--Wind Tunnel Experiments and Mathematical Model Simulations to Study Heat Transfer from a Flat Surface to a Cold Nitrogen Cloud in a Simulated Atmospheric Boundary Layer," Topical Report for Gas Research Institute, November 1996.

Spicer, T. O., J. A. Havens, and H. L. Walker, "Evaluation of Mitigation Methods for Accidental LNG Releases: Volume 5/5--Using FEM3A for LNG Accident Consequence Analysis," Topical Report for Gas Research Institute, April 1997.

Havens, J. A., Tom Spicer and Kent Perry, "LNG Technology: Volume 8/4-New Models Predict Consequences of LNG Releases," Topical Report for Gas Technology Institute, Fall 2002.

Havens, J. A., Tom Spicer, Wendy Sheppard, and Kent Perry, LNG Terminal Siting: FEM3A Vapor Dispersion Model Development, Proceedings AICHE Annual Spring Meeting, Houston, Texas, 2004.

106a Kinetic Aspects of Boosting Hydrogen Production by Means of Methanol

Charlotte V Ovesen, John Bogild Hansen, Niels Christian Schiodt, P E Hojlund Nielsen

The demand for hydrogen in refineries is steadily increasing. Even refineries with relatively new hydrogen plants often experience continuous or temporary shortage of hydrogen only few years after start-up. The need for simple methods for boosting hydrogen capacity is therefore considerable. A typical, modern, large-scale hydrogen plant consists of four sections: a purification section with sulphur removal from the feed; a steam reforming unit in which hydrocarbons is processed over a Ni-based catalyst into a mixture of hydrogen, carbon monoxide, carbon dioxide and methane; a water gas shift section where hydrogen is produced according to the water gas shift reaction over a Cu-based catalyst; and finally a purification section.

Methanol is known as an excellent hydrogen carrier, and from steam reforming of one mole of methanol, three mole of hydrogen are obtained: $CH_3OH + H_2O <=> CO_2 + 3H_2$. Steam reforming of methanol is carried out over a Cu-based catalyst. Cu-based catalysts are also used for methanol synthesis and water gas shift process. Thus Cu-based catalysts are capable of performing both the water gas shift reaction and the methanol reforming reaction. This recognition has led Haldor Topsoe A/S to introduce a new process to boost hydrogen production in hydrogen plants by means of methanol [1]. The invention consists of adding methanol to the water gas shift section. By simultaneously running the shift and methanol reforming reactions, the hydrogen production can be increased by more than 25% compared to a conventional hydrogen plant. The new process is called the Methanol-to-Shift^{^{TM}} Technology. The shift reaction is exothermic:

$CO + H_2O <=> CO_2 + H_2$ -DeltaH^{280o}C = 39 kJ/mole

whereas methanol reforming is endothermic:

$CH_3OH + H_2O <=> CO_2 + 3H_2$ -DeltaH^{280o}C= -60 kJ/mole

The methanol reforming reaction gets the necessary heat of reaction primarily from the released reaction heat from the water gas shift reaction. Thus it is an energy-balanced process which may become thermoneutral. A kinetic model is essential for analysing this aspect. In this paper we discuss the kinetics of the shift and methanol reforming reaction on the basis of a global kinetic model for Cu-based catalyst. The model is based on 13 elementary reactions taking place on Cu, and the kinetic parameters are derived from fundamental studies on Cu-catalysts [2]. It can simulate the water gas shift reaction, the methanol synthesis reaction and the methanol reforming reaction. We show how we from this basic model can derive a simple kinetic model applicable for studying the aspects of the Methanol-to-Shift^{TM} Technology. The analysis of pilot data with respect to reaction rate of the water gas shift reaction and methanol reforming will be presented. We discuss different scenarios for operation of the shift converter with and without addition of methanol, and we analyze the likelihood of getting either a too hot or a too cold catalyst bed. The example highlights the importance of considering reaction networks rather than single reactions in the developing and refining of catalytic processes.

References

[1] P. E. H. Nielsen, J. B. Hansen, N. C. Schiodt, patent no. WO 2004087566, October 14, 2004.

P. E. H. Nielsen, J. B. Hansen, N. C. Schiodt, patent no. WO 2004087567, October 14, 2004.

[2] C. V. Ovesen, B. S. Clausen, J. Schiotz, P. Stoltze, H. Topsoe, and J. K. Norskov, J. Catal., 168, 133 (1997).

106b A Novel Thermochemical Conversion Process for Distributed Hydrogen Production from Natural Gas and other Fuels

James L Aderhold, Jr, Rachid B Slimane

Small-scale hydrogen production faces significant technical and economic challenges. GTI has been developing a novel thermochemical conversion concept that promises to provide the basis for developing a cost-effective process for distributed hydrogen production, characterized by high efficiency, reduced CO2 emissions, lower energy requirements, and other advantages over traditional steam methane reforming (SMR), partial oxidation (POX), and autothermal reforming (ATR) systems. The key feature of GTI's process is the superadiabatic reactor, a new type of chemical reactor in which a slowly-propagating, self-sustaining reaction zone provides very favorable conditions for high conversion of natural gas into syngas, i.e., hydrogen (H2) and carbon monoxide (CO). GTI's envisioned process is quite simple, consisting of the superadiabatic POX reactor with commercially-available shift conversion, separation, and purification (e.g., PSA) technologies.

In this paper, based on preliminary bench-scale test results and thermodynamic and kinetic modeling, an assessment will be made of the techno-economic merits of the superadiabatic POX concept as a basis for a cost-effective process for small-scale distributed hydrogen production from natural gas and other fuels for refueling stations or stationary power facilities.

106c Developing Structural Property Relationships of Pt/Ceria Catalysts for Low Temperature Water Gas Shift

Gary Jacobs

Knowledge of the mechanism and the nature of the active site are important keys to development of low temperature water gas shift (WGS) catalysts employed to generate CO-free H2 for fuel cell applications. An important synergism between noble metal promoter and partially reducible oxide components has led to the postulation of at least two mechanisms, one involving surface formates as intermediates, having its roots in the catalysis of formic acid decomposition and another relying on a support-mediated redox process, whereby ceria is reduced by CO, generating CO2, with subsequent reoxidation of the reduced ceria by water, liberating hydrogen in the process.

In this discussion, a link was established between a normal kinetic isotope effect (KIE) observed for low temperature WGS (CO + H2O and CO + D2O) and that of water-assisted formic acid decomposition (HCOOH + H2O and DCOOH + H2O). The KIE for WGS is discussed in terms of the dynamic response of H and D labeled surface formate, as monitored by DRIFTS. The results suggest that the rate limiting step involves C-H bond rupture of formate. Isotopic tracers were employed to monitor the exchange of surface species at steady state reaction conditions. Isotopic transients were followed by DRIFTS for the forward shift, underlining the importance of the autocatalytic role of water in the decomposition of surface formate species.

Based upon these mechanistic insights, structural property relationships were developed. The active site is strongly suggested to be bridging OH groups associated with Ce3+ in the surface shell. Increasing ceria surface area led to important increases in catalyst activity. Pt promoter loading was also explored. Increasing Pt loading facilitated ceria surface shell reduction, accompanied by removal of surface carbonates, such that the active bridging OH groups formed at lower temperature. DRIFTS results suggest that Pt may also assist in improving the formate decomposition rate.

Partial Oxidation of Methane on Ni-MgO Catalysts Supported on Metal Foams

Abolghasem Shamsi, U.S. DOE/National Energy Technology Laboratory, Morgantown, WV 26507-0880[1] and James J. Spivey Dept. Chemical Engineering, Louisiana State University, Baton Rouge, LA 70803

Abstract. The catalytic partial oxidation of methane on Ni-MgO solid solutions supported on Fecralloy® metal foams has been studied at temperatures from 600-800°C, 1 atm, 3000-19,000 mL/g-h, CH_4/O_2 = 2/1. The catalysts were prepared in two ways: (a) oxidized foams were dipped directly into a citrate/ethylene glycol solution containing nitrate precursors and (b) oxidized foams were primed with a bohemite primer and washcoated with Ni-MgO powders, derived from the above solution, in an acidic solution. Results show that the foam-supported catalysts reach near-equilibrium conversions of methane and H_2/CO selectivities at the indicated space velocities. Rates of carbon deposition differ greatly among the catalysts, varying from 0.24 mg C/g cat h for the dipped foams to 7.0 mg C/g cat h for the powder-coated foams. Both of these rates are much lower than for the calcined, unsupported powder (57 mg C/g cat h). The results suggest that exposed Fe and Cr on all of the foam samples interact with the Ni-MgO catalyst to kinetically limit carbon formation.

Introduction. The partial oxidation of methane (POX) has been widely investigated as a way to produce synthesis gas[1,2]. The reaction produces synthesis gas with the 2/1 H_2/CO ratio that is required in methanol and Fischer Tropsch synthesis, among other processes:

$$CH_4 + 1/2\ O_2 \rightarrow CO + 2\ H_2 \qquad \Delta H_{r,700^\circ C} = -22.8\ \text{kJ/mol}\ ;\ \Delta G_{r,700^\circ C} = -214\ \text{kJ/mol}$$

Despite the advantages of this reaction, it is not practiced industrially because of the difficulty in preventing the highly exothermic complete oxidation reaction, which leads to unselective products[1]. Most catalysts that have been investigated for the POX reaction have been based either on noble metals[3,4] or Ni[5,2]. Ni-MgO of particular interest because the formation of a Ni-MgO solid solution over a wide range of compositions leads to a catalyst in which the Ni atoms can be reduced by hydrogen to active metallic N_i^o [2]. Unlike other conventional Ni-based catalysts on oxide supports, reduction in hydrogen forms small, strongly held metallic Ni particles because of the strong interaction of NiO with MgO in the solid solution[6,2].

Metal foams are one of a number of alternatives for high temperature monolithic catalyst supports for reforming reactions. Typically, the metal foams have an open, reticulated structure with a void volume greater than ~85%. Their high mechanical strength and thermal conductivity make these materials important for a number of industrial applications, including catalyst supports[7].

Experimental. Fecralloy® metal foams obtained from Porvair, Inc. (Hendersonville, NC) were oxidized in air at 900°C for10 h to produce an alumina layer on the surface of the alloy to which washcoats and catalysts would adhere[8]. The calcined samples were primed with a boehmite solution prepared by dispersing 10 wt% AlOOH (Alcoa) in a 0.4 wt % aqueous HNO_3 solution and mixing it for 10 min. A solution containing Ni and Mg cations was used to washcoat the foams[9], which were then were dried 90°C 2h, 500°C, 2h and calcined at 800°C for 2 h. Partial oxidation (POX) was carried out in a fixed bed microreactor at a CH_4/O_2 feed ratio = 2/1. The carbon deposited on the catalysts was characterized by TPO to 900°C at 20°C /min.

Results and Discussion. Methane conversions on the catalyzed foams approached equilibrium at 3000 h^{-1}, but not at 19,000 h^{-1}, and are comparable to conversions reported elsewhere[5]. Carbon deposition as measured by TPO was significantly lower on the metal foams than on a Ni-MgO powder catalyst of the same nominal composition. The reasons for the significantly lower carbon deposition rate on the dipped foams compared to the impregnated foam may be related to the exposed surface of the foam. SEM/EDX (not shown) before reaction shows relatively large Ni-MgO particles on the surface of the calcined foam. Large portions of the foam surface, which contains Fe and Cr in addition to alumina, are exposed to the gaseous reactants. Both of these elements have been shown to be active for the POX reaction[10]. The results suggest that the greater exposed foam surface on the dipped samples minimizes deactivation of the Ni-MgO catalyst by interaction with the surface Fe and Cr present on the foams, and that the greater the exposed area of the foam, the less the rate of carbon deposition.

---

[1] York, A.P.E., T. Xiao, M.L.H. Green, Topics Catal., 22(3-4), 2003, 345-358.
[2] Hu, Y.H. and E. Ruckenstein, Catal. Rev. Sci. Eng., 44(3), 2000, 423-453.
[3] Mallens, E.P.J., J.H.B.J. Hoebnik and G.B. Marin, Catal. Lett. 33, (1995) 291.
[4] Pena, M.A., J.P. Gomez and J.L.G. Fierro, Appl. Catal. A: General, 144 (1996) 7.
[5] Barbero, J., M.A.Pena, J.M.Campos-Martin, J.L.G. Fierro, P.L. Aras, Catal. Lett., 87(4-5), 2003, 211.
[6] Ruckenstein, E. and Y.H. Hu, Ind. Eng. Chem. Res., 37, 1998, 1744-1747.
[7] Banhart, J., Prog. Material Sci., 46(6), 559, 2001.
[8] Valentini, M. G. Groppi, C. Cristiani, M. Levi, E. Tronconi, P. Forzatti, Catal. Today, 69, 2001, 307-314.
[9] Lessing, P.A., Am. Ceram. Soc. Bull., 68 (1989), 1002.
[10] Gonzalez, M.G., N.N. Nichio, B. Moraweck, G. Martin, Materials Lett., 45(1), 2000, 15-18

Non-oxidative Conversion of Methane to Hydrogen in a DC Plasma Reactor

Christopher L. Gordon, Lance L. Lobban, and Richard G. Mallinson

School of Chemical Engineering and Materials Science
Institute for Gas Utilization Technologies
University of Oklahoma
Norman, OK 73019

ABSTRACT

Natural gas is being looked at as an energy source, as well as a basis for the production of many chemicals. New technologies are needed for smaller scale, remote and niche applications to utilize the natural gas resources that will not be suitable for large-scale production. Additionally, distributed production may be desirable for developing a hydrogen transportation infrastructure. Low temperature plasma reactors have been shown to be an effective method for the conversion of methane, and under some conditions have low power consumption. Cold plasmas have energetic electrons that exist at high temperatures while the bulk gas temperature remains relatively low (~65°C). This allows for reactions that would not normally occur at these low temperatures, and removes the necessity to pre-heat feed streams. For the conditions studied, the majority of the energy that is input into the system is in low-level electron energies that are directed toward the excitation of the methane. This excitation of the methane is an effective method of overcoming the activation barrier required for the conversion of methane. The dc plasma reactor is capable of producing hydrogen and acetylene single pass yields as high as 50%. To date, the lowest energy consumption of the dc plasma system is 7.0 eV/molecule hydrogen produced and 7.5 eV/molecule of methane converted using a non-optimized power system. The target goal, according to the Department of Energy, is approximately 2.0 eV/hydrogen produced. Also, typical industrial processes use about 5 eV/molecule of methane converted, benefiting from substantial energy recovery that may not be feasible for small scale systems. However, the efficiencies of the dc plasma system are based solely on the power input to the system. The cost of the process could be reduced by selling the valuable hydrocarbon products produced, since little carbon and hydrogen is wasted in combustion.

INTRODUCTION

It is thought that methane will play an important role in the energy and chemical industries of the future. Methane is the "greenest" carbon based source of energy. The conversion of methane via steam reforming can result in four molecules of hydrogen for every molecule of carbon dioxide produced (including the water gas shift reaction). However, typical methane conversion processes require an intensive energy input to drive the endothermic reaction.

Plasma reactors have shown to be an effective method for the conversion of methane. Acetylene and carbon black have been produced via an arc discharge for many years. The primary drawbacks to these reactors is the high energy input and "low–value" heat that leaves the process, as well as the excessive carbon formation. It is desired to have a process with less thermal waste, as well as production of products with a greater end-value.

Cold plasmas have shown to be capable of converting methane to a variety of different products. Cold plasmas are characterized by energetic electrons that exist at high temperatures while the bulk gas temperature remains low (less than 100°C) Thanyachotpaiboon [1] et al. showed that an ac dielectric barrier discharge is capable of converting pure methane directly to higher hydrocarbons, such as ethane, ethylene, propane, and butane. Larkin [2] et al. have shown that an ac dielectric barrier discharge is capable of the partial oxidation of methane to organic oxygenates. However, the low energy efficiency of the dielectric barrier discharge reactors used in those cases have hindered their progress.

The ac tubular reactor and dc tubular reactor have significantly better energy efficiency, as well as high selectivities toward desired products. The dc plasma system has been shown to be an effective process for the conversion of methane to acetylene. The dc plasma system operates around 65°C which reduces the amount of heat that is wasted as unrecoverable energy. In addition, the primary product is acetylene and not other undesired products, such as carbon dioxide and water. The initial studies of the process used helium and oxygen to help stabilize the process. Helium is impractical as a feedstock in industrial applications. In addition, it is desired to operate without oxygen because oxygen is expensive to generate and its conversion results in the production of less desired products: carbon dioxide, carbon monoxide, and water. Unfortunately, with oxygen removed, the tendency to form carbon increases. Carbon is an undesired by-product and carbon deposits have been found to destabilize the dc plasma system.

It has been known that the creation and stabilization of the streamer discharge is a key component of the dc plasma system. Liu [3] et al. found that the presence of a metal oxide catalyst enhanced the conversion and selectivity within the plasma system. It was hypothesized that the OH groups within the metal oxide catalyst played an important part in the mechanism of conversion and selectivity. In addition, the effect of various zeolites was studied within the dc plasma system. It was thought that the zeolites, contain numerous hydroxyl sites, would lead to the enhancement of the plasma. However, it was concluded that only certain zeolites, basic catalysts, were effective in sustaining the discharge under those conditions.

Understanding the role of the material as it helps create and stabilize the discharge is a crucial piece in the development of the dc plasma discharge. In general, the desired results occurred while there was a presence of both a material and a streamer discharge. Likewise, the arc discharge and the absence of a material resulted in low conversion and low product selectivities. It is therefore desirable to determine what role, if any, the material and/or the type of discharge had on the dc plasma system.

EXPERIMENTAL PROCEDURE

The experimental apparatus is similar to the system that has been described previously[4]. The feed gases consisted of a combination of methane, oxygen, hydrogen, and helium. Helium was only used in initial experiments and for characterization studies of the catalyst. The feed gas flowrates were controlled by Porter mass flow controllers, model 201. The feed gases flowed axially down the reactor tube. The reactor is a quartz tube with a 9.0 mm O.D. and an I.D. of either 4.5 mm or 7.0 mm. The configuration of the reactor can be seen in Figure 2.2.

The reactor's electrode configuration consists of a point to plane geometry, meaning that the top electrode is a wire point electrode and the bottom electrode is a flat perforated plate that also serves as a support for the catalyst. The top electrode is positioned concentrically within the reactor, and the gap between its tip and the plate is 8.0 mm and both are made of steel.

As mentioned before, this system operates at low temperatures. A furnace around the reactor is used to heat the system to the desired temperature. The temperature is measured by an Omega K-type thermocouple that is attached to the outside of the reactor near the catalyst bed. The temperature measured on the outside has been calibrated against the internal temperature of the reactor, and has been discussed elsewhere[5]. However, when the desired operating temperature is below 373 K it is necessary to use cooling air to control the temperature since the plasma itself does heat the gas to some extent. The system pressure is varied between 0.1 and 0.3 kPa, depending on the experiment. The pressure is controlled using a GO back pressure regulator.

The product gases are passed through a dry ice/acetone bath that allows for any condensable organic liquids to be separated from the product gases. It should be noted that the dc system produced no measurable liquids, including water. The effluent gases can be analyzed on-line by either a gas chromatograph or a mass spectrometer. The gas chromatograph is a CARLE series 400 AGC (EG&G) gas chromatograph equipped with a hydrogen transfer system to separate the hydrogen; a HayeSep column to separate carbon dioxide, ethane, ethylene, and acetylene; and a molecular sieve column to separate oxygen, nitrogen, methane, and carbon monoxide. A thermal conductivity detector (TCD) was used to quantify the separated products. Also, a MKS mass spectrometer is used for on-line analysis of the products and for temperature programmed oxidation of carbon deposited on the catalyst.

RESULTS

The dc plasma system uses small concentrations of oxygen (2/1H_2/CH_4 with 2% O_2), but it is highly unlikely that the 2% oxygen could contribute to yields of acetylene upwards of

50%. Therefore, it is likely that another process is responsible for the increased production of acetylene as the E/P is reduced for both the ac cylindrical reactor and the dc plasma reactor. The oxygen excitation increases as the E/P is decreased; however, the amount of energy directed toward the excitation of methane increases over the same range. It is believed that the mechanism of acetylene production and high conversion is through the excitation of the methane, because varying the oxygen concentration on the feed (from 0.5 to 5% oxygen) does not result in an increased methane conversion.

The total energy may then be summed and normalized to provide an energy distribution of the amount of energy directed toward different processes as a function of reduced electric field. The reduced electric field is dependent on the breakdown voltage, gas gap, and pressure of the system; and the reduced electric field has units of volts/cm/torr, given in Equation 3.3.

$$E/P = \frac{breakdown\ voltage}{electrode\ gap * system\ pressure} \qquad \text{Equation 3.3}$$

The reduced electric field plays an important role in the average electron energy, as well as the amount of energy that is deposited into the various collision processes. For the range of reduced electric fields encountered in the lab (E/P = 5 to 70), over 95% of the energy is in the form of the inelastic group process. Thus, the processes of elastic collisions (no transfer of energy), ionization (increases with E/P, though still small in highest E/P), and attachment may be ignored. Within the inelastic group processes the excitation and dissociation are the two major processes that consume the energy.

Figures 1, 2, and 3 show the effect of the E/P on the amount of energy that goes into the inelastic methane processes, as well as the acetylene and ethane selectivity data from the dc plasma system for a 1/1/8 $CH_4/H_2/He$ and 2% O_2 feed, 1/1 H_2/CH_4 with 2% O_2 feed, and a 2/1 H_2/CH_4 with 2% O_2 feed, respectively. The E/P was adjusted in two different ways; by varying the gap height and by changing the operating pressure of the system. The three feed conditions vary significantly yet all three follow the same trend. As the E/P is reduced, the acetylene selectivity increases as the fraction of energy directed to methane excitation increases. Conversely, as the E/P is reduced, the selectivity toward ethane decreases since the energy directed toward methane dissociation also decreases. The lowest E/P yielded a product distribution that is almost entirely acetylene, while the highest E/P resulted in an ethane dominated product. This data correlates very well with the 'ethane only' production seen in the ac planar system and the C_2 trends seen in the ac cylindrical system.

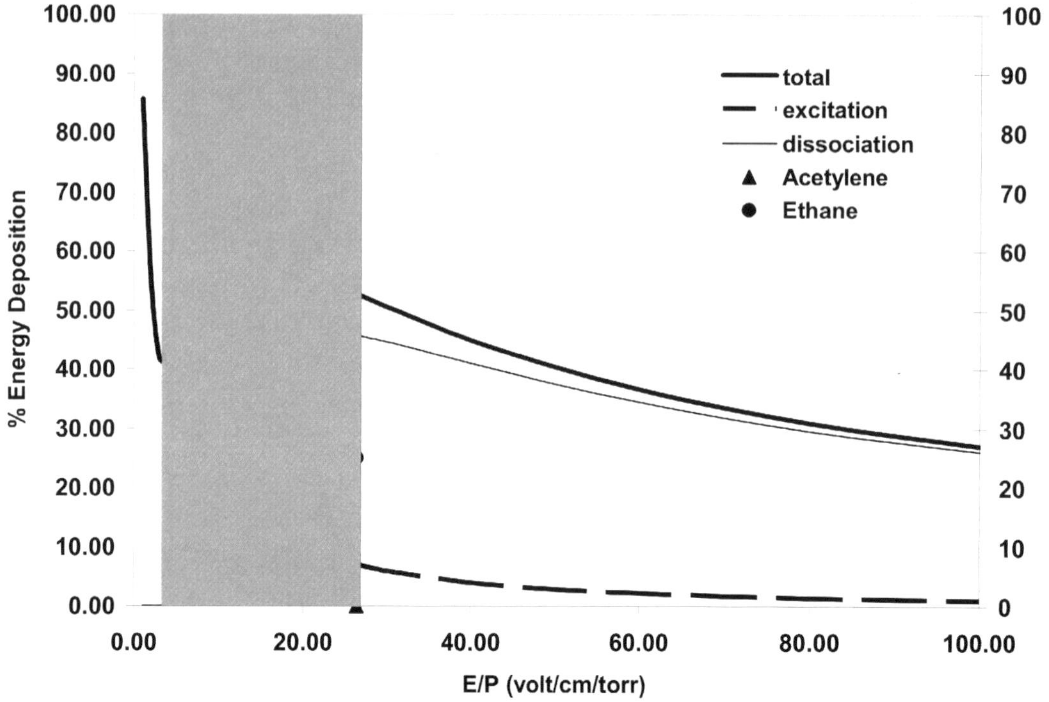

Figure 1. Effect of the reduced electric field (E/P) on the percentage of energy deposited into the inelastic group process (excitation and dissociation) of a 1/1/8 $CH_4/H_2/He$ with 2% O_2 system. Overlaid is the C_2 selectivity data of dc plasma system.

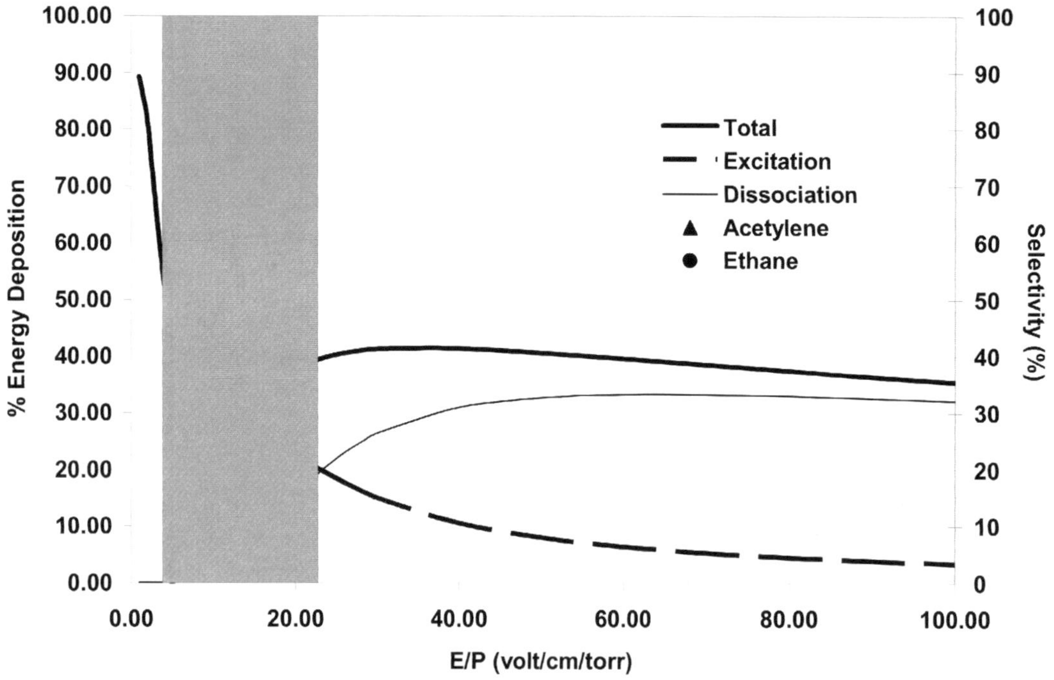

Figure 2. Effect of the reduced electric field (E/P) on the percentage of energy deposited into the inelastic group process (excitation and dissociation) of a 1/1 CH_4/H_2 with 2% O_2 system. Overlaid is the C_2 selectivity data of dc plasma system.

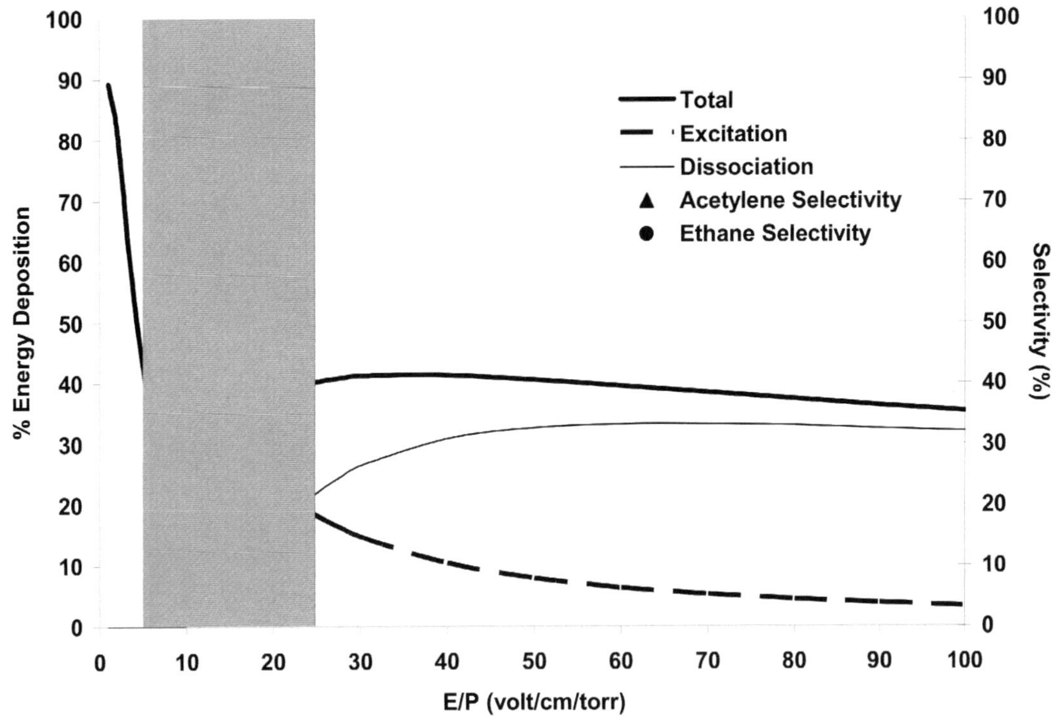

Figure 3. Effect of the reduced electric field (E/P) on the percentage of energy deposited into the inelastic group process (excitation and dissociation) of a 2/1 CH_4/H_2 with 2% O_2 system. Overlaid is the C_2 selectivity data of dc plasma system.

A mechanism for the production of ethane through the dissociation of methane can be seen in Equation 1 and 2. However, the production of acetylene through the excitation of methane is more complicated. The excitation of methane corresponds to the two levels of vibration and transitional excitation which require energies of 0.16 and 0.36 eV, respectively. It is believed that the excitation of the methane molecules allows them to overcome the activation energy to proceed to subsequent reactions. When an electron collides with a gas molecule it excites the molecule to a higher energy level (Equation 3), and the electron loses the corresponding amount of energy[6]. The excited molecule can then initiate some reactions due to its higher internal energy. So, a reaction that may typically only occur at very high temperatures, may be initiated by an excitation collision.

$$CH_4 + e \rightarrow CH_3 + H \qquad \text{Equation 1}$$

$$2CH_3 \rightarrow C_2H_6 \qquad \text{Equation 2}$$

$$CH_4 + e \rightarrow CH_4^* + e \qquad \text{Equation 3}$$

The energy transferred to the methane molecule by the inelastic collision with the electron leads to the excitation of the methane molecule. This excitation of the methane molecule raises its energy to a level that may allow it to undergo other reaction pathways. Then the reaction may follow a pathway somewhat similar to the pyrolytic conversion of methane to acetylene. The energy required to drive the reactions would proceed through the excitation of the molecules by the low-level energy electrons.

The methane conversion and selectivity toward acetylene is higher at lower values of the reduced electric field where the majority of the energy is input into the system via numerous low energy electrons. On the other hand, at higher reduced fields the majority of the energy goes into electrons that have much larger electron energies, while the methane conversion is much lower and ethane is the predominate product. It is believed that these numerous excitations lead to the higher conversion of methane, as well as the subsequent reactions that lead to the formation of acetylene. And the higher energy electrons associated with the higher reduced field are less efficient in the conversion of methane. For each molecule there is an optimum electron energy for the ionization of the species. If the electron energy is too high, there is less probability that the electron will result in the ionization of the molecule. It is believed that this same theory applies to the excitation of a molecule. That is, the electrons have sufficient energy to excite the methane, but are not as efficient as the lower energy level electrons.

ACKNOWLEDGEMENTS

The authors would like to express their appreciation to ChevronTexaco Inc., the United States Department of Energy (Contract DE-FG21-94MC31170), and the National Science Foundation for sponsoring a Graduate Research Traineeship in Environmentally Friendly Natural Gas Technologies.

REFERENCES

1. Thanyachotpaiboon, K., T.A. Caldwell, S. Chavadej, L.L. L. L. Lobban, and R.G. Mallinson, *Non-oxidative conversion of methane to higher hydrocarbons in low temperature plasmas.* AIChE Journal, 1998. **44**: p. 2252.

2. Larkin, D.W., L.L. Lobban, and R.G. Mallinson, *Product selectivity control and organic oxygenate pathways from partial oxidation of methane in a silent electric discharge reactor.* Industrial Engineering & Chemical Research, 2001(40): p. 5496.

3. Liu, C.-J., A. Marafee, R.G. Mallinson, and L.L. Lobban, *Methane Conversion to higher hydrocarbons in a corona discharge over metal oxide catalysts with OH groups.* Applied Catalysis A: General, 1997. **164**: p. 21-33.

4. Gordon, C.L., L.L. Lobban, and R.G. Mallinson, *Ethylene production using a Pd and Ag-Pd Y-zeolite catalyst in a dc plasma reactor.* Catalysts Today, 2002.

5. Liu, C.-J., A. Marafee, B.J. Hill, G. Xu, R.G. Mallinson, and L.L. Lobban, *Oxidative coupling of methane with ac and dc corona discharge.* Industrial & Engineering Chemistry Research, 1996. **35**(10): p. 3295-3301.

6. Eliasson, B. and U. Kogelschatz, *Nonequilibrium volume plasma chemical processing.* IEEE Transactions on Plasma Science, 1999. **19**(6): p. 1063-1077.

Biological Processing of Carbonaceous Sources Using Extremophiles for CO_2-Neutral Ultra-Pure Hydrogen Production

Mojgan Anjom[1], Bin Dong[2], Suellen A. Van Ooteghem[3], Daniel van der Lelie[4], Safiyh Taghavi[4], and Devinder Mahajan[1,2]*

[1]*Materials Science and Engineering Department, Stony Brook University, Stony Brook, New York 11794*
[2]*Energy Sciences & Technology Department, Brookhaven National Laboratory, Upton, NY 11973*
[3]*BioHydroGenesys Inc., Covington, GA 30014*
[4]*Biology Department, Brookhaven National Laboratory, Upton, NY 11973*

Keywords: *Hydrogen production, biological hydrogen, catalytic water-gas-shift, thermotoga neapolitana, low temperature catalysis*

[*]Corresponding author: dmahajan@bnl.gov

Abstract

Presently, natural gas derived-synthesis gas is the main commercial source of hydrogen (H_2) for PEM based fuel cells that effectively utilize platinum catalysts under an ambient pressure and at 80°C. Since synthesis gas invariably contains copious amounts of carbon monoxide (CO), an expensive step involving staged water-gas-shift (WGS) reactions are used to remove CO to acceptable levels (< 50 ppm) and thus prolong the Pt catalyst life. We recently reported a liquid phase, low temperature (< 150°C) approach to WGS reaction in which a single-site catalyst was used to efficiently process impure synthesis gas containing several %CO (Mahajan, U.S. patent 6,596,423 (2003)). An alternate biological approach is being developed in our laboratory that involves the use of extremophilic microorganisms such as *Thermotoga neapolitana* that efficiently mediates H_2 production from glucose under ambient microaerobic atmosphere and 70°C (Van Ooteghem et al. Biotech. Lett. 26: 1223 (2004)). Itaconic acid was successfully used as a buffer to overcome pH-induced limitations of cell growth and to facilitate enhanced production of CO-free H_2, an inherent benefit of using the biological approach.

Introduction

One of the most recently identified orders of H_2-producing bacteria is Thermotogales that invariably produce H_2 as a product of their catabolism (Belkin et al. 1986, childers et al. 1992, Van Niel et al. 2002, Van Ooteghem et al. 2002). However, contrary to what has been described, one particular member of this order, *T. neapolitana*, efficiently produces H_2 gas under microaerobic conditions (Van Ooteghem et al. 2002). This occurs despite Tga. Neapolitana being described as a heterotrophic, extremely thermophilic, obligate anaerobic (Belkin et al. 1986). We defined optimal growth and hydrogen production conditions for Tga. neapolitana and systematically identified non-catabolized substrates. The substrate utilization data were used to select non-metabolized carbohydrates, such as Itaconic Acid, as physiological buffers to overcome pH-induced limitations of cell growth and enhance H_2 production.

Materials and methods

Cultures were grown in serum bottles containing 50 ml growth medium and headspace approx. 110 ml. The strain was grown in ATCC1977 medium which contained per liter: 1 g NH_4Cl, 0.3 g K_4HPO_4, 0.3 g KH_2PO4, 0.2 g $MgCl_2.6H2O$, 0.1 g $CaCl_2.2H_2O$, 10 g NaCl, 0.1 g KCl, 1 g Cysteine hydrochloride, 2 g yeast extract, 2 g trypticase, 5 g glucose, pH 8.5. The medium was supplemented with 10 ml filter sterilized trace element and vitamin solution. Aliquots of modified ATCC 1977 medium (50 ml) were pipetted into serum bottles, placed in an oil bath at 98°C, heated for 10 min, and then sparged for 5 min with N_2. Sparged bottles were immediately capped with a rubber stopper, sealed, and sterilized by autoclaving. To provide a 2% (v/v) inoculum, 1 ml culture was withdrawn from a freshly grown culture and injected directly into the recipient bottle through the septum.

Cell growth rates and population densities were determined by direct counting using a Leitz phase contrast microscope (40x magnification). Direct counts were correlated with turbidity measured at 600nm. The cultures, incubated at elevated temperatures, were allowed to cool to room temperature prior to gas sampling and analysis. Gas concentrations were measured as volume% (v/v) of total gas in the serum bottle headspace with gastight syringes and assayed by gas chromatography using thermal conductivity detectors.

Results and discussion

Hydrogen production. H_2 and CO_2 production with *T. neapolitana* were determined under microaerobic conditions. T. neapolitana, showed 20% (v/v) or higher H_2 levels. These data confirm earlier observations (Van Ooteghem et al. 2002) that *T. neapolitana* and other members of the order Thermotogales, believed to be strict anaerobes, were able to produce significant amounts of H_2 gas when the cultures were inoculated under microaerobic conditions (6-8% (v/v) O_2, see Figure1). Thus, it appears that none of the organisms tested are obligate anaerobes as claimed in earlier studies (Childers et al. 1992, Huber et al. 1986). Baughn and Malamy (2004) recently made a similar observation that strict anaerobe *Bacteroides fragilis* grows in and even benefits from nanomolar concentrations of O_2. With species belonging to the order Thermatogales, we have observed that the O_2 tolerance range could be extended to the micromolar level. *T. neapolitana* was found to repetitively yield a H_2/CO_2 ratio of approximately 2:1 suggesting that this species uses a different metabolic pathway for H_2 production compared to other strains. It was observed the O_2 concentration at the beginning of the experiment rapidly decreased over time as more CO_2 was produced. Interestingly, tested culture didn't produce detectable amounts of CO or methane (CH_4) based on the sensitivity of our GC measurements. The high-purity H_2 produced (CO < 50 ppm) contrasts commercial H_2 production from fossil fuels using chemical catalysis, in which the occurrence of CO contamination adversely affects the platinum catalyst in the proton exchange membrane (PEM) fuel cell (Larminie & Dicks 2003). H_2 purification technologies to remove CO to < 50 ppm including one based on a single-site catalyst are the subject of a recent patent from our group (Mahajan 2003). Our present result is noteworthy in that it signifies the applicability of biologically produced CO-free H_2 as a direct feed to PEM fuel cell.

Cell Growth and Density. In order to link physiological data such as H_2 production to cell numbers, we determined the relation between cell counting data and the optical density of a T. neapolitana culture as a function of time. The results are presented in Figure 2. In agreement with the previously reported data (Childers et al. 1992), we observed that *T. neapolitana* could vary in morphology and, depending on the growth phase, two different stages were noted. The

Figure 1: H_2 and CO_2 production by T. neapolitana under microaerobic conditions

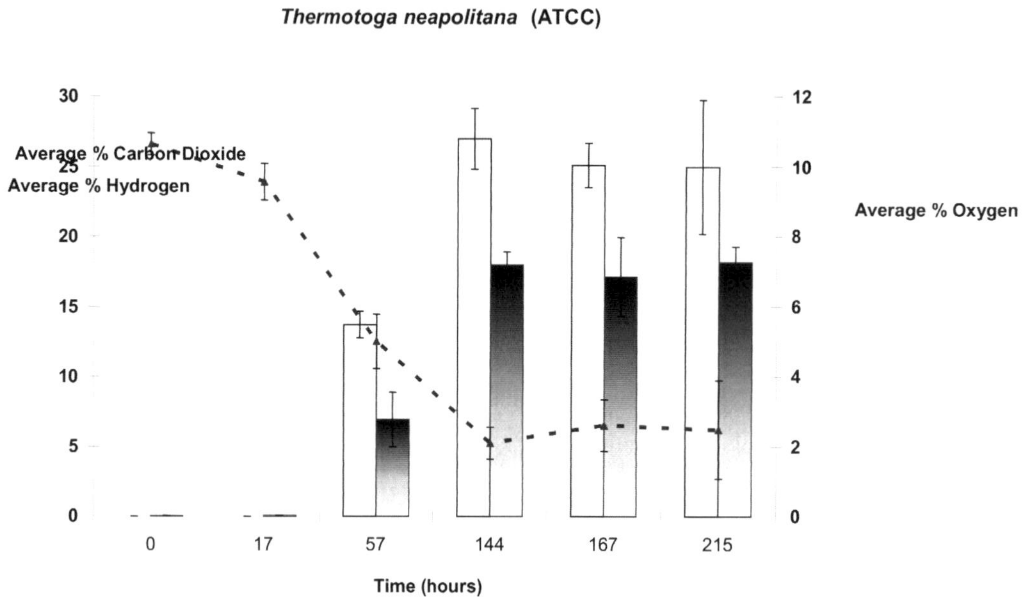

Figure 2: Cell counts and optical density for T. neapolitana

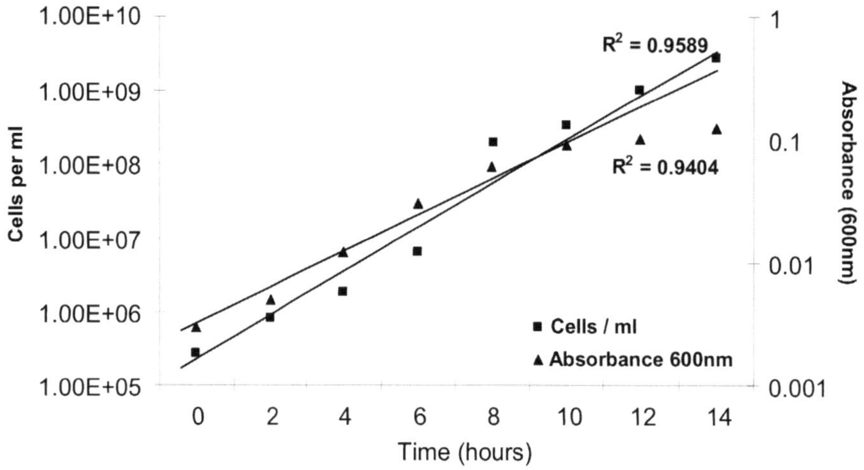

bacteria appeared as small, single rods in the early hours of the experiment but after approximately 6 hours, the culture entered the log growth phase with accompanying morphological changes wherein the cells became larger and were sometimes present in pairs. The culture achieved a maximum cell density of approximately 3.0×10^9 cells/ml after 14 hours of incubation at 70ºC. Thereafter, the absorbance continued to increase with no increase in the corresponding cell numbers

Improved growth and H_2 production. The amount of growth of *T. neapolitana* and H_2 production in a small batch reactor is often limited by a rapid decrease of pH. Based on the Biolog results, it was possible to determine carbon sources that could serve as simple, non-toxic physiological buffers to overcome this pH limitation. Figure 3 compares the H_2 production results with glucose as carbon source in the absence and presence of itaconic acid as a buffering substrate. Residual growth in the absence of glucose was also determined. The pH of the medium that only contained glucose dropped within 20 hours from 7.4 to 4.5. However, in the presence of itaconic acid, the pH decreased more slowly, from 7.4 to 5.9, after 20 hours with H_2 production reaching a maximum plateau value of 27% (v/v). In addition, the agitation of cultures at 75 rpm greatly improved H_2 production: 20 hours after the beginning of growth, an agitated *T. neapolitana* culture had approximately 23.5% H_2 (v/v) gas in its headspace. Thus, pH control and moderate agitation are critical factors for stimulated growth to enhance H_2 production by *T. neapolitana* under microaerobic conditions.

Figure 3: Improved H_2 production by T. neapolitana in the presence of itaconic acid as buffering substance.

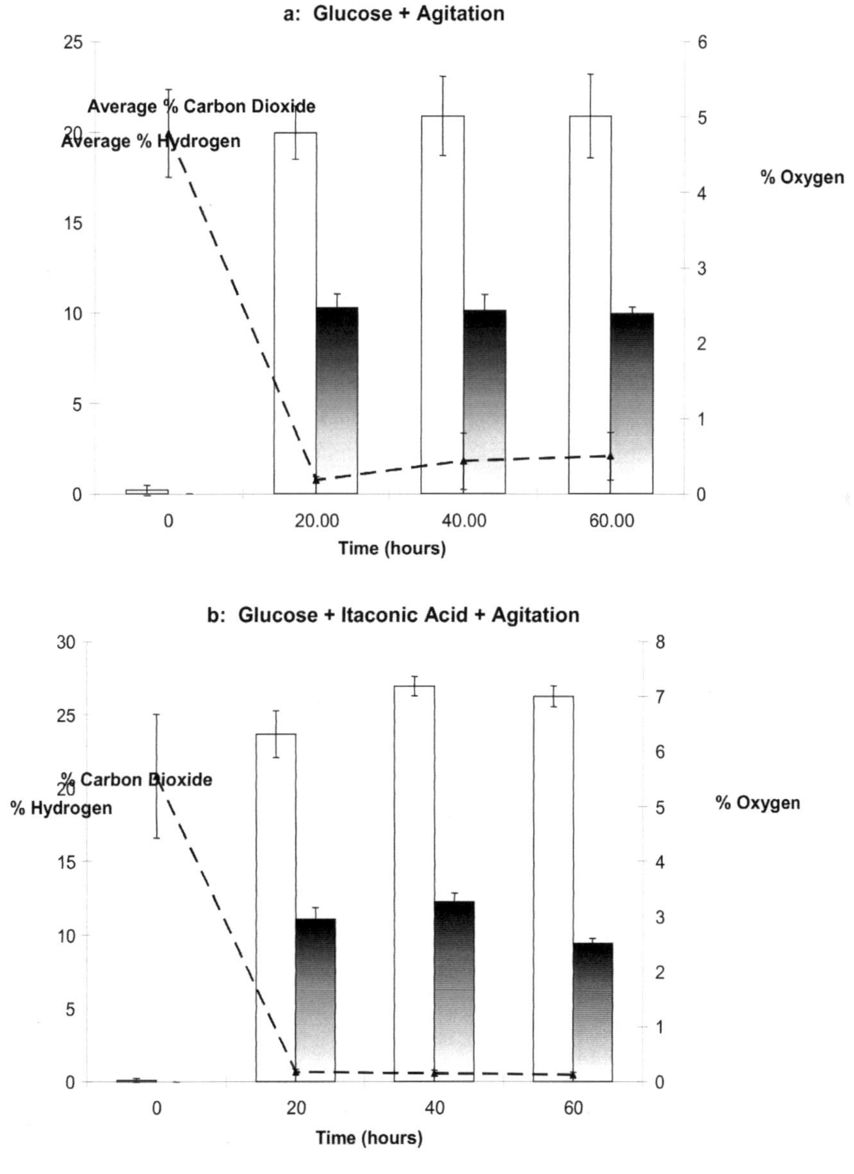

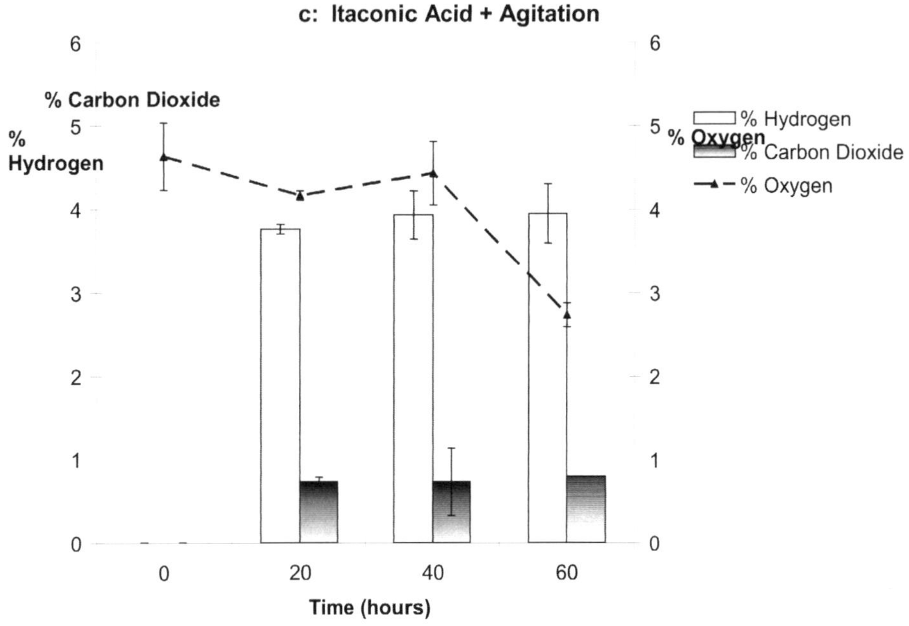

Figure 4: Effect of glucose on hydrogen production

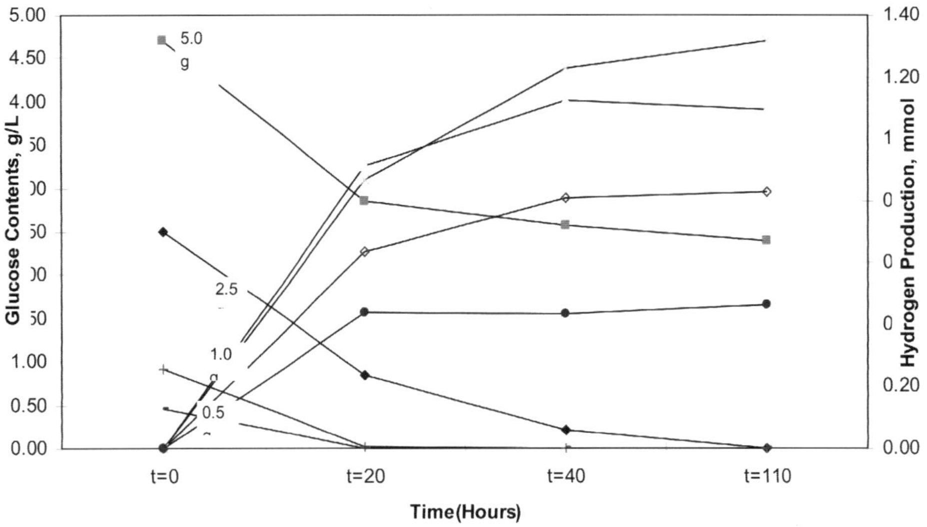

Subsequently, the ATCC1977 medium was modified: 2 g NH4Cl, 0.3 g KH2PO4, 0.3 g H2HPO4, 0.2 g MgCl2.6H2O, 0.1 g CaCl2.2H20, 10 g NaCl, 0.1 g KCl, 0.5 g cysteine hydrochloride, 2 g yeast extract, 2 g trypticase, 5 g glucose, PH 8. It was observed that 5g/l glucose was remains excess in the solution and does not have any effect in hydrogen production. (Figure 4).

Acknowledgements

We thank Dr. Karen Nelson and her team at TIGR for 16S rRNA gene sequencing and analysis. The initial phase of the work was funded through the U.S. DOE (S.A.v.o). D.v.d.L. is presently being supported by the BNL Laboratory Directed Research and Development funds. D.M. thanks Stony Brook University for a start-up grant.

References

Adams MW (1990) The metabolism of hydrogen by extremely thermophilic sulfur-dependent bacteria. FEMS Microbiol. Rev. 75: 219-238.

Balch WE, Wolfe RS (1976) New approach to the cultivation of methanogenic bacteria: 2-mercaptoethanesulfonic acid (HS-CoM)-Dependent growth of Methanobacterium ruminatium in a pressurized atmosphere. Appl. Environ. Micribiol. 32: 781-791.

Baughn AD, Malamy MH (2004) The strict anaerobe Bacteroides fragilis grows in and benefits from nanomolar concentrations of oxygen. Nature 427: 441-444.

Belkin S, Wirsen CO, Jannasch, HW (1986). A new sulfur-reducing, extremely thermophilic bacterium from a submarine thermal vent. *Appl. Environ. Microbiol.* **51**: 1180-1184.

Childers SE, Vargas M, Noll, KM (1992) Improved methods for the cultivation of the extremely thermophilic bacterium Thermotoga neapolitana. Appl. Environ. Microbiol. 58: 3949-3953.

Dobranik JK, Zak JC (1999) A microtiter plate procedure for evaluating fungal functional diversity. Mycologia 91: 756-765.

Huber R, Langworthy TA, Konig H, Thomm M, Woese CR, Sleytr UB, Stetter KO (1986) Thermotoga maritima sp. nov. represents a new genus of unique extremely thermophilic eubacteria growing up to 90^{0}C. Arch. Microbiol. 144: 324-333.

Larminie J, Dicks A (2003) Fuel Cell Systems Explained. John Wiley & Sons, 2nd ed., New York, New York, USA.

Mahajan D (2003) A Method for Low Temperature Catalytic Production of Hydrogen. U.S. patent # 6,596,423.

Preston-Mafham J, Boddy L, Randerson PF (2002) Analysis of microbial community functional diversity using sole-carbon-sources utilisation profiles—a critique. FEMS Microbiol. Ecol. 42: 1-14.

Schroder C, Selig M, Schonheit P (1994) Glucose fermentation to acetate, CO_2, and H_2 in the anaerobic hyperthermophilic eubacterium Thermotoga maritima: involvement of the Embden-Meyerhof pathway. Arch. Microbiol. 161: 460-470.

Van Niel EWJ, Budde MAW, De Haas GG, Van der Wal FJ, Claassen PAM, Stams AJM (2002) Distinctive properties of high hydrogen producing extreme thermophiles, Caldicellulosiruptor saccharolyticus and Thermotoga elfii. Int. J. Hydrogen Energy 27: 1391-1398.

Van Ooteghem SA, Beer SK, Yue PC (2002) Hydrogen production by the thermophilic bacterium Thermotoga neapolitana. Appl. Biochem. Biotechnol. 98-100: 177-189.

The Importance of Antisurge Contriol Valve Performance

John Wilson
Business Development Manager
Emerson Process Management – Fisher Valve Division

AIChE Spring Meeting, April 2005
5th Topical Conference on Natural Gas Utilization
Atlanta April 10-14, 2005

Whether in petrochemical facilities, gas transmission pipelines or liquefied natural gas (LNG) facilities, the compressor systems are some of the most critical pieces of the equipment in the entire process. For example, in an LNG facility, the refigerant compressors can cost on the order of $50M each. This is a sizeable investment that requires the latest technology in order to gain the optimum efficiency from the compressor and turbine driver while also protecting the compressor from failure during an upset condition.

Whether looking at an axial or a centrifugal compressor, one of the biggest areas of concern is the potential for surge conditions. To fully understand the effects that this condition can have on a system, it is best to review the surge condition itself.

Imagine that a compressor is turning at a speed to maintain a desired downstream pressure. If there is a sudden decrease in downstream demand, the downstream pressure will spike and the compressor can not maintain the flow due to the increased head across the compressor. If the pressure spike is large enough, flow can then move back through the compressor. As the head across the compressor bleeds down, flow will then go back through the compressor in the forward direction. This event can occur in as little as half of a second.

Surging of the compressor can lead to catastrophic failure after only a short amount of time. Outside of the worst consequence, other effects include upsetting of operations, altering internal compressor clearances and overstressing of the seals. The costs alone of replacing the compressor seals is in the order of $20K to $50K.

In order to prevent the system from experiencing a surge event, the speed of the compressor can be altered to an operating point away from the surge line, but this is a very inefficient way of controlling the event. Another way to prevent a surge event from occurring is to install an antisurge system. The antisurge system consists of a surge controller, an antisurge valve and a variety of additional accessories.

Figure 1 shows a typical antisurge control schematic. The surge controller takes into account the temperature (TT), flow (FT) and pressure (PT) upstream and downstream of the compressor, and the compressor speed (ST). Other variables may be included depending upon the constituency of the gas. The surge controller will operate the valve as close to the surge line as possible to maximize compressor and gas turbine efficiency without jeopardizing the performance of the compressor.

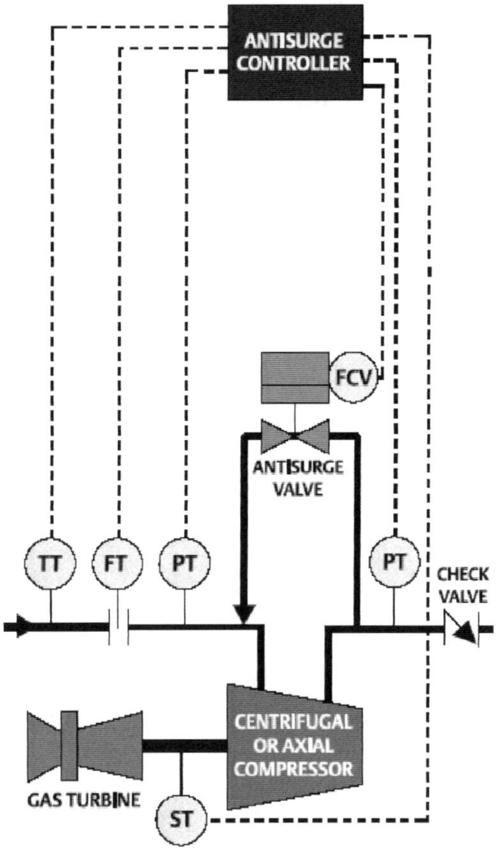

Figure 1: Typical Compressor Antisurge System

In the event of a potential surge condition, the antisurge controller will send a signal to the antisurge valve to open to a required set point to pass a given amount of flow back to the inlet of the compressor. Depending upon the process upset, this may require a small amount of valve opening or full valve opening. Whether for small or large steps, the valve must move to set point quickly (i.e. less than two seconds). If the valve does not respond quickly, it can expose the compressor to additional surges.

Not only must the antisurge valve open quickly, it must possess enough capacity to pass the required amount of flow to protect against a surge event. Because the high differential pressure between the inlet and outlet of the compressor (1000 psi or greater), there is the potential for noise generation as flow is bypassed around the compressor. To eliminate the potential for noise and subsequent vibration, the valve must possess noise attenuating trim.

These factors are critical to valve operation, but specifying a valve only by these two factors can lead to inefficient operation while potentially affecting reliability of the unit. This was not an issue with early antisurge systems as the sole function of the valve was to bypass flow to prevent a surge condition. However, not only is today's antisurge valve used to prevent surge conditions, it is also used to adjust compressor suction and discharge pressure to improve efficiency. The performance of the antisurge controller and the efficiency of the compressor is heavily dependent upon the performance of the antisurge control valve. If the performance of the control valve is not stable, the antisurge controller is detuned, which will reduce efficiency.

Understanding the dynamic performance requirements is one factor that is often overlooked is the dynamic performance of the valve when not experiencing the need for full stroke capability. In the last five years, dramatic improvements have been made to the design and selection of antisurge valves. The following text will highlight the new capabilities and the effect that these improvements will have on valve selection and performance.

Advancements in Noise Control

Many advancements have been made in noise control in the past few years. Traditionally, there have been only a handful of noise control trims on the market. These noise abatement trims provided up to 30 decibels of noise attenuation and relied upon two common methods; frequency shifting and velocity control.

The concept of frequency shifting has been used for over 35 years. The principle behind frequency shifting relies upon shifting the peak frequency of the fluid exiting the valve trim. This increases noise transmission loss from the piping system and lowers fatigue stress in the pipe. International Electrotechnical Commission (IEC) standard 60534-8-3 allows one to calculate the peak frequency exiting a given trim set. Equation 1-1 shows the calculation method.

$$f_p = 0.2\, u_{vc} / D_j \qquad [1\text{-}1]$$

Where f_p is the peak frequency and u_{vc} is the velocity at the vena contracta and D_j is the jet diameter exiting the trim. As one can see by the relationship the peak frequency exiting the trim is a direct function of the diameter of the trim outlet passages. Therefore, the smaller the hole, the greater the frequency shift and the more noise attenuation available.

However, the standard goes on to state that the distance between the holes shall be enough to prevent jet interaction as the fluid exits the holes. If the holes are spaced too close together, the combination of the outlet jets will generate turbulence and noise. This is a critical function that is generally overlooked in noise attenuation.

The other method of noise attenuation commonly used is velocity control. By lowering the trim exit velocity, the acoustical efficiency is lowered leading to lower sound pressure levels.

The most common manner of controlling velocity is by using a series of fixed restrictions in the flow path or more commonly referred to as tortuous path technology. This type of technology relies upon a series of 90-degree turns to minimize fluid velocity. This type of approach works well to control velocity across a given area on average, but requires a fair amount of space to incorporate the necessary number of turns leading to larger valve space requirements.

Because of the small outlet passage normally associated with this type of trim, jet separation is critical to the best noise attenuation. Research, however, has shown that this is one of the downfalls of this type of technology.

The latest technology available combines the velocity control and frequency shifting methods. From a velocity control standpoint, the difference lies with the appropriate recovery area downstream of a restriction where a pressure drop occurs. Because of the change in density and increase in volume once a pressure drop is taken, it is key to allow a sufficient amount of area for the fluid to expand. If this does not occur, high internal velocity is possible leading to turbulence and noise generation.

After the fluid has fully recovered, it can then go through additional staging to maintain velocity control while also maintaining jet separation.

Figure 2 compares the velocity profiles of two noise abatement trims. The models are prepared using computational fluid dynamic (CFD) software to model the behavior of flow through the trim. In this case, low velocities are shown in blue and high velocities are shown in red.

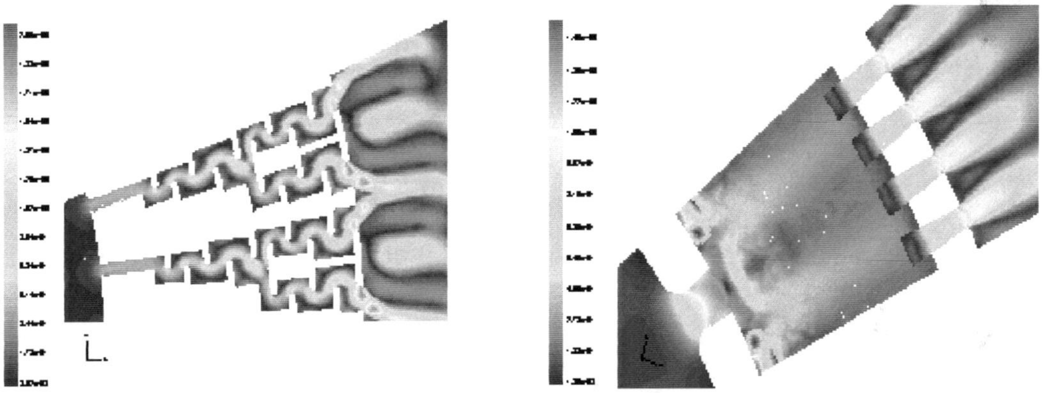

Figure 2: Tortuous Path Trim (Left) vs. Staged Pressure Drop Trim (Right)

Looking at both models, one can see that both control internal velocities well. However, the trim outlet velocity and jet independence are suspect with the tortuous path design. First off, the high velocities noted at the trim outlet signify the generation of significant turbulence, which will lead to valve noise generation. This is caused by not allowing for the appropriate amount of recovery area as the fluid passes through the turns. The other issue that comes out is that the outlet jets interact thus reducing the peak frequency at the trim outlet leading to reduced pipe transmission loss and additional noise that can be measured outside of the valve.

Exhaustive testing has shown that the staged pressure drop trim can yield up to 40 dBA of noise attenuation. This is 10 dBA more attenuation than drilled hole or tortuous path technologies are capable of providing. Not only does the trim yield better noise reduction it does so with 30% of the area that would be taken up by a similar tortuous path design. This allows the trim selection to be optimized with the valve body surrounding the trim. If the trim outlets are too close to the valve body wall, turbulence can be generated again leading to further noise generation.

IEC 60534-8-3 also calls out the potential for noise generation caused by valve outlet velocity. If the valve outlet velocity is too high, additional noise can be generated. In the past, noise calculations were limited to outlet velocities of 0.3 Mach and below. However, the standard includes formulas to account for additional noise generation with outlet velocities above 0.3 Mach.

In order to address the potential for additional noise generation due to valve outlet velocity, the valve outlet size must be increased. However, this can lead to additional valve and piping related expense.

Many instances have been found where vendors have ignored the additional noise generated by high valve outlet velocities. In certain cases, ignoring this has led to severe pipe vibration and subsequent damage to the piping support, the valve and accessories and the pipe itself.

In compressor antisurge applications, it may be possible to live with the additional noise generated when the valve is responding to a surge condition as this condition will last only a short time and will not likely cause serious damage. By thoroughly reviewing the surge control system and the noise requirements, it is possible to reduce the valve outlet size and the piping size to optimize the system for both noise control and installed cost.

Improved Valve Internals Eliminate Axial and Radial Plug Motion

The valve trim selection goes far beyond just selecting a noise abatement trim. The primary function of the low noise trim is to prevent damaging vibration from being introduced into the system. While the noise abatement trim addresses noise generated through the throttling area of the valve, there are other sources of noise and vibration that must be considered.

One area that has created issues with antisurge valves and other valves used in compressible fluids has been unstable operation and vibration at certain valve travels. Two different mechanisms can cause these issues with vibration and control.

The first mechanism is created by pressure imbalances that can be experienced with standard balanced valve plugs. If there is a sudden change in upstream pressure, the difference in pressures cannot be registered on both sides of the plug immediately and can cause the valve to experience repeated axial motion. This motion can drastically affect control and can lead to issues with compressor efficiency and operation. Repeated axial plug motion can also cause pressure pulsations in the downstream piping. This can cause excessive vibration in the piping system.

The other mechanism that can create issues with control and vibration is when the valve is first lifted off of the seating surface. At this point, the plug can begin to move radially against the valve cage. This action can cause extreme valve and piping vibration.

There are two improvements that have been made to eliminate both of these issues. A spoked plug is used to eliminate the potential for pressure imbalances on the balanced valve plug. This design reacts to transient pressure changes and pulsations in the fluid better than standard balance constructions by changing the natural frequency of the plug assembly. Now, any pressure changes that occur in the fluid are quickly registered from the bottom side of the plug to the top side due to less restriction thus equalizing the forces.

To eliminate the potential radial vibration a lower metal piston ring is installed into a machined groove at the lower end of the valve plug. This ring effectively eliminates any flow that can move between the plug and cage and any associated vibration. By controlling the clearance flow, the overall turndown of the valve can also be increased.

Noise and vibration control are key components of antisurge valve selection, but they are not the only issues that must be explored. The ability to provide fast, accurate control is as important if not more. The next section will look at the advancements made in actuation and accessory selection.

The Importance of Turndown and Controllability

Not only must the valve trim be able to provide the necessary level of noise attenuation, it must also provide adequate turndown for the application. Traditionally, a valve with 25:1 turndown was adequate for antisurge control. However, in applications where efficiency gains are critical, the industry is moving to valves that can achieve 100:1 turndown and greater. Since antisurge valves are being used more in closed loop control, the valve must be able to provide stable flow control at the lower ends of travel while also providing the necessary capacity in case of a surge event.

Selecting the Right Actuator and Accessories

As mentioned previously, because of the fast stroking requirements for antisurge valves, one could easily make the mistake that the only requirement for selection of the actuator and accessories is to make the valve move quickly. This couldn't be further from the truth. It is key to fully understand the step response requirements for every application. Certain fast stroking applications may have requirements detailing the amount of overshoot that can occur with a certain sized step response.

Let's begin by looking at the requirements around the actuator itself. In the past, because of the fast stroking speed requirements for this application, the use of electro-hydraulic actuators was pursued. This was because of the fast acting nature of the device and its ability to yield accurate step responses. Over time, it was determined that not only was the system extremely expensive from an initial capital expenditure, the maintenance and cost surrounding the operation made them very unattractive.

As pneumatic technology has improved, this type of actuator is today's common choice for antisurge applications. Most commonly used are double acting piston type actuators that have a fail-safe action. With antisurge valves, the fail-safe is in the open position and is accomplished by use of a trip system.

With larger antisurge systems, the actuators can become very large in order to provide enough force to obtain the required shutoff and to overcome any packing and internal valve friction. With the larger size comes the potential for damage when the actuator is stroked from closed to full open. For example, the valve plug for a 30" valve can weigh as much as 750 pounds. For long stroke applications, the plug will be moving up to 24 inches in less than a second. This is a great deal of force acting on the actuator travel stops. If the actuator is operated in this manner, the piston can hammer against the top of the actuator casing causing vibration and the potential for actuator damage.

In order to overcome this possibility, the actuator can be installed with an air-brake system. This system engages during the last three to five percent travel to slow down the piston to prevent it from hammering against the top of the actuator casing. If the actuator is not supplied with an air-brake system, damage can occur to the actuator, positioner and accessories while also potentially inducing damaging vibration into the downstream piping.

It is also important to review friction that can occur inside of the actuator itself. Chrome plating the actuator cylinder reduces friction between the casing and the piston. Also coating the piston with a wear band will eliminate issues with galling in the actuator cylinder.

When it comes to the accessories, the common approach to gaining fast stroking speeds is to incorporate volume boosters into the system. The booster amplifies the air volume enabling the valve to open or close quickly. In order for the boosters to operate properly, the air on the other side of the piston must be relieved quickly. Quick-exhaust valves have met this need in the past.

Quick-exhaust valves are three-way valves that have an elastomeric disk that switches flow between ports. A few inches of water column pressure differential will cause the disk to switch, making quick-exhaust valves fast-acting, on-off devices. To reduce sensitivity, a pneumatic bypass valve is often piped around the quick-exhaust valve.

Quick-exhaust valves have been applied successfully on low pressure spring and diaphragm actuators as well as high pressure piston actuators. However, when used on large volume actuators, they can introduce large overshoots that are difficult to attenuate by adjusting gains in the positioner or by adjusting the bypass valve around the quick-exhaust valve. The performance of these devices is also dependent upon age and environmental conditions. As temperature and humidity change, so does the performance of a quick exhaust valve requiring repeated tuning. In antisurge valves, quick-exhaust valves should not be used on piston actuators where tight throttling control is required or on those applications where large overshoots cannot be tolerated. Instead, for these applications, dynamic performance objectives should be achieved using multiple volume boosters.

By using multiple volume boosters, not only will performance improve, the ability to tune the accessories will be that much simpler. Over time, the elastomers in a quick-exhaust valve will degrade requiring repeated tuning with the bypass valves. Elimination of these devices in lieu of boosters will minimize the need to constantly tune the valves during scheduled shutdowns.

The actuator and accessory selection is critical to the proper operation of the antisurge system. It is also necessary to ensure that this selection is coupled with a control element that facilitates easy tuning and the ability to monitor performance on-line in real time.

Tuning & Control

The control valve positioner holds the greatest potential to improve actuator control and response time. In the past, spool type positioners with fast response were typically used in these applications because of their ability to respond quickly to step changes. However, these devices require constant tweaking and do not allow the user any way to monitor the performance of the valve.

As the performance of smart positioners has improved, so has their ability to be used to facilitate tuning of the valve and the ability to predict potential problems before they occur. This has been brought about by improvements in the positioner gain settings and in performance monitoring equipment.

The smart positioner recommended for this application is a high gain, proportional-plus-derivative controller that has three adjustments; forward path gain, minor-loop feedback gain and velocity feedback gain.

The forward path gain is used to set the speed of response where a higher gain value yields a faster response. The velocity feedback gain controls the secondary damping, which is used to attenuate slight overshoots in the response. The minor-loop feedback controls the primary damping and is used to reduce the cycling in the response and allow higher forward path gains to be used.

The primary tuning parameter is the forward path gain. Increasing the forward path gain will produce fast travel responses, low dead bands and a host of other desirable characteristics. However, if the forward path gain is set too high, the system may start to overshoot and enter into a limit cycle.

To dampen the effects of a high forward path gain, two derivative feedback elements can be incorporated in the control algorithm. The primary damping element is relay motion feedback, often referred to as the 'minor-loop'. Increasing the minor-loop feedback will dampen the response, allowing higher forward path gains to be placed in the positioner. A secondary damping element is velocity feedback. Increasing the velocity feedback gain will also dampen the response

Figure 3 shows the results of a step response performance of a 30" antisurge valve with 24" of travel beginning with 10 percent steps up to 70 percent steps. Notice that the higher gain setting allows the valve to react quickly to the step response change with minimal overshoot. It should be noted that the valve met the stroke time requirements while dramatically improving the ability for accurate closed loop control.

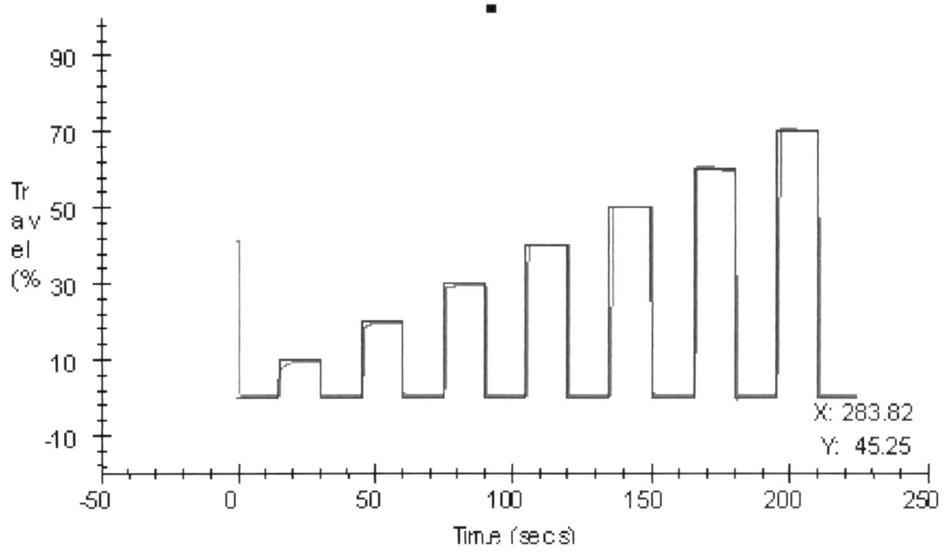

Figure 3: Step Response for 30" Compressor Antisurge Valve

This type of positioning system, coupled with the applicable accessories has yielded extraordinary results. Analysis has found that by using this type of system, the resolution is a mere one-quarter percent with a one-percent deadtime when looking at one-eighth percent input increments. Linearity of the signal to valve performance is also less than 1.0 percent for any given point of travel.

With traditional antisurge systems, it could take an experienced technician up to twelve hours to properly tune one valve. By eliminating the quick exhaust valves and including the latest in smart technology, this process can be reduced down to 15 minutes per valve.

Monitoring Valve Performance to Ensure Reliable Performance

The majority of this paper has focused around improvements in hardware that improve performance of the antisurge system. Accurate, repeatable performance is important, but ensuring that the valve will operate when called upon is of equal importance.

Control valve diagnostics have improved since their inception in the late 1980's. Until recently, diagnostics could be performed only when a valve was not in operation. Online diagnostics have made it easy to diagnose potential valve problems while the valve is operating. These diagnostics are run in the background without any interruption to the process. While this works well for valves that normally throttle, it presents challenges for valves that are normally closed, i.e. antisurge valves.

To improve reliability of the antisurge system, additional diagnostic capabilities have been identified. These include triggered and on-seat diagnostics and partial stroke testing.

Triggered diagnostics work well for antisurge valves since the diagnostic data is collected as soon as the valve opens. Once the valve receives a signal to open, the digital valve positioner begins to collect data. The data is stored in the positioner and can be retrieved at a later date for review.

On-seat diagnostics and partial stroke testing are key for ensuring reliable antisurge valve performance. While the valve is on the seat, diagnostics can ensure that the valve positioner is ready to respond to any changes in setpoint. Partial stroke testing and diagnostics open the valve to a given travel (i.e. two or three percent) and then close the valve in a controlled fashion. This test ensures that the valve is not obstructed by debris in the line. The diagnostics can then be used to identify potential areas for concern.

Accurate diagnostic information is key for maintaining system reliability and availability. Only by monitoring the valve in it's normal operating states can one be sure that the valve will act as desired when called upon.

Conclusions

The proper selection of antisurge valves relies on more than just meeting noise and stroking speed requirements. Elimination of the potential for damaging vibration in the system is normally the biggest concern, which goes beyond just selecting a low-noise valve trim. Advances have been made in overall trim selection that provides better control while eliminating the potential for axial or radial vibration.

Meeting the stroking speed requirement is only the first piece of the puzzle. Ensuring that the valve can perform accurately and quickly when operating in a closed loop mode is an important point to consider. By eliminating the need for unreliable quick exhaust valves, the performance of the entire system can be improved. When coupled with the latest in smart positioner technology, the gains can be increased allowing for fasting closed loop response while preventing dramatic overshoot in the system.

By combining the latest technologies from the valve to the actuator to the positioner to the accessories, one can be sure that they will achieve the optimal performance of the antisurge system.

Biography of Speaker

John Wilson is the Business Development Manager at Emerson Process Management – Fisher Valve Division, Marshalltown, Iowa. He received a degree in chemical engineering from the University of Nebraska. John works extensively on severe-service control valve applications in the power and hydrocarbon industries. Prior to joining Fisher, John worked for Omaha Public Power District and Nebraska Boiler Co.

Non-Linear Oscillations in LNG Liquid-Vapor Two-Phase Flows

Andrew Kimmel
University of Nevada Reno
Department of Physics
Reno, Nevada

AIChE Spring Meeting, April 2005
5th Topical Conference on Natural Gas Utilization
Atlanta April 10-14, 2005

Synopsis

Expanded two-phase LNG fluid flows are often subject to pressure induced oscillations. Due to its compressibility the vapor bubbles act as a spring with an asymmetric non-linear characteristic. Increasing pressure fluctuations cause less displacement of the liquid mass than decreasing pressure fluctuations.

The volume of the vapor bubbles increases or decreases differently if the pressure fluctuations are increasing or decreasing the average pressure value. Consequently increasing pressure fluctuations in a two-phase pipe flow cause less displacement in the direction of the pipe flow than decreasing pressure fluctuations.

The displacement depends on the ratio of LNG liquid to vapor, the ratio of pressure fluctuations over average pressure and on the exciting frequency of the pressure fluctuations.

The paper describes the governing non-linear equation for two-phase fluid oscillations and presents practical solutions for the steady state under forced pressure induced oscillations. Resonance criteria and effects of LNG two-phase flows on plant safety and operation are discussed.

Process Understanding and Control Approach of Separator Column coupled with Overhead Compressor: Lessons learnt from Malaysia LNG Tiga

S. Jamaludin
Process Control Engineer Malaysia LNG Tiga Startup Team
Sarawak, Malaysia

S. Manaf
Senior Manager Malaysia LNG Tiga Startup Team
Sarawak, Malaysia

AIChE Spring Meeting, April 2005
5th Topical Conference on Natural Gas Utilization
Atlanta April 10-14, 2005

ABSTRACT

Unstable distillation or separator column operation will leads to unnecessary flaring and off-specification product. The need for reprocessing and excess heating/cooling is a waste of energy and can disrupt the operation of the plant as a whole. In a highly competitive oil and gas industry, total plant energy management and plant thermal efficiency is fast becoming critical; hence any unnecessary flaring and wastage of energy cannot be tolerated. Separator column designed for bulk separation of component without reflux vessel behaves differently on disturbances when coupled with overhead compressor. The arrangement gives rise to process behavior that is almost totally the opposite of what can be normally expected from a distillation column. Designed control system based on the wrongly understood process behavior leads to highly unstable column operation requiring constant operator intervention to prevent flaring, tripping of overhead compressor and tripping of compressor backflow prevention. Malaysia's PETRONAS LNG complex in Bintulu, Sarawak, comprises of MLNG Satu (1983) and MLNG Dua plants (1995). Each of these MLNG plant facilities consists of three independent process trains with total LNG deliverable capacity of 7.6 mtpa and 7.8 mtpa for MLNG Satu and MLNG Dua respectively. In 2003, the complex was expanded with two more trains for MLNG Tiga Sdn. Bhd, bringing the total capacity to about 23 mtpa. The three plants is being operated as one integrated site, all being part of the Gas Sector of PETRONAS. This paper is written with the intention of sharing the process and process control knowledge learnt from the startup and operational experience of condensate stabiliser column and demethaniser column in MLNG Tiga plant for better design of future installation.

1 INTRODUCTION

1.1 PETRONAS LNG COMPLEX

Situated in the island of Borneo, in a small town of Bintulu, Malaysia; Petronas LNG complex consist of 3 LNG plants namely Malaysia LNG (MLNG), Malaysia LNG Dua (MLNG Dua) and Malaysia LNG Tiga (MLNG Tiga). MLNG, which consist of 3 LNG trains was incorporated in 1978 and delivered its maiden cargo to Japan in January of 1983. Growing demand of LNG and vast reserve of natural gas from the central Luconia basin open ways for further establishment of MLNG Dua Train 4, 5 and 6 in 1994 and MLNG Tiga Train 7 and 8 in 2003.

With the completion of MLNG Tiga, Petronas LNG complex became the world largest LNG plant in a single location with total capacity of 23 million tonne per annum. To date, Petronas LNG complex had successfully delivered more than 4000 cargoes of LNG to its buyers without a single failure. With impeccable delivery record and world's class safety achievement; Petronas LNG complex sets the industry standard.

1.2 PROCESS DESCRIPTION

The plant can be divided into 4 main sections, which are the upstream facilities, gas treating, liquefaction and storage/terminal.

The upstream facilities receive gas via 125km pipeline from the sea in the Central Luconia basin and separate the condensate from the gas. The gas is metered and enters the individual train gas treating section. Condensates are sent to the stabilization unit for removal of light components before it is sent as stabilized gasoline to crude oil terminal. Recovered light hydrocarbons are reinjected back to the LNG trains.

The first part of the gas treating section deals with the acid gas in the feed. Amine-based solvent is used to remove CO_2 and H_2S to meet product specification and to prevent freezing in the liquefaction unit. The natural gas feed also need to dried to 1 ppm of water concentration to prevent freezing. This is done in thermal swing molecular sieve adsorber beds.

Before the feed gas is liquefied, heavy component need to be stripped off to meet product specification and to prevent freezing. In a scrub column, hydrocarbons that are heavier than pentane are removed and sent to the fractionation unit for further separation to recover valuable methane, propane and butane.

The natural gas is then ready to be liquefied by using Air Products and Chemicals Inc. (APCI) license process. APCI process uses 2 types of refrigerants, which are propane and mixed component refrigerant (mixture of nitrogen, methane, ethane, propane and butane). Natural gas is first pre-cooled by propane before it is finally chilled to –151 deg C in the Main Cryogenic Heat Exchanger (MCHE) by the mixed component refrigerant. The final cooling down to –161 deg C is by Joule-Thompson expansion through control valve.

Finally, LNG is piped to the tank farm for storage at atmospheric pressure. Once the LNG ship is ready, LNG is loaded to the ship from the tank through the loading arm at the jetty head. That completes the process, and the LNG is now ready to be used as fuel for power generation upon gasification at the customer's LNG receiving terminal.

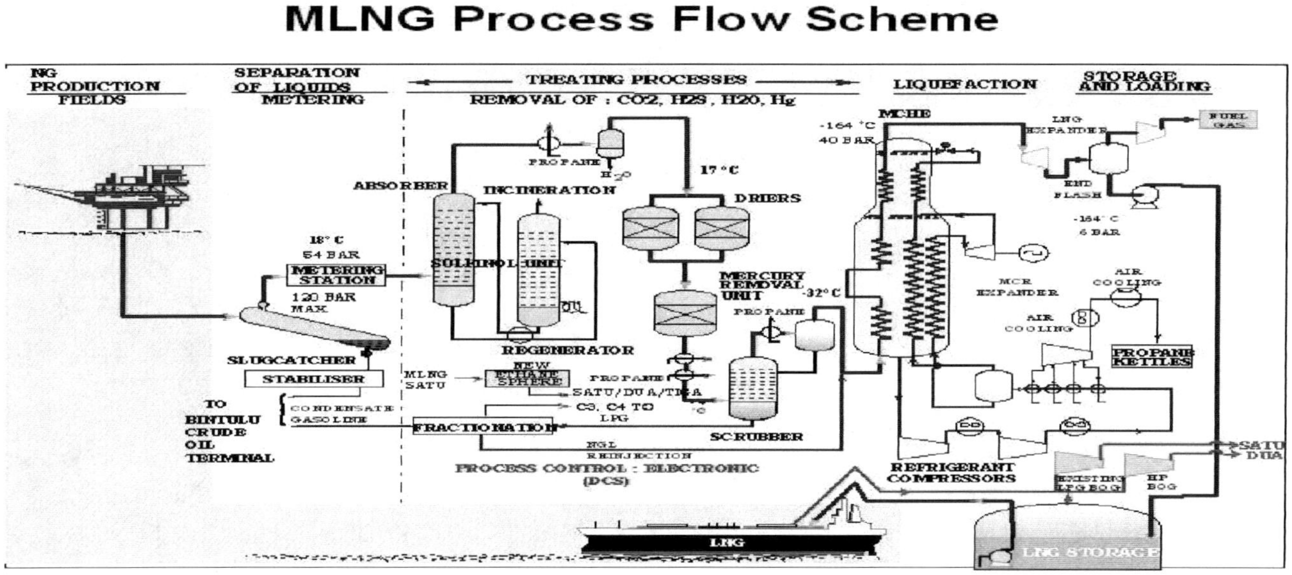

Figure 1: Simplified Flow Scheme of LNG Process

2 PROCESS INTERACTION OF SEPARATOR COLUMN OPERATION WITH OVERHEAD COMPRESSOR

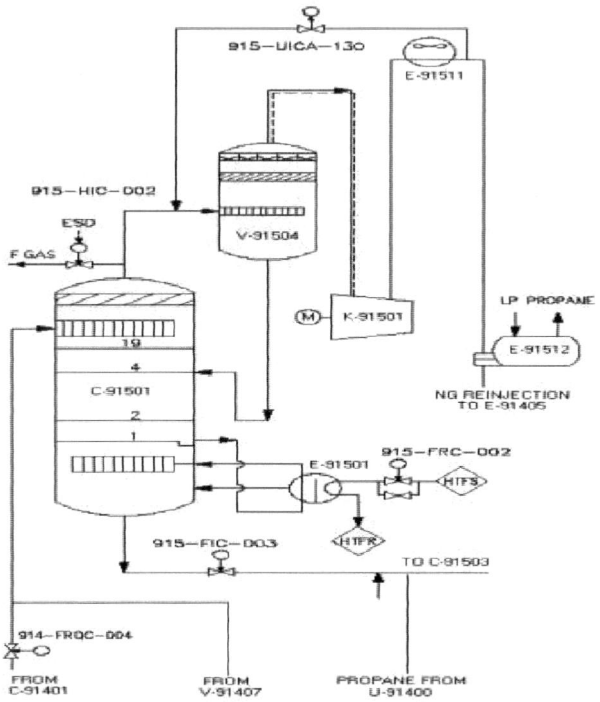

Figure 2: Separator Column With Overhead Compressor

Operations of separator column become more sensitive to disturbances when coupled with overhead compressor. This is especially true for separator column of simple separation for e.g. stripper column or stabiliser where there is no reflux vessel. For distillation column with reflux vessel, process interaction with overhead compressor will be dampen as compositional change in the top of the column will not have immediate impact on the composition of gas going to the overhead compressor. Hence, the column has more time to recover before the disturbances is seen and made worst by the overhead compressor.

With overhead compressor, separator column need to recover fast from disturbances in order not to be too far outside the operating envelope, which will results in flaring. The difference in process behavior compared to ordinary separator column without overhead compressor can be explain as the compressor flow act as secondary disturbance on the onset of the first disturbance. Furthermore, the impact of the secondary disturbance on the column operations is much faster than the impact of the first disturbance. Hence, there is a large process behavior difference between the two columns on transient during disturbances.

For example, when the column pressure drops below setpoint, an uncompensated column tray temperature controller setpoint will be too high. As a result, composition at the top of the column will be richer with heavier components of hydrocarbon. Due to the presence of overhead compressor, the pressure in the column will get even lower as the compressor flow increases due to higher density gas at the top of the column. In contrary, for ordinary separator column without overhead compressor, the increase in the density of the gas at the top of the column will not further reduce the column pressure. Therefore, in such situation, heat to the bottom of the column can be increase to assist in bringing up the column pressure. Similar action cannot be done for separator column with overhead compressor because the increase in heat will further increase the density of the gas at the top of the column, which in this case will further decrease the column pressure.

The reverse will happen when the column pressure or the tray temperature drops below setpoint. The absence of reflux vessel means that the change in the column top composition will immediately change the composition of gas at the overhead compressor suction.

3 PROCESS BEHAVIOUR AND INSTABILITY IN STEADY OPERATION

There are 2 separator columns in MLNG Tiga that were designed with overhead reinjection compressor and without reflux vessels. Conventional control scheme with tray temperature control on reboiler and pressure control on compressor capacity had been applied in both units. Since startup until steady operation, both columns have been displaying process behavior as described above. Disturbances in the columns cause changes in compressor flow, which further upsets the columns operation. Normally, manual operator intervention are required to bring back the column into the operating envelope as control system fails or takes too long to do so. Flaring, trip of overhead compressor on motor thermal overload and trip of overhead compressor backflow prevention are common to occur as results of disturbances in the columns.

3.1 DEMETHANISER COLUMN

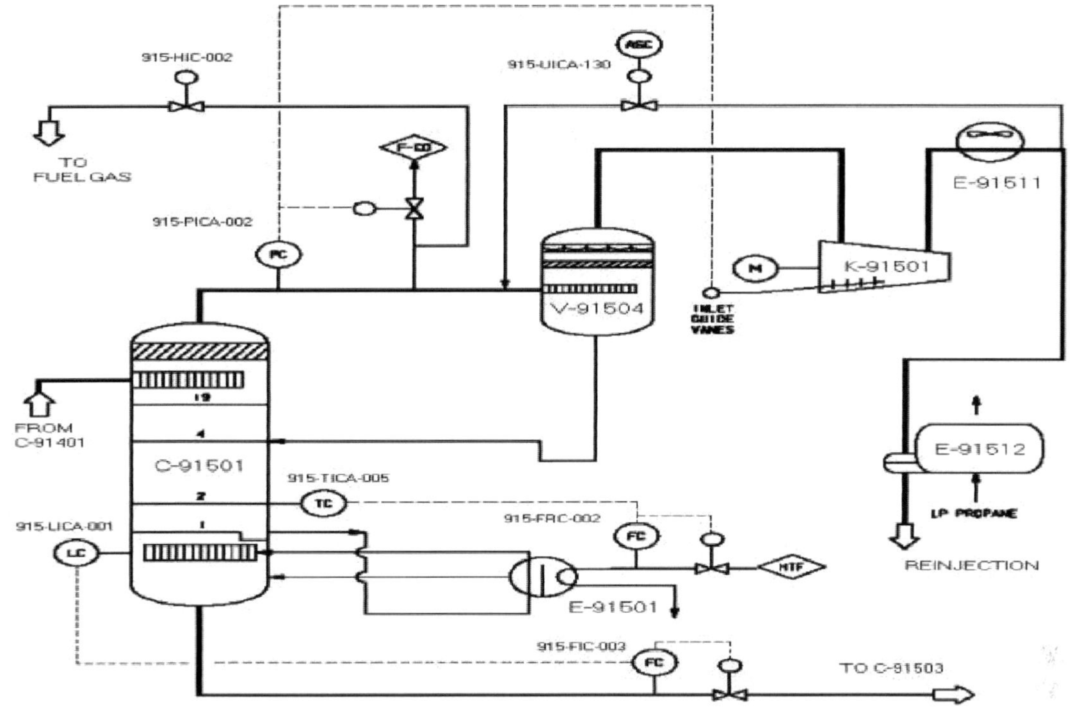

Figure 3: Demethaniser Column and Its Control Scheme

The demethaniser column receives its feed from the bottom of the scrub column where hydrocarbons heavier than pentane are separated. Methane is recovered from the top of the demethaniser to be reinjected back to the liquefaction unit. The bottom product flows to depropaniser column and debutaniser column for recovery of propane and butane.

The demethaniser column is normally not subjected to large disturbance, as the feed composition is steady and the feed flow only changes by small amount. Hence, the instability in the column is not very severe. The control scheme is as such that the effective cut point control is via tray #2 temperature control, which is cascaded to flow controller on the heating medium and the pressure control is a split range of the overhead compressor inlet guide vanes (IGV) and opening to flare. Since the separation is easy without tight specification for the top product, there is no control for the column separation index.

To add to the inadequacy of the designed control scheme, the selected control handle for both pressure and temperature control do not have sufficient gain. The overhead compressor IGV does not have a large effect on the pressure of the column and tray #2 temperature is not sensitive to column disturbances. On large pressure increase, the IGV will be fully open and the pressure will still be rising until the pressure control starts opening the valve to flare. The opposite happens when the pressure is decreasing, where the IGV will be fully closed with the column pressure continue to decrease.

At higher pressure than setpoint, before the flare valve opens to eventually bring back the column pressure down, the temperature setpoint becomes too low. Composition of hydrocarbon at the top of the column becomes leaner as indicated by the drop in the column top temperature. At lower molecular weight and lower temperature, the gas density decreases thus the overhead compressor flow decreases. Decrease in gas flow from the top of the column further increases the column pressure, which further escalate the disturbance. Compressor recycle valve will open as the operating point move closer towards the surge control line. Eventually the backflow prevention of the compressor trips and compressor goes to full recycle. Valve to flare immediately open and stay open until the backflow prevention is reset and the overhead compressor resumes reinjection.

On the other hand, when column pressure drops below setpoint, the column top temperature will increase as the composition of the overhead gas gets heavier. Compressor flow then increases with the increase in the gas density. The increase of gas density can be large enough that the overhead compressor motor trip on thermal overload. Immediately the valve to flare will open as the pressure in the column increases.

Both of the scenarios described above are common to happen on large disturbances as the pressure control is too slow and has too small process gain. Column pressure control is unable to bring the operating point back to setpoint fast enough before the composition at the top of the column changes. Once the top composition changes significantly, flaring is inevitable.

3.2 STABILISER COLUMN

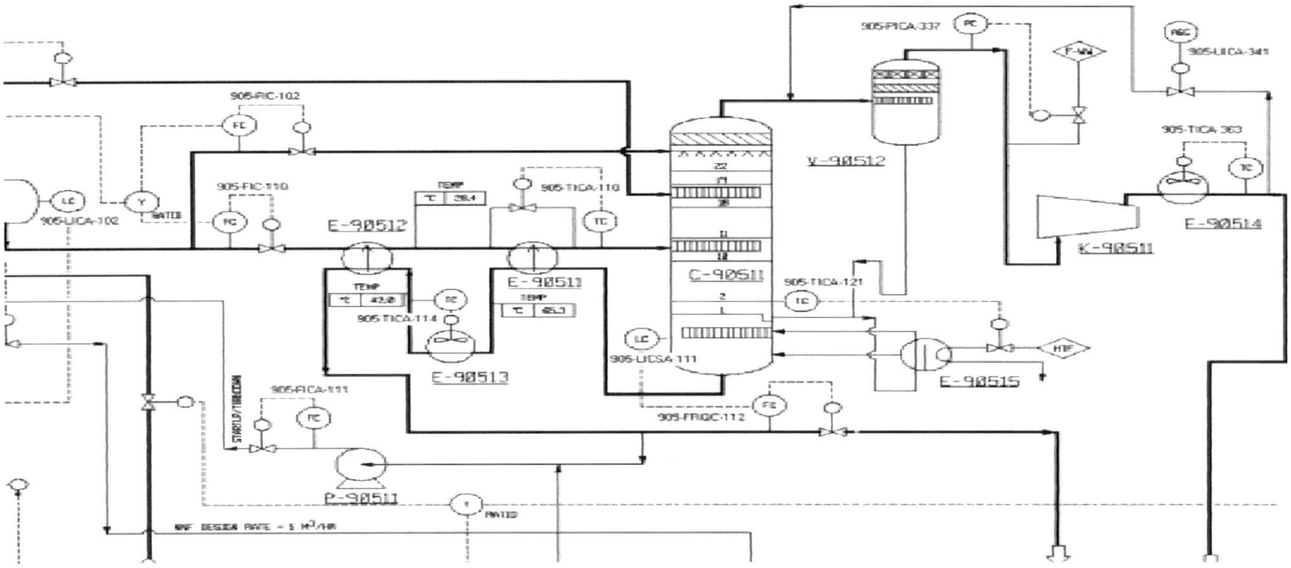

Figure 4: Stabiliser Column and Its Control Scheme

Stabiliser column functions to process and stabilizes the condensates received from upstream facilities. Light hydrocarbons are removed from the top of the stabiliser column to be reinjected back to the LNG train. Stabilized gasoline leaves the bottom of the column and pumped to gasoline storage tank.

Unlike the demethaniser column, disturbances to the stabiliser column are larger and more frequent. The composition and flow rate of the feed to the column can change drastically within a short period of time. The effective cut point is controlled by tray #3 temperature control cascaded to flow control of the heating medium. Separation index control is by varying the reflux flow to feed flow. Varying the compressor capacity via discharge control valve controls the column pressure.

Pressure control of the stabiliser column acts by varying the discharge control valve rather than by using compressor IGV like the demethaniser, which is a much better control handle. The process gain is larger and the response is also faster but unfortunately it is still not fast enough for such application.

Similar process interaction between the overhead compressor operating point and the column operating conditions persist and even to a larger extent compared to the demethaniser column due to larger disturbances experienced by the stabiliser column. Just like the demethaniser column, runaway pressure in the column causes flaring, trip of overhead compressor back flow prevention and trip of overhead compressor motor on thermal overload.

4 NEW CONTROL APPROACH AND IMPACT ON COLUMN STABILITY

One of the most important factors for successful column operation is good pressure control. This is because the boiling points of all components depend on pressure. Normally for separator column control, "averaging" pressure is required, that is the pressure does not need to be controlled extremely tightly but it needs to be kept around an average value. However, for separator column with overhead compressor, pressure need to be controlled faster and tighter, in order to prevent the top composition from changing significantly.

The main problem with conventional control approach on separator column with overhead compressor is that the pressure control is too slow. Whether the pressure control is controlling the compressor IGV, compressor suction valve or compressor discharge valve; it cannot be tuned any faster because pressure change is not a very fast process for large column.

A different approach is needed as a standard control scheme for separator column with overhead compressor. An easiest solution is to cascade the column pressure control to a slave flow controller. The slave flow controller can be tuned much faster and will be able to react immediately on changes of compressor flow before the master pressure controller realize of the change in column pressure.

However, a better and more complete approach is to integrate the column pressure control with anti surge control of the overhead compressor. Integrating compressor capacity control with anti surge control is normally done for a different reason. It is particularly useful when interaction between capacity and anti surge controller can drive the compressor into the operating range of the anti surge control system. For compressor system with a rather flat process system resistance line, reduction in speed by the capacity controller can cause the compressor to suddenly go into full recycle.

An anti surge controller calculates the current compressor operating points relative to the compressor surge control line. This is done by calculating the surge parameter, which is a combination of process variables that when the value is constant, represents a curve in compressor performance diagram. The surge parameter is calculated by taking into accounts the suction pressure, discharge pressure and the mass flow rate through the compressor. The design of the surge parameter is such that its sensitivity for changes in compressor operating condition is minimized. Even though other process operating variables such as molecular weight, suction temperature, compressibility factor, ratio of specific heat and polytropic exponent are not measured or calculated, its influence on the actual operating point of the compressor is already covered by measuring pressures and flow.

The advantage of integrating the column pressure control with the overhead compressor anti surge control compared to column pressure and compressor flow cascade, is that the control system will not only react to changes in the compressor flow but also to changes in the molecular weight and suction temperature at the suction of the compressor. This will make the column pressure control faster and more robust. There are a number of ways how the integration of column pressure control and overhead compressor anti surge control can be done but basically the column pressure control will not only act as the compressor capacity control; adjusting IGV, discharge valve, suction valve of compressor speed but it will also act on the compressor anti surge recycle valve. Column pressure control will be the master pressure control acting on the slave anti surge controller. The compressor anti surge control is not affected and the anti surge controller will be able to override the pressure controller whenever it needs to open the anti surge recycle valve to prevent surge. Furthermore, there is also the advantage of not needing a separate flow meter on the discharge of the compressor to be used as the slave flow controller.

In addition to the integrated column pressure control and overhead compressor anti surge control, it will also help if the tray temperature controller is pressure compensated. Therefore, the temperature controller will also be able to bring the column back to the operating envelope on disturbances and not only depends on the action of the pressure controller. Furthermore, uncompensated temperature controller can act on the wrong direction on pressure variation. For example, if column pressure controller is in manual or during large disturbances, where column pressure decreases far below the setpoint, not only that the top temperature of the column will increase but the tray temperature will also increase. The temperature controller will then decrease its output in effort to decrease the tray temperature, which is an act in the wrong direction that will further reduce the column pressure and hence increase the column tray and top temperatures.

5 FUTURE EXPECTATIONS AND CONCLUDING REMARKS

Experiences in MLNG Tiga have shown that there are significant process interactions between the operations of separator column with its overhead compressor that warrants a more integrated control scheme for the units to ensure column stability. Failures to do so will results in frequent flaring, tripping of overhead compressor's motor on thermal overload and tripping of the compressor back flow prevention.

The column pressure is highly affected by the loading of the overhead compressor. Since the compressor loading can vary in a short period of time, column pressure control needs to be fast acting, which is not made possible by conventional column pressure control where pressure control acts directly on the speed, IGV, suction or discharge valve of the compressor to vary its capacity. Column pressure needs to be controlled tightly to ensure that the top composition does not change as changes in the column top composition and hence the overhead compressor suction will quickly change the compressor loading.

A better control approach towards operational stability of separator column coupled with overhead compressor is to integrate the column pressure control with the overhead compressor anti surge control system. The integrated control system will be able to react fast to changes of molecular weight and temperature at the top of the column as well as changes to the overhead compressor loading. Another improvement that can be done on the control scheme is to compensate the column tray temperature controller for pressure changes. The temperature controller can then assist the pressure controller in disturbances or when the pressure controller is in manual to quickly bring the column back into the operational envelope. This is also to avoid reversing of the temperature controller gain when the pressure control is in manual or during large disturbances where the pressure has drifted away from the setpoint.

Even though the unstable separator column coupled with overhead compressor might only be a small part of the whole plant, every effort towards reducing flaring deserve priority attention. For the panel operator, stability of the column will mean less alarms and operator attention can be focused towards optimizing the process. With no additional cost for the modification of the control scheme as all that it takes are software changes, and the potential benefits that it offers; the change towards integrated compressor capacity and anti surge control system is well worth considering for this particular application.

6 REFERENCES

Jonk, H., Overstegen, B., Jones, K.A., Snoek, J. and Solkesz, E.J. (1985). 'Guide to Anti-Surge Control Systems For Centrifugal Compressors', Report MF 85-0260.

Shell Global Solutions International B.V. (2001). 'M111 Distillation and Process Control Course Manual'.

7 BIOGRAPHY OF SPEAKER

Mr. Saifullah Jamaludin is an Associate Member of Institution of Chemical Engineering UK. He graduated from University of Edinburgh, Scotland with BENG (HONS) Chemical Engineering in 1999 and joined Malaysia LNG Sdn. Bhd., a subsidiary company of PETRONAS.

Saifullah first assignment was as a Process Technologist in the Technical Services Department with responsibility of troubleshooting and monitoring of MLNG and MLNG Dua process including storage and loading.

In 2001, Saifullah was transferred to Operations Department to join a newly formed section of MLNG Tiga Startup Team as Process Control Engineer. At that point in time, construction of MLNG Tiga plant had just begun. MLNG Tiga Startup Team was formed early in the construction stage of the project with the intention of integrating it with the MLNG Tiga Project Team to ensure safe, timely and efficient startup of the plant in 2003.

At the end of 2001, 2 Senior Process Control Engineers from Shell joined the group and together they formed Process Control Group of the MLNG Tiga Startup Team. Early in the project, Process Control Group was deeply involved with review of Process Engineering Flow Scheme, Operating Manual, Operational Procedure Guide, Instrumented Protective Function, Process Control Narrative and Operational Functional Test Procedure. During commissioning, all control parameters and especially complex control loops in Distributed Control System (DCS) and compressor anti surge control was checked and simulated. When MLNG Tiga Train 7 and 8 were finally started up in March 2003 and August 2003 respectively, fine-tuning and minor troubleshooting was done. Both of the trains were successfully started up safely and smoothly.

With the end of the MLNG Tiga project, Saifullah is currently transferred to MLNG Process Control Group responsible for monitoring and troubleshooting of process control issues in MLNG, MLNG Dua and MLNG Tiga. Saifullah is also currently involved with the planning and implementation of Advanced Process Control in MLNG Tiga.

Modeling and Dynamic Simulation as Design Tools for LNG Plants

Carlos Yengle
Jaleel Valappil
Ramachandra Tekumalla
Mehrotra Vibhor
Dave Messersmith

Bechtel Corporation
3000 Post Oak Blvd
Houston, Texas

Hahn, Paul
ConocoPhillips Company
3000 Post Oak Blvd
Houston, Texas

AIChE Spring Meeting, April 2005
5th Topical Conference on Natural Gas Utilization
Atlanta April 10-14, 2005

Synopsis

During the EPC stage of a LNG plant project, design team engineers face several challenges with process-related development, planning, and design required for economical and effective completion of the project. Process engineers must have the necessary technical skills to transfer the process into appropriate equipment and control system lay out in order to accomplish the process requirements. Being able to visualize the flow and the processing of fluids through the entire system and within each process component, as well as being able to recognize what will emerge during the process and upset conditions are critical skills in understanding the process dynamic responses and equipment mechanical specifications. Unfortunately, many design techniques are merely based on general industry standards that can sometimes result in over/under design of the equipment and process control. Fortunately, dynamic simulation software tools currently available are an extremely valuable mechanism that support engineers in their quest for successful completion of the project.

This paper is divided into two sections. The first section describes the development of dynamic models of the Phillips Optimized Cascade LNG Process including data validation to an existing LNG operating plant. Dynamic models were built for various individual projects. Each model replicates the entire process within each facility.

The second section of this paper discusses various applications of the models. It includes a description of four separate validation studies. High fluctuations in the inlet feed gas pressure due to hydraulics of the inlet pipeline and uncertainties in the flow regime were of concern in the first study. The stability and controllability of the proposed pressure and flow control combination in the Inlet Feed gas unit was successfully validated. In the second study, the stability and controllability of the Nitrogen Rejection Unit during transient conditions and process upsets was evaluated. In the third study, core and shell exchangers were forced to minimum core sub-mergency in order to evaluate their warm end temperature approach. In the fourth study, upset conditions were created in the heavies unit in order to identify the minimum exposure temperature.

Developing a dynamic model for each individual LNG project has proven to be a critical tool for the process design team. Having a close-to-real plant engineering tool through dynamic simulation has given Bechtel engineers a better understanding of the process dynamics. Additionally, it has enabled Bechtel engineers to effectively design the equipment and process control system.

Biography of Speaker

Carlos M. Yengle is an Engineer Specialist with Bechtel Corporation in Houston, Texas. Carlos' fields of expertise include dynamic simulation, virtual plant simulators, and process modeling. He is actively involved in building and validating dynamic plant models and utilizing simulation for operator training, design improvements and optimization/control, start-up support and evaluating safety and operating procedures of LNG plants. Carlos received a Bachelor of Engineering degree in Chemical Engineering from Universidad de Trujillo-Peru and a Masters of Engineering Degree in Chemical Engineering from University of South Florida.

Controlling Complex Turbomachines in LNG Plants

Krishnan Narayanan
Compressor Controls Corporation
Des Moines, Iowa

AIChE Spring Meeting, April 2005
5th Topical Conference on Natural Gas Utilization
Atlanta April 10-14, 2005

Introduction

The design of a Liquefied Natural Gas (LNG) plant is based on one of several commercially licensed process technologies. Although the choice of refrigeration cycle and a turbomachinery configuration are dependent on the process design, a compression system is always the fundamental constituent of any world-scale LNG plant. These compression systems are typically comprised of large, complex turbomachines. Therefore, integrated control of turbomachines is of utmost importance for safe, efficient, and reliable operation.

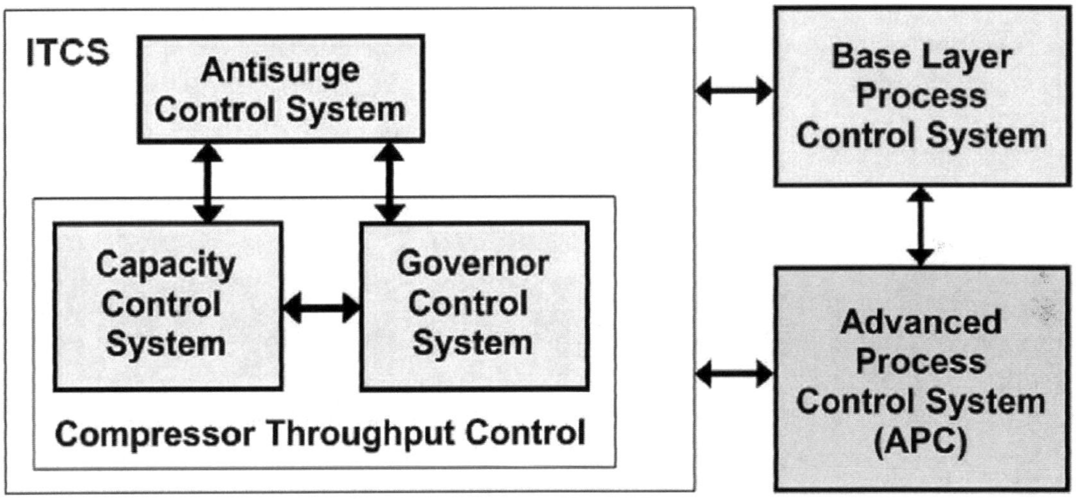

Figure 1. Integrated Turbomachinery Control System (ITCS)

Compressor Controls Corporation (CCC) has developed and successfully implemented Integrated Turbomachinery Control Systems (ITCS) for several world-scale LNG plants around the globe. ITCS (Figure 1) is a multivariable control system that provides antisurge control, capacity/throughput control for compressors, and governor control for turbines. This paper addresses turbomachinery control system requirements for major compressors in different LNG processes.. Additionally, control system design for various operating modes will be highlighted along with considerations for turbomachinery performance testing. Although the paper does not address specific control requirements for end-flash gas or boil-off gas compressors, in general the requirements stated here are equally applicable to these machines.

LNG Process Layout

The control system design outlined in this paper is relevant to all commercially licensed LNG processes. However, we will focus our discussions on Propane (C3) precooled – mixed refrigerant (MR) liquefaction process from Air Products & Chemicals, Inc., and the Optimized Cascade Process developed by Phillips Petroleum. For LNG plants utilizing a propane (C3) precooled, mixed refrigerant (MR) refrigeration cycle, the natural gas feed after pre-treatment is delivered to the main cryogenic heat exchanger (MCHE) at a given pressure and temperature. A C3 cycle precools natural gas and MR vapors (from compressor discharge) prior to their entry into the MCHE. Adjusting the throughput of compressors in both refrigeration cycles (C3, MR) liquefies the natural gas. On the other hand, at LNG plants designed using Phillips process, the feed gas supply undergoes a standard pre-treatment process to remove carbon dioxide, water and mercury. The Phillips Process then utilizes three refrigeration cycles – propane, ethylene and methane to cool the gas in steps to -161 degrees Celsius.

Operating Modes

Base-load LNG plants have world-scale capacity and operate continuously. On occasion, base-load plants are forced to operate at reduced rates as a result of inclement weather (resulting in tanker delays), upstream gas-plant trips, machinery or process problems, and seasonal factors. Various operating modes are described below.

<u>Start-Up Mode</u>

Cold Start – Rotating machines have been shutdown for extended periods and the heat exchangers are not cold (say temperatures above –10 °C). The duration for a cold start-up is long because the heat exchangers must be cooled down to a desired temperature (at a specified rate) to avoid thermal shocks to its structure.

Warm Start – Rotating machines are operated at minimum governor speed with compressors in full recirculation. The duration for a warm start-up is much shorter because the heat exchanger is still relatively cold.

<u>Part-Load Mode</u> – The plant is operated at reduced production rates attributed to restrictive conditions similar to those mentioned above. Compressor capacities are reduced to match the refrigeration load imposed by lower production rates. Part-load mode, in general, tends to be unstable owing to loop interactions. Moreover, interactions between series-compressor stages render the process to become very difficult to control.

CCC's ITCS has proven to be capable of eliminating the aforementioned controllability problems.

Base-Load Mode – The plant is operated at production rates equal to or exceeding the design ratings. The goal is to maximize production by running turbomachines at their power limits.
Shutdown Mode – Compressors are unloaded in a controlled fashion during normal shutdowns and unloaded abruptly during emergency conditions.

Integrated Turbomachinery Control System

LNG plants meet their production objectives through one of two control philosophies. The one taken by most plants is to maintain product temperatures within tight margins. As a result, these plants do not have direct control of production rates. Their goal is to maximize flow rates by adjusting the available refrigeration-cycle duty while maintaining desired product temperatures. The second philosophy involves maintaining a desired flow rate in addition to maintaining product temperatures within tight margins.

Irrespective of the choice of control philosophy, integrated turbomachinery control is essential for meeting a plant's production commitments. The fundamental objective is to efficiently operate the turbomachinery within a safe operating envelope by properly integrating the compressor and driver controls. This envelope is defined in Figures 2 and 3 for a gas turbine driven compressor.

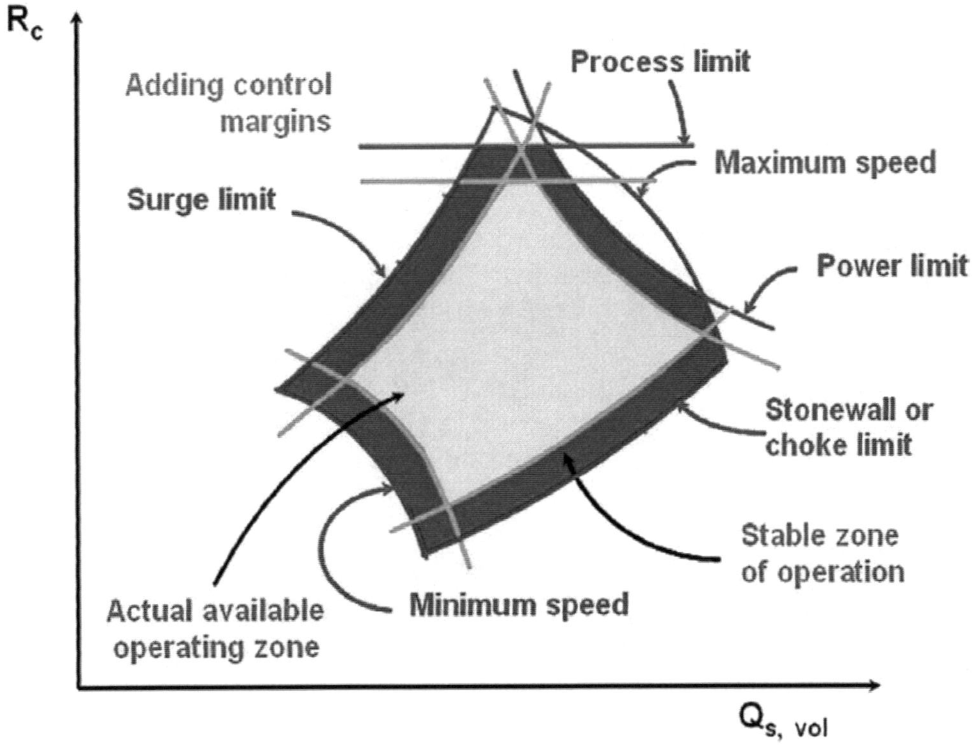

Figure 2. Compressor map with defined boundaries of operation

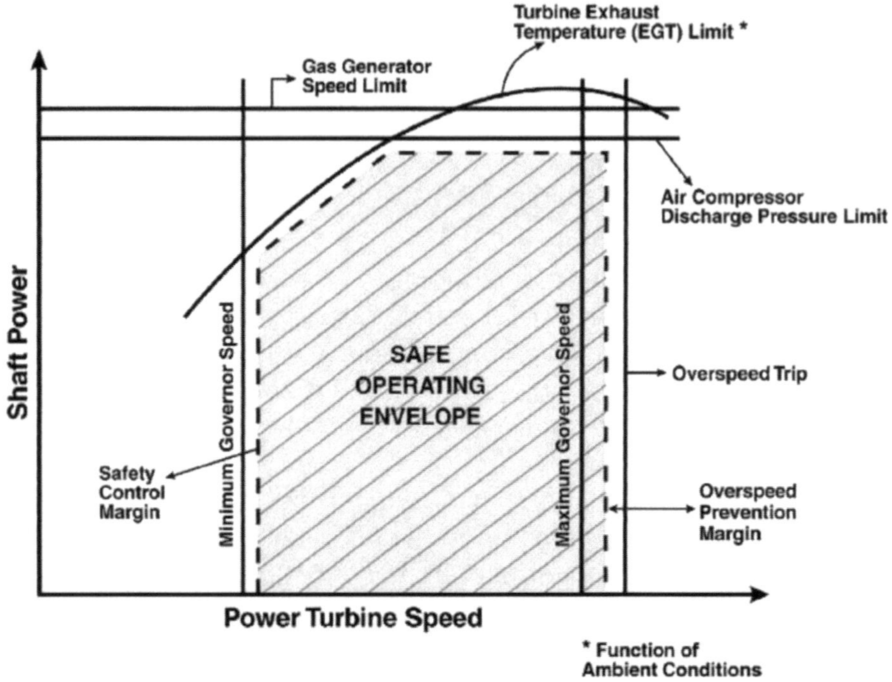

Figure 3. Operating envelope of a Two-shaft gas turbine

An ITCS should be implemented by applying state-of-the-art microprocessor technology. Advanced control algorithms using adaptive open- and closed-loop techniques are employed to prevent compressor surge. Advanced constraint-control schemes (such as predictive set-point and feed-forward rate-of-change calculations) are important for maximizing the performance of control system.

An ITCS must be versatile enough to adapt to diverse turbomachinery train configurations (parallel or series), such as driver (fixed versus variable speed driver), helper-driver, condenser, sidestream loads, and stage designs for refrigeration compressors.

To ensure optimum performance, control algorithms should be executed in a maximum of 40 ms for an antisurge system, and 20 ms for a governor system. The control system must be designed for fail-safe operation with built-in strategies that allow safe and reliable on-line operation upon failure of transmitter signals. A minimum of duplex, fault-tolerant, control architecture is recommended to increase system reliability and availability.

Antisurge Control System Requirements

Algorithm Invariance and Surge Limit Surface

Suction conditions for refrigeration compressors change continuously. A surge limit line, defined by an antisurge algorithm, must not vary with operating conditions [1]. MR compressors typically have a family of surge limit lines; that is, a surge limit surface [2] instead of one curve. The varying parameters are speed or inlet guide-vane (or stator-vane) position, gas composition, suction temperature, and suction pressure.

Inaccurate implementation of a surge limit surface in the antisurge controller, results in loss of turndown, or operating in the unstable zone, which can be highly damaging due to surge.

Surge Parameter for Sidestream Machines

Sidestream flows of C3 compressor present challenges to an antisurge control design. A compressor's internal design prevents the measurement of flow in the sidestream-stage inlet. As a result, special algorithms (using principles of mass balance) are required to calculate this flow. CCC implements a unique algorithm to define surge parameters for a sidestream compressor by using dynamic similitude theory [2].

Dynamic Decoupling between Series Stages

During plant start-up and part-load operation, compressor recycle valves are modulated under antisurge control action. For compressors operating in series (for example, a 3-stage axial/centrifugal MR compressor configuration of C3-MR process or multi-staged methane compressors in Phillips Cascade process), opening the low pressure (LP) stage recycle valve reduces flow to the downstream stage and pushes the high pressure (HP) stage towards surge. Conversely, closing the LP stage recycle valve will drive the downstream stage away from surge.

In order to prevent the HP stage from rapidly approaching surge, it is imperative that its recycle valve be opened by a necessary amount (as a feed-forward response) that is a function of the LP stage's recycle valve opening. This feed-forward (decoupling) action stabilizes the interacting antisurge control loops.

For a multistage C3 compressor, the downstream stage (for example, the third stage) does not have a major influence on the upstream stage (first stage); these decoupling (feed-forward) gains must then be configured according to varying degrees of interaction between compression stages.

Automated Start-Up and Shutdown Algorithm

An antisurge control algorithm must be energy efficient for all operating modes of the plant. Automated recycle valve response ensures that the compressor is automatically loaded and unloaded in an efficient, safe, and controlled manner during start-up and normal shutdown. In case of major, process upsets or emergency conditions, the compressor is unloaded rapidly by quick opening of the recycle valves.

Constraint Control

An antisurge control system may be required to open the recycle valve even when operating conditions are away from the surge limit. Process and equipment design constraints may require the recycle valve to be opened to prevent structural damage or process trip.

Suction pressure – It may be detrimental to allow the compressor inlet drum suction pressure to drop below atmospheric conditions, owing to process or machinery constraints.

Inter-stage or final stage discharge pressure – Process and equipment design constraints may require an antisurge control system to limit compressor inter-stage or discharge pressure levels thereby preventing structural damage to compressor casings; or preventing opening of the PSVs during over pressure conditions.

Suction temperature – On losing feed gas to the MCHE, suction temperature of the LP stage compressor will drop. In some cases, the machinery and process safety requirements may require the MR compressor to be tripped when suction temperature goes below an acceptable level.

Irrespective of the upsets in the process, the design should guarantee that the compressor's operating point stays within the safe envelope shown in Figure 2. Recurrent opening of the recycle valve results in diminished production caused by excessive power usage and reduction in refrigeration capacity. An inefficient antisurge control system will prevent a plant from operating at the required production rates.

Requirements for Capacity Control System

A capacity control system's primary function is to adjust flow rates through the compressor in accordance with the refrigeration load imposed by production requirements.

Main Control Variable

Refrigeration compressors – The main process variable is typically the suction pressure at the first-stage drum.

MR compressor – The variable to be controlled for an MR compressor is typically the ratio of pressures between the HP stage separator-drum and the LP stage knockout-drum.

For variable speed drivers, flow through the compressor is adjusted by varying the driver speed. Adjusting the stator or inlet guide vane position regulates the performance of MR compressors with fixed speed driver configuration.

Constraint Control

Constraint control algorithms will be required to limit turbomachinery performance within the boundary of a safe operating envelope, as illustrated in Figures 2 and 3. This implies that a capacity control system must guarantee that safe operation is never sacrificed (during increased production demands) if the operating point of the turbomachinery is close to any performance limits. An ITCS ensures that performance is optimized within the safe operating envelope; thus allowing maximum possible compressor throughput for a given constraint.

Process and equipment design constraints may require a capacity control system to restrict compressor throughput, thereby preventing structural damage to compressor casings or preventing sudden opening of the flare valves or overloading the drivers if throughput exceeds the driver power limit.

Load Sharing of Series or Parallel compressors

For multi-stage compressor trains operating in parallel (such as methane, ethylene or propane compressors in Phillips process or boil-off gas compressors), the flow should be distributed between the trains in an optimum fashion. This ensures that one train is not at its power limit, with the other train operating on surge control line with recycle flow. It is essential that the parallel compressor trains approach the surge control line simultaneously resulting in optimum operation of the recycle valves [3]. For series compressors with independent drivers (such as MR compressors in earlier generation of C3-MR process), the overall pressure ratio should be distributed between stages in an optimum fashion. This ensures that the LP stage is not at its power limit, with the HP stage close to its surge-control line, or vice versa. Similarly, it is important that the LP and HP compressor stages approach the surge control line simultaneously; resulting in optimum operation of the recycle valves.

Decoupling with an Antisurge Control System

Antisurge and capacity control systems for C3 and MR compressors in Propane-MR process and for methane, ethylene and C3 compressors in Phillips Cascade process must be integrated through dynamic decoupling. At reduced rates, recycle valves are in operation and conflict with the capacity control system. This, in turn, results in unstable operation if the antisurge and capacity control systems are not integrated. For example, if production were to be reduced, an independent capacity controller would decrease flow through the C3 compressor by reducing throughput. This could result in the recycle valves being opened by the antisurge controllers. Inlet-drum pressure would then increase and the capacity controller would react by increasing the speed. Antisurge controllers would close the recycle valves as a consequence of the increase in throughput, thereby causing suction pressure to drop, and so on.

This cycle would repeat itself with potential for causing unstable operation unless the operator runs the capacity control in manual, or the capacity and antisurge control system gains are reduced. The reduction in gains is termed *loop detuning* and prevents unstable cycling (hunting) of control loops. However, loop detuning is undesirable because it results in loss of tight control during normal operation. Loop decoupling in an ITCS alleviates this problem and results in stable operation without having to detune the offending loops. Figure 4 depicts improvements in control system response brought about by dynamic decoupling for a C3 compressor.

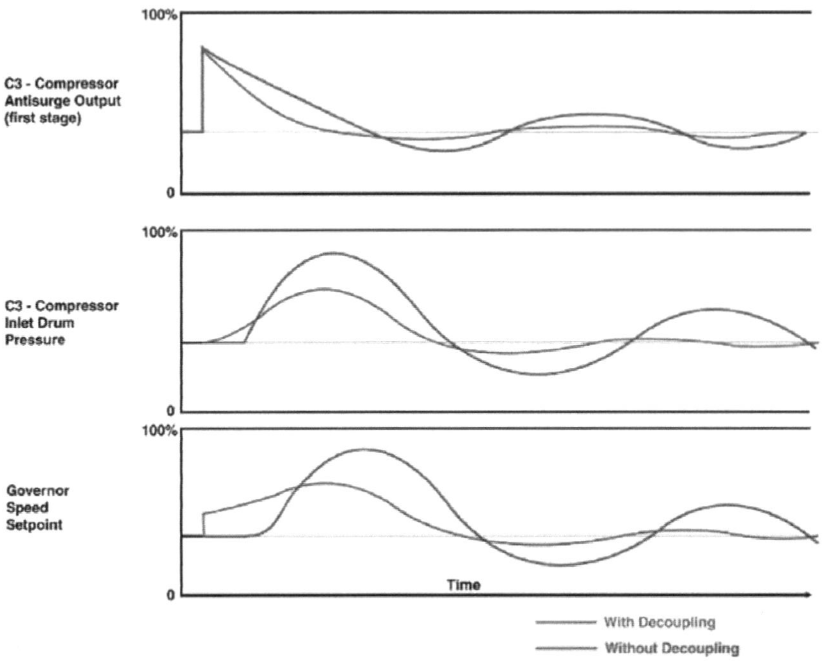

Figure 4. Dynamic decoupling and integration between the antisurge and capacity control system

Automated Loading and Unloading

Refrigeration compressors are generally brought up to minimum governor speed in LOCAL mode by employing a turbine's governor system. An operator can load the compressor by changing the governor speed set point, or by putting the governor control system in REMOTE mode and letting the capacity control system load the compressor automatically (in a controlled manner) using the remote speed set point to the governor.

Successful loading of compressors in MANUAL mode relies heavily on operator experience and skill. Loading of compressors operating in series or bringing a compressor on line with a parallel running train (at elevated header pressure) is not a trivial task. The interaction between the series or parallel stages and recycle valves makes manual loading extremely challenging. Even simple mistakes can prove costly and severe, resulting in delays of process start-up and possible compressor damage from surge.

It is important that the capacity control system employ automated loading and unloading algorithms. This will ensure that compressor throughput is regulated in a controlled manner and integrated with the antisurge control system to bring the compressor on-line smoothly, or take the compressor off-line to operate in a standby mode.

ITCS must load the compressor such that thermal cooling rates (recommended by the heat exchanger manufacturer) are not violated. Rapid adjustments of recycle valves or fast changes in compressor throughput can produce sharp temperature fluctuations causing thermal shocks and the potential for structural damage to the cryogenic equipment.

Governor Control System Requirements

A governor control system must be able to operate in either a LOCAL or REMOTE set point mode contingent upon plant operating conditions. LOCAL mode is used to bring compressors up to minimum governor speed at plant start-up.

For plants employing variable speed drivers, the capacity control system then interacts with the governor (in REMOTE mode) by adjusting the speed set point in accordance with the refrigeration duty requirements. In order to maximize production, the capacity control system will drive the governor to push the turbine to its power limit. The governor and capacity control system should be able to detect a reduction in refrigeration load and drive the turbine away from a power limit condition. This demonstrates the need for tighter integration between the capacity and governor control systems.

Governor system for C3-MR based LNG plants with constant speed compressor operation should account for bog down of gas turbine caused by helper motor overloading or trip. Integration with capacity control system will ensure loading of compressor (using inlet or stator vanes) within the performance limits of MR compressor [Figure 2].

Advanced Process Control (APC) System

Integrating the ITCS with the APC can potentially maximize the profitability of LNG plants. An advanced control system runs in a traditional, distributed control system (DCS) to optimize production. The function of an advanced control system is to generate set points to the ITCS and to the regulatory process control system in response to the optimization routines employed in a DCS. For example, the optimum set-points for methane, ethylene and C3 compressors in a Phillips optimized cascade process can be determined by the APC system and integrated with ITCS to enhance the plant profit margins.

Turbomachinery Performance Testing

The surge-limit parameters configured in the antisurge control system should, at the least, be verified for safe operation of the compressor within the vicinity of theoretical surge limit line. For retrofit projects, performance data (supplied by OEMs) usually deviate from actual data because of degradation in machinery performance over time. Consequently, it may be required to test a compressor in the "as installed" condition to confirm an acceptable stable regime of operation.

Conducting compressor performance tests is the prerogative of the plant operator/owner and is determined from suggestions and requirements of the OEM and the control system vendor, plant demand, and operation schedules.

Conclusions

Various challenges in the design of turbomachinery control system for commercially available LNG processes were highlighted. Figures 5 and 6 show the overview of ITCS implementation for APCI's typical C3-MR process and Phillips' cascade design. The ITCS developed by CCC based on the requirements outlined in the paper has been successfully commissioned in various LNG plants around the world. Seamless exchange of information among the process licensor, EPC Company, LNG producer & operators, OEM, LNG technology advisor, and control system vendors is highly essential for successful execution of ITCS projects.

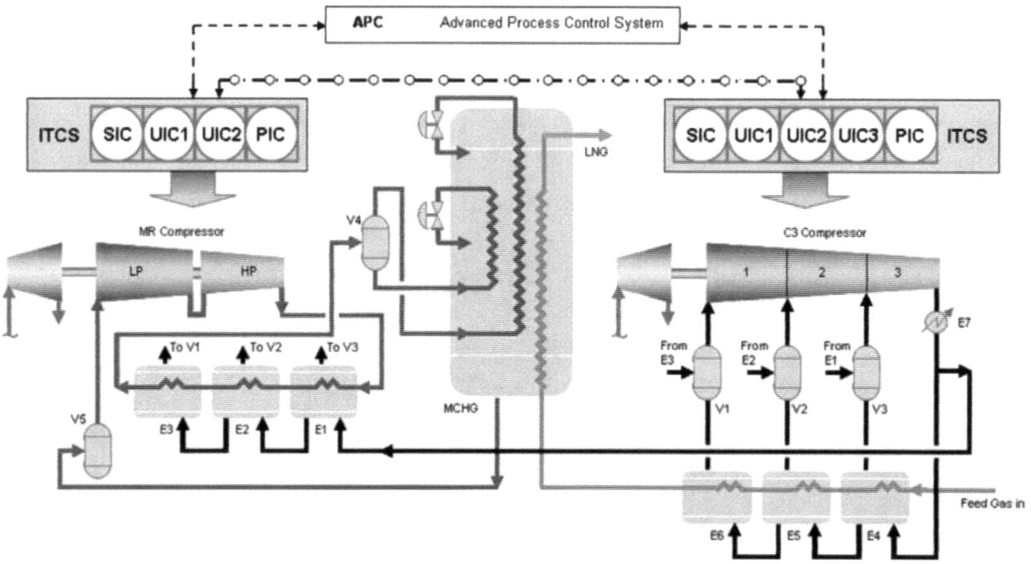

Figure 5. Integrated Turbomachinery Control System (ITCS) for APCI's typical C3-MR LNG process

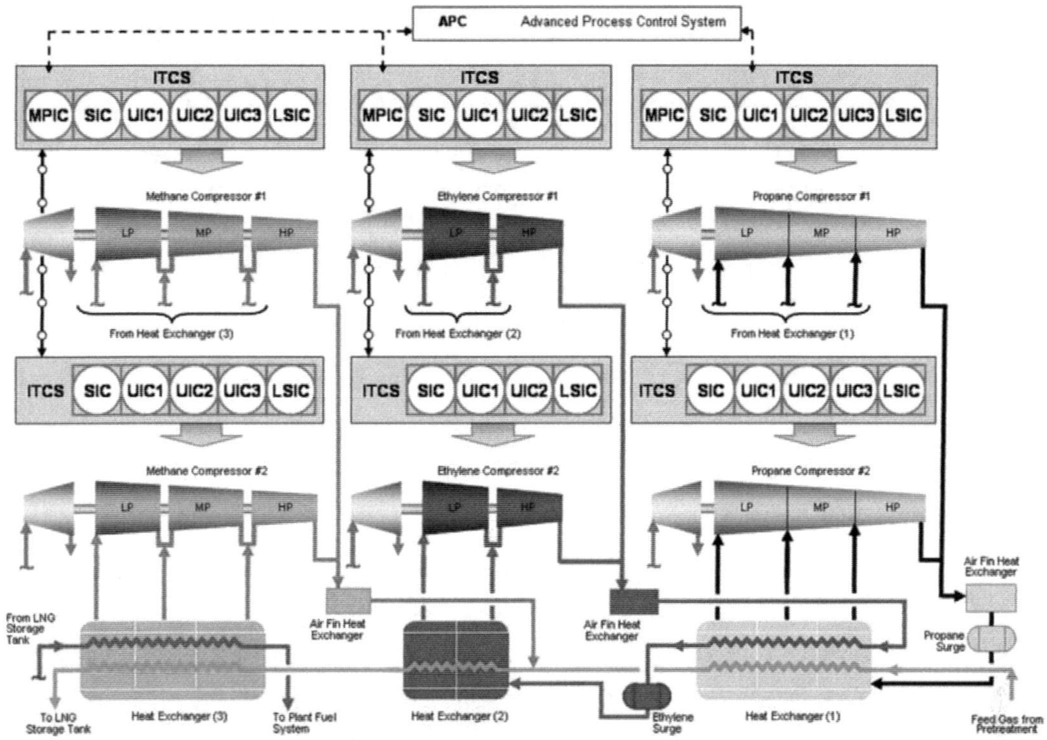

Figure 6. Integrated Turbomachinery Control System (ITCS) for a typical Phillips Cascade LNG process

ITCS will ensure that turbomachines in a LNG plant are controlled within a safe operating regime in a reliable and efficient manner by properly integrating the compressor and driver controls. ITCS will also simplify compressor start-ups and restarts; ensure that control valves are automatically positioned; compensate for changes in ambient conditions and gas compositions; and efficiently share the load between parallel operating compressors. These automations result in smooth start-ups, considerable decrease in downtime, and the prevention of excessive damage to rotating machines and to the cryogenic equipment; in addition to providing high-quality control of machinery and of the process during normal operations.

Due to ever-increasing global demand for LNG, the opportunities from continuous operation of an LNG plant with uninterrupted productions are greater than before. With several new construction and expansion projects being planned around the globe, this is not a good time for an LNG plant to experiment with unproven turbomachinery control system vendors and risk meeting delivery commitments due to unplanned shutdowns. As LNG technology licensors are striving to develop innovative designs, turbomachinery controls should not be the inhibiting factor in promoting creative process schemes involving complex turbomachines.

References

[1] Compressor Controls Corporation Patent; No. 5,508,943. *Method and Apparatus for Measuring the Distance of a Turbocompressor's Operating Point to the Surge Limit Interface*. April 1996.

[2] Compressor Control Corporation. *Series 3 Plus Antisurge Controller for Centrifugal and Axial Compressors*. Instruction Manual, Publication IM301, February 2001.

[3] Compressor Control Corporation. *Series 3 Plus Performance Controller for Centrifugal and Axial Compressors*. Instruction Manual, Publication IM302, February 2001.

Biography of Speaker

Krishnan Narayanan, Compressor Controls Corporation's Application Marketing Director, is author of several technical papers and a regular speaker on turbomachinery control systems in various industry forums. Mr. Narayanan has over seventeen years of experience in the Oil & Gas industry with expertise in control system design, field engineering, dynamic simulations, and advanced control applications. In addition, he continuously evaluates technological issues in the market place and leads product development initiatives for new and existing products. His work has awarded him several patents in advanced control methods for turbomachinery systems. Mr. Narayanan is a registered Professional Engineer (P.E) and holds MSME degree from University of Minnesota, Minneapolis and BSME from Indian Institute of Technology (IIT), Madras, India.

Optimizing LNG Expander Performance across the System Resistance

Christian Fischer
Shell Sakhalin Energy SE-LNG
Yokohama, Japan

Hans Kimmel
Ebara International Corporation
Sparks, Nevada

AIChE Spring Meeting, April 2005
5th Topical Conference on Natural Gas Utilization
Atlanta April 10-14, 2005

Abstract

The performance of LNG expanders operating at different flow rates is subject to system resistance. Applying Gilbert Habets' established economics criteria for LNG expanders (Dec.1999, Hydrocarbon Engineering) the paper presents optimization rules to maximize enthalpy reduction across the system resistance curve.

Offshore LNG Receiving Terminal Facilities that Optimize LNG Shipping Operations

Craig Taylor
HNG Storage Company

Capt. Craig Jackson
Golar Shipping

James Davis
AMEC Paragon Inc.

AIChE Spring Meeting, April 2005
5th Topical Conference on Natural Gas Utilization
Atlanta April 10-14, 2005

ABSTRACT

The LNG shipping industry thrives on set routes with established originations and destinations over long contractual periods to assure the maximum return on the investment made in modern LNG carriers. Also, these long-term contracts assure the liquefaction facilities and LNG producers of off-take capacity and in turn their own economic viability. The fully integrated Owner would have both the LNG production and loading facility as well as the receiving terminal selected and contracted as part of their multi-billion dollar investment in the LNG supply chain from gas well to eventual market.

Some LNG producers choose to not own receiving terminals as assets but rather to contract capacity from independent developers. Several recent announcements have indicated that this may be an emerging trend for U.S. LNG imports. Coincidently, the independent LNG import terminal developer desires to sell his terminal capacity to one or more Capacity Holders to achieve his business objectives and investment returns. These facilities follow the traditional LNG terminal model of receiving sufficient LNG on a "just-in-time" basis to satisfy the send-out demand with enough storage to meet the anticipated interruptions to normal service due to weather and traffic.

However, if a particular LNG receiving terminal would offer significant financial advantages / incentives to LNG producers in the form of significantly lower shipping, storage and processing fees, could that terminal attract long term contracts? Furthermore, could the added flexibility of large-volume, low-cost storage boost the emergence of a spot market and the interest in maximizing the buying and trading cargoes?

This paper discusses a terminal concept that is optimized with respect to the shipping operations and that could provide preferential financial terms to capacity holders with both long-term contracts and for any cargoes purchased on a 'spot' market. This is achieved by integrating the shipping operations into the design of the receiving terminal.

INTRODUCTION

Traditional LNG-trade economic models may be impacted by the emergence of a new LNG receiving terminal concept that focuses as much upon the hydrocarbon storage aspects as it does LNG receiving and gas send-out. An offshore terminal design is discussed within this paper in which large volumes of gas storage can be provided at very low cost and which offers direct access to berths / moorings from shipping fair-ways without the need to transit The benefit is a significantly lower cost of operations and a unique capability to de-couple LNG receiving from gas send-out by means of large volumes of low cost gas storage.

LNG SHIPPING

The economic viability of a typical \$1.5 - \$2.5 billion USD liquefaction plant depends upon a relatively constant volume of LNG daily that must be moved to market. Project financing will mandate the Owner to establish: 1) a long term gas supply contract at agreed rates; 2) long term LNG shipping contracts to assure the LNG reaches market; 3) re-gasification and storage capacity at one or more import terminals connected to end-use markets. To secure the shipping, long terms charters will dedicate a suitable number of the proper size ships to service the LNG product flow. The numbers of ships and their sizes are selected for a particular trade as part of the overall LNG supply chain. Without assured transport of the LNG and unloading and re-gasification capacity at receiving terminals secured, the liquefaction plants would not be funded or built.

LNG Tanker Fleet Size

The numbers and sizes of ships (reference Table 1) vary considerably with the newer ships becoming ever larger to achieve economies of scale. There are about 175 ships currently in charter service operation worldwide with a combined capacity of over 20 million m^3 and another 100+ ships in various stages of design and construction.

Table 1: LNG Carrier Fleet

Fleet Composition	Vessels	Total Capacity Cubic meters	Avg. Size
Existing Fleet	175	20,700,000	119,000
LNG Carrier Orders	112	16,300,000	145,000

Some of these existing ships are small 75,000 m^3 short-haul and coastal designs. At the other extreme, there are new world-scale 151,000 m^3 tankers by

Teekay in early construction. There are also reports of tenders issued for 200,000+ m^3 mega-ships from international oil companies that are designed for large loads and long hauls to take advantage of economy of scale.

This added shipping capacity comes at a time when over 304 million cubic meters per year of new and expanded LNG facilities will be coming on stream from 2004 to 2009. Each LNG liquefaction plant has limited storage and must load ships at a regular pace to avoid production impacts. The carriers on order, if actually built, will be expected to transport that new volume annually. That will require about 19 trips annually for each ship in this new "fleet" and each trip must be completed on average within 18.5 days[1].

To assure financing of traditional LNG projects, each vessel, whether existing or new-build, is in service under a long-term (up to 25 years) hire serving a specific shipping route between a specific LNG liquefaction facility and designated receiving terminal(s). With this in mind, the charter rate is established based upon experience, economic models and simulations using information such as:

- The cost of the ship – about $180 mm USD for a new 145,000 m^3 vessel
- Annual operating costs – fuel, crew, insurance, port fees, maintenance
- Annual volume of LNG to be carried – number of shipments per year
- The number of ships required - distance and time for round trip
- The shipping distance from berth to berth including the need to navigate locks, channels or rivers
- The expected annual delays due to weather forcing alternative transit routes, reduced steaming speed or delays in transiting restricted waters, e.g., fog
- Expected delays in transiting rivers, channels or harbors due to other shipping traffic
- Demurrage and Day Rates – about $75,000 /day

Numbers of Ships - Example

These ships are expensive to build and operate. Therefore it is advantageous to minimize the number of ships dedicated to a trade. A very simple example of determining the number of ships required to service a given LNG liquefaction plant and a selected LNG receiving terminal is provided in Table 2. In this hypothetical example, the LNG fleet required to service this single train LNG plant is 6 ships of 138,000 m^3 capacity. Note the amount of time allocated for unplanned delays. Each LNG carrier's whereabouts is carefully monitored for steaming speed, course, location, sea state, weather and other critical data

[1] Using 15 annual days downtime

points. Weather may force use of an alternate route or delay in transiting a channel due to fog.

Just as the LNG liquefaction plant needs regular load out, the receiving and gasification terminal requires the same. An inland 1.0 bcfd re-gasification terminal will require about 48,000 m^3/day that in turn requires a 138,000 m^3 carrier (carrying about 125,000 m^3 net = 3.1 bcf gas equivalent) to arrive about every 2.5 days. This is about 146 cargoes per year. As is shown in Table 2, the number of equivalent size shipments from a single LNG liquefaction train is 85.

This highlights the challenge of balancing a LNG train with given capacity to a receiving terminal with perhaps different capacity. These results in a balancing act of finding sufficient LNG to meet the terminal demands and the terminal must find sufficient supply to service its markets. There is little room for deviations from LNG production or receiving and gasification planning or any delays in transit. It is not unexpected then to see LNG suppliers arranging for capacity at multiple receiving terminals and even buying a portion of the capacity. They wish to assure themselves of getting the LNG somewhere.

Table 2: Example of Project Fleet Sizing

FACTOR	VALUE	COMMENTS
Annual LNG Production	6 million tonnes / yr[2]	Avg. single LNG train
Size of LNG carriers	138,000 m^3	Net capacity
Annual Number of Shipments to be Made	85.12	= 6 mm tpa x 2.27 / 138,000
Shipping Distance to Port	3,000 nm	
Time to Load	24 hours	@ 12,000 m^3/hr
Time to Transit – one way	166 hours	7 days @ 18 knots avg.
Allowed Delay, Slow Down	6 hours	
Time to Unload	20 hours	@ 10,000 m^3/hr
Time to Traverse Channel [3]	14 hours	River – total both ways
Total Terminal "Port time"	34 hours	Unload + traverse time
Total Time – Full Cycle	16.6 days	
Number of Annual Trips per Ship	15.53 = 15	=365 x 24/563
Number Of Ships	**6**	**85.12/15.53 = 5.48**

[2] 1 metric tonne = 2 cubic meters (actual factor is 2.27 m^3/tonne)

[3] The planned time for this transit from point at which Pilot boards till he leaves at departure

New Trends in Shipping?

Several interesting trends are emerging. First, the current shipping negotiations for new LNG plants and even some renewal contracts are favoring shorter terms of 12-15 years with options for renewal. Second, some Asian markets are seeking seasonal cargoes to better balance winter/summer extremes without the need to build expensive cryogenic storage tanks. Third, some of the newer LNG plants can produce more than nameplate capacity during certain times of the year. This could be several extra cargoes producing additional revenue if the transport is available and if the market can accept a "spot" cargo, i.e., the cargo can be sold. 1-3 extra cargoes could have a value of $30 - $40 mm USD net. If the LNG plant financing is based upon nameplate, this amount could be almost entirely bottom line.

At the same time it is interesting to see the numbers of ships being planned and built at the various shipyards around the world that have no dedicated routes or charters. At least 11 of the 140,000+ m^3 size have been announced. Typically the shipping and the destinations are agreed in advance as part of the LNG plant financing plan. These unassigned ships may be place holders in the construction queue hoping that someone will pick up the options as a means to transport additional LNG from expansions or de-bottlenecking

The traditional "tight" LNG business model is to make a given amount of LNG, match the size and number of ships to transport that amount of LNG to a designated receiving and gasification terminal that has enough storage to maintain gas send-out between ship arrivals. The entire system is "optimized" to that business model and there is only a pre-determined amount of flexibility built in.

[0]In this current LNG market the more significant aspect seems to be the receiving terminal rather than the shipping. Ship building is ahead of the terminals at this time and is limited only by shipyard space that eventually becomes available. Not so of the terminal – each will be contracted for long periods.

TERMINAL INTERACTION WITH SHIPPING

There are two key factors governing the interaction of the shipping and the terminal. They are: 1) the LNG storage capacity provided at the terminal; 2) the ability to receive ships of all sizes, moor them when they arrive, and offload them as quickly as possible. The current design practice is to minimize capital cost and operating costs by adopting the "just-in-time" LNG delivery by ship and managing everything else to that delivery.

LNG Cryogenic Storage

The cryogenic storage tank provides a modest buffer that allows variance in either the receiving (tank refilling time) or the gas send-out rate due to changes in demand. Cryogenic storage is very expensive (reference Table 3) with a typical full-containment storage tank of 160,000 cubic meters costing $65 - $75 mm USD for a Gulf Coast location resting on a piled foundation. A similar size single containment tank would be less (about 70%). Both are sized to match the ships to be used. Typically tanks serving a 138,000 m3 vessel would be designed at 160,000 net[4] to assure sufficient space for a full tanker load yet with a minimal "heel" in the tank bottom for carrying operations until the next ship arrives. These tanks require considerable ground space and safety systems.

Table 3: Relative Cost for LNG / Gas Storage

Storage Type	LNG Gross Vol.	Net Gas Cap.[56]	Cost	$ per bcf
Single Containment	160,000 cu. mtr.	3.0 billion scf	$45 mm	$15 mm
Full Containment	160,000 cu. mtr.	3.0 billion scf	$65 mm	$17.3 mm

A "tight" design with little spare storage provides operational challenges or lost opportunity for the unplanned, unknown and unforeseen. For example, a lower gas demand or vaporization downtime may leave too much LNG in the tanks to offload an on-time ship. This might either cause a "dumping" of the LNG at a lower cost per Btu to make room or cause a delay in the offloading. This is a readily observable event in the current US gas markets. The price of gas at terminals follows the shipping arrival schedules due to inadequate tank storage.

Some terminals delay mooring and offloading until the terminal and harbor are clear of all traffic and until there is sufficient tank capacity to receive all the cargo at full rate. This delay is done in order to minimize the time the vessel is alongside and thus the risk/exposure of the vessel.

---

[4] Working liquid volume; excludes heel. A 138,000 m3 vessel will net about 125,000 m3 at the terminal after boil-off of 0.15%/day burned as turbine fuel in transit.

5

[6] Net operating volume assumed to be 90% of gross; accounts for heel

Similarly, a delay in transit due to weather or ship traffic could cause the terminal to run low on LNG. Or a major change in weather could create a demand peak that could be opportunistic if sufficient LNG / gas could be made available without jeopardizing the contracted gas sales volumes.

Current Receiving Terminal Design Issues

With few exceptions, as discussed above, every LNG receiving terminal built to date has a "tight" design. Also, its berths are located in a harbor or water-way requiring transit of a channel or river frequented by other ship traffic and with maneuvering restrictions for weather.

Not every vessel can operate on every shipping route as most LNG receiving terminal locations have restricted water depth due to dredged channels making ships larger than 125,000 or 138,000 m^3 impassible with full loads. Inland locations require special safety and security precautions to be implemented that can disrupt other ship traffic or delay the LNG carrier while other critical ship traffic clears.

If the trend is toward more spot or seasonal cargoes, then the ability to serve a broader range of ship sizes would be an asset. This means not only serving the vessels chartered by the major capacity holder but any that might have opportunistic cargoes.

Given the NIMBY feeling in most of the USA, an offshore terminal has significant advantages in that little time and money is lost in consulting and debating with the local community. With any onshore terminal there is significant risk that a given project could be delayed or cancelled as a result of socio-political influences and overt opposition.

NEW "ENERGY" TERMINAL DESIGN

A new offshore concept is being implemented at the Freedom[7] LNG Receiving Terminal. It will accommodate virtually any vessel and would provide sufficient storage for accepting spot and seasonal cargoes. The design utilizes the patented Bishop Process™ in which LNG is offloaded, pumped to pressure and immediately vaporized to be injected into salt caverns for storage. Concurrently, a portion of the gas is pumped to pipeline pressure and vaporized to service the send-out demand. Between ship arrivals, the send-out demand is serviced from the caverns without compression.

The unique aspect of this design is the "de-coupling" of the LNG receiving / and the gas being sent-out to market. The enabling technology is the vast amount of

[7] Managed by Freedom LNG Receiving Terminal L.P.

inexpensive storage in the form of multiple salt caverns mined in an offshore salt dome – of which there are many in the Gulf of Mexico.

Design Concept

The Freedom terminal is comprised of jackets, platforms, pipelines and wells – very typical of the Gulf of Mexico. Artificial land will not be built. Two or three mooring berths (simple breasting and mooring dolphins) in several orientations in less than 100 feet of water to accommodate ships of up to 250,000 m3 capacity. Extended movement LNG arms will provide offload rates of 10,000 to 12,000 m^3/hr.The vaporization scheme is comprised of four parallel trains of 1.4 bcfd each with vaporized gas in the dense phase being sent to the caverns or to the pipelines. The average send-out rate is designed for 3 bcfd into at least three pipelines. The parallel trains offer dispersion of assets and built-in redundancy, making standby spares un-necessary.

The four process units will each be built on standard 8-pile jackets and two-level decks that can be fabricated in several yards along the Gulf Coast. The heaviest lift is limited to 4,000 tons enabling several lift barges to make the installation. The other Freedom platforms include a 4-pile quarters platform and a 3-pile flare minimal structure – all of which will be connected with bridges.

The send-out pipeline laterals are typical of Gulf of Mexico buried marine gas pipelines placed with conventional lay barges and installation procedures. There will be high-pressure (ANSI 1500) gas lines from the vaporization trains to the 8 salt caverns that will flow 750 mm scfd. The caverns will operate in uncompensated mode with an operating range of 1,200 to 3,000 psig. Each cavern will have two wells, each with a dry tree on a minimal structure, and each well will have a subsurface shutoff valve.

The initial design calls for 60 bcf of total gas storage of which 30 bcf is working gas. The remainder is cushion gas, which is used to maintain the caverns at the pipeline pressure. This cushion gas is fully recoverable by injecting seawater or brine. However, this recovery is not considered part of the normal operations.

The construction of this facility is typical of any offshore Gulf of Mexico jacket, platform, pipeline and drilling. The fabrication and installation can be accomplished in 28-30 months once the fabrication begins and the cavern drilling commences. First gas can be achieved in 32 months using either an LNG cargo or gas from the connecting pipelines to provide the cushion gas. An equivalent onshore terminal with two 160,000 tanks would be on-line in 38-40 months.

The Freedom terminal is environmentally benign. The power generated is with LM2500 gas turbine units (commonly used offshore) with ultra low NOx combustors. Any supplemental heating will also have ultra clean burners. There is no LNG inventory at site as the LNG is vaporized upon offloading. There is no diesel fuel other than emergency generators and pumps. Bunkering will not be provided to the carriers. LNG carriers have an excellent safety record and are clean in operation.

Safety in the design is enhanced by the dispersed assets (platforms will be a minimum of 200 feet apart) and redundancy from parallel trains. Personnel are typically not on the process platforms during unloading operations. Between unloading operations, the system is idle and depressurized. There is very low inventory of LNG or gas at any time. As soon as the LNG is taken off the ship it is pumped to pressure, vaporized and either sent to pipeline or to the caverns. The LNG pumps will be equipped with a HIPPS system to assure safety. Upon eventual decommissioning of the facility, the accepted Gulf of Mexico procedures for removing and decommissioning of offshore platform and for pipelines will be utilized. Similarly, the caverns can be filled with brine and closed.

Salt Cavern Storage Flexibility

The Freedom terminal will have over 30 billion scf of working gas storage capacity. It will service the gas market with up to 3 bcfd of gas and a peak capability to reach almost 5 bcfd as demand warrants. The hydrocarbon is stored underground as pressurized dense phase gas enabling the terminal to dispatch large volumes with minimal response time. As seen in Table 4, the cost per billion cubic feet of gas stored in salt caverns is vastly less than the cost of storing the hydrocarbon as a cryogenic liquid in tanks.
Normal operations do leave storage capacity for early arrivals and non-chartered spot cargoes purchased opportunistically. In a similar manner, delays in arrivals would not disrupt send-out as there can be up to 10 days of gas send-out stored. Another aspect is to use the terminal for long-term strategic gas storage. As a point of interest, the incremental cost to add 10 bcf cavern storage would amount to $50 mm USD. There is no practical upper limit to storage capacity as additional caverns can be mined during normal operations.

Table 4: Salt Cavern Storage Cost Comparison

Storage Type	LNG Gross Vol.	Net Gas Cap.	Cost	$ per bcf gas
Single Containment	160,000 cu. mtr.	3.0 billion scf	$45 mm	$15 mm
Full Containment	160,000 cu. mtr.	3.0 billion scf	$65 mm	$17.3 mm
Salt Cavern	*1.6 mm cu. mtr.*[8]	24[9] billion scf	$117 mm	$4.9 mm

[8] This is the LNG equivalent of the gas stored; LNG cannot be directly injected into salt caverns

The cost of gas storage in salt caverns becomes a minor cost factor in the design. This enables the shipping operations to be flexible to receive and offload chartered vessels and any spot cargoes or seasonal cargoes that might be acquired. The volume stored would enable the facility to weather storms while continuing to meet send-out obligations.

Ships Servicing Freedom

With 3 bcfd being delivered to market through as many as five pipelines, it is advantageous to utilize bigger carriers and to provide multiple docks to assure sufficient maneuvering time and docking time to minimize non-productive offloading activity. The offshore location facilitates ship movements, maneuvering and docking without the encumbrance of being in closed channels or rivers. The terminal and its multiple berths would provide mooring for two ships concurrently while offloading one at a time virtually eliminating any vaporization downtime between ships.

The design basis of 3 bcfd send-out rate will require 243 shipments annually of 215,000 m^3 net vessels. These will come from various locations with differing transits times. The operating flexibility of the Freedom terminal with its offshore location, vast storage, and easy ship access will enable this number of shipments to be serviced in somewhat of an un-balanced mode, i.e., ships can arrive when they need to. The terminal will have two dedicated 7,000 hp tractor tugs to facilitate movements.

New Paradigm

The design of Freedom will establish a new paradigm in LNG terminals. Some of the major aspects are:

1. It will become an "energy terminal" with as much attention to the storage of gas as to the receiving and send-out.
2. Gas can be stored to meet varying demands with rapid response to demand changes.
3. Opportunistic purchases of spot cargos can be leveraged during higher seasonal demand because there will be always a place to put the gas[0]
4. In periods of low demand the gas can be purchased cheaply and stored cheaply keeping the vessels in steady employment
5. Reducing exposure to increasing purchase prices in times of high demand by taking advantage of the high selling price by selling the previously purchased and stored cheap gas.

---

[9] 34 bcf total gas stored = 24 bcf working gas

6. With the addition of compression, gas can be purchased from the pipeline at low demand and low prices to be stored in the caverns for later sales when prices are better
7. The ability to accommodate CNG carriers with adequate storage to offload their entire cargo.

Low Cost Services

The Freedom LNG Import Terminal unit capital cost is approximately 70% of that of a similar-sized land-based terminal. The operations unit cost is slightly more than that for a shore-based terminal due to offshore transport of crews for 7-day turns, dedicated vs. shared tugs, power generation vs. power grid, offshore accommodations, catering, etc.

The energy unit cost for the Freedom terminal is significantly less than existing land-based terminals. Land-based terminals typically use submerged combustion vaporizers to gasify LNG. The Freedom terminal will use sea-water-based re-gasification energy, which is significantly lower cost that submerged combustion.

The cost data for the Freedom terminal has been examined from two perspectives: (1) US Government study and (2) Project sponsor engineering. First, the Bishop Process™ concept that Freedom uses was studied extensively under sponsorship of the National Energy Technology Laboratory ("NETL") of the U.S. Department of Energy beginning in 2002. (The Freedom terminal is a licensee of the Bishop Process™.) That NETL work culminated in a report finalized in November of 2004. That report includes extensive cost analysis data for salt-cavern-based receiving terminals.

Second, the sponsors of the Freedom terminal took the "lessons-learned" from that NETL effort and developed a clean design basis that minimizes the overall capital cost and operating cost while maximizing flexibility, "optionality", and security. That objective has been achieved to a great extent with further optimization having been identified in the pre-FEED design basis. The benefit to the Owner will be a service fee that is lower than that of conventional onshore terminals and offshore gravity based structure ("GBS") terminals. The Freedom terminal should be the most cost-effective terminal in the US.

Offshore Natural Gas Liquefaction Process Selection

Michael Barclay, Senior Process Engineer
Noel Denton, Process Engineer
Foster Wheeler Energy Limited
Reading, Berkshire, UK

AIChE Spring Meeting, April 2005
5th Topical Conference on Natural Gas Utilization
Atlanta April 10-14, 2005

ABSTRACT

Offshore natural gas liquefaction has different process requirements to the traditional on-land baseload plants. While thermodynamic efficiency is arguably the most important process selection criteria for large onshore natural gas liquefiers, other factors become more important for offshore projects. Thermodynamic efficiency is likely to remain critically important. However, for offshore applications criteria such as compactness and process safety become more significant considerations. The high efficiency pre-cooled mixed refrigerant cycles that dominate onshore LNG installations will likely not meet the needs of the future mobile and offshore liquefaction projects.

The authors review natural gas liquefaction cycles in the context of, compactness, ease of operation, process safety, and efficiency for offshore liquefaction. Particular attention is paid to the lower-efficiency single and dual expander processes. These cycles offers several advantages over traditional cascade and mixed refrigerant (MR) liquefiers for offshore, on-board and mobile applications. The authors share their insights gained through their unique design and operating experience on small-scale N_2 cycle LNG plants and marginal field developments.

1 INTRODUCTION

Increasing global demand for natural gas is supporting the rapid growth of worldwide LNG production capacity. As demand continues to grow and the value of natural gas remains high, the impetus to monetise non-traditional gas resources also grows. Offshore floating LNG production has generated interest because it offers the potential to avoid flaring or reinjection of associated gas and to monetise smaller or remote fields of non-associated gas. With the realisation of large floating production, storage, and off-loading (FPSO) facilities for oil production and more recently LPG production, an LNG FPSO project appears to be increasingly more likely in the future.

1.1 Global LNG Trade and Emerging Markets

Worldwide LNG trade in 2002 was approximately 110 Mtpa and is expected to increase by over 70% to 190 Mpta in 2010 (Smaal, 2003). In 2003 there were a total of 20 liquefaction facilities in operation with a total of 66 trains and a combined production capacity of 135 Mtpa. In the same year there were 14 liquefaction trains under construction with a combined additional capacity of 58.4 Mtpa (DOE/EIA, 2003). LNG demand, driven by both existing and emerging markets is expected to play an increasingly important role in the natural gas industry and global energy markets. Three key emerging markets are China, India, and the US.

China and India have large potential coastal markets, and high domestic growth, and both have either just completed or are currently building their first receiving terminals (DOE/EIA, 2003). With two receiving terminals in various stages of development, China's economic growth will be fuelled partially by LNG. China's first LNG receiving terminal (758 Bcf per year (3.3 Mtpa)) in Guangdong is scheduled for completion in 2006/2007. India has LNG receiving terminals in various stages of completion in Dabhol, Dahej, and Hazira. These receiving terminals have a combined import capacity of over 485 Bcf per year (10.0 Mtpa).

The US is re-emerging as a large LNG importer and this trend is likely to be continued as domestic gas production growth fails to keep up with demand. With two receiving terminals reopening in the past three years the US now has four import terminals on the Atlantic and Gulf coasts with a combined peak capacity of 1.2 Tcf per year (26.0 Mtpa) and a baseload capacity of 0.88 Tcf per year (18.5 Mtpa). Receiving capacity is expected to increase dramatically in the next five years with four (of the over two dozen proposed) new LNG receiving terminals expected to be constructed on the Atlantic and Gulf coasts from 2007 through 2010 to meet LNG imports projected to increase nearly 60% during that period (DOE / EIA, 2004). Import terminals are also planned on the West Coast to support high demand from power generation and industry in the California and along the Mexican boarder.

1.2 Traditional Onshore Liquefaction

The logical starting point for any new LNG production scheme should be the existing industry and processes. The base-load LNG industry now has over 40 years of history starting with permanent operations of the Camel plant in Algeria in 1964. The earliest plants consisted of fairly simple liquefaction processes based on either cascaded refrigeration or single mixed refrigerant processes and train capacities were less than 1 Mtpa. In 1972 Brunei Lumut 1 utilised the first two-cycle process using a propane pre-cooled mixed refrigerant (C3MR) developed by Air Products and Chemicals Int. (APCI). This process became the dominant liquefaction process technology by the late 1970s and continues to be the workhorse of the LNG industry today. During this period APCI and others have made significant improvements on the original C3MR process. Economies of scale, improved process simulation tools, and improved equipment performance (i.e. liquid expanders and gas turbine drivers) have all dramatically decreased installed liquefaction plant costs, improved performance, and increased the capacity of liquefaction trains. The continued development of traditional LNG plant design can be seen by comparing recently commissioned plants to current and planned facilities. Less than five years ago Foster Wheeler and Chiyoda completed the engineering, procurement and construction (EPC) contract for Oman LNG. At the time of start-up (Feb. 2000 for Train 2), this plant had the largest trains in operation at 3.3 Mtpa and set a benchmark for process efficiency with a reported average specific power of 10.15 kW per tonne per day of LNG (McLachlan et. Al., 2002). Five years later, installed train capacities are over 5 Mtpa with projects in development for 7.8 Mtpa.

The liquefaction process typically accounts for 30-40% of the capital of the overall plant (Shukri, 2004) and has a large impact on utilities and operating costs. Selection of the appropriate cycle is critical to cost effective LNG projects. Historically, liquefaction cycle selection was an easy choice to make: the APCI C3MR. Table 1 shows the baseload liquefaction trains currently operating, in various stages of construction, and planned (in the case of AP-X).

Table 1. LNG trains by liquefaction process (DOE/EIA, 2003, Shukri, 2004, Mayer, 2004)

Liquefaction Process	**Licensor**	**Number of Trains**			**Startup**	**% of**
Nov-04		Running	Constr.	Planned	**Year**	**Market**
Propane Precooled MR	**APCI**	**55**	**9**	**?**	**1972**	**72%**
Optimised Cascade	Conoco-Phillips	3	4	?	1999	8%
Single Refrigerant MR	APCI	4	-	-	1970s	4%
Classic Cascade	Marathon/Phillips	4	-	-	1969	4%
Teal Dual Pressure MR		3	-	-		3%
Prico Single Stage MR	Black & Veatch	3	-	-		3%
MR Processes (C3MR & Dual-MR)	Shell	-	3	?	2005	3%
Multifluid Cascade	Linde-Statoil	-	1	?	2006	1%
AP-X Process	APCI	-	-	3 (4)	2007/2008	0%

The table illustrates two key points: first the APCI C3MR process dominates the industry; secondly there has been a considerable diversification of liquefaction processes in the last 5-7 years. This increased competition has led to increased train capacity, improved driver integration, and decreased capital costs.

Four trains of APCI's new AP-X liquefaction process are planned in Qatar for Qatargas II trains 4 and 5 and RasGas II trains 6 and 7(LNG Express, 2005). These planned plants represent the state-of-the-art in land-based liquefaction featuring single train liquefaction capacities of 7.8 Mtpa using a N_2 expander cycle to affect the sub-cooling of the LNG.

1.3 State of Development for FPSO

The use of floating production systems within the offshore oil and gas industry is maturing technology with over 119 currently operating FPSOs and FPSs, a further 22 under construction or conversion, and 139 prospects being considered from 2004 onwards (Infield Energy Data Analysis, 2005). The offshore market is dominated by FPSO facilities with, as of year-end 2002, more FPSO deployments worldwide than all the other floating production systems put together (Spars, Tension Leg Platforms and Semi-Submersibles) (Harbinson et. al., 2003). These FPSO production systems range from Aframax based systems processing ~ 30,000 BPD of crude to ULCC (Ultra Large Crude Carrier) based systems processing 200,000 BPD of crude. The use of FPSOs provides a number of benefits as the production facility also provides storage for the crude. This eliminates the requirement for local infrastructure to transport the crude to shore.

The floating production industry has recently moved to processing gas and condensate as well as crude. Traditionally, associated gas has been re-injected and remote gas reserves were left untapped due to the difficulty in delivering gaseous (low energy density) products to market. How can we benefit from the associated gas or stranded gas reserves? This challenge has led to the development of LPG FPSOs such as the Sanha for ChevronTexaco situated offshore Angola (ChevronTexaco, 2005). This project helps to eliminate routine flaring and provides marketable products but still requires residue gas to be re-injected.

Floating offshore LNG production offers the potential to avoid such re-injection and captures the other benefits associated with crude installations. It eliminates the need for infrastructure and provides storage to enable tankers to be utilised to transfer the product to market. This concept is now being studied by numerous organisations such as Shell with their Floating LNG (FLNG) and Floating Oil and Natural Gas (FONG) concepts for processing gas and associated gas respectively (Faber et. al., 2002). The floating production concept has not yet been extended to LNG production offshore but Foster Wheeler and others are prepared to engineer such projects (Raine and Kaplan, 2003; Raine et.al., 2003).

2 Offshore Liquefaction

Different floating LNG production process selection criteria have long been recognised with early studies on the topic occurring over 25 years ago (Kennett et. al., 1981). N_2 expander cycles for offshore liquefaction were discussed and studied by Foster Wheeler and others in the 1980s (Finn et. al., 2000). During the twenty years since initial conceptual design, the development of hydrocarbon processing on FPSOs has matured with several production facilities currently operating in a variety of environments and configurations.

2.1 Offshore Liquefaction Process Criteria

Offshore liquefaction facilities have different technology selection criteria than their onshore counterparts leading to different optimal processes. Offshore facilities must be compact, light, and offer high inherent process safety. They must also consider the additional constraints placed on the system in the marine environment such as vessel motion and offer a high degree of modularity, ease of operation, low equipment count, quick start-up, and high availability. Additionally, because FPSOs will be processing gas from marginal fields they must be tolerant of a variety of process conditions and have a high degree of inherent process robustness. High process efficiency remains an important selection criterion because, even with inexpensive feed gas, poor efficiency must be paid for with increased utilities, compressors capacity, and other major CAPEX items.

The technology in use on existing FPSOs such as turbines, compressors, towers and separators has already set the groundwork for installing machinery on floating facilities. This development over the past 20 years allows the step to be taken to offshore LNG with a large number of already proven components.

Other factors which must be considered are storage and offloading. The transport of LNG in tankers is well established. However partial fill conditions in the LNG FPSO will occur as the LNG is being processed prior to offtake. This may result in sloshing which is of particular concern in membrane tanks. The consideration of loss of containment must also be addressed when considering hull fabrication. The use of concrete for the hull provides benefits in the storage of cryogenic fluids as it retains its structural integrity when in contact with the LNG; however, this must be measured against potential cost reductions if traditional steel ship designs could be utilised.

If offloading is considered with a typical spread moored configuration such as might be found offshore West Africa, then side by side offloading could be considered. This provides the benefit in that typically LNG tankers load at midships therefore providing more flexibility. However in less benign seas, weather vaning configurations are often used possibly with tandem offloading being required. To facilitate this a number of technology suppliers have looked at flexible loading arms for transfer of LNG between the production vessel and the tanker such as the SBM soft yoke mooring and offloading (SYMO) system. These factors have all been examined in various studies such as Project Azure and the Shell development work on FLNG and FONG (Sheffield and Mayer, 2001; Faber et. al., 2002).

2.2 Expander Processes

The process criteria outlined in the previous section suggest that gas turbo-expander based processes are well suited for offshore liquefaction. A simple expander refrigeration cycle, the reverse-Brayton, is seen below in Figure 1.

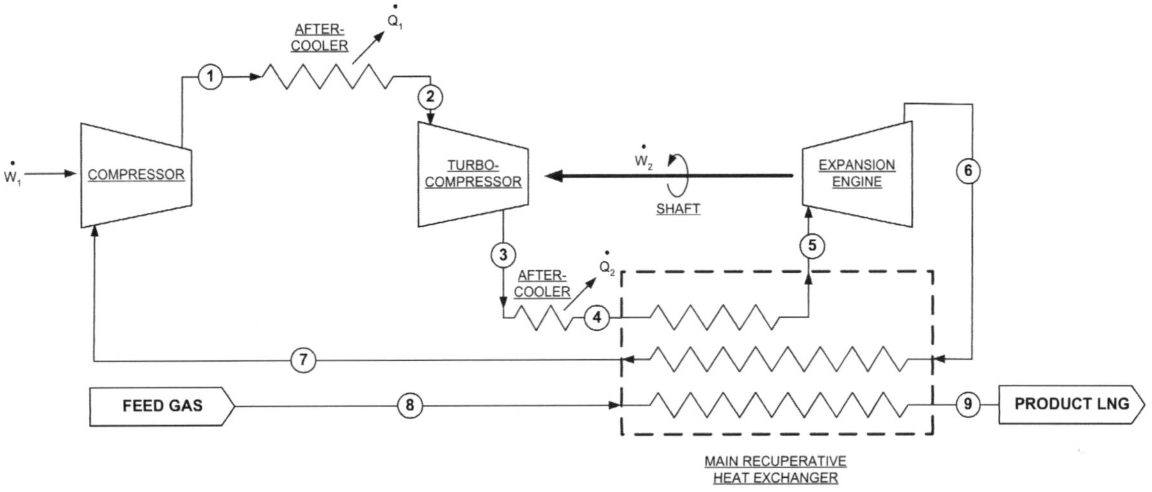

Figure 1. Modified-Brayton cycle refrigerator.

The working fluid in the refrigeration system, typically N_2, is compressed in the main compressor with *Stream 1* as its discharge. The heat is rejected from the cycle to the environment as the heat of compression in the after-cooler prior to a second stage of compression in the compressor loading the turbo-expander. The *Stream 3* has the highest pressure developed in the cycle. *Stream 3* then flows through a second after-cooler to reject additional heat to the environment. Next, S*tream 4* enters the recuperative heat exchanger where the refrigerant is cooled to well below ambient temperature. After leaving the heat exchanger, the cooled gas undergoes an isentropic expansion in the turbo-expander, which causes a large temperature drop in the refrigerant and produces shaft work. The lower pressure cold refrigerant in *Stream 6* is warmed in the recuperative heat exchanger with simultaneous pre-cooling of the incoming high-pressure N_2 refrigerant and removal of the sensible and latent heat in the process stream producing LNG. The warm, low-pressure refrigerant in *Stream 7* is then recompressed in the primary compressor to complete the refrigeration cycle.

While the process described above has been used to liquefy natural gas it has low efficiency because a gas with uniform flowrate through the cycle cannot closely match the cooling requirements in the process gas. Several variations on a single expander reverse-Brayton cycle may improve efficiency significantly (Foglietta, 2002). These variations include using two expanders (with or without the same working fluid), pre-cooling the feed gas with a MR or propane chiller, and expanding a saturated LNG product in controlled stages.

2.3 Comparison of Conventional Processes to Expander Processes

As discussed above, all large-scale LNG production processes use either a MR or pure component cascaded refrigeration cycle. Expander cycles use all or mostly gas refrigerants and offer lower efficiency but many benefits for offshore liquefaction. The following paragraphs compare how each of the offshore process selection criteria is satisfied by traditional baseload and turboexpander liquefaction cycles.

Compact and Light

The lack of compactness for the MR cycles is potentially their largest disadvantage. MR cycle liquefiers require extensive plot space for refrigerant storage and management. Two-way two-phase flow requires specific equipment (heat exchanger) and piping layout. The flammable nature of the refrigerant also places additional constraints on process and piping layout, offset distances, and restricted area classifications to ensure adequate process safety and code compliance. Using dual mixed refrigerant processes may be acceptable for offshore liquefaction as seen in concepts using either Shell or APCI processes (Sheffield and Mayer, 2001).

Expander cycles using N_2 as the refrigerant have the potential to be extremely compact because they feature:

- All gas service so there is no large refrigerant storage and management system decreasing plot requirements and weight

- Decreased offset distances because the refrigerant is non-flammable and places fewer constraints on equipment positioning

- Most of the surface area is dedicated to gas-gas service with flexible orientation. Heat exchanger cores can be arranged as needed. This allows the design of conformal coldboxes and modularized plant layout

Note that although the refrigerant circulation rate and main heat exchanger duty are significantly decreased in expander cycles, the required heat transfer surface may not decrease because the refrigerant heat transfer coefficient is also much lower.

High inherent safety

The LNG industry has built an excellent safety record that must continue to be aggressively protected as existing plants age and novel processes and production schemes are commercialised. Transitioning established MR processes to offshore production facilities has an advantage in that their risks and hazards are well understood, documented and mitigated.

MR and Cascade processes have large flammable refrigerant inventories, high refrigerant circulation rates through process lines, and extensive overpressure potential and flare requirements. This makes these processes inherently less safe than expander processes.

Expander processes have higher inherent safety because the refrigerant is inert. Recent studies suggest that although MR liquefaction cycles have significantly increased risk, the effect on overall facility risk may be small. Risk analysis studies completed for project AZURE suggest that the total process risk for the FPSO is not dramatically increased by the large increase in liquefaction process risk associated with a dual mixed refrigerant process (Sheffield and Mayer, 2001).

Marine Environment

Hydraulic design of the liquefaction process should consider the special constraints created by the marine environment. Previous articles have highlighted the impact of moving floating production on various LNG process equipment items (Waldie, 2002; Hatanaka et. al., 2001). Appropriate design of the process equipment and plant layout can ensure either expander or MR processes are fit for purpose.

Ease of operation, low equipment count, quick start-up

Turboexpander-based processes have the advantage in all three of these process selection criteria. Turboexpander processes are extremely easy to operate and control. The high equipment count and start-up time for MR processes is a function of refrigerant management and the inherent properties of the process. Storage tanks, separators, valve manifolds, and instruments and controls required to adjust and control the refrigerant charge and composition in the MR processes is high.

Cold production in a turboexpander process is largely independent of the process gas. Mixed refrigerant processes are more complicated because refrigerant composition, process pressures and temperatures, and the feed gas conditions are all coupled.

Start-up times may be limited by either thermal shock and equipment cool down times or by non-mechanical issues. Heat exchangers and piping must be cooled down in a controlled fashion to ensure thermal stresses associated with the change in temperature do not damage the equipment. Non-mechanical issues relate to refrigerant charging and tuning in both processes.

High Efficiency

There are only incremental improvements to be made in the existing MR cycle liquefiers. They were initially chosen because they offered the highest efficiency and have been refined for the past 30 years. The high efficiency on Oman LNG and other recent plants are a testament to this fact.

Expander-based processes on the other hand have been steadily improving with advances in plate fin heat exchangers, turbomachinery, and process configuration. Expander cycles shift the key efficiency determining element of the process from the LNG heat exchanger to the expander. Expander efficiency is critical to efficient operation as is compressor efficiency and process configuration. A dual-expander processes for LNG liquefaction may be either open art or proprietary and offer the potential to greatly increase efficiency by better matching the natural gas cooling curves. Efficiency of several expander plants is seen below in Table 2.

Table 2. Expander plant performance relative to Oman LNG

Plant	Liquefaction Process	Status	Licensor	Efficiency kW*day/ton	Relative Power
OMAN LNG, Trains 1,2 (*1*)	**C3 Precooled MR**	**Operational**	**APCI**	**12.2**	**100%**
Wildwood LNG Plant (*2*)	Single N_2 expander, closed	Operational	CFS	40.5	332%
LNG Jamal BOG Reliquefier (*3*)	Single Expander, N_2	Built, des. Cap.	-	37.8	310%
Kryopak EXP-Typical (*4*)	Single expander, process fluid	Operational??	KryoPak, Inc.	20.4	167%
Predicted / Patented (*5*)	Dual Expander C1 / N_2	Simulated	ABB	16.5	135%
Predicted / Patented (*5*)	C3 - Dual Expander C1 / N_2	Simulated	ABB	**13.5**	**111%**

1. *G. McLachlan, C. Ayres, K. Vink, M. Al Mukhainy, Efficient Production of LNG from the Oman LNG Project, Gastech 2002, Doha, Qatar, Oct. 2002.*
2. *M. Barclay, D. Gongaware, Distributed Liquefied Natural Gas Infrastructure: Design of the Next Generation of Liquefiers, Proc. of 21st Int'l. Congress of Refrigeration, Washington D.C., Aug. 2003.*
3. *N. Hatanaka et. al., A Challenge to Advance LNG Transport for the 21st Century- LNG Jamal: New LNG Carrier with Reliquefaction Plant, LNG-13, 2001.*
4. *Kryopack, Inc. website, www.salofompanies.com/html/kryopak/process_exp.htm, Jan. 2005.*
5. *J. Foglietta, Production of LNG using Dual Independent Expander Refrigeration Cycles, Presented at AIChE Spring Meeting 2002, New Orleans, March 2002.*

Data from the tables comes from a variety of sources. The Wildwood liquefaction plant is a distributed-scale LNG facility operating near Stockton, California on a single N_2 expander cycle. The plant processes a low-BTU non-associated gas containing about 20% N_2 at a feed pressure of 280 psig (19 bar) and produces a 97%+ C1 LNG stream. The apparent low efficiency can be contributed to a variety of features. First, the work associated with lower pressure liquefaction and N_2 rejection is included in the provided value. Secondly, initial CAPEX for the project was a key factor in the project. This supported the choice of an expander process and other decisions that further decreased efficiency including the use of modular stainless steel heat exchangers. Thirdly, the plant has a small production rate and some equipment is inherently less efficient.

The second N_2 expander plant is onboard the LNG – Jamal, an LNG carrier built for the Oman LNG project. Kryopak provides details for a "typical" actually operating single expander plant using the process gas as the working fluid at a capacity of 125 ton of LNG per day. The final two values from an ABB reference are simulated for a dual-expander process using both N_2 and flammable refrigerants in separate circuits without and with propane pre-cooling respectively.

Table 2 shows a wide range of performance for commercial expander natural gas liquefiers. As in all such comparisons, scrutiny should be paid to what exactly is being measured and how data is reported as they come from various sources. From the table it is fair to say that:

- All expander cycle plants are less efficient than the C3MR benchmark

- Dual-expander processes offer higher efficiency than single expander processes
- Propane pre-cooled dual expander processes may be approaching the efficiencies of state-of-the-art C3MR cycles but have surrendered their inherent advantages of low flammable refrigerant inventory, low equipment count, and simplicity

Regardless of reporting and site specific differences, it is clear that some small expander-based processes have been selected even though they were considerably less efficient than competing technologies. This is a testament that factors other than efficiency, notably project economics, are ultimately the drivers in project progression. While these projects undertaken to date are relatively specialized, it is apparent that some expander-based processes offer potential for the more general offshore market.

Based on the ABB and Kryopak reported efficiencies, expander processes may offer adequate efficiency for offshore liquefaction on FPSOs in either single or double expander configurations depending on process specific details.

3 Conclusions

Demand for LNG is increasing due to growing traditional and emerging markets. This growing demand coupled with evolving market conditions is creating the impetus for additional production including that from floating production. Consideration of many relevant process selection criteria for onshore and offshore natural gas liquefaction would suggest that expander cycles are better suited to offshore liquefaction on FPSOs than traditional liquid refrigerant processes. The expander cycle's primary disadvantage remains that it has an inherently lower efficiency than common onshore processes such as C3MR and Cascade processes.

When considering marginal field developments, either onshore or offshore, the use of non-proprietary turbo-expander based processes should be considered because they are offer ed the potential for a compact plant, simple robust operation, lower initial CAPEX, and a high degree of process safety. Higher operating costs must be considered with other costs for the shorter, less defined marginal project life. The ultimate choice of which process to select will remain dependent on project specific variables and potential development state of novel processes.

References

DOE/EIA, The Global Liquefied Natural Gas Market: Status and Outlook, Energy Information Administration, U.S. Dept. of Energy, DOE/EIA-0637, Dec. 2003.

DOE/EIA, Annual Energy Outlook – 2004, Energy Information Administration, U.S. Dept. of Energy, DOE/EIA-0383(2004), Jan. 2004.

F. Faber, L.R. Resweber, P.S. Jones, A.E.J. Bliault; Floating LNG Solutions from the Drawing Board to Reality, Offshore Technology Conference, Houston, Texas, May 6-9, 2002.

A J Finn, G L Johnson, T R Tomlinson, *LNG technology for offshore and mid-scale plants*, 79th Annual GPA Convention, Atlanta, March 2000.

J. Foglietta, *Production of LNG using Dual Independent Expander Refrigeration Cycles*, Presented at AIChE Spring Meeting 2002, New Orleans, March 2002.

N. Hatanaka, H. Yoneyama, T. Mihashi, M. Oka, M. Ohmori, *A Challenge to Advance LNG Transport for the 21st Century – LNG Jamal: New Carrier with Reliquefaction Plant*, LNG-13, 2001.

A. Kennett, D. Limb, B. Czarnecki, *Offshore Liquefaction of Associated Gas- a suitable process for the North Sea*, Offshore Technology Conference, May, 1981.

G. McLachlan, C. Ayres, K. Vink, M. Al Mukhainy, *Efficient Operation of LNG from the Oman LNG Project*, Gastech 2002, Doha, Qatar, Oct. 2002.

M. Meyer, *LNG Liquefaction Process- Why the Big Fuss about Selection*, IChemE London SONG Meeting, Nov. 9, 2004.

D. Harbinson, S. Robertson, R. Knight, Extract from "World Floating Production Report 2003 - 2007" as published in Offshore Magazine, Aug. 2003.

B. Raine, A. Kaplan, G. Jackson, *Making the Concrete Case*, Offshore Engineer, Dec. 2003, pp. 35-40.

B. Raine, A. Kaplan, *Concreter-based Offshore LNG Production in Nigeria,* LNG Journal, Sept./Oct. 2003, pp. 28-30.

J.A. Sheffield, M. Mayer, *The Challenges of Floating LNG Facilities*, Proc. of the GPA Spring Meeting 2001.

T. Shukri, *LNG Technology Selection*, Hydrocarbon Engineering, Feb. 2004.

A. Smaal, *Liquefaction Plants: Development of Technology and Innovation*, Working Committee Contribution, 22nd World Gas Conference, Tokyo, 2003.

B. Waldie, *Effects of Tilt and Motion on LNG and GTL Process Equipment for Floating Production*, GPA Europe Annual Conference, Rome, Italy, Sept. 2002.

LNG Remains Keystone to Development of Qatari Gas Resources, LNG Express, Vol. 15, No. 1., Jan 2005.

Global Perspectives Floating Production Market Update Report 2004/08, Infield Energy Data Analysis, http://www.infield.com/floating_production_market_reports.htm, Jan. 2005.

Construction of Sanha LPG-FPSO Continues Safely and on Schedule, ChevronTexaco Website, http://www.chevrontexaco.com/news/spotlight/25april2003_sanha.asp, Jan. 2005.

Gas Commercialization Costs – on the Way up

Devendra Agrawal, PE
Marathon Oil Company
Houston, TX

AIChE Spring Meeting, April 2005
5th Topical Conference on Natural Gas Utilization
Atlanta April 10-14, 2005

ABSTRACT

The impressive gains made by the LNG industry in the last decade in reducing the costs of gas commercialization by increasing unit capacities are in recent times being eroded by a variety of world dynamics. As technological improvements continue to stretch the size of each element of the LNG chain, trying to achieve further economies of scale, factors outside of technology continue to reduce benefits that could be achieved. This paper highlights some of these factors in play at each element of the LNG value chain, gas production, liquefaction, transportation and regasification.

Most of the world's easy to commercialize gas reserves are either already commercialized or on the way to being commercialized. Producers must now look to commercialize gas in difficult places and harsher environments, for example gas reserves in the permafrost regions. This increases the cost of gas production. Several factors are in play in raising LNG liquefaction costs such as, rising price of steel, drop in the value of Dollar, limited contractor competition, less favorable feed gas compositions, cost of CO_2 sequestering and so on.

In transportation, technology improvements are pushing for larger LNG tankers, however, LNG must now be transported over greater distances to reach favorably priced markets and the net result is increase in transportation costs. On the regasification end, public outcry, NIMBY, BANANA are forcing the plants to be located away from markets, increasing investment and adding pipeline transportation cost for gas delivery to market. And further more, environmental and interchangeability considerations are forcing higher investments in regasification plants.

Biography of Speaker

Devendra Agrawal is Principal LNG Technology Engineer with Marathon Oil Company in Houston, Texas, USA and provides technology support to projects in gas utilization areas. He is involved in Marathon's integrated gas initiative and participates in the entire LNG value chain from production to shipping and regasification. His present responsibilities include technical support of Marathon's Equatorial Guinea LNG plant under construction.

Devendra Agrawal holds a Bachelors degree in Chemical Engineering from the Institute of Technology, India. Prior to joining Marathon in 2003, he spent 17 years with various firms in the Oil & Gas industry. He is a registered professional engineer in the state of Texas and a senior member of AIChE.

LNG Plant Costs – Past and Present Trends and a look at the Future

Charles C. Yost
Houston, TX
Merlin Associates

Robert N. DiNapoli
Atlanta, GA
Merlin Associates

AIChE Spring Meeting, April 2005
5th Topical Conference on Natural Gas Utilization
Atlanta April 10-14, 2005

ABSTRACT

Over the long term, since construction of the first LNG plants in the early 1960's, LNG plant capital costs have benefited from economies of scale both in terms of individual train sizes and larger total plant sizes. This trend, however, has not been uniform with technology related problems impacting costs in the early 1970's followed by significant cost increases in the mid-1970's through the 1980's as a result of word-wide inflation and an overly conservative contracting and design strategy. Unit capital costs peaked in the mid-1970's through the mid-1980's at over US$ 600/mtpa, however, capital costs have steadily decreased to around US$ 200/mtpa for LNG plants currently under construction.

Despite current efforts to optimize LNG plant design utilizing even larger train sizes, the US$ 200/mtpa unit cost for all except very large, mega -size trains (8 mtpa and larger), may represent a low water mark for new projects as demand for such critical materials as specialty steels and other alloy metals used extensively in LNG projects soar and EPC contractors are further exposed to price risk in lump-sum bidding. This paper explores these issues citing critical milestones in LNG plant development leading to more cost effective plant designs and the potential future impact of cost increases for materials and services on the capital cost of new projects that may completely offset traditional cost improvement efforts such as economies of scale.

Several examples of expected future LNG plants costs are given showing the impact of location and plant capacity.

LNG PLANT COST RELATIONSHIPS

One of the main driving forces behind the rapid growth in worldwide LNG trade has been the substantial reduction in the unit capital investment required to produce a tonne of LNG. Over the past three plus decades worldwide LNG capacity has grown from a few million tonnes per year (mtpa) to more than 150 mtpa. Over the same period the capital investment to produce an annual tonne of LNG has decreased from over US$ 400/tpa (on a 2004 US dollar basis) to somewhat below the US$ 200/tpa level.

This trend, however, has not been uniform with a number of interrelated factors working both for and against lowering LNG plant costs. Factors instrumental in driving LNG capital costs lower include.

- Economies of scale both in terms of individual train sizes and larger total plant sizes.
- Improvements and refinements in liquefaction technology.
- Better and more sophisticated computer-related design tools.
- An accumulation of successful design and operating experience leading to more efficient and optimized design of plant facilities.
- Increased competition among liquefaction process vendors, major equipment suppliers, and EPC contractors.
- Adoption of lump sum engineering-procurement-construction (EPC) bidding strategy.
- Reduction in over-design and safety factors following successful demonstration of long term reliability.

At the same time LNG projects must contend with a number of other factors that can counteract, to some degree, these cost reduction efforts. Examples include:

- Substantial capital investment required for projects most often constructed in a remote location and/or in a less developed country requiring substantial infrastructure and support facilities, both temporary and permanent.
- The need to import a large number of skilled construction workers to supplement the limited skilled indigenous work force with its attendant social and political implications.
- The international nature of LNG projects makes plant investment subject to worldwide inflationary pressures and currency fluctuations.
- The need for sufficient backup and redundant facilities to insure security of supply to sole source LNG buyers.

- Reluctance of project sponsors, buyers and lenders to accept generally unproven technology and related inexperienced equipment suppliers and contracting firms given the large capital investment and criticality of the product to the buyer.
- Possible restrictions in worldwide procurement bidding tied to export credit agency financing.
- Limited number of qualified EPC contractors with skills and resources necessary to execute an LNG project.
- Competing demands in periods of high worldwide construction activity for the large quantities of high quality stainless steel and high nickel steel and large rotating equipment and gas turbine drivers required to build an LNG plant.

LNG PLANT COST OVERVIEW

An important point in understanding LNG plant costs is the distribution of project costs among facilities and services required to implement an LNG export project and the influence designers and contractors have on these costs. Table 1 shows the general distribution of major component costs as a percentage of total project costs for a typical grassroots LNG plant installation. Also shown are the important project variables that influence these costs.

TABLE 1

	% Cost Distribution	Major Project Variable			
		Design Capacity	Site Labor	Site/Project Related	Project Execution
Liquefaction Trains	34 - 38	X	X		
Utilities	12 - 16	X	X		
LNG Storage & Loading	10 - 15		X	X	
Building & Misc.	3 - 5			X	
EPC Contractor	14 - 16				X
Marine Related	3 - 6		X	X	
Infrastructure	0 - 6			X	X
Other Project Related	10 - 12				X

Facilities included in the EPC contractor's scope of work typically involve feed gas pretreatment, liquefaction, utilities and offsites, LNG storage and loading, and the marine and harbor facilities that, altogether, represent about 80 - 85% of the total plant capital cost. Permanent infrastructure (housing, community infrastructure, and support for operating personnel and marine support services) accounts for an additional 3 - 4% of the costs. The remaining 12 - 16 % is for Owner project management and start-up costs, Construction/Builder All Risk insurance and project venture costs.

Table 1 also illustrates the fact that for a world scale size LNG plant less than 50% of the total LNG plant project cost is capacity related. These capacity-related costs involve the liquefaction trains and supporting cooling system and a portion of the plant utility requirements.

LNG storage costs are a function of storage tank design (the choice of containment system) and storage volume which is set primarily by shipping considerations (ship parcel size, distance to and number of Buyer locations, and potential shipping schedule/delays etc.).

Although emphasis is rightly given to processing and design related improvements as a means of reducing LNG plant costs, much of the cost of an LNG liquefaction project is beyond the influence of the design engineer and is mainly a function of site related conditions and project development and project execution efforts. Costs for the loading jetty and related marine facilities, plant infrastructure, owner project management and venture costs are basically fixed for a given location regardless of the project size.

HISTORICAL TREND OF LNG PROJECT COSTS

Greenfield Projects

The influence of the factors discussed above is evident on examination of historical LNG plants costs. Figure 1 shows the historical trend, on a unit capital cost basis (US$/tpa), of the 17 grassroots LNG projects currently in operation. Capital cost is expressed on a US dollar-of-the day basis at the year of startup or the year of completion if startup was delayed more than a year. The dotted line represents Merlin Associates' estimate of unit capital cost for projects under construction or in an advanced engineering phase.

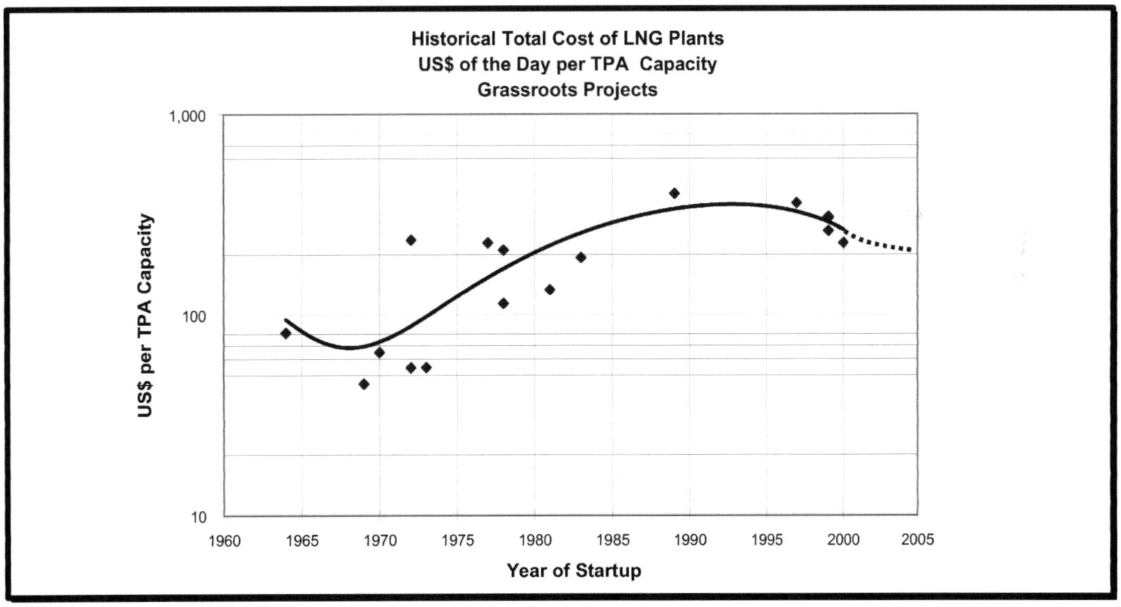

FIGURE 1

A similar plot of grassroots LNG project cost on an inflation-adjusted basis is given in Figure 2. Inflation adjustments are derived from the Nelson-Farrar Refinery Cost index published quarterly in Pennwell's Oil & Gas Journal.

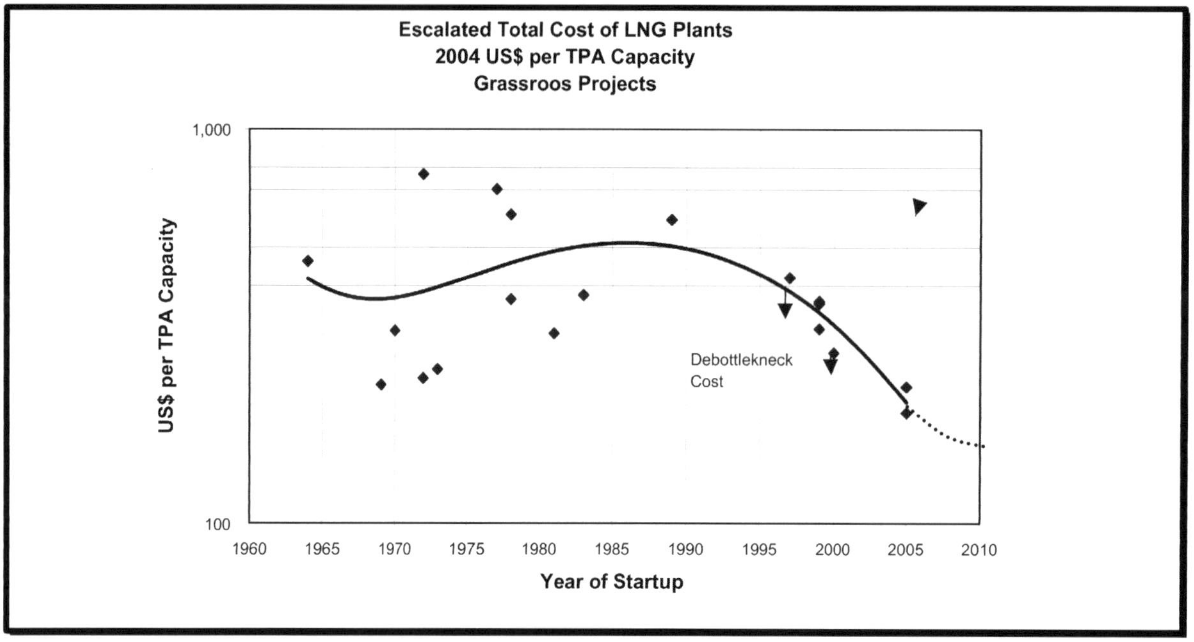

FIGURE 2

These cost trends reflect both the evolution in plant design philosophy and influences of external economic and monetary forces. The reduction in costs for the very recent LNG projects is clearly more evident when projects are compared on an inflation-adjusted basis.

The LNG plant cost - technology relationship is further shown in Figure 3 which superimposes several critical technology milestones in LNG processing technology on the inflation-adjusted historical cost trend.

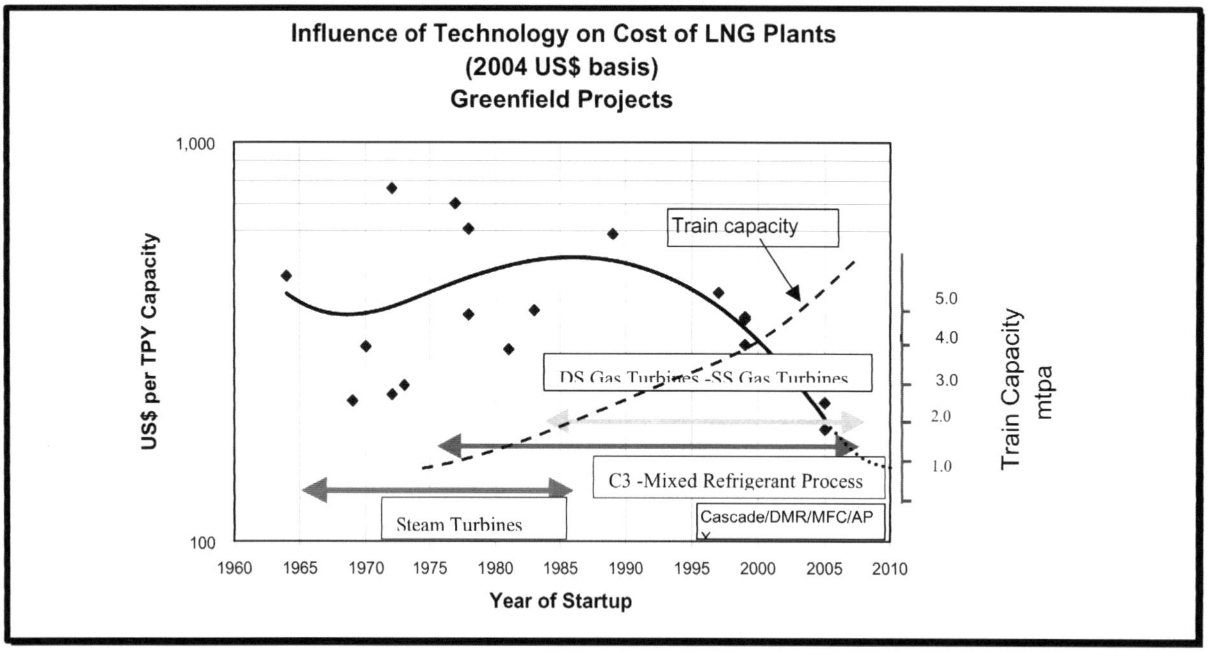

FIGURE 3

Major stages in the evolution in LNG plant costs over the years shown in Figure 3 can be summarized as follows:

- The earliest LNG plants used variations of the classic cascade refrigeration cycle for natural gas liquefaction. Engineering/design technology was borrowed from experience in air separation, helium recovery, ethylene plant refrigeration systems, and oil refinery design. Train capacities were small, on the order of 1 mtpa. Total costs were low for these plants which, for a variety of reasons, not representative of succeeding LNG projects.

- Plants built in the early 1970s (Libya, Brunei, and Algeria) were considerably larger than the first plants in Arzew and Alaska and utilized variations of the mixed refrigerant process in an effort to reduce project capital costs. Because of design and operating problems several of the projects failed to achieve design LNG production capacity. Many of the design problems resulted from extending previous design parameters beyond their reliable limits – projects were being designed and constructed using new technology at (or beyond) the edge of the envelope. In several cases it was not apparent until the project was commissioned that design parameters had been exceeded. Design problems also led to significant cost overruns on a number of projects.

- The late 1970s and very early 1980s saw the emergence of the Air Products & Chemicals, Inc. (APCI) developed propane pre-cooled, multi-component refrigerant (P-MR) process as the preferred liquefaction technology for use in large, multi-train processing configurations. Refrigeration compressors were steam-turbine driven and basic plant design had substantial built-in over capacity to insure security of supply to the typical critical-service LNG Buyers.

- From the early 1970s through the end of the 1980s, inflation adjusted unit LNG plant costs reached a peak due, in large part to high, worldwide inflation, a reimbursable EPC contractor bidding strategy restricted to contractors with proven LNG experience, and a conservative design philosophy to insure supply integrity.

- Further increases in train capacity using the now proven APCI P-MR liquefaction technology occurred in the mid- to late-1980s. A major shift in plant design technology in an effort to drive down capital costs also occurred during this period with gas turbines becoming the preferred choice as drivers for the main refrigeration compressors. Air-fin cooling at Australia's Northwest Shelf Project proved to be a viable option to once-through seawater cooling or a closed loop cooling water system.

- The results of cost reduction efforts began to be realized in the mid-to-late 1990's due to competition not only among EPC contractors but also among liquefaction process vendors as well as a more realistic assessment of basic LNG plant design margins. When compared on an inflation-adjusted basis new LNG projects are nearly an order-of-magnitude lower in unit cost than peak costs experienced in the mid-1970's.

- Current projects reflect full use of the power output of the large, single shaft Frame 7 gas turbine with demonstrated LNG capacities approaching 5 mtpa for the P-MR liquefaction process. Design capacities for P-MR process projects now under construction exceed 5 mtpa using supplemental power available from 8 - 20 MW starter/helper motors. Similar production rates are expected for new projects using Shell's Dual Mixed Refrigerant (DMR) process. Current Phillips Optimized Cascade (POC) process projects now operational achieve outputs of 3.3 mtpa with new projects expected to produce more than 5 mtpa using multiple compressor-drivers in parallel. Norway's Snohvit project will achieve production of 4.1 mtpa using Linde's Multi Fluid Cascade (MFC®) process.

LNG Expansion Projects

Figure 4 presents similar historical cost data for the major expansions at existing base load LNG plant installations. Expansion facilities typically include one or more liquefaction trains (gas pretreatment, dehydration, liquefaction and associated refrigeration systems) and necessary utility support systems as well as additional LNG storage capacity and LNG loading facilities where required in several projects.

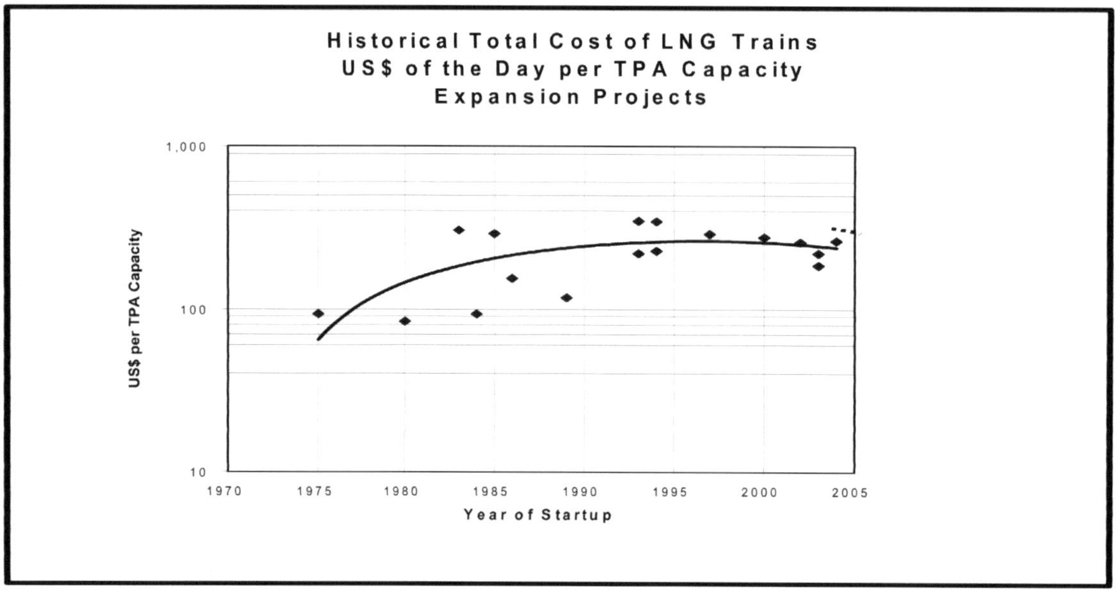

FIGURE 4

Inflation-adjusted LNG expansion cost data are shown in updated Figure 5.

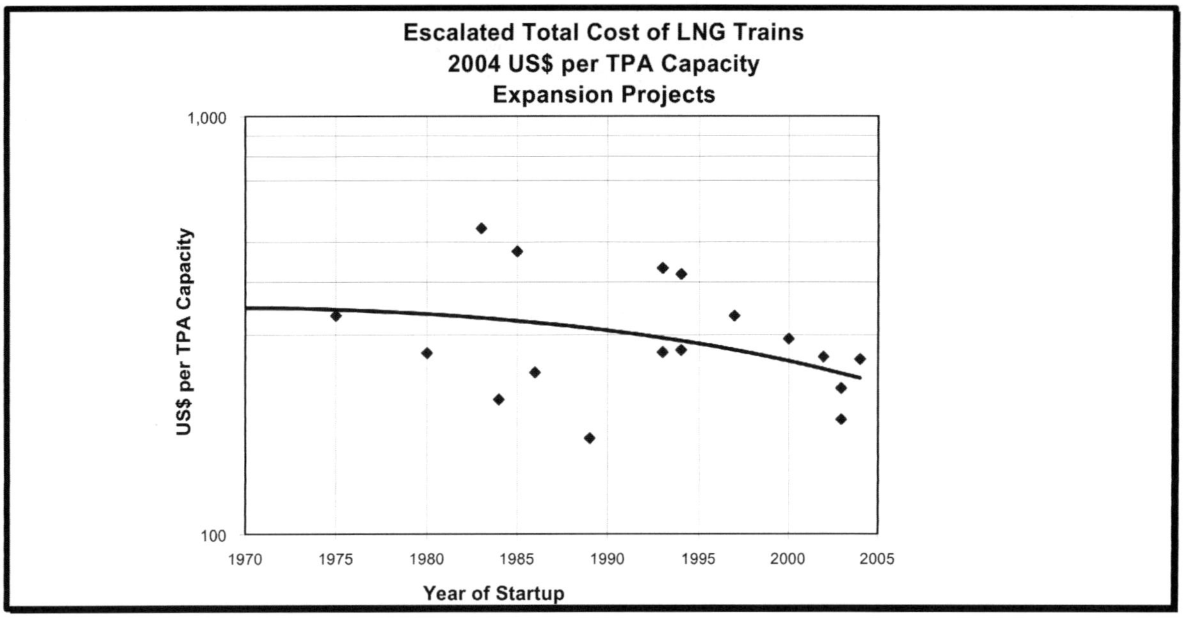

FIGURE 5

Expansion projects show a similar trend in cost reduction as the grassroots projects on an inflation-adjusted basis although not as large as for grassroots projects. This is due, in large part, to the restricted scope of supply for expansion projects with less room for cost and engineering innovation.

EVOLUTION IN TRAIN CAPACITY

Existing Projects and Current Projects

Economies of scale as a result of both larger individual train capacities and larger total plant sizes have had a major impact on reducing LNG plant capital costs. Figure 6 shows the historical trend in LNG Greenfield project costs (on an escalated unit basis) and corresponding project liquefaction train capacity since installation of the first projects in the early 1960s.

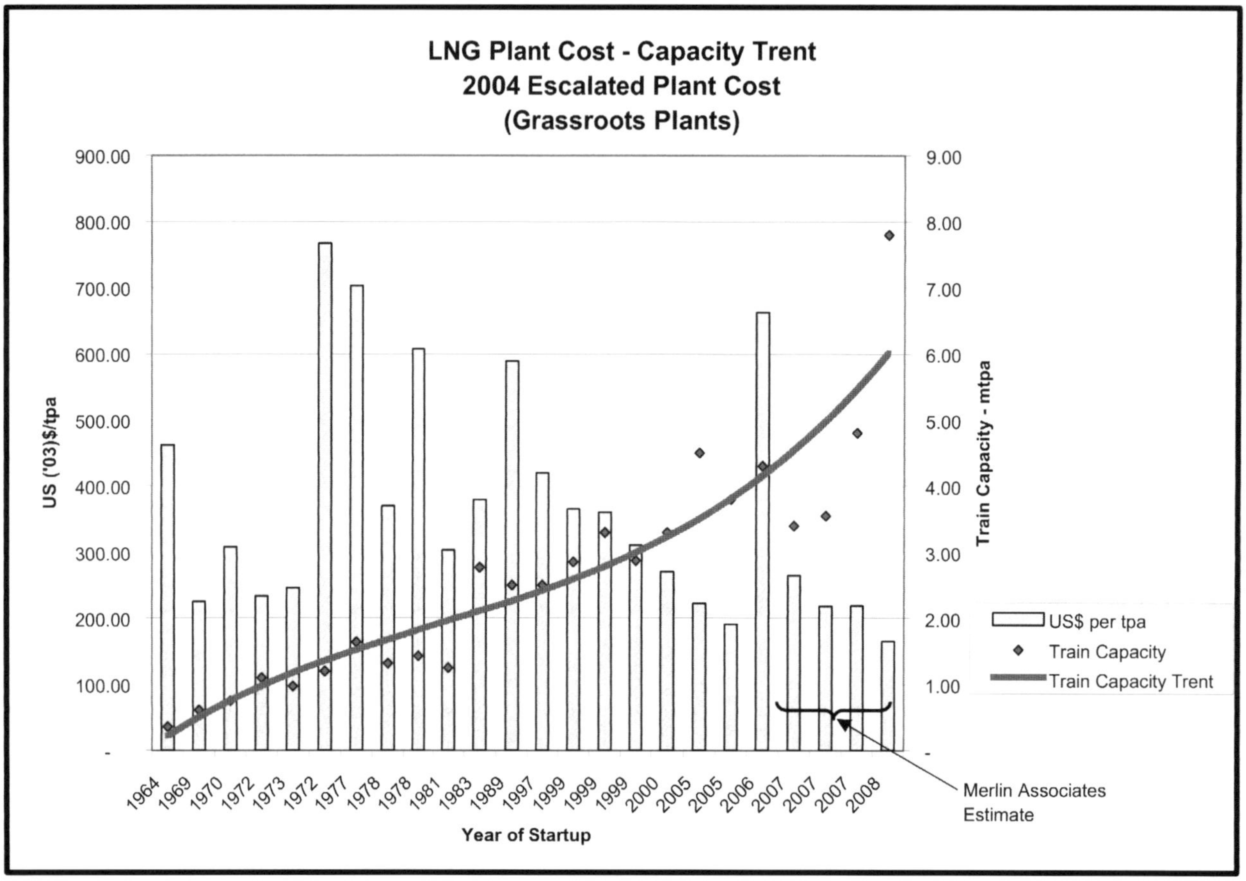

FIGURE 6

Train LNG capacities have steadily increased from less than 1 mtpa for the first LNG plants in the 60's and 2.0 to 2.5 mtpa for plants built in the 1980's to about 4.1 mtpa for the recently committed Nigeria LNG plant expansion project. The largest LNG train currently under construction is the 5.2 mtpa capacity Train 4 Atlantic LNG project in Trinidad.

New Projects and Future Trends

The announced 7.8 mtpa train capacity for the QatarGas II, QatarGas 3, and RasGas Trains 5 and 6 expansion projects are the largest train size selected for an active project to-date. Startup of the first train of the initial large train is scheduled for early 2008 and will utilize APCI's new AP-X™ technology. While this technology is essentially the basic P-MR process with the addition of a nitrogen refrigerant cycle, it does utilize equipment, principally the Frame 9 gas turbine, not previously proven in LNG applications. Merlin Associates estimates of the unit cost for an 8 mtpa capacity train plant are just under US$ 150/tpa.

Vendors and process licensors for the DMR, POC, MFC® and Prico® processes, in addition to the AP-X™ process, have also proposed liquefaction train designs with capacities in the 8 - 10 mpta range. At these very large train capacities the economies of scale begin to diminish as the physical size of many items of equipment and related piping and valving reach practical limits. All large train processes utilize some combination of parallel refrigeration compressor-driver arrangement and/or liquefaction cold box arrangement or a third refrigerant circuit (in the case of the AP-X™ process) to overcome limitations in proven centrifugal compressor design and heat exchanger sizing.

Figure 7 shows the authors' estimates of the trend in LNG plant cost for the larger train sizes now being considered for new projects. Costs have been developed using Merlin Associates' proprietary Capital Cost Estimating Model© for a generic mid-east location using once-through seawater cooling and are expressed in constant mid-2004 US dollars. Costs are based on use of conventional propane precooled-mixed refrigerant technology (APCI P-MR process) for trains below 8 mtpa and APCI's AP-X™ technology for the 8 mtpa train capacity cases. APCI technology has been assumed for consistency purposes since it offers the largest data base of existing plant costs over the widest range of train capacities. The authors believe plant costs utilizing other, competitive liquefaction technologies are generally consistent with these costs.

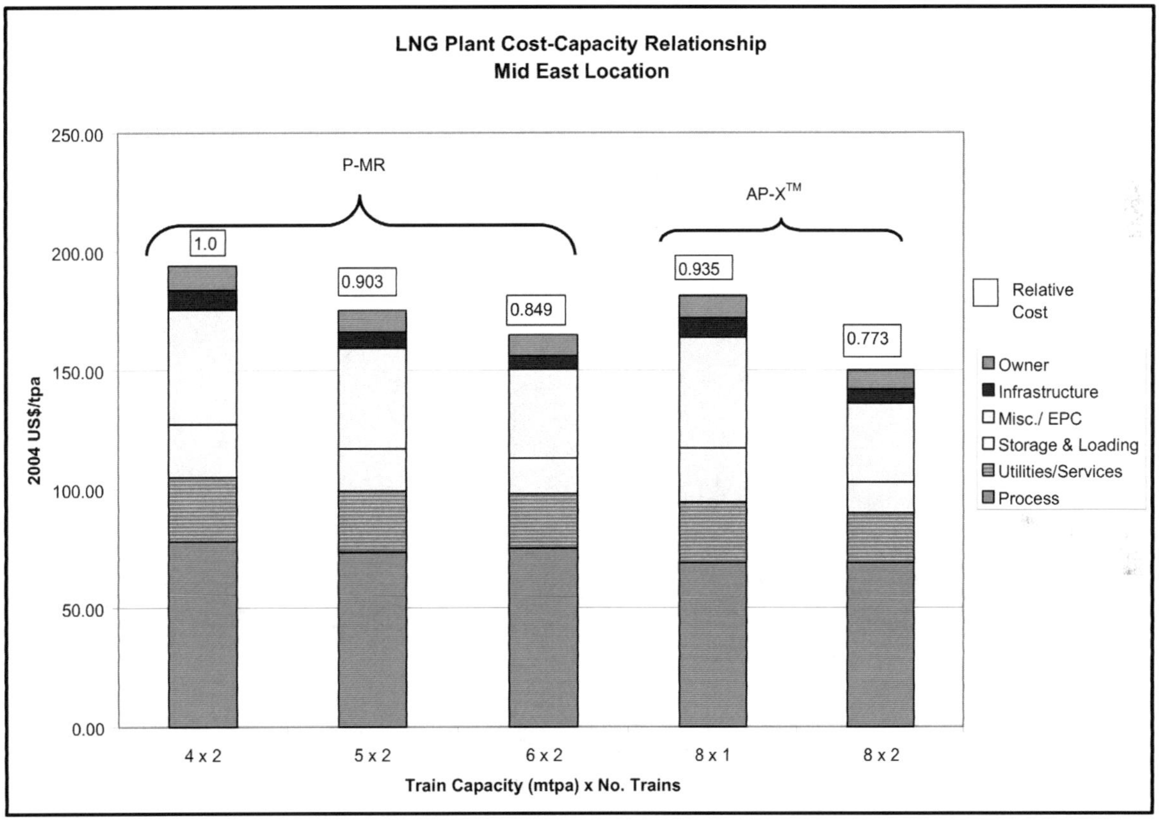

FIGURE 7

The cost advantage of large train sizes is illustrated by the fact that total project costs on a US$/tpa basis for a plant utilizing a single 8 mtpa capacity liquefaction train is about 15% less than a comparable capacity facility using two smaller, 4 mtpa capacity trains. Further economies on the order of nearly 25% are realized using two 8 mtpa capacity trains. The large train size offers on the order of a 12% decrease in cost for the process systems portion (gas pretreatment, liquefaction and fractionation/stabilization) of total plant cost. Greater savings, on the order of 33 - 50%, for utilities, storage and loading (including harbor facilities) and EPC costs result with larger trains and larger capacity projects as these plant infrastructure costs are prorated over larger production rates.

IMPACT OF PLANT LOCATION AND LABOR COST

Beyond plant capacity and related train size, plant location and the cost of site labor, in particular, has the most significant impact on LNG plant cost. As noted earlier in this paper LNG export plants are often located in remote areas with a minimum of supporting infrastructure and a shortage of skilled construction labor. Frequently this requires importing large numbers of low paid and low productivity skilled craft personnel. The project must then include the cost to train and develop sufficient numbers of skilled workers from the local work force and the imported work force. This has often been the case for plants built in Southeast Asia and in the Middle East. Alternatively the plant location may be in a developed country with a large resource of relatively higher cost but high productivity skilled workers. This has been the case for plants built in Australia and more recently in Norway.

Figure 8 illustrates the impact of unit labor rate and related productivity on the installed cost of various capacity LNG projects utilizing conventional P-MR or AP-X liquefaction technology and once-through seawater cooling. Three labor cases are shown for each capacity case: a low labor case assuming an all-in average labor rate of US $8.00/hr with a productivity of 0.4 (relative to a USGOM productivity of 1.0), a medium labor case assuming a US$ 20/hr labor rate with a productivity of 0.5, and a high labor case assuming a labor rate of US$ 48/hr with a productivity of 0.85.

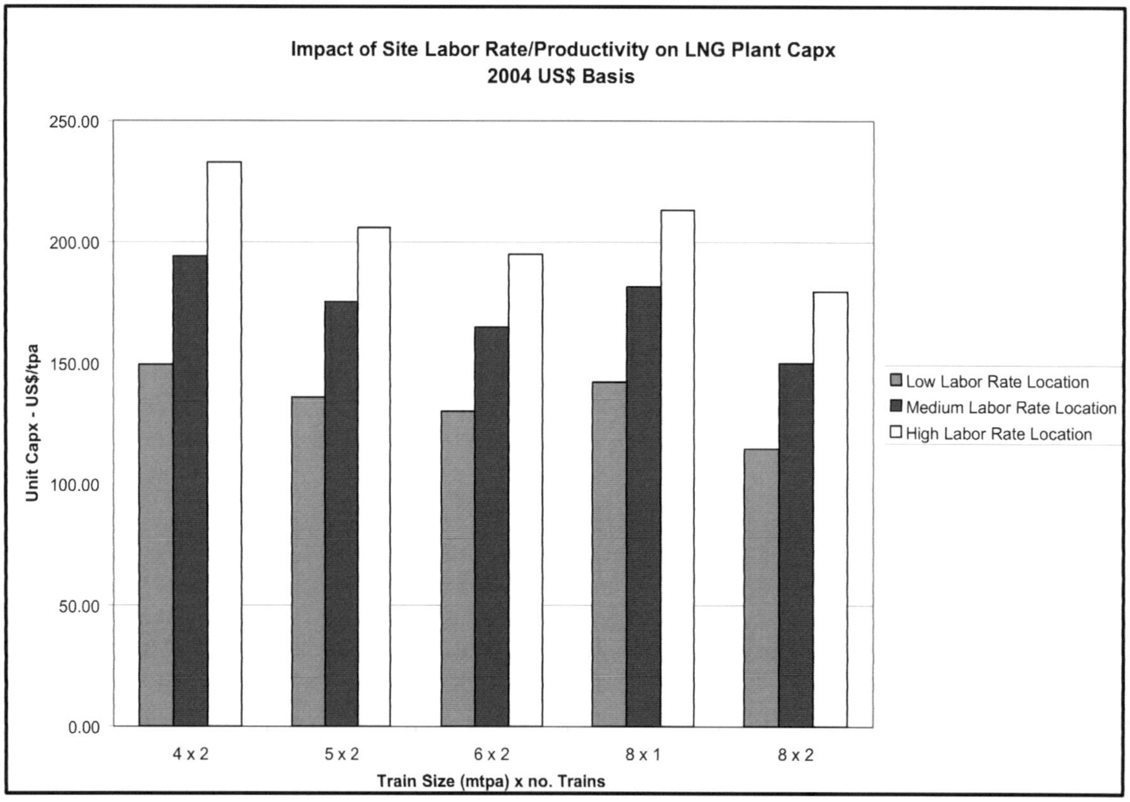

FIGURE 8

Since construction labor typically constitutes from 40 - 50% of total LNG project costs it is not surprising that total unit cost can vary by as much as US$ 30/mtpa for the cases shown in Figure 8. It is interesting to note that the cost of an 8 mtpa, two train project built under the low labor scenario is comparable wit that of a larger 16 mtpa, two train plant built under the medium labor scenario.

IMPACT OF COMPETITION FOR GOODS AND SERVICESE

Like all process plant construction LNG plants are subject to equipment, materials and services price and supply risk. Since LNG plants utilize large quantities of stainless and other specialty alloy metals, EPC lump sum contracts are exposed to a significant risk during periods of competing worldwide construction activity. As more LNG projects move from the planning stage to the execution stage, competition for goods and services develops among projects and further demands are placed on the finite, limited number of EPC contractors with the resources and financial strength to undertake an LNG project. The same holds true for the sourcing of skilled technical and construction labor. Increased demand for large items of rotating equipment available from only a limited number of vendors may also substantially increase the lead time and costs for delivery of such equipment impacting the duration of engineering-procurement -construction schedule. All result in the potential for major increases in LNG plant costs.

Figure 9 examines the impact of these variables on unit installed plant cost for the five capacity cases described earlier in this paper. The base comparison is made for conditions of medium site labor costs. Increases assumed for this analysis included 50% to selected S & E and bulk material costs, a 40% increase in EPC home office cost and a 25% increase in EPC site management and services costs plus a 5% increase in labor cost to account for schedule extension.

Cost increases similar to this scenario are consistent with several current projects that are in the process of negotiating final EPC contracts and having to deal with cost increases over the last 18 months for alloy metals and increasing vendor and fabrication shop costs. EPC costs are more recently beginning to increase as the EPC contractor backlog has increased and the contractors have difficulty in finding sufficient numbers of experienced personnel for design, construction supervision, and project management.

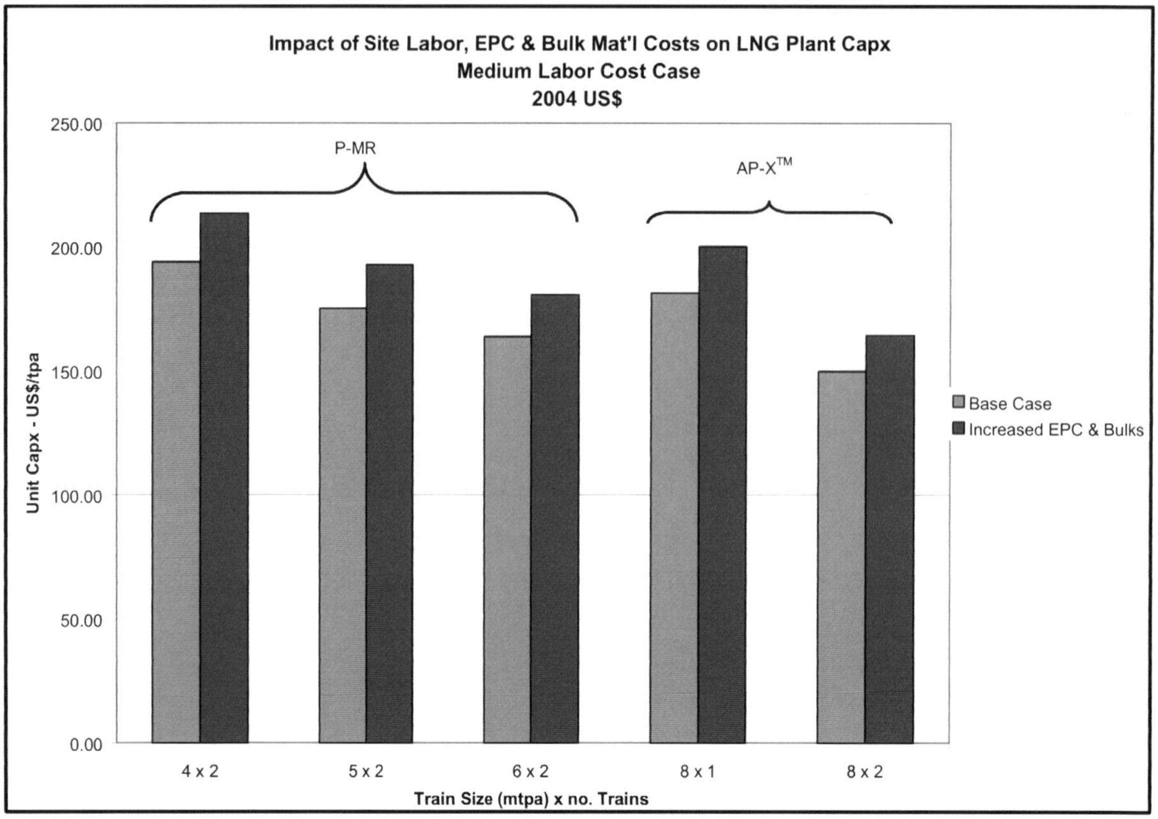

FIGURE 9

Cost increases as a result of these stresses are on the order of 6 - 10% of the total project cost and can negate a significant portion of the savings in plant cost achieved through economies of scale. For example, approximately 95% of the savings in unit cost between an 8 mtpa two train project and a 10 mtpa, two train project can be lost through cost escalation in materials, schedule completion delays and EPC contractor services. Similarly, over 50% of the cost saving between an 8 mtpa, two train project and a 12 mtpa two train project can be lost.

A LOOK AT THE FUTURE

Base load liquefaction of natural gas is now a mature industry. The technology is well developed and no significant breakthroughs leading to substantial cost reductions are anticipated. Compression power - the critical element of liquefaction technology - is fixed at minimum levels by the laws of thermodynamics, a fact that tends to even out advantages/disadvantages of competing liquefaction processes. Equipment selection then becomes a critical element in optimizing process designs. The bottom line is that further innovations in process design are not expected to have a major impact on reducing costs of new LNG projects.

The question remains as to whether there will be further cost reductions through economies of scale. As discussed in this paper economies of scale has been a major factor in reducing unit plant capital investment over the past twenty years. Further gains, however, appear to be approaching a point of diminishing returns as critical items of equipment reach their physical and design maximum size. This necessitates installing more items of equipment in parallel and/or increasing the complexity of the processing sequence, all tending to offset advantages of large capacity trains.

Another factor limiting eventual train size is the sheer volume of LNG produced. Depending on the processing configuration, shutdown of a single item of equipment can result in the loss of a substantial quantity of LNG relative to the total volume imported to a particular country. Also, to build a two train LNG plant using 8 mtpa capacity trains requires the marketing of some 16 mtpa of LNG as a single block. This seriously limits potential markets and buyers. Furthermore, a very large gas reserve on the order of 16 TCF (460 BCM) is required to support such a project over a 20 year contract period.

Finally the construction of large, mega size trains as an industry standard presents numerous equipment, construction and logistic problems. Cost escalation is likely under these conditions due to limited sources of specialty equipment, materials, construction labor, and EPC contracting capabilities to build multiple projects of this size.

The authors believe that for the immediate future train sizes in the 4 - 5 mtpa size range will continue to be the industry norm with unit costs averaging in the US$ 175 - 200/tpa range (current dollar basis) with some forward escalation. Large, mega-size trains of 8 - 10 mtpa capacity can reduce unit cost to under US$ 150 mtpa for certain locations such as Qatar with ample gas reserves to support such projects.

Cryo-Energy from LNG Regasification to Pipeline Specifications

Ari Minkkinen and Alexandre Rojey
IFP

AIChE Spring Meeting, April 2005
5th Topical Conference on Natural Gas Utilization
Atlanta April 10-14, 2005

Abstract

Today conventional LNG terminals use techniques such as open rack seawater and submerged combustion vaporizers to regasify LNG making no effort to recover the inherent cryogenic refrigeration energy of the LNG. The challenge presented in the recovery of this "CryoEnergy" is how to achieve it most effectively and how best to use it in-situ since refrigeration cannot be exported long distances. Additionally, as most fluids have freezing points well above LNG storage temperature they would potentially clog an indirect surface heat exchange regasification process.

This paper presents novel LNG regasification processes being developed by IFP which make use of the inherent "CryoEnergy" provided by LNG. These processes are considered mostly for applications within a reasonable proximity to a gas fired power plant. Common to all is that warming for regasification takes place in a very compact heat exchanger configuration while cooling and/or condensing non-freezing liquids and/or vapors. In a future application when CO_2 capture for sequestration becomes a widely accepted practice to mitigate global warming, an ideal heat transfer fluid for LNG regasification is the captured CO_2 vapor itself being liquefied or solidified for easy transport and sequestration. This is the basis for what is to be called the " green LNG supply chain ".

Introduction

The energy thirst of gas deficient nations is increasingly being satisfied by importing liquefied natural gas (LNG) from faraway natural gas liquefaction plants to LNG receiving terminals. The USA and UK both of which face a growing domestic natural gas supply deficit will probably import the largest volumes in the future. This dynamic growth of global LNG trade is expected to expand from 125 million tons per year (Mtpy) to 235 Mtpy by year 2010 and to 380 Mtpy by 2030 [Chabrelie 2004].

At the LNG terminal, the landed liquefied gas, stored at atmospheric pressure, is pumped to gas pipeline pressure then re-gasified at a constant contractual volumetric flow rate. Today, most terminals regasify LNG at the final pipeline send-out pressure which is often well above the cricondenbar of the PT phase envelope of the raw LNG composition. Since phase separation cannot be achieved at such pressures, there is no simple way to control the compositional properties of the re-gasified raw LNG, lest it be blended with other gases. Often the raw LNG in these cases becomes the product pipeline gas and its quality does not necessarily meet existing pipeline gas specifications, most of which impose a maximum gross calorific value (GCV) and hydrocarbon dew point. Today, this factor is being construed as a serious obstacle, especially in the USA, already reluctant to accepting imported LNG to its existing pipeline distribution network [Fatica 2004].

Most conventional LNG terminals make no effort to recover the inherent refrigeration potential of LNG regasification and use the heat available in seawater to regasify LNG. Environmental and ecological considerations limit the temperature decrease of the outlet seawater to around 5°C calling for enormous flow rates of water to vaporize LNG in conventional seawater open rack vaporizer (ORV) terminals [Habibullah et al. 1999]. In these ORVs skin ice formation is somewhat tolerated by the washing action of such massive water flows. However, the size and cost of piping and pumping is very high and the need for chlorination to combat algae growth adds complexity. A simpler, more compact and less expensive configuration is presented by the submerged combustion vaporizer (SCV); but this is at the expense of fuel gas consumption and air pollution. Other types of vaporizers using intermediate fluids are sometimes employed to have a compact layout; especially for shipboard regasification; but there is generally no thought to recover cryogenic energy from such vaporizers [Campbell et al. 1999].

IFP has been actively involved in LNG for a number of years. It has developed the natural gas liquefaction technology using a simple dual mixed refrigerant process [Fischer 2002] which is trademarked Liquefin ™. Today, this technology is being licensed worldwide by Axens, a fully IFP owned subsidiary. More recently, to complete its LNG technology portfolio, IFP has undertaken a process development effort focusing on the LNG regasification process. The aim of this work is to bring to the industrial gas energy market cost effective, ecological and less CO_2 emissive technologies. An emphasis is placed on using commercially available equipment in a novel manner and/or configuration. In some of these patent pending processes, a normal primary alcohol is the preferred choice for the fluid transferring heat to LNG. A primary alcohol, such as methanol, in liquid state can be easily pumped considerable distances and thus the recovered refrigeration doesn't need to be used in-situ.

Maritime LNG supply chain and its terminal link today

Linked to the end of the maritime LNG supply chain is the LNG terminal with its regasification process. This process has become a crucial element in the LNG chain and how it is accomplished can impact significantly on the economics and environmental friendliness of the overall LNG project.

Liquefaction for transport and storage

Today there are many available natural gas liquefaction technologies and these can be reviewed in many trade publications and is not the subject of this paper. Needless to say, those technologies using brazed aluminum plate fin heat exchangers (PFHEs) to liquefy natural gas support the premise of this paper since PFHEs can also be advantageously used to regasify LNG. To this end, Liquefin ™ technology already mentioned is notably of these.

Most liquefaction plants today produce LNG for the Japanese and Korean markets which alone absorb over 62% of the LNG trade [Chabrelie 2004]. In these markets, LNG imports mostly feed the power generation sector and there is little incentive to limit the LNG energy content, composition nor GCV to consumer specifications. Because of this, liquefaction processes and LNG trains have, up to now, been basically designed around the Japanese and Korean appetite. As a consequence, most LNG cargoes floating around the world are too hydrocarbon rich for the USA and other mature gas producing infrastructures which traditionally handle differently treated leaner wellhead pipeline gas.

Transport of LNG cargoes are conventionally carried out today with tanker ships of from 90 000 to 135 000 m^3 capacity but the tendency is for ever bigger volumes exceeding 200 000 m^3 to capture the economy of scale advantages. Most ships use boil-off gas as supplemental fuel and thus loose the most volatile LNG and lowest GCV constituents during the transport time. Obviously the longer the transport distance time the more is the boil-off loss. This factor further aggravates the landed LNG composition and final regasified LNG pipeline gas quality by increasing the GCV above specification limits.

Not much needs to be said about LNG storage except that it is largest single cost in the LNG terminal [Habibullah et al 1999]. LNG tanks are needed both at the LNG liquefaction plant as well as the LNG terminal to provide reserve capacity and smoothen the intermittent transport by ships to assure a constant contractual flow rate of gas to the pipeline distribution system. The landed LNG is stored at cryogenic temperature at nearly the atmospheric boiling point of methane (minus 160°C). It is lifted from the storage tank by submerged first stage low head pumps to a pressure of around 8 bar and effectively the lifted raw LNG is sub-cooled. This sub-cooling provides more than enough thermal absorption capacity for re-condensing some boil-off vapors. In general the raw LNG taken to the suction of the final send-out pumps at around 7 bar would have less than saturation bubble point temperature of roughly minus 150°C.

Regasification to pipeline condition

The conventional LNG terminal regasifies LNG to send-out condition without regard for recovery or beneficial use of the inherent cold energy equivalent of LNG. Moreover, in an attempt to avoid gas compression, the raw LNG received at the suction of the main multi-stage send-out pumps is boosted immediately to above the final send out pressure. Often, as mentioned, this pressure is well above the cricondenbar of the raw LNG's compositional phase envelope as illustrated in Figure 1 and as such no vaporization per se takes place. The LNG regasification path in this thermodynamic sense is simply an isobaric heating of a dense phase "super cricondenbaric" fluid to ambient temperature at pipeline pressure. As such there is no possibility to employ a simple upstream vapor liquid separation step to control the quality of the raw LNG and the landed LNG composition becomes the pipeline sales gas.

Sometimes, inert gas or other third party lean gas can be admixed to achieve a given compositional property such as the GCV for the final sales gas product. Otherwise, more complex integrated downstream gas liquids extraction techniques to remove C_2 plus NGLs can be incorporated within the LNG terminal [Mak et al. 2004]. Sometimes, the marginal sales revenue from co-production of a fractionated LPG and/or stabilized NGL justifies the increased cost of such a scheme [Price 2004].

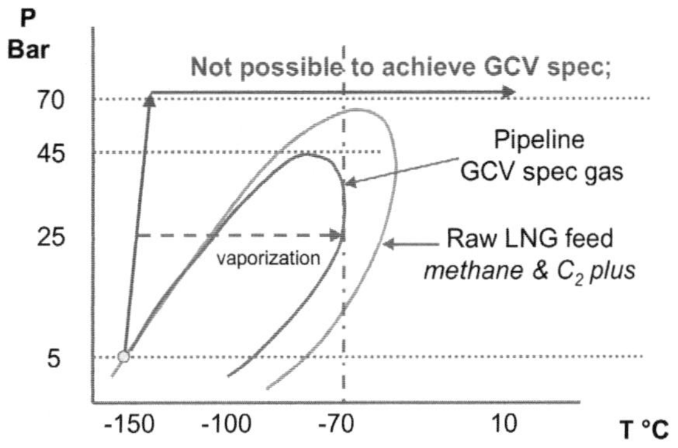

Figure 1 PT Phase Envelopes

When the regasified LNG from the terminal is destined solely for its consumption in a dedicated power plant, the send-out pressure condition can be significantly lower than pipeline gas requirements. In this case, it needs to be just enough to feed an industrial gas turbine combined cycle power plant at around 20 bar. Generally, in these situations there is no particular specification for GCV; the higher the better is the rule.

Pipeline gas quality specifications

A word on pipeline gas quality specifications is in order as this has apparently become a serious matter; especially in the USA as the composition of landed LNG varies considerably with traditional pipeline gas quality and interchangeability factors [Fatica 2004]. Table 1 hereafter illustrates some typical values reported in literature.

Table 1 Pipeline Gas and Typical LNG Import Specification

Component vol%	USA Specification	Typical Landed LNG
Methane	88 min	86 to 95
Ethane	6 max	4 to 14
C_3 to C_5	3 max	3 to 7
C_6 plus	0.2 max	0.5 to 1.0
$N_2 + CO_2$	1.4 to 3.5	0.5 to 1.0
Gross Calorific Value (GCV)		
Btu / SCF	960 to 1 050	1 050 to 1 200
MJ / *s* m^3	35 to 45	45 to 52

Suffice it to say the typical landed LNG composition is far from the leanness required to meet the North American pipeline standard.

CryoEnergy recovery technology

New, patent pending, LNG regasification technology conceived by IFP for near imminent application to grass roots LNG terminal projects or for revamps of existing terminals programmed for expansion is called "Gasifin". This is a natural follow-up and the completion link to the commercial Liquefin™ technology. Processes within the Gasifin technology suite make use of the inherent potential cryogenic level refrigeration energy "CryoEnergy" provided by the regasification of LNG. This is used to cryogenically chill a circulating heat transfer liquid consisting of an anhydrous normal primary alcohol. This cryogenic alcohol or "Cryohol " loop in turn indirectly provides by sensible heat transfer various practical and beneficial chilling services within a gas turbine combined cycle (GTCC) power plant. In one obvious application, the refrigeration duty can be applied to chilling gas turbine (GT) intake air to significantly increase its power output. However, since the benefit due to chilling GT intake air is dependent upon ambient conditions which are site specific and vary seasonally day to night, a preferred application, presented as an illustrative example, is condensing propane in a simple closed loop propane expander power recovery cycle. This power recovery cycle is advantageously added to the tail end of the GTCC power plant heat recovery steam generation (HRSG) convection section just before the flue gas stack. Residual waste heat energy content from industrial GTCC power plant flue gases is nearly as constant as LNG flows need to be to a gas pipeline. Thus heat is always available to regasify LNG.

Although both ethanol and n-propanol have somewhat lower freezing points, methanol, is the preferred choice for the "Cryohol" transferring heat to LNG as it has a low enough freezing point, the lowest viscosity, as high a specific heat and density and other favorable transport properties at cryogenic temperatures. It is also a relatively inexpensive common bulk chemical readily available from many suppliers at high purity. However, to avoid localized very low skin temperatures below minus 97°C and potential plugging in the chilling of methanol with indirect heat exchange surfaces a novel co-current heat transfer configuration is used upstream of the main brazed aluminum plate and fin heat exchanger (PFHE). This upstream co-current heat exchange step incorporates a shell and tube (S&T) heat exchanger designed with appropriate mechanical expansion joint to compensate the thermal stresses built-up from high inlet temperature difference (delta T) imposed by co-current warming and chilling flows. After the co-current heat exchange step, the bulk regasification duty using a compact PFHE can be designed to exchange heat in a more efficient, purely counter-current mode without exceeding a 50°C delta T between adjacent heating and cooling streams. According to specialists in the field, this is widely recognized as the limit of present brazed aluminum PFHE design [ALPEMA 1994].

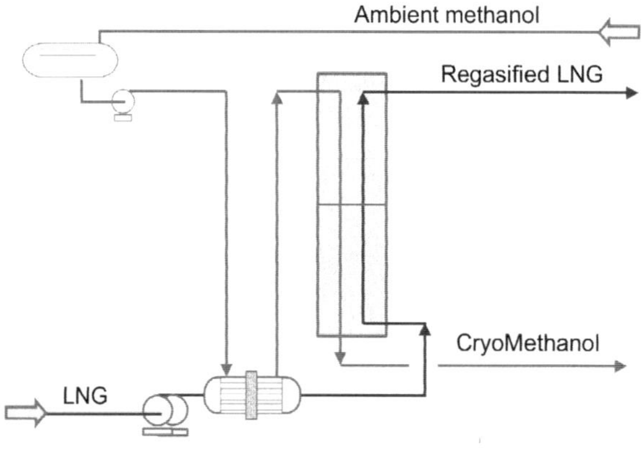

Figure 2 Type 1 Gasifin Process Flow Diagram

The Gasifin Process

In the most simple LNG terminal concept, labeled Type 1, where the regasification is conducted at pipeline send-out pressure above the cricondenbar of the raw LNG feed, the process scheme can be as depicted in Figure 2. A pure, dry, circulating methanol stream at ambient temperature pumped to some pressure is passed to the shell-side of a fixed tube sheets single pass shell and tube (S&T) heat exchanger. LNG feed, pumped to slightly above send-out pressure, at around minus 150°C, is taken to the tubes side of the heat exchanger where it flows co-currently to the warm methanol. The tubes are preferably made of aluminum to resist the cryogenic temperature and are designed to withstand the send-out pump shut in pressure. The shell side can be made of any suitable low temperature carbon steel or aluminum and is equipped with an appropriate shell-side expansion joint to take up the thermal expansion stresses developed.

The co-current temperature profiles of the S&T heat exchange are depicted in Figure 3. Due to the high temperature difference at the inlet the intermediate skin temperature outside of the tubes is maintained well above the freezing temperature of methanol. The pre-chilled methanol is then passed to the top entry channel of a PFHE mounted vertically inside a cold box. The methanol is split into a multitude of parallel finned plate passages to flow downward counter-currently against adjacent parallel finned plate passages of rising cold LNG preheated in the S&T heat exchanger and entering at the bottom. The delta T between adjacent plates in the PFHE never exceeds 50°C.

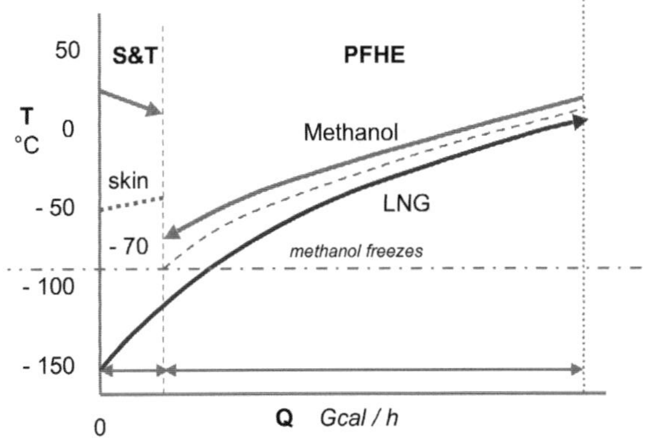

Figure 3 Heat Exchange Temperature Profiles Type 1

The cryogenically cold methanol stream exits at the bottom of the PFHE while the re-gasified LNG exits at the top at pipeline send-out condition. The CryoMethanol is then pumped to the GTCC power plant and is used in beneficial refrigeration services within the power plant. Once reheated to ambient temperature in the power plant the methanol is returned at some positive pressure to the Gasifin process at LNG terminal and passed to a surge drum to be re-circulated through the heat exchangers to close the loop. This scheme allows some reasonable kilometers distance between the LNG terminal and the power plant and thus pressurized cryogenic LNG doesn't ever need to physically pass through the power plant. The terminal and power plant can be separate operating entities if business interests dictate so.

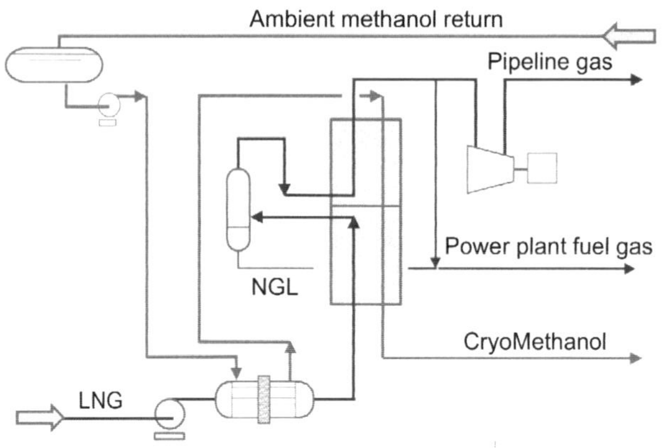

Figure 4 Type 2 Gasifin Process Flow Diagram

Gasifin with GCV control

In an alternative scheme, labeled Type 2, where dew point and GCV control are required, the raw LNG feed is regasified in several sequential steps conducted at a pressure below the cricondenbar of the raw LNG composition. In this mode of operation depicted in Figure 4 the hydrocarbon dew point and GCV of the raw LNG feed can be controlled simply by the withdrawal of a higher molecular mass and thus higher gross calorific value (GCV) liquid representing a raw non-stabilized NGL.

The warm dry circulating methanol stream is passed to the shell-side of the S&T heat exchanger as before. In this alternative, however, raw LNG feed is pumped with a less powerful lower head send-out pump to some intermediate pressure, but well below the cricondenbar, and below the final send-out pressure. It is also taken to the tubes side of the S&T heat exchanger to flow co-currently to the warm methanol. In this heat exchanger, the raw LNG feed can be preheated up to its' bubble point temperature at the prevailing pressure of the inter-stage separator drum. Again the tubes are preferably made of aluminum to resist the cryogenic temperature and designed to withstand only the pump discharge pressure; not send out pressure.

The co-current temperature profile of the S&T heat exchange is depicted in Figure 5. As before, the high delta T at the inlet maintains the skin temperature outside of the tubes above the freezing temperature of methanol. The pre-chilled methanol is passed to the top entry channel of the top plate fin heat exchanger PFHE mounted vertically inside a cold box on top of the bottom PFHE. The methanol flows downward counter-currently against rising cold vaporized LNG originating from the inter-stage vapor liquid separator drum which can be located inside the cold box. The delta T between the cold LNG vapor and the warmer methanol stream at the bottom of the top PFHE is maintained below 50°C as shown in Figure 5.

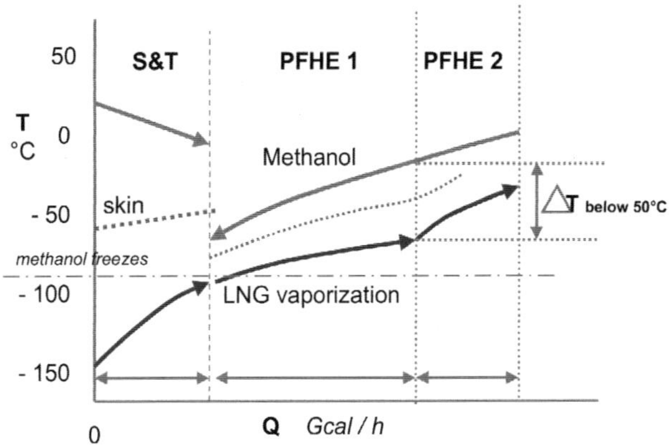

Figure 5 Heat Exchanger Temperature Profiles Type 2

The cold liquid methanol stream continues its downward passage through the bottom PFHE against rising colder vaporizing LNG which was preheated to bubble point in the S&T heat exchanger and entered at the bottom of the bottom PFHE. The induced upwards sloping warming curve of the LNG is made possible by operating below the cricondenbar of the LNG mixture. This curvature is characteristic of multi-component vaporization and if it was pure methane would be a horizontal line. The vaporization of LNG stream in the bottom PFHE is favorably adjusted under controlled heat input with flow and temperature controllers not to reach completeness. The exiting two phase vapor / liquid stream is then passed to the separator drum already mentioned. The smaller liquid fraction, consisting of a non-stabilized hydrocarbon mixture having a higher GCV and molecular mass than greater vapor fraction, is removed. This stream is ideally mixed with the gas slip stream sent to the power plant as fuel. No further fractionation nor stabilization is required as the liquid can be consumed by the power plant. This affords simplicity and cost savings for an LNG regasification terminal with GCV control compared to integrated NGL extraction and export schemes described by others [Price 2004] .

Finally the superheated LNG vapor exiting the top of the top PFHE is taken to the suction of a gas compressor where by compression the final send-out pressure an temperature is attained. In a preferred mode the exit temperature of the vapor stream is kept low enough to obviate the need for cooling the compressor discharge before entry to the pipeline gas system. The compression power required is also minimized by the low suction temperature but obviously this is a trade off with CryoEnergy recovery as will be shown in an illustrative example.

CryoMethanol use in a revamped GTCC power plant

In an advantageous and relatively easy to realize add-on revamp to a GTCC power plant, the "CryoMethanol" refrigeration duty can be used to condense propane in a low level waste heat to power recovery cycle. This obviates the mentioned problems and deficiencies associated with chilling only ambient gas turbine intake air. Moreover it improves the overall efficiency of the GTCC on a continuous basis by making more power available to the overall credit of the power plant and rejecting flue gas to atmosphere at significantly lower but still buoyant temperatures.

Brief description of a simple power recovery cycle

The CryoMethanol from the Gasifin process can be indirectly used to condense propane at around 0°C temperature level making possible a condensing back pressure of around 4 bar. Cold condensed propane is then pumped to heat transfer coils at the "tail-end" of the GTCC flue gas HRSG exchanging waste heat to preheat, vaporize and superheat the propane before expanding it to the achievable condensing pressure in a propane expansion turbine generator. A simple closed loop propane cycle is achieved with a net power export attained from low temperature level flue gas as depicted in Figure 6.

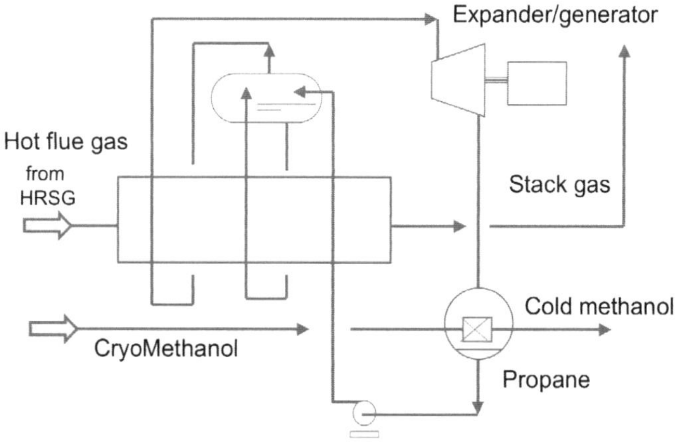

Figure 6 Propane Power Cycle Process Flow Diagram

Before returning to the Gasifin process, the still cold methanol stream exiting the propane condenser is further heated to near ambient temperature by giving obvious beneficial heat transfer to other ancillary utility streams within the power plant; such as chilling cooling water used for the surface condensers within the steam turbine power generators.

An illustrative example of Gasifin integrated with a GTCC power plant

Take an LNG terminal, associated with a **3.4 million tpy LNG** project, which is regasifying **500 MMSCFD** or roughly **400 t/h** of LNG to a gas pipeline at **70 bar** (1 000 psig) send-out pressure continuously. A summary of the order of magnitude power and capacity parameters at the LNG terminal are given in Table 2:

.

Table 2 Summary of LNG Terminal Parameters

Parameter	Type 1	Type 2
Send-out pump Mw	5.4	0.5
CryoMethanol pump Mw	0.6	0.5
Gas compressor Mw	0	12
Total power Mw	**6**	**13**
Regasification duty Gcal/h (MMBtu/h)		
❑ S&T HE	12 (48)	19 (75)
❑ PFHE	67 (265)	48 (190)

Total duty Gcal/h (MMBtu/h)	**79 (313)**	**67 (265)**
Relative total heat exchange surface	**2.6**	**1.0**
CryoMethanol		
❑ Outlet temperature °C	minus 70	minus 70
❑ Return temperature °C	20	20
❑ **Circulation flow rate** t/h (GPM)	**1 700 (9 000)**	**1 450 (7 700)**

At the power plant, which consumes roughly **60 MMSCFD** fuel gas to gas turbines in a combined cycle configuration rated at **400 MW,** around **3 000 t/h** of hot flue gas is available from the HRSG convection section. The order of magnitude performances, which are attained with the CryoMethanol assumed to arrive at minus 65°C to the power plant are summarized in Table 3.

Table 3 Summary of GTCC Power Plant Parameters

Parameter	with Type 1	with Type 2
CryoMethanol circulation flow rate t/h	**1 700**	**1 450**
Refrigeration made available Gcal/h (MMBtu/h)		
❑ for condensing propane	56 (220)	48 (190)
❑ for ancillary chilling services	19 (75)	16 (63)
Total chilling Gcal/h (MMBtu/h)	75 (295)	64 (253)
Propane cycle		
❑ circulation flow rate t/h (GPM)	560 (4 900)	480 (4 200)
❑ extracted waste heat duty Gcal/h (MMBtu/h)	66 (263)	57 (226)
Net recovered electric power Mw	**10**	**8**

The GasifinPlus process

Due to an ever growing environmental concern over "green-house" gases contribution to global warming, IFP, together with other international integrated energy companies, firmly believe that CO_2 capture for its sequestration will eventually become a significant bulk business operation in fossil fueled power plants and other CO_2 emitting industries. IFP also believe that the future LNG terminal can play an important role to facilitate the processing required for CO_2 collection and capture to sequestration. To this end, IFP has conceived an extended version of the Gasifin concept to include CO_2 capture to storage for eventual export to sequestration destinations. This is called the GasifinPlus process. A future green LNG terminal configuration incorporating the GasifinPlus process integrated with a GTCC power plant is conceptualized in the a simplified block flow diagram in Figure 8. The main features of the this integrated LNG terminal are:

- LNG to be regasified is split into two Gasifin cold boxes in separate parallel cryogenic refrigeration recovery services
- First Gasifin cold box cryogenically chills methanol as previously described
- The cryogenic methanol is used in an open instead of a closed loop
- Methanol serves not only as a cryogenic heat transfer fluid but also as a refrigerated physical solvent to dehydrate and capture CO_2
- Heating of the CO_2 laden open methanol loop with GT air (as shown) or by condensing propane, as previously described, provides the flash regeneration heat needed to liberate CO_2 as a dry vapor above the CO_2 triple point pressure

- Second Gasifin cold box uses captured CO_2 vapor as a heating medium to regasify LNG whilst being liquefied.
- Liquefied CO_2 is stored in pressurized spheres for ready export

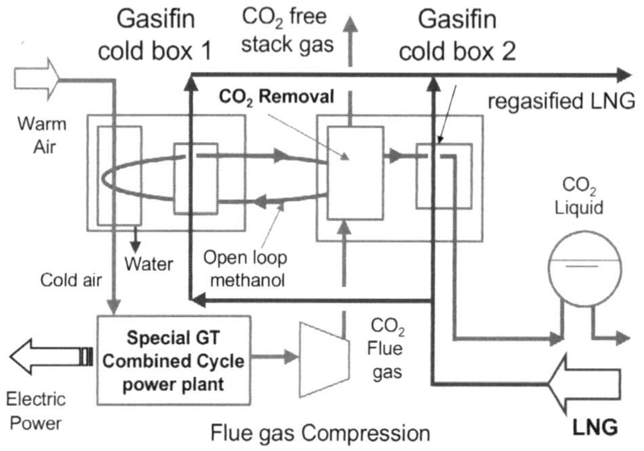

Figure 8 Integrated GTCC with GasifinPlus Process

A brief description of the process

In the patent pending GasifinPlus process the use of the CryoMethanol is extended to include not only condensing propane in a waste heat power recovery cycle already described or chilling gas turbine intake air, as shown in Figures 8 and 9, but also for chill dehydration and CO_2 removal. Refrigerated methanol exhibits a high physical solvency for CO_2 as attested by the Rectisol^{TM} process [Ranke et al 1984] and by IFP's IFPEXOL^{TM} process [Rojey et al 1989][Minkkinen et al. 1992]. Cold methanol is also more resistant to the presence of oxygen in flue gas than are CO_2 reactive amines. While methanol, as a physical solvent, tends to co-absorb hydrocarbons, this is not an issue since they are absent in flue gases. However, CO_2 partial pressure is important and a good deal of compression and/or concentration of CO_2 is needed to boost the flue gas to an optimum absorption pressure. With a judicious heating of the bulk remaining CO_2 free flue gas after the methanol wash followed by expansion a significant amount of the compression energy can be recovered making this a very feasible scheme.

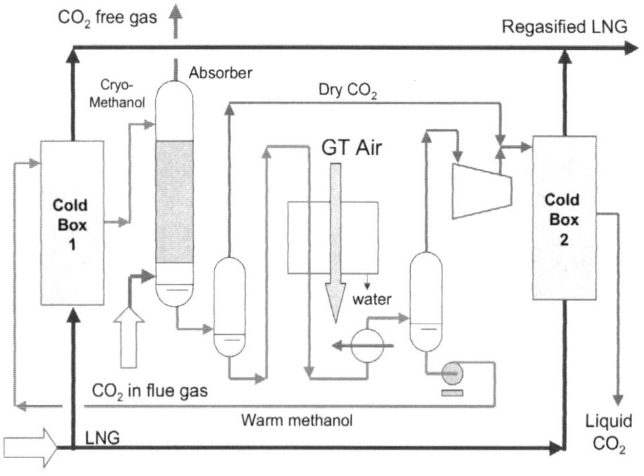

Figure 9 GasifinPlus Process

Pressurized methanol, chilled by LNG in Gasifin cold box 1 as shown in Figure 9, is first passed to an absorption column to physically absorb CO_2 from chill dried and pressurized flue gas from the upstream compression step. The rich methanol is flashed to liberate part of the CO_2. Then, the still extremely cold methanol, containing remaining dissolved CO_2, provides by sensible heat transfer, the beneficial chilling duties within a GTCC power plant whilst being warmed to above ambient regeneration temperature which upon flash at a lower pressure releases dry CO_2. The CO_2 so captured when compressed above its triple point pressure can easily be liquefied in the second Gasifin cold box 2 by regasifying LNG.

The second Gasifin cold box uses an indirect intermediate fluid, such as ethane, to liquefy CO_2 vapor. The intermediate fluid condenses and vaporizes at around minus 60°C to avoid CO_2 solidification by preventing its exposure directly to skin temperatures below minus 60°C induced by LNG regasification temperature profile. The CO_2 which is stored as a minus 50°C liquid in spheres can be pumped and transported by pipeline for export or transported in railway cars to users such as oil fields practicing EOR miscible CO_2 injection.

The Green LNG supply chain with carbon return loop

Finally, the future evolution of the green LNG supply chain will not only include capture of CO_2 but could be modified to return captured CO_2 for sequestration in the producing gas field which provided the LNG. This will perhaps require a revolutionary design modification of the LNG maritime transport system. In a glimpse to this end, a conceptual view of this ecological LNG supply chain with a return carbon loop is depicted in Figure 10.

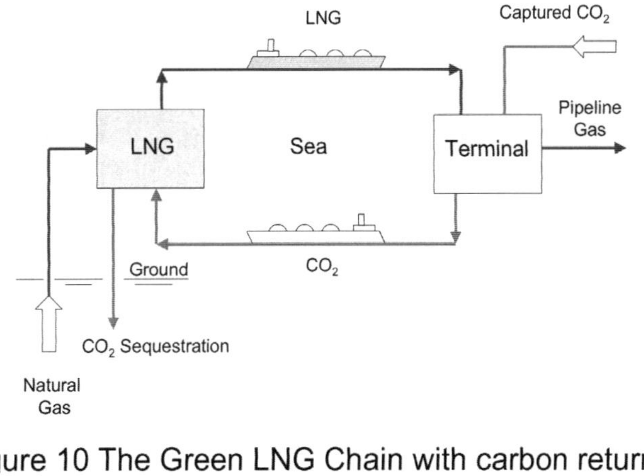

Figure 10 The Green LNG Chain with carbon return loop

In this future green LNG supply chain, the LNG terminal not only achieves CryoEnergy recovery to the credit of the power plant but it also deal with the CO_2 emitted from its flue gas stack. Valuable third party CO_2 traded for emissions credits can also be conveniently processed at the terminal. Finally, to loop the green chain to its origin, it includes a maritime return transport of the captured CO_2 in the liquid or solid state to the natural gas liquefaction plant and its re-injection to a suitable underground reservoir. Since the LNG tanker no longer returns to the point of origin empty; a much better way to run a tanker fleet is realized. Moreover, the returned liquid CO_2 could be beneficially re-vaporized in the natural gas liquefaction process as a once-through pre-cooling refrigerant before it is re-injected in the vapor state underground. This would reduce by an equivalent energy the external refrigeration power load on the LNG process.

A final word on the LNG tanker fleet is in order. Present day LNG tanker holds are not designed to carry higher than LNG density fluids at vapor pressures of above 5 bar. Liquid CO_2 transport is more compatible today with service of refrigerated LPG tankers. However, recent interest postulated by some integrated energy majors in the advantages of pressurized LNG facilities could be instrumental to inciting change in the LNG maritime transport system. Presumably, new ideas could be captured and concretized to future LNG tanker construction ultimately making them more compatible to the this green concept with a carbon return loop.

Conclusion

The LNG regasification process is the terminal link of the maritime LNG supply chain. It is a crucial processing step which needs to be more compact, less obtrusive and less costly. Today it also needs to respond to energy conservation, if not outright cryo-energy recovery. In the USA and industrialized countries where a mature pipeline gas distribution network exists, LNG regasification at the terminal needs also to achieve interchangeability and meet existing gas pipeline specifications in a cost effective, if not simple, manner.

The Gasifin technology meets all of these objectives. It meets them simply, compactly and therefore more cost effectively than conventional LNG regasification techniques practiced today. Moreover, it affords an elegant integration of the LNG regasification terminal with a gas fired electric power plant without imposing that the two to be right on top of one another.

In the future, when green business practice makes absolute sense and CO_2 emissions credits and/or penalties are factored in, the LNG terminal can become the ideal focal point for collection, capture, storage and shipping of CO_2. Then an already existing Gasifin LNG terminal extended to the GasifinPlus processing scope by including CO_2 capture and liquefaction will find an ecologically best overall synergy within the LNG supply chain and gas power industry.

References

ALPEMA *The Standards of the Brazed Aluminum Plate -Fin Heat Exchanger Manufacturer's Association* First Edition 1994 Mechanical standard 5.8 page 33

Campbell, G. and Zednik, J. "*Shipboard Regasification Terminal* " preprint 78th Annual GPA Convention 1999

Chabrelie, M-F. "*The New LNG Industry* " Hydrocarbon Engineering Volume 9 Number 12 December 2004 p. 13

Fatica, A. "*Gas quality debate heats up as more US LNG imports loom*" Oil & Gas Journal June 14, 2004 p. 69

Fischer, B. "*A New Process to Reduce LNG Cost*" presented at the AICHE Spring National Meeting March 10, 2002

Habibullah, A. and Weldin, F. "*Cost Reduction Ideas for LNG terminals*" preprint from 78th Annual GPA Convention 1999

Mak, J., Nielsen, D.,Schulte, D. and Graham, C. "*A New and Flexible LNG Regasification Plant*" preprint form 83rd Annual GPA Convention 2004

Minkkinen, A., Larue, J., Levier J.F. and Patel, S. "*Methanol gas treating scheme offers economics, versatility* " Oil & Gas Journal June 1, 1992 p. 65

Price, B.C. "*Optimize energy integration for LNG terminals*" Hydrocarbon Processing Vol. 83 No.7 July 2004 p. 43

Ranke, G. and Mohr, V. "*The Rectisol Wash, New Developments in Acid Gas Removal from Synthesis Gases*" Newman, S. Editor Acid and Sour Gas Treating Processes 1985. Gulf Publishing Company. p. 80

Rojey, A., Larue, J. 1989 US Patent 4 979 966

Biography of Speaker

Ari Minkkinen is a principal process advisor at IFP near Paris in France. In this capacity he is actively involved in all aspects of oil and gas process technology development from conception to commercialization. His previous affiliations have been with Total in Paris as chief process engineer and with ABB Lummus both in France and the USA as a process design manager.

He holds a post graduate degree in Chemical Engineering from the Polytechnic Institute of New York, and is cited as sole or principal inventor in over 25 US and foreign patents in the hydrocarbons processing, refining and petrochemicals field. He was a past chairman of the Gas Processors Association Europe and is a member of the American Institute of Chemical Engineers

Challenges of LPG Supply to Nigerian Domestic Economy

Dr. N. John Erinne, FNSChE
Chex & Associates
Lagos, Nigeria

AIChE Spring Meeting, April 2005
5th Topical Conference on Natural Gas Utilization
Atlanta April 10-14, 2005

1. INTRODUCTION

Situated on the West Coast of Africa, a zone prolific in oil and gas reserves, Nigeria is the leading oil producing country in Africa. Proven reserves of oil are estimated at 35b bbls while current output is about 2.3m bbls per day. On the other hand, total natural gas reserves amount to about 180 tscf while current production rate is about 5.1bscfd[1]. With these, estimated depletion times of the oil and gas reserves are about 40 and 100 years respectively. Evidently therefore, Nigeria is increasingly regarded as a predominantly gas territory, with significant oil.

The population of Nigeria is estimated at about 125m, making it the most populous nation in Africa and the dominant entity in West Africa. To put this is perspective, Nigeria is about as big as UK and France combined and half of USA, in terms of population.

2. GAS SECTOR

2.1 RESERVES

About 53% (95tscf) of the natural gas reserves is locked in with oil as Associated Gas (AG), while the balance (85tscf) is Non-Associated Gas (NAG). This however may be a gross understatement of the NAG potentials of the country as virtually all the gas discoveries so far have been coincidental on oil exploration. There is as yet very little or no specific focus on gas exploration[2].

2.2 PRODUCTION

Of the 5.1bscfd production, about 3.9bscf (76%) are obtained from AG while 24% comes from NAG. Besides, only about 57% of the total produced gas is utilized, while 43% is flared in the oilfields. The 2.2 bscfd of natural gas flared is derived essentially from AG and accounts for about 56% of total AG production[3]. In effect therefore, only 44% of the AG production is currently being utilized gainfully.

2.3 TREND

Worrisome as this situation may be, it represents very considerable progress in gas utilization in the last decade and half. For instance flare proportion has been reduced approximately by half from nearly 80% of total gas production in the early 1990s to about 43%

in 2003. Conversely, proportion of gas utilized has nearly tripled from just over 20% to 57% during the same period **(Table1)** [4].

Besides, the last decade and half has also witnessed rapid growth in total gas production, from about 3.0 bscfd in 1990 to 5.1 bscfd in 2003, representing an average annual growth of about 4.5%[4].

2.4 DEVELOPMENT PROJECTS

The progress achieved is substantially a response to deliberate government policy of encouraging gainful utilization and monetization of natural gas[5]. In particular, current policy requires all producers to end flaring by 2008.

By consequence, the industry has witnessed development of several new major gas utilization projects in the last decade **(Table 2),** mostly in the areas of:

- Liquefaction (LNG)
- NGL/LPG Extraction
- Power Generation
- Process Feedstock
- Gas Export

3. LPG SUPPLIES

3.1 REFINERIES

Traditionally, all LPG supplies to the Nigerian economy have been produced from the existing 3 local Refineries, with a total installed capacity of 380,000 tonnes per annum **(Table 3).** However, in the last decade and half performance of these government owned and operated Refineries has become rather erratic and output very low, leading to serious shortfalls in supplies to the domestic economy. For instance, in the early 1990s, average production was about 120,000 tonnes per annum, representing barely 32% of available capacity. Between 1994 and 2002 output dropped to 40 – 80,000/year, with an average of about 65,000 tonnes/year. Annual production between 1998 and 2002 is shown in **Table 4**[6].

3.2 IMPORTS

Owing to the difficulties in domestic production from the Refineries imports have had to be arranged, to augument at least part of the shortfall. It is estimated that in the 5 years from 1998 to 2003 imports accounted for about 40-50% of total supplies per annum, on the average.

3.3 EXTRACTION FROM NATURAL GAS

Production of LPG from natural gas commenced in Nigeria only about 7 years ago with the commissioning of the CHEVRONTEXACO and AGIP plants. Subsequently the EXXONMOBIL and NLNG plants have also come on stream **(Table 5).** Incidentally, all these are focused exclusively on exports, with the exception of the AGIP plant which supplies propane to the local Petrochemicals Complex at Port Harcourt[5,7,8].

Presently, total exports of LPG from these plants amount to over 2m tonnes per annum.

4. LPG DISTRIBUTION CHAIN

4.1 MARKETERS

Distribution of LPG across the country has traditionally been dominated by the Marketers. 3 broad categories of marketers may be identified in this regard:

- Majors
- Independent Majors
- Independents

i. Majors

These are major marketers of petroleum products, including LPG. They are either presently or were once subsidiaries of multinational petroleum companies (EXXONMOBIL, CHEVRONTEXACO, TOTAL, SHELL, BP, AGIP etc). They own and operate large storage and bottling facilities at key locations across the country, especially the refinery locations and major industrial/commercial centers (Lagos, Warri, Port Harcourt, Kaduna etc).

ii. Independent Majors

There are also a few sizeable marketers solely focused on LPG marketing. They also own and operate sizeable storage and distribution facilities at key locations, as well.

iii. Independents

These are the numerous independent, indigenous marketers who operate small storage and bottling plants all over the country. Some are in wholesome marketing of petroleum products while others are limited to LPG solely. Recent counts put the total number of such mini-filling plants at over 150.

Essentially, the Marketers lift bulk LPG from the Refinery Depots to their locations in tank trucks, for subsequent filling into cylinders for distribution or delivery in bulk in smaller tank trucks to commercial consumers. Some, mostly the majors and independent majors, also lift bulk LPG from the coastal Refinery Depots at Warri and Port Harcourt by marine vessels, mainly to Lagos.

The only facility for piped delivery to marketer's storage is between the Refinery Depot and the Storage Depot of OANDO, a former subsidiary of ESSO, at Warri. There is no capability for distribution by railroad.

4.2 NATIONAL DEPOTS

In the early 1990s the government authorities, through the Nigerian National Petroleum Corporation (NNPC), embarked on an ambitious program of constructing storage depots at 9 selected locations across the country, with combined storage capacity of 12,000tonnes, as a means of facilitating the distribution of LPG in the country **(Table 6 & Fig.1).** Most of these were completed in the mid 1990s but were not fully commissioned as this coincided with the steep downturn in the performance of the Refineries and hence shortfalls in product supplies. Only a few were indeed fully commissioned. Deliveries from the Refinery LPG Depots were designed to be mostly by tank trucks, as these Depots were not connected to the Refineries by pipeline. This also constituted an additional constraint owing to a number of factors:

- Long distances to be traversed
- Poor state of the roads
- Limited availability of tank trucks
- High cost of delivery

A few of the Depots, notably Calabar and Lagos, are however located on the coast and are designed to be fed from marine vessels.

4.3 IMPORTS

LPG imports since the late 1990s have usually been discharged into either the Refinery LPG Depots in coastal locations (Warri, Port Harcourt) or the National LPG Depots in Lagos and Calabar. Marketers then lift from these facilities to their own facilities for distribution.

4.4 LPG GADGETS/EQUIPMENT

The Marketers, Storage Facilities, Filling Plants, Tank Trucks, Marine Vessels and Pipelines have been highlighted above as necessary factors in the distribution network. In addition, LPG utilization gadgets and equipment also constitute a necessary factor for sustenance of the distribution chain. These include such items as: cylinders, cooking stoves, hoses and regulators.

There are 2 known manufacturers of cylinders in the country located in Lagos and Ibadan in the South West and several assemblers of cooking stoves, mostly in Lagos.

5. LPG UTILIZATION / CONSUMPTION PATTERN

5.1 UTILIZATION

LPG utilization in Nigeria may be classified broadly as follows:

- Household Consumption
- Commercial / Industrial Consumption

i. Household Consumption

This refers to LPG used in households as a cooking fuel. Over 90% of the demand is in this segment.

LPG competes mostly with wood and kerosene as a domestic fuel. Popularity of wood is however waning, with increased urbanization, in view of the difficulties of supply and distribution, as well as convenience of handling, in the urban centers. Besides, and perhaps more importantly too, exploitation of forest resources for fuel wood is actively being discouraged by Government owing to environmental considerations, especially as pertains to desert encroachment.

Kerosene on the other hand tends to be very popular with the urban class, especially the lower income wrungs, and is as well becoming increasingly popular in the rural areas also. Nevertheless, kerosene suffers disadvantage relative to LPG in terms of cleanliness of flame, speed of heating and overall convenience of handling.

ii. **Commercial / Industrial Consumption**

- Commercial Cooking/Heating
  - Hotels, Restaurants, Hospitals, Schools, Institutions, Laundries/Dry Cleaning, Bakeries etc

- Industrial Heating
  - Process Heating, Ovens, Furnaces, Dryers etc

- Feedstock
  - Petrochemical, Aerosol Propellant, etc

LPG is preferred in industry when a clean fuel is particularly desirable either in moderate quantities or at locations remote from natural gas distribution lines. These applications are estimated to account for less than 10% of the total LPG consumption.

5.2 DEMAND / CONSUMPTION PATTERN

Recent studies indicates that in 2002 only about 2.8% of Nigeria households used LPG, down from about 4.4% in 1992[9].

Furthermore, other studies also indicate that per capita total consumption of LPG in the Household and Commercial Sectors in Nigeria is barely 0.8kg, evidently the lowest in West Africa and indeed one of the lowest in the world **(Tables 7&8).** This is obviously a paradox, considering that Nigeria is a leading oil and gas country. To put it further into perspective, it represents barely 24% of the West African average per capita consumption of about 3.4kg and only 8% of per capita consumption of Senegal, a non oil and gas country[6,10].

5.3 NATURAL GAS

Ordinarily, natural gas presents strong competition against LPG. However, in view of the complete absence of piped natural gas distribution infrastructure to households natural gas poses no challenge in the Household Segment in Nigeria. Besides, natural gas distribution to industry is only limited and therefore poses only limited competition to LPG in the Industrial Segment.

6. LPG PRICES

6.1 PRODUCT PRICES

Available LPG pricing data for the period 1992 – 2002, illustrating Ex-Refinery Prices and Retail Prices, are presented in **Table 9.** Estimated Differentials to International Spot Market Prices, as well as Retailer Gross Margins, are also presented[9].

Evidently, between 1992 and 1997 Ex-Refinery and Retail Prices remained fairly stable and low, reasonably below International Spot Market Prices. With increasing supply shortfalls and consequent price deregulation, by 1998 both Ex-Refinery and Retail Prices began to climb, rising to about $400/tonne and $1,200/tonne respectively by 2002. This also reflects over $150/tonne differential above world spot prices.

The current Retail Price in Nigeria in particular is out of step with similar figures from other countries as shown in **Table 10.** Nigeria is the most expensive relative to the countries of Latin America and USA. For illustration, USA Retail Prices are about 50% of Nigeria but USA GDP per capita is over 500 times greater than Nigeria's.

6.2 GADGET COSTS

Product prices apart, the cost of the gadgets for LPG utilization, especially in the Household Sector, could also be an issue. Estimated cost of a complete set of basic gadgets required (cylinder, cooking stove, hose & regulator) amounts to about $30. This represents about 10% of the annual GDP per capita of $310 or over 1 month earning per person. The implication is that these gadgets are hardly affordable to the average households and are likely to be available only to the highest income households in Nigeria.

7. POTENTIALS

Evidently, LPG is grossly under-utilized and potentials gravely unfulfilled in Nigeria owing to several constraints, as outlined below.

7.1 CONSTRAINTS

These constraints may be categorized broadly as follows:

- Supply Constraints
- Over – Pricing

i. Supply Constraints

These include:

- Production Shortfalls from the Refineries
- Reliance on Imports
- Limited Distribution Infrastructure and Logistics Facilities

ii. Over – Pricing

- High Ex-Refinery and Retail Prices of LPG

- High Cost of LPG Gadgets / Equipment

7.2 PROJECTIONS

A recent World Bank Study indicates that if these constraints are reasonably addressed it is possible for per capita consumption of LPG in Nigeria to rapidly attain a level of 3.74kg per annum. This would translate to a total annual demand of about 460,00 tonnes, over 350% above current demand levels[10].

Beyond this, further growth of the industry would depend on basic economic indices such as:

- GDP Growth Rate
- Population Growth Rate

Based on an assumed GDP Growth Rate of about 4.5% and Population Growth Rate of 2.6%, it is estimated that demand could rise to 1.1m tonnes per annum in the next 15years, by 2020. Other forecasts based on the West African Regional average consumption of 3.4kg per capita per annum arrive at approximately the same conclusion indicating a potential of 1.0m tonnes per annum by about 2020 – 2022[9,10].

8. STRATEGIES

Possible strategies for elimination of the noted constraints and achieving the projected potentials are examined.

8.1 PRODUCTION

A number of measures are proferred which will facilitate rapid recovery and growth of LPG production for the domestic market.

i. Refineries

All the 3 Refineries which are presently public enterprises need to be privatized and preferably sold to companies with proven competence in operation of refineries. In principle, privatization has been agreed by the Federal Government of Nigeria since about 5 years ago but not much progress has been recorded in this regard. It is strongly suggested that the process be put on a fast track if the plants are to return to meaningful operation in the near future, in order to restore supplies of petroleum products to the economy, including LPG.

ii. Production from Natural Gas

Evidently, the existing natural gas based LPG plants/projects are sited in remote coastal locations, for ease of exports. This however leads to serious logistical problems in diverting even a fraction of their production to the domestic market. Nevertheless, the scope exists for extraction of LPG from associated natural gas flared in the oilfields further inland and subsequent injection into the domestic market, with less logistical difficulties. Already, one of such projects has been proposed in South East Nigeria by OWEL GAS, based on feedstock from SHELL, CHEVRONTEXACO and ADDAX operations. It is hoped that this project will be driven to conclusion and also that similar other projects would be initiated in the near future.

With viable technology for even smaller sized extraction plants now available this option is considered increasingly attractive and is strongly recommended.

iii. **Cryogenic LPG**

LPG from the LNG operations is not very readily applicable in the domestic economy as it is basically cryogenic, whereas the requirement is for pressurized product. A measure of flexibility would therefore be necessary in the design of further expansions of the plant, in line with the new LNG trains, to produce some pressurized LPG, for the domestic market, as may be necessary. It may also be worthwhile to consider retrofit of the existing plant at Bonny, to adapt portion of the output to pressurized LPG.

8.2 LOGISTICS

i. **Marine Vessels**

From a policy stand-point, the Federal Government of Nigeria has recently intervened and approved the diversion of up to 300,000 tonnes of LPG from the EXXONMOBIL facility at Bonny Island to the domestic economy. This is however yet to make the desired impact in view of inadequate marine vessels to lift the product to the main LPG Coastal Depots at Lagos in the South West and Calabar in the South East. A window of opportunity therefore evidently exists for adventurous enterprenuers in the Marine Industry to invest in marine vessels suitable for supporting LPG distribution. They are well encouraged to explore such opportunity actively.

ii. **National Depots**

It is obviously very disappointing that after a decade most of the Depots, especially the 7 located inland, have still not been fully commissioned and are therefore not operational. It is strongly recommended therefore that the Depots be included in the package for privatization of the national petroleum products distribution network and that the process of unbundling will be accelerated. This is considered the quickest and surest way to get these Depots to be operational.

iii. **Distribution Pipelines**

It is expected that consequent on privatization of the LPG Depots and Refineries serious consideration would be accorded the question of connection of the inland Depots by pipeline from the Coastal Depots or Refineries. For instance, the following configuration may be reasonably anticipated:

Eastern Axis: Calabar/Enugu/Makurdi; from Calabar Coastal Depot or PH Refinery
Western Axis : Lagos/Ibadan/Ilorin; from Lagos Coastal Depot
Northern Axis: Kano/Gombe/Gusau; from Kaduna Refinery

These investments will not only enhance the operations of the Depots and the entire Distribution Chain, they will also contribute substantially towards the realization of the national butanization goals.

iv. **Tank Trucks**

In the interim, opportunity also evidently exists for specialist logistic companies to set up tank truck fleets for bulk movement of LPG from the Coastal Depots to the other 7 National Depots located inland. Till date no haulage companies dedicated to LPG transportation exist.

v. **Rail Haulage**

On a longer term basis, serious consideration also requires to be given to distribution of bulk LPG by rail. Incidentally, several of the inland Depots are sited in locations serviced by the national railway network. This should enhance the economics of such a project.

8.3 PRICING

i. **Product Price**

Evidently, the Ex-Refinery price of bulk LPG set by Government through NNPC is rather high, even by world commercial standards, and requires to be reviewed downwards urgently, if demand for domestically produced LPG is to be stimulated. It is also expected that in the future if the Refineries are duly privatized and subsequently operate commercially the combination of improved efficiency, enhanced output and competition will drive the prices down.

Furthermore, Retail Gross Margins have remained very high, exerting pressure on the Retail Prices. If the supplies bottlenecks are eased it is expected that competitive pressure will also squeeze the Retail Gross Margins and drive the Retail Prices down.

ii. **Equipment Prices**

Given the commitment of the Federal Government of Nigeria to promoting LPG as the preferred household fuel, a complete review of the fiscal regimen applicable to manufacturing of cylinders and stoves is called for. Enhancement of the incentives will assist drive down the prices which are presently uncomfortably high.

Besides, it will be realistic to expect that if the various constraints associated with Production, Logistics and Product-Pricing are addressed, there will be a steady rise in demand for these items, which will enable the manufacturers reduce unit production costs and subsequently unit prices.

9. CONCLUSION

Evidently, the LPG Sector in Nigeria is gravely under-achieving, despite the enormous gas resources of the country, owing to a combination of under-supply and over-pricing problems. The avowed government policy of actively promoting the use of LPG and making it the preferred household fuel is at grave risk of collapsing. Various suggestions have been proffered for elimination of the identified problems and constraints, in the areas of production, distribution/logistics and pricing. If these are implemented it is realistic to envisage an early recovery and rapid growth in the industry.

REFERENCES

1. Igbatayo, S. A.:'Meeting, Global Energy Demand: The Emergence of Nigeria's Gas Industry', Proceedings of Nigerian Gas Association International Conference, Abuja, Nigeria Nov. 2004

2. Erinne, N. J: 'Natural Gas: Nigeria's Energy for the Future', Proceedings of AIChE Spring Conference, Houston, Texas, April 2001.

3. Dankoru, E: 'Deregulation as a Catalyst in Creating Opportunities for improved Gas Utilization', Proceedings of Nigerian Gas Association International Conference, Abuja, Nigeria, Nov. 2004

4. Nigerian Gas Association: 'Nigerian Gas: Quarterly Journal of NGA', Jan. 2004

5. Oladele, A. O. R:'Challenges of Sourcing Finance for Gas Process Industries in Nigeria',Proceedings of 33rd Annual Conference of Nigerian Society of Chemical Engineers (a Division of Nigerian Society of Engineers), Abeokuta, Nigeria, Nov. 2003

6. PL Consulting:'Nigerian Oil & Gas Industry Outlook', 2003

7. Caldwell, P.L:'Nigerian Oil & Gas Projects, Prospects, Financing and Incentives on Reducing and Eliminating Gas Flares', Proceedings of 1st Annual Workshop of Nigerian Gas Association International Conference, Abuja, Nigeria, Nov. 1999.

8. Issa, U. A:'Diversifying Nigeria's Petroleum Exports–The Nigeria LNG Experience', Proceedings of Nigerian Gas Association, Abuja, Nigeria, Nov. 2004

9. Purvin & Gertz Inc:'West Niger Delta LPG Study', 2002

10. Osezua, C. A.'Opportunities and Challenges of Extracting LPG from Natural Gas for the Domestic Market', Proceedings of Nigerian Gas Association International Conference, Abuja, Nigeria, Nov. 2004.

NATURAL GAS PRODUCTION & UTILIZATION TREND : TABLE 1

YEAR 1992	TOTAL PRODUCTION (bscfd)	% UTILIZED	% FLARED
1992	3.14	2.17	78.3
1993	3.20	21.1	78.9
1994	3.16	20.6	79.4
1995	3.19	21.0	79.0
1996	3.54	27.2	72.8
1997	3.13	29.8	70.2
1998	3.67	37.7	62.3
1999	3.64	39.9	60.1
2000	3.90	46.5	53.5
2001	5.03	49.0	51.0
2002	5.08	53.8	46.2
2003	5.12	56.9	43.1

MAJOR GAS UTILIZATION PROJECTS: TABLE 2

PROJECT LNG	PROMOTER	LOCATION	START DATE
NLNG Trains 1& 2	NNPC/Shell/Total/Agip	Bonny Island	1999
NLNG Train 3	NNPC/Shell/Total/Agip	Bonny Island	2003
NLNG Trains 4 & 5	NNPC/Shell/Total/Agip	Bonny Island	2005
NLNG Train 6	NNPC/Shell/Total/Agip	Bonny Island	2007
Brass LNG	NNPC/ChevronTexaco/ Agip/Phillips	Brass	2007
OK LNG	NNPC/ChevronTexaco/ British Gas	Olokola	2009
Mobil LNG	NNPC/ExxonMobil	Bonny Island	2010
NGL/LPG			
OB/OB NGL Plant	NNPC/Agip/Phillips	Omoku	1997
Escravos LPG Plant	NNPC/ChevronTexaco	Escravos	1997
Oso NGL Plant	NNPC/ExxonMobil	Bonny Island	1998
NLNG LPG Plant	NNPC/Shell/Total/Agip	Bonny Island	2003
Global Gas Project	Global Energy	Cawthorn Channel	2006
Izombe LPG Project	Owel Gas	Izombe	Proposed
POWER GENERATION			
Egbin Power Plant	NEPA	Ikorodu, Lagos	1990
Afam Power Plant	NNPC/Shell/NEPA	Afam, PH	2005
Kwale IPP	NNPC/Agip	Kwale	2005
ELF IPP	NNPC/Total	Omoku	2007
Mobil IPP	NNPC/ExxonMobil	Bonny Island	2010

FEEDSTOCK			
NAFCON Fertilizer Complex	NAFCON	Onne, PH	1990
Eleme Petrochemicals Complex	NNPC	PH	1995
Escravos GTL Project	NNPC/ChevronTexaco/	Escravos	Proposed
Sasol	Lekki Methanol Project	Lagos	Proposed
GAS EXPORT			
WAGP	ChevronTexaco/Shell/ NNPC/Ghana/Togo/Benin	West African Coast-Nigeria/ Benin/Togo/Ghana	2006
Trans Sahara Gas Project	NNPC/SONATRACH	Trans Sahara - Nigeria/Niger/ Algeria	Proposed

REFINERIES LPG PRODUCTION CAPACITIES: TABLE 3

REFINERY	LPG PRODUCTION CAPACITY *(000 tonnes/yr)*
Warri	70
Kaduna	200
Port Harcourt	110
	380

LPG PRODUCTION FROM REFINERIES 1998 – 2002: TABLE 4

YEAR	PRODUCTION *(000 tonnes)*
1998	70
1999	42
2000	60
2001	60
2002	68

LPG FROM NATURAL GAS PLANTS/PROJECTS: TABLE 5

PLANT/ PROJECT	OPERATOR	LOCATION	NATURAL GAS CAPACITY (mscfd)	OUTPUT CAPACITY (bpd)
OB/OB NGL Plant	Agip	Omoku	50	3,200
Escravos LPG Plant	ChevronTexaco	Escravos	285	14,000
OSO NGL Plant	ExxonMobil	Bonny Island	600	50,000
NLNG LPG Plant	NLNG/Shell	Bonny Island	1,500	45,000

PROJECTS

Global Gas Project	Global Energy	Cawthorn Channel	-	-
Izombe LPG Project	Owel Gas	Izombe	-	-
NLNG LPG Project	NLNG \ Shell	Bonny Island	2,000	35,000
Escravos GTL Project	ChevronTexaco	Escravos	300	2,000
Escravos Gas Project Phase 3	ChevronTexaco	Escravos	365	33,000
East Area Gas Project Phase 2	ExxonMobil	Bonny Island	630	30,000

NNPC NATIONAL LPG DEPOTS / CAPACITIES: TABLE 6

LOCATION	CAPACITY *(000 tonnes)*
Lagos	4,000
Ibadan	1,000
Ilorin	1,000
Calabar	1,000
Enugu	1,000
Makurdi	1,000
Kano	1,000
Gombe	1,000
Gusau	1,000

WEST AFRICA PER CAPITA LPG CONSUMPTION: TABLE 7

COUNTRY	PER CAPITA LPG CONSUMPTION *(kg/yr)*
Cameroun	1.8
Nigeria	0.8
Ghana	1.8
Ivory Cost	3.2
Senegal	10.3
Congo – Brazzaville	1.3
Angola	3.8
Gabon	12.8

DEVELOPING WORLD PER CAPITA LPG CONSUMPTION: TABLE 8

COUNTRY	PER CAPITA LPG CONSUMPTION *(kg/yr)*
China	9.6
Brazil	40.0
Chile	58.3
Indonesia	3.3
Vietnam	2.8
India	5.8
Pakistian	2.4
Honduras	6.6
Nicaragua	6.1
Egypt	32.7
Philippines	10.1
Nigeria	0.8

COMPARATIVE INTERNATIONAL LPG RETAIL PRICES: TABLE 10

COUNTRY	RETAIL PRICE *($ / tonne)*
Mexico	564
El-Salvador	325
Guatamela	325
Venezuela	321
Argentina	1,050
Uruguay	780
Brazil	600
Peru	>900
Chile	>900
USA	625 – 750
Nigeria	1,280

NIGERIAN DOMESTIC LPG PRICES: TABLE 9

YEAR	EX-REFINERY PRICE GROSS *($ / tonne)*	DIFFERENTIAL TO INT. SPOT *($ / tonne)*	*RETAIL PRICE *($ / tonne)*	RETAIL *($ / tonne)*
1992	55	(133)	139	84
1993	43	(129)	127	84
1994	43	(119)	243	200
1995	49	(128)	106	57
1996	49	(173)	359	310
1997	49	(155)	835	786
1998	298	155	1,025	727
1999	380	193	1,143	763
2000	357	50	1,281	924
2001	376	105	1,052	676
2002	390	152	1,280	890

* Based on retail price of 12.5kg cylinder of LPG

LNG Expander for Extended Operating Range in Large-Scale Liquefaction Trains

Joel Madison
Hans Kimmel
Ebara International Corporation
Sparks, Nevada

AIChE Spring Meeting, April 2005
5th Topical Conference on Natural Gas Utilization
Atlanta April 10-14, 2005

Abstract

The size of LNG liquefaction trains is steadily increasing due to economical reasons. As a consequence of the increase in the size of trains most of the equipment used in the liquefaction process has also to be increased in size. Particularly compressors, heat exchangers, gas expanders and liquid expanders have to be increased either in size or quantity. The paper presents the concept and design of LNG expanders for extended operating range in large-scale liquefaction trains.

Consequences of Underwater Releases of LNG

Robin Pitblado, Dennis Butts, John Baik
Det Norske Veritas (USA), Inc, Houston

AIChE Spring Meeting, April 2005
5th Topical Conference on Natural Gas Utilization
Atlanta April 10-14, 2005

Introduction

There has been substantial debate in the USA over the potential consequences of a marine accident involving an LNG vessel at or approaching one of the 4 current USA import terminals or one of the up to 45 proposed new terminals in North America. This debate has occurred at public meetings associated with the approval process, in conferences, or published technical papers. Some recent publications on this topic include: Fay (2003), ABS (2004a,b), DNV (Pitblado et al 2004a,b), Sandia (Hightower et al 2004), Cornwell (2004). Sandia and DNV both argue that a risk-based approach taking account of event likelihoods and the contribution of mitigation measures is a superior decision basis to a purely consequence-based approach (as in CFR 193-2051 and NFPA49a). There are significant issues and uncertainties that remain regarding underwater spills and this is the focus of this paper.

Events above the water level, such as those due to collision with another vessel, striking a fixed object above the waterline (also termed allision), or intentional act (e.g. explosion or sabotage) are relatively easy to conceive initially, although there are many possible outcomes and the modeling of these is highly variable amongst practitioners. Underwater releases have not received as much attention. These could be due to collision (e.g. with the bulbous bow of another large vessel), grounding, or intentional act (e.g., underwater explosion).

Past Events

The LNG marine industry has an excellent safety record with no loss of LNG cargo from a breach of cargo containment accident since delivery of the first cargo in 1958 by the Methane Pioneer sailing from Lake Charles. Thomas (1998) reported cumulative LNG traffic data to the end of 1997 and projecting this forward shows that by end 2004 there would have been in excess of 42,000 loaded voyages (equating to 84,000 loaded port transits as each voyage passes loaded through the source port and the reception port), 92 million loaded miles (184 million total voyage miles), and 3.6 billion m^3 LNG delivered.

In this time there have been a number of marine incidents. Mostly these have been small events such as collision / contact with a support vessel (e.g., tug), hard impact with piers, and groundings. Analysis of these events shows an incident rate similar to that reported for crude oil vessels, however it is believed that there is better reporting of LNG events and this suggests a somewhat lesser rate of occurrence. This might be expected given the generally higher level of crew standards and training for LNG vessels than the general crude oil fleet.

The authors are aware of three serious grounding events that have occurred to LNG vessels (two laden and one unladen). These are: El Paso Paul Kayser (1979), LNG Taurus (1979), and MISC Tenaga Lima (2004), respectively. The laden cases involved a membrane vessel (Paul Kayser) and a Moss Spherical vessel (LNG Taurus). The Paul Kayser struck pinnacle rocks off Gibraltar under full load and at full speed (19 knots) and may be regarded as a worst case grounding. The damage affected the full length of the hull, with the double hull and the insulation taking most of the deformation, although the forward LNG membrane was indented by 100-200mm no leak occurred. The LNG Taurus ran aground in storm conditions while trying to maneuver in restricted port waters. No release occurred. The unladen Tenaga Lima case occurred in Korean waters due to fishing nets fouling the propeller, and while these were successfully cleared, the vessel ran aground while trying to return to deeper water.

Two well known crude carrier groundings were the Exxon Valdez in 1989 and the Sea Empress off Milford Haven in 1996. Both of these involved navigation errors and released substantial oil. The two inquiries that followed both claimed that no significant spill would have occurred if the vessels had been double hulled. Some other grounding events (e.g. the Braer off Shetland in 1993) resulted in sinking and a double hull would not have been prevented loss of cargo. Thus double hulls have a demonstrated significant benefit in preventing loss of LNG and other cargoes, however extreme accidents are still possible. The USCG and Captain of the Port transit restrictions make such grounding or collision events highly unlikely, and the double hull provides probable another factor of ten reduction in spill likelihood.

Collisions are a greater source of energy, but probably a lesser risk than grounding based on historical data. Paik et al (2001) and others have published LNG vessel collision studies and these provide good confidence on minimum necessary collision speeds. Spherical vessels offer some advantage in that except for a small fraction of the hull length where the sphere is at its closest approach to the hull, the tank is well separated from the hull and collisions may not impact the tank. A further factor is that all collision studies to date currently assume that touching the LNG tank will result in leakage. In fact the LNG container (membrane of sphere) is ductile and Moss Rosenberg (LNG vessel constructor's) believe that it can withstand substantial deformation before failing. They suggest this adds a 20% margin onto the necessary collision speed. The ullage space above the liquid is typically 500m^3 at port arrival and thus the tank may be indented by 6m or more before the tank goes liquid full and a leak is forced. This would significantly increase the collision or grounding speed necessary for breach of containment.

Barriers

Pitblado et al (2004a) presented likely grounding and collision accidents and identified a maximum credible hole size of 0.75m. Intentional events could give larger hole sizes, but due to the progressively increasing effort required, the maximum credible hole size was identified as 1.5m. It was argued that other targets would assume higher priority in the mind of terrorists as the effort increased and potential outcome remained uncertain as to damage potential – a significant number of outcomes will result in zero harm to the public and the environment. Sandia (Hightower 2004) used somewhat larger holes (up to 2.5m diameter for accidents and 3.5m for intentional events). They also considered risk factors, but mainly in the form of mitigation measures. In practice, dispersion distances predicted by DNV were longer than those of Sandia and the predicted consequences were thus not that different although the event holes were larger in the Sandia report.

Table 1. compares the barriers that must be breached to obtain a significant LNG release below the waterline. Aside from the geometry, the key difference between the spherical and membrane designs is that the sphere is a self supporting pressure vessel attached to the inner hull by its skirt, while the membrane is a flexible shell supported by the insulation / inner hull.

Table 1. Barriers protecting against a below waterline release

BARRIER TYPE	MOSS SPHERICAL VESSEL	MEMBRANE VESSEL
External hull	Yes: normal vessel steel	Yes: normal vessel steel
Internal hull	Yes: normal vessel steel, 2+m separation (where the sphere curves away from the inner hull, separation from the sea can be 8+m)	Yes: normal vessel steel, Nearly 4m separation from sea
LNG sphere support skirt	Yes: Structural steel	n/a
External double insulation	n/a	Yes:1+m thick (perlite / wood)
Secondary barrier	n/a	Yes: additional membrane (nickel steel or plywood)
Direct thermal insulation	Yes: several materials (e.g. polyurethane 0.2-0.3m thick)	Yes: 1+m thick (perlite / wood)
LNG Tank wall	Yes: Aluminum or nickel steel	Nickel steel membrane
Pressure relief	Yes: fire rated	Yes: fire rated

In total these provide 5 – 6 barriers between the LNG cargo and the outside seawater. Older crude carriers have only 1 barrier, double bottom vessels 1½, and modern double hull tankers have 2. The hull and skirt barriers are effective through the structural strength of the steel, the insulation barriers are effective in cushioning and spreading the load. The LNG container walls mainly provide ductility as a barrier, with significant deformation likely before breach of containment. Other barriers not listed in the table are probabilistic, between 10-80% effective typically, as striking spheres away from their closest approach to the hull and striking membranes at the separating caisson would result in no direct impact onto the tank.

Ingress of Water

a) Spherical vessel

A collision, grounding or other external hole penetrating the double hull of a Moss-type spherical carrier will allow water into the LNG tank area. Spherical tanks might appear to offer very large spaces for water entry and circulation – were they simply located in cubic spaces. However, in practice designers ensure the double hull uses the spherical void spaces for fuel oil or ballast water and the air gap between the sphere insulation and the inner hull is relatively small. This limits the volume of water entry and resultant hydrodynamic stresses to the hull due to the loss of buoyancy.

The inner and outer hulls and transverse members are constructed of normal marine plate and these steels are not resistant to cryogenic embrittlement. Embrittlement typically occurs below -25 to -40°C while the LNG is at -162°C. Spills of LNG onto unprotected deck plates have historically led to rapid cracking. In underwater cases, seawater ingress affords that part of the hull thermal protection from LNG contact and embrittlement. In order to cause a breach of 0.75-1.5m in the LNG tank, the external breach will be significantly larger, 3 – 4m at least and water ingress will be immediate. Parts of the hull above the water level would be subject to embrittlement, but there are also mitigating factors here:

- the leak will be immediately adjacent to the external hole and thus the vulnerable plating above the hole, subject to cryogenic cooling and failure, is not providing much direct support as the load will have redistributed onto either side
- LNG carriers are equipped with water spray systems that would be utilized in such an event to provide thermal protection to the hull and deck plating
- the supporting sphere skirt is very strong and has been calculated by DNV to be able to support the LNG tank even with approximately 90° of arc ($1/4^{th}$ of circumference) damaged

Thus through load redistribution and design redundancy substantial loads can be sustained and water ingress actually protects the structural members. This will afford the vessel protection against break-up and escalation of the initiating event to involve other tanks. This is a qualitative argument and a full finite element analysis would be necessary to resolve the uncertainties. It is not possible currently to offer definitive conclusions without such analysis.

b) Membrane Vessels

Ingress of water into a membrane vessel is in some ways simpler as the double insulation and LNG containment system is fixed to the inner hull plating. While there are several designs, many membrane systems use perlite (a non-combustible insulation) contained in wooden boxes. A hole punched through the insulation will wet the remaining perlite in the boxes directly affected. While the insulting properties will be lost, the boxes are not large and the zone affected will be limited to the region around the immediate damage area. The loss of thermal insulation will be much less than the heat input from the sea directly to the LNG.

Underwater Scenarios

Typical factors associated with an underwater release of LNG from a collision, grounding or explosion event are listed in Table 2.

Table 2. Comparison of factors affecting LNG release

RELEASE FACTOR	MOSS SPHERICAL VESSEL	MEMBRANE VESSEL
Single Tank capacity	27,450m^3	27,450m^3
Tank dimensions	37.5 m inner diameter	W=34.5m L=32.m H=24.6m
Typical vessel draft	11.5m	11.5m
Bottom of tank below waterline (full LNG load)	9.5m	7m
Initial LNG hydrostatic head (at base)	33.8 m LNG	23.6 m LNG
Water backpressure head	0.87 barg	0.63 barg
Assumed initial water entry = (98% tank fill level and 500m^3 gas space) to create a 0.2barg positive pressure	90 kg	90 kg
Pressure differential LNG at hole to seawater at hole	0.71 barg (positive LNG – side)	0.54 barg (positive LNG–side)
Initial LNG discharge rate (Average rate about 50% less)	2100 kg/s – 0.75m hole 8300 – 1.5m hole	1800 kg/s – 0.75m hole 7200 – 1.5m hole
Initial LNG discharge velocity	11.1m/s	9.6m/s
Further water entry	Nil for a time, see below	Nil for a time, see below
Semi-steady state?	Yes	Yes

The LNG release scenario can be almost semi-steady-state. Thus it can be explained in this sequence:

a) initiating event creates a significant hole in the external hull (e.g. 3-5m diameter) below the waterline and progresses through all remaining barriers to cause a serious breach of LNG containment (e.g. 0.75-1.5m diameter);
b) water immediately pours into the hull and makes contact with the hole;
c) simultaneously LNG starts to flow out of the hole due to its hydrostatic head;
d) assuming the hole is not precisely horizontal in orientation some water will flow into the hole at its lower edge due to relative density differences (as shown by Fay, 2003);
e) very rapid heat transfer will occur at the interface and large quantities of natural gas will evolve;
f) through the combination of the previous two steps natural gas vapor will collect in the vapor space above the LNG in the tank;
g) normally the pressure is maintained very close to atmospheric, but the relief system is set to 0.2 barg and DNV believe this pressure can be reached rather quickly (equivalent to 90kg of water entry to the tank or equivalent contact through the hole);

h) this pressure is sufficient to cause the discharge flow to increase to 9-11 m/s and water will not be able to flow freely into the tank against this high discharge velocity – further vapor formation will occur away from the hole and mostly rise to the water surface outside of the vessel;
i) as water ingress to the tank slows the pressure stabilizes at 0.2barg (the relief setting) and it will remain at the level for some time;
j) as LNG is discharged under pressure, and the level of LNG in the tank declines, the vapor space pressure will also decline;
k) at various points in time, not easily estimated by calculation, water will again be in contact with the LNG in a manner that natural gas vapors can rise up into the tank generally keeping the vapor space pressure near the limit of 0.2 barg, although it may drop below that level at several points in the process;
l) during this process the release rate will be declining with the decreasing head of LNG in the tank, until an effective state of equilibrium is reached between the LNG and water where the release rate of LNG is minimal compared with the initial rate;
m) the point at which the equilibrium between LNG and water is reached is at 43% of the tank level using the assumption of a hole 0.5m above the bottom of the tank. Estimation of flow rates after this is difficult but should be substantially less than the initial rate and the resultant hazard zones will decrease.

The two LNG vessel designs, spherical and membrane, are approximately equivalent in this regard.

Effect of RPT

Underwater scenarios are complicated due to the direct contact with seawater and the potential for Rapid Phase Transition (RPT) pressure impacts. Nedelka (2003) describes current knowledge of RPT's. These are often given a TNT equivalence, based on longer distance impacts (e.g. window breakage) that correlate to TNT explosions; however, RPT's are at much less velocity (maximum sonic velocity vs supersonic velocity for TNT detonations) and their destructive potential will be correspondingly less. The situation here is subsurface – with extra pressure. One significant factor underlying RPT is the type of heat transfer – film boiling (where a film of methane forms between the water and the LNG) reduces the heat transfer relative to nucleate boiling and thus reduces the RPT effect. Theoretically, higher pressure will tend to reduce film boiling and thus somewhat increase the potential for RPT. The cause of the pressure is the rapid vapor formation – 600:1 (LNG : natural gas) at sealevel and 300-400:1 at the depth here. The reduction in volume increase will significantly affect the RPT pressure rise. Finally, the depth of the hole and the lesser backpressure towards the sea surface than into the tank will tend to direct RPT effects outwards. Overall, RPT effects to the LNG tank may be less pronounced than initially thought.

RPT effects could impact the space between the inner and outer hull. These are partitioned but otherwise open spaces as LNG vessels carry little water ballast when fully loaded.

Amount of LNG Released

Below waterline releases in a membrane carrier will discharge in a relatively quick manner about 57% of the cargo before hydrodynamic forces balance slowing to a much smaller release rate, although it is possible that eventually all cargo would be lost through seepage. For consequence modeling purposes, the 57% figure is used as the cutoff for maximum effects. In the case of above waterline cases, once the LNG has flowed out of the tank above the hole, the remaining LNG may be released very slowly as a vapor due to heat leakage, with the possibility of releasing about 69% of a single tank (Pitblado 2004a).

Effect of Fire

The Sandia report (2004) identified that in collisions the primary outcome would be fire due to the high energy release and thus chance of ignition. That study identified hole sizes of up to 1.4m diameter for collisions or grounding. They mainly argue that collisions are the primary risk as grounding events tended to be smaller and that most collisions will ignite the released LNG, creating primarily a pool fire risk. DNV note that the frequency of grounding is likely to be higher than collision given a) the historical record of 3 serious groundings and no serious collision and b) the USCG escort regime which should minimize the potential for interactions. Many groundings in US ports will be onto muddy bottoms and at less than full speed (often 12 kts of less) and relatively smaller hull damage would be expected. It is considered unlikely that a grounding event at less than 12 knots would result in breach of containment. For worst case grounding events, DNV has used 0.75m as a representative hole size.

Grounding events would be less likely to ignite as the damage and energy dissipation would all be beneath the surface. The LNG released will rise very quickly to the surface with some boil-off and then spread to a large pool where most of the vaporization will occur. These may give rise to longer distance dispersion events that may drift, find remote ignition, and flash back to the source. The initially large but thin LNG surface pool, will burn at a high rate ($0.35 kg/m^2s$) and it will shrink to a sustainable size (about 45% of its initial unignited dimension) where the prolonged release rate matches the pool burning rate.

Authen et al (1976) examined the potential effects of a major water surface fire adjacent to the LNG vessel for both membrane and spherical designs. They note that most analyses have shown that LNG carrier hulls should withstand very large fires, without hull breakdown, so long as the fire is only on one side of the vessel. However further analysis shows that there are mechanisms for thermal insulation degradation and inner LNG tank failure under several circumstances. In extreme circumstances, this could lead to hull failures. These are strongly mitigated by the relatively short duration expected of LNG fires, which would be less than one hour even at reduced burn rates. They suggest further detailed analysis is required, however prevention of collisions and groundings is the focus of their conclusions due to the uncertainty of fire outcome. It is noted that a large sea fire burned around the Yuyoh Maru LPG and naphtha tanker in Tokyo Bay for many days in the 1970's with no failure in the LPG cargo containment system.

Dispersion

Dispersion calculations have been carried out by many groups to support LNG applications and some comparisons have been provided by Sandia (2004) and Pitblado et al (2004b). These currently display a considerable variance. Dispersion results obtained using

DNV's PHAST model for the discharge cases here are shown in Table 3 for two weather conditions.

Table 3. Dispersion Results

CASE	DISPERSION DISTANCE TO LFL Weather and Hole Size (diameter – m)			
	D5 weather		F2 weather	
Vessel type and release location	0.75m	1.5m	0.75m	1.5m
Membrane – surface	870m	1700m	1800m	3400m
Membrane – sub-sea	790m	1600m	1200m	3300m
Spherical – surface	960m	1800m	2000 m	3700m
Spherical – sub-sea	840m	1700m	1800m	3500m

Note: LFL – Lower Flammable Limit (5% vol) calibrated in PHAST against trials
The 1.5m cases are shaded grey as DNV believe that these cases associated with intentional acts should almost always result in immediate ignition and pool fire rather than long distance dispersion. These results are reported here to allow comparison with others.

Uncertainties

There remain significant uncertainties in how major accidents would develop for LNG vessels. An extended finite element analysis may provide additional understanding of how the hull loads are redistributed. However, given the range of possible assumptions, ultimately some experimental work would assist first responders in planning their best emergency response. Several LNG vessels are reaching the end of their commercial life are scheduled for breakup. Ideally one of these vessels should be used for experimental purposes, perhaps partially filled with a non-flammable and non-toxic cryogenic liquid (e.g. liquid nitrogen). A collision or grounding scenario can be simulated and measurements taken on the vessel response.

Conclusions

Sub-sea releases of LNG have been less well treated than surface releases. This paper has provided an analysis of possible sub-surface holes and demonstrates that a pseudo-steady state release can result that is mostly a little less in consequence than similar sized surface releases from a similar type of LNG tanks. This is due to hydrostatic forces leading to a lower discharge rate and less LNG released, partially offset by higher initial boiloff.

References

ABS / FERC (2004a) Consequence Assessment Methods for Incidents Involving Releases from Liquefied Natural Gas Carriers, GEMS 1288209, May 13 2004, contract FERC04C40196.

ABS / FERC (2004b) Staff Responses to Comments on Consequence Assessment Methods, Docket AD04-6-000, 39pp, June 18, 2004.

Authen, Skramstad, Nylund (1976) Effect of Fire on the Cargo Containment, GasTech Conference, GasTech 76.

Cornwell J (2001), Letter to Mr. Don Juckett, United States Department of Energy, October 2, 2001 (and and reported in Vajello EIA)

Cornwell J and Johnson D (2004), Modeling LNG spills on water, 4th Conf Natural Gas Utilization, AIChE Spring Meeting, New Orleans, April 25-29

Fay JA (2003), Model of Spills and Fires from LNG and Oil Tankers, J Haz Matls B96, p171-188.

Nedelka D, Sauter V, Goavic J, Ohba R (2003). Last Developments in Rapid Phase Transition Knowledge and Modeling Techniques. 2003 Offshore Technology Conference, Houston, OTC 15228.

Netherlands (1999) Guidelines for Quantitative Risk Assessment – Purple Book, Committee for the Prevention of Disasters, CPR 18E.

Paik JK, Choe IH, Thayamballi AK (2001). On Accidental Limit State Design of Spherical Type LNG Carrier Structures Against Ship Collisions. International Collision and Grounding of Ships Conference. Copenhagen.

Pitblado R, Baik J, Hughes G, Ferro C, Shaw S (2004a) Consequences of LNG Marine Incidents, Center for Chemical Process Safety Conference, Orlando, Jun 30-July 2.

Pitblado R, Baik J, Rachunathan V (2004b) LNG Decision Making Approaches Compared, Mary Kay O'Connor Process Safety Center Conference, Texas A&M Univ, Oct 26-27.

Hightower M, Gritzo L, Luketa-Hanlin A, et al (2004) Guidance on Risk Analysis and Safety Implications of a Large Liquefied Natural Gas (LNG) Spill Over Water, Sandia National Laboratories, SAND2004-6258, December.

Thomas Wd (1998) Cumulative LNG Deliveries through 31 Dec 1997, SIGTTO Report LNG Log 23, Society of International Gas Tanker and Terminal Operators, London.

Large LNG Fire Thermal Radiation – Modeling Issues & Hazard Criteria Revisited

Phani K. Raj
Technology & Management Systems, Inc
Burlington, MA

AIChE Spring Meeting, April 2005
5th Topical Conference on Natural Gas Utilization
Atlanta April 10-14, 2005

Abstract

Current models for evaluating the exclusion (hazard) zones around LNG fires in both US Regulations and NFPA-59A standard are prescriptive and require the consideration of large LNG releases. These models do not consider the effects of the combustion dynamics associated with large size pool burning. Oxygen starvation in the core of LNG fires of diameters larger than about 35 m leads to the formation of non-luminous, cold soot (smoke) resulting in a reduction of thermal energy radiated by the fire to the surroundings. The net effect is a smaller (calculated) thermal hazard distances for people exposure (by factors of 2 or 3 compared to results ignoring this phenomenon). Available large scale LNG fire test information is reviewed to quantify the effect of this phenomenon. This paper also discusses the common mistakes made in calculating the thermal radiation hazard distances around large fires by using, for the energy radiated from the fire, a constant percent of energy generated by combustion. The criteria for setting thermal radiation hazard zones around large hydrocarbon fires are also reviewed.

Introduction

In the past three years there has been considerable interest among regulatory agencies and the general public in Liquefied Natural Gas (LNG) fires, thanks to a number of applications submitted to the regulatory agencies to develop and operate LNG import terminals in the US [FERC, 2005]. The public is concerned about the potential consequences of release of a large quantity of LNG, either by accidents or due to acts of terrorism. While the public concern has been expressed for releases from both terminals and LNG ships, significantly more attention has been focused on large quantity releases from ships [Fay, 2003; ABS, 2004; Sandia, 2004] and public safety assessments been performed. In many of these evaluations, the principal hazard scenario postulated is the occurrence of a LNG pool fire on water caused by releases from ships and thermal radiation from such fires. Calculations have been performed using simplified description of the geometry, physics of fire and the effects of thermal radiation on people (and structures). Results presented by some researchers indicating that thermal hazard zone radii extending to 1,500 meters or greater have alarmed the public and the emergency response personnel. Unfortunately, these researchers have not considered many physical phenomena that occur in large fires, which tend to reduce the intensity of thermal radiation emission. Also, they have ignored many of the mitigating factors of short-term exposure of an urban population to fire thermal radiation.

This paper presents a summary review of data from most recent large scale LNG fire tests and discusses the many important phenomena that occur in large fires and the parameters that need to be considered in evaluating the hazard potential of LNG fires. Data from large LNG fire tests that have not been publicized in the literature are indicated. Also, an approach to considering these data in a newer generation LNG fire models is discussed. The focus of the paper is limited to LNG pool fires.

LNG Experiments & Results

A number of organizations, worldwide, conducted LNG fire experiments in the 1970s and 1980s. These tests have included LNG spill tests on land (contained within a dike) and a limited number on water. Tests ranged from immediate ignition upon release, delayed ignition of dispersed vapor, dispersion without ignition of vapors generated by the boiling LNG, and in some cases, energetic ignition of stoichiometric concentrations of LNG vapor and air in large volume balloons (simulating unconfined energetic ignitions). Details of LNG fire field tests and their results are indicated by May & McQueen [1973], AGA [1974], JGA [1976], Raj, et al., [1979], Raj [1982], Mizner and Eyre [1982], Nadelka, et al., [1989]. Some general information on the experimental parameters and results from these tests are indicated in Table 1.

The set of experiments conducted in China Lake, CA in mid 1970s remains as the only ones designed and conducted to understand the characteristics of unconfined LNG (pool fire) burning on water and measure the thermal radiation field in the surroundings external to the fire [Raj, et al, 1979]. The volume of LNG spilled varied from 3 m^3 to 5.7 m^3 with spill rates varying from 0.02 m^3/s to 0.11 m^3/s. The observed fire base diameters (averaged over the near steady burning duration) varied 8.5 m to 16.8 m. The observed duration of steady state burning diameters ranged from 42 s to 195 s. The best-fit correlation of the observed steady state fire base diameter and the spill rate indicated an average LNG pool regression rate of 5.7 x 10^{-4} m/s (corresponding to a mass loss rate of 0.242 kg/m^2s, or 12.1 MW/m^2 thermal release rate). The least square curve fit of the plot of the ratio of photographically measured average height of visible fire plumes (with the error bars connecting the maximum and minimum heights) to steady state diameters of the fires indicated a linear variation with 2/3 power of the dimensionless burning rate (also termed "modified Froude number"), similar to the correlations of Thomas (1963, 1965); see later discussions on the flame height correlations. Plate 1 shows a typical fire plume during the mid part of the steady state burning period. The fire was very columnar and generally yellow in color. The flame height variation relative to the mean height was about $\pm$20%. Narrow angle radiometers were directed at 4.6 m and 6.2 m above the base of the fire (location of very radiative part of the fire). The narrow angle radiometer data indicated the atmospheric absorption corrected flame emissive power of 220 $\pm$30 kW/m^2. The wide-angle radiometer readings showed considerable scatter over the duration of burning. The computed mean emissive powers computed from measured radiometer readings and applying correction for both the view of the radiometers and the atmospheric absorption ranged from 200 kW/m^2 for initial stages of fires to 225 kW/m^2 for later stages. This computation of emissive power was prone to significant calculation uncertainties because of difficulties in evaluating the atmospheric absorption, instantaneous flame heights and shapes and corrections for radiometer filed of view.

Table 1
Summary of LNG Fire Experiments

#	Details of Field Tests			Type of Tests	# of Tests	Fire Dimensions	Principal Results			Technical Reference(s)
	Year	Sponsored by	Conducted in				Liquid Regression Rate (m/s)	Wide-angle Radiometer based, Mean Emissive Power (kW/m^2)	Avg fraction of combustion energy radiated (%)	
1	1962	US Bureau of Mines	Lake Charles, LA	LNG spill on diked ground	NA					Burgess, and Zabetakis (1962)
2	1969	Esso	Libya	LNG fire in a land diked area (trench): Continuous LNG feed	6	70 m long x 25 m widest x 5 m depth (avg). Eq diam = 18 m	1.6×10^{-4}	92	12 - 16	May and McQueen (1973)
3	1973	AGA	San Clemente, CA	Diked pool on land	7	Diameter = 1.8 m	1.5×10^{-4}	100	20	AGA (1974)
					8	Diameter = 6.1 m	2.2×10^{-4}	143 – 178 (max)	25	Raj & Atallah (1974)
4	1974 - 76	USCG	China Lake, CA	Unconfined pool on water: Continuous spill	5	Steady state fire diameters from 8.5 m to 15 m	4×10^{-4} to 10×10^{-4} with a mean of 5.7×10^{-4}	185 to 224	12 to 32 (depending on spill rate)	Raj, et al (1979)
5	1976	JGA	Japan	Diked pool on land	3	2 m x 2 m square	NA	58	13	JGA (1976)
6	1980	British Gas	england	Diked pool on land	29	Square and rectangular (2.5:1) dikes. Equivalent diameters 6.9 m to 15.4 m	NA		NA	Moorhouse (1982)
7	1980	Shell Research	Maplin Sands, England	Diked pool on land	1	Diameter = 20 m	2×10^{-4}	150 to 220	NA	Mizner and Eyre (1983)
				Unconfined pool on water		Mostly vapor fires resulting from delayed ignition	NA	NA		
8	1981	Tokyo Gas	Japan	Diked pool on land	8	Square pools of 2.5 m x 2.5 m	NA		NA	Kataoka (1981)
9	1987	Gaz de France	Montoir, France	Diked pool on land	3 of dia 35 m	1.8, 6.1, 10.2, 20 and 35 m diameter shallow dikes	3.3×10^{-4}	257-273	NA	Nedelka, Moorhouse & Tucker (1989)

Source: Raj, et al (1979)

Plate1: LNG pool fire (d= 15 m) on Water in a pond (50 m x 50 m x 1 m)

Source: Gaz de France (2003)

Plate 2: Snap shot of 35 m dia LNG Fire on insulated concrete dike

Source: AP Photo, 9th June 2004, Iraq.
(*Reprinted with permission from the Associated Press)*

Plate 3: Fire on an oil spill from a pipeline rupture

A series of three tests conducted in Montoir, France in 1987 involving LNG spill on to an insulated concrete floor dike of 1000 m^2 area (35.7 m diameter) and its ignition constitute the largest land spill fire tests to date. Nedelka, et al. (1989) have presented limited information on these tests; the bulk of data remains unpublished.

The Montoir tests show that large LNG fires burn with the production of copious amounts of black soot (smoke), contrary to the conventional assumption that LNG fires burn clean. The effect of smoke is to obscure the inner burning core of the fire thereby reducing the amount of total radiation that is emitted by the fire. In addition, the fire displays considerable similarity in its dynamics with other hydrocarbon fires of the same size, including the "puffing" type of burning in which large fireball puffs are released at a frequency determined by the fire size. Plate 2 is a clip from a video film of one of the Montoir tests showing the fireball type burning and considerable smoke production in the lower layers of the fire. Plate 3 is from a news clip of an oil pool fire resulting from a pipeline rupture[1].

The results provided by Nedelka, et al., can be summarized as follows: The average burning rate is 0.14 kg/m^2s (i.e., liquid regression rate of 3.3 x 10^{-4} m/s or 7.1 MW/m^2), spot emissive powers measured by narrow-angle radio meters ranged from 290 to 320 kW/m^2, the average surface emissive power (computed from the wide-angle radiometer data corrected for atmospheric absorption and the projected view of a specific area of fire as seen by the radiometer) ranged from 230 to 305 kW/m^2. The projected image of fire seen by the radiometer was smaller than the projected image of the average dimensions of the visible fire. If the entire visible fire area is used the equivalent surface emissive power calculated resulted in values between 257 and 273 kW/m^2.

LNG Fire Hazard (Evaluation) Models

Two general approaches have been in use to determine the distance to specified level of heat flux hazard from fires (Raj, 1977; Raj 1979; Considine, 1984; Moorehouse & Pritchard, 1982; McGrattan, et al., 2000). These are the so-called "solid flame" model and the "point source" model. The application of each of these models and the modifications/enhancements needed in the models, in the light of data from large-scale LNG fire experiments, are discussed below.

The **Solid Flame Model** is based on the equation

$$\dot{q}'' = E\,F\,\tau \tag{1}$$

where, $\dot{q}''$ is the radiative heat flux received by an object located at a specified distance from and orientated in a specified angle to the fire, E is an average value of thermal radiation flux emitted by the fire over a defined surface of a specified geometrical fire shape (generally referred to as the "surface emissive power – SEP"), F is the geometrical view factor between the radiation receiving object and the radiation emitting parts of the fire and τ is the transmissivity of the atmosphere to thermal radiation from the fire over a mean distance between the object and the geometrical shape of the fire ($F\tau$ can also be evaluated by

1 The diameter of the fire is unknown but guesstimated to be between 30 and 50 m.

determining their product values for several locations on the fire surface and summing the individual contributions for all surfaces that can be "seen" by the object).

Fire Shape In most "solid flame" models the shape of the fire is chosen to be a circular cylinder of diameter equal to the base diameter of the fire and axial length representing the visible plume of the fire. The axis of this cylinder is assumed vertical in low wind speeds (wind velocity below a critical value) and tilted in the wind-direction by an angle wrt the vertical, which depends on the wind speed, the diameter of the fire and the burning rate. (In some models, the drag of the base of the fire at ground level due to the wind is also considered. Other modifications include considering the horizontal cross section of the tilter cylinder to be elliptical rather than circular).

Fire Plume Length (L) Correlations of the following type due originally to Thomas (1963, 1965) have been used in the models to calculate the fire plume length for a fire of diameter D.

$$\frac{L}{D} = A\, F_C^p\, (U^*)^q \tag{2}$$

where, A, p and q are correlation constants,

$$F_C = \frac{\dot{m}''}{\rho_a \sqrt{g\, D}} = Combustion\ \ Froude\ \ Number = Dimensionless\ burning\ rate \tag{3}$$

and

$$U^* = \frac{U_{wind}}{\left[\frac{\dot{m}''}{\rho_a} gD\right]^{\frac{1}{3}}} = Dimensionless\ Critical\ wind\ speed \tag{4}$$

Thomas (1963) proposed a value A=42, p=0.61 and q=0, which was modified in a later report (Thomas, 1965) to A=55, p=2/3 and q = -0.21. Many models, including the models required by regulations (LNGFIRE3, 2004) use a constant correlation (A=42, p=0.61, q=0, with a wind tilt angle that depends upon U*). Using a single value for the exponent on the combustion Froude number to determine the fire plume length over all values of F_C is incorrect. For large LNG fires (D>30 m) the correlations used in such models as LNGFIRE3 over predict the length of flame and, hence, the radiation hazard distance. In fact, Thomas (1963) has provided data that indicate different values for the exponent (p) in different value ranges of F_C; p = 0.4 for $F>10^{-1}$; p= 0.61 for $10^{-2}<F<10^{-1}$ and p= 2/3 for $F<10^{-2}$. In fact, it can be shown from a simple analysis of air entrainment (Raj, 2005) that p = 2/3, if the fire length (L) is defined as the location by which the fuel is completely consumed within the plume by burning. The same p=2/3 correlation has also been indicated in a book by Murgai (1976) based on analysis of forest fire data. Moorhouse (1982) has proposed a correlation with p=0.254. Haskestad (1983) shows data for a large span of F ($10^{-3}<F<10^{0}$) indicating varying values for the exponent (p). The experimental results of Cox and Chitty (1985) indicate that p=1 for $10^{-4}<F<10^{-3}$ and p = 2 for $10^{-5}<F<10^{-4}$.

Figure 1 shows the plot of measured visible flame length to diameter ratios from a number of LNG experiments. The Moorhouse correlation is unacceptable on the grounds that it is not physically justifiable (it has the wrong slope of L/D ratio variation with Froude number), and the Cox & Chitty correlation is not applicable in the range of experimental combustion Froude numbers. For postulated LNG fire diameters (330 to 512 m) due releases from ships from acts of terror ([Sandia, 2004) the estimated combustion Froude number (F_C) is in the range 1.83×10^{-3} to 1.47×10^{-3}. The values for the flame height-to-diameter ratios (L/D) used by Hightower, et al., (based on Moorhouse correlation) for the above ranges of Froude numbers are 1.28 and 1.35, respectively. These are considerably higher than those predicted by Thomas' correlation (viz, L/D = 0.71 and 0.822). All other things being equal the radiation hazard distances predicted by Hightower, et al. are, therefore, higher than what should be the case based on Thomas' correlation for flame heights which are based physical principles and experimental measurements.

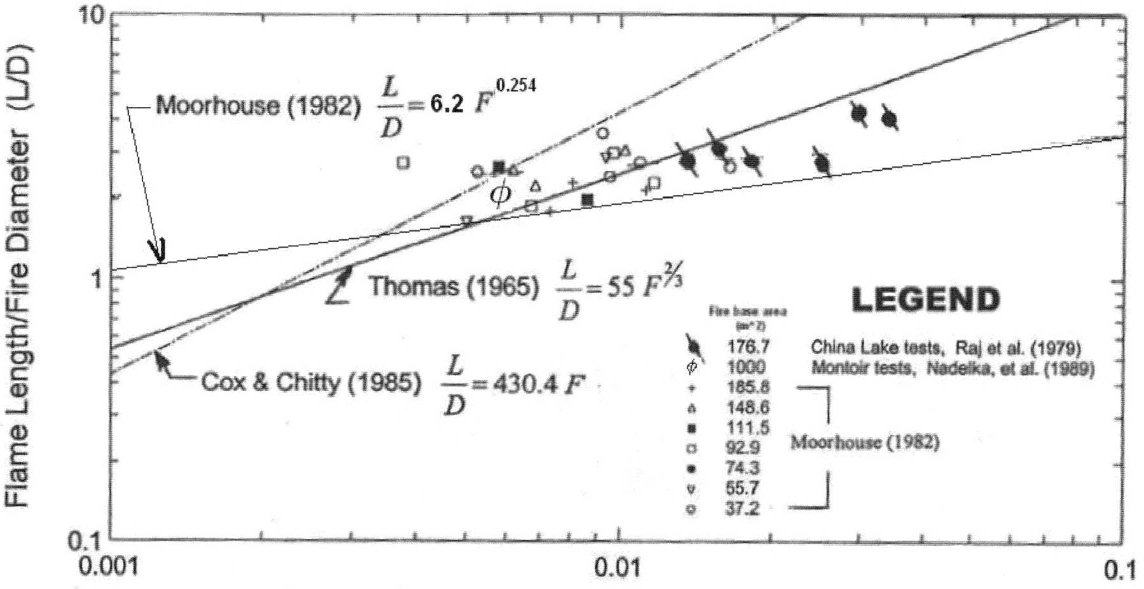

$$F = \left[\frac{\dot{m}''}{\rho_a \sqrt{gD}} \right] = \textit{Dimensionless burning rate or Froude number}$$

Figure 1: Fire height to diameter ratio data from LNG fire experiments

The models to calculate the fire radiation hazard should also consider the different regimes of combustion within the height of the visible flame. Current generation models do not take these into consideration. It has been recognized in the fire literature that as the of fire diameter increases the burning region is no longer represented by a columnar, cylindrical plume but consists of distinct zones. This can be seen in photographs in Plate 1 and Plate 2. McCffrey (1983) proposed a three-zone model for a turbulent diffusion fire. The first zone consists of a fuel rich core region. In the second zone the flame is still anchored to the base but pulsates in size both radially and axially due to the effects of large scale eddies in the entrained air. The third zone, termed "the intermittency region" is where peeled off blobs of fuel burn in irregular clumps. This model for burning is shown schematically in Figure 2. While no specific calculation procedures are available to determine the heights of each zone, it is known that the first zone generally extends to about 10% of the visible flame height, the second zone varying between 10% and 40% of the flame height.

Plate 2 and Plate 3 also show the distinct "puff" type of burning as the diameter of the fire increases. A large diffusion fire can be considered to be made of a series of connected starting thermals, which exhibit the classic mushroom formation. Malalsekera, et. al., (1996) have described the physics of formation of such burning puffs. It is obvious that the "top" of the burning flame ("flame height") is not a unique position in space but changes with time. Haskestad's (1983) data indicates a correlation for the length of the intermittent zone (L_I) with the combustion Froude number F_C (for $7.5 \times 10^{-4}<F<2.5 \times 10^{-1}$) as follows:

$$\frac{L_I}{L} = 0.167 - \log_{10}(F_C^{\frac{1}{4}}) \qquad (5)$$

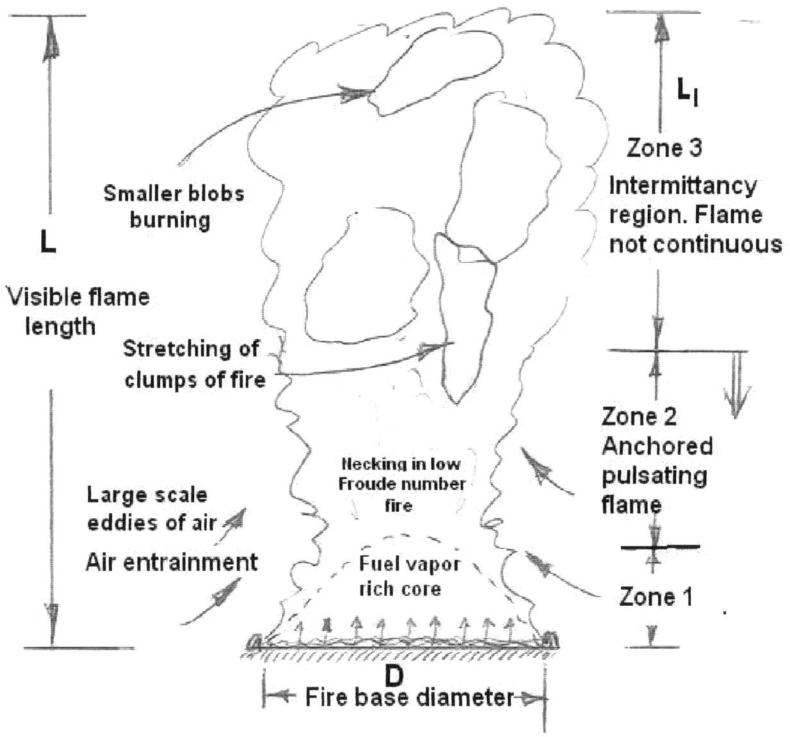

Figure 2: Schematic representation of the intermittency in buoyant diffusion flame height

In the intermittent fire region the flame is not coherent and, therefore, cannot be considered as a continuous emitter of radiation over the entire length of the intermittency length region. New generation of LNG fire models must consider the effect of this phenomenon. It may be appropriate to consider different values for the surface emissive power for the three regions. McGrattan, et. al.,(2000) have used a variation of this theme based on assumptions of the percent of combustion energy released, where all of the energy released is assigned to the first zone (and from which an effective height of uniform emissive power fire is determined). The application of the approach of McGrattan, et al., to LNG fire radiation assessment is questionable.

Surface Emissive Power-SEP (E) Several values for the parameter E (or SEP) have been reported in the literature for LNG fires based on calculations from heat fluxes measured by radiometers. (Some of these results, obtained by using Thomas' correlation for flame heights are indicated in Table 1). These results show considerable scatter. The increase in the value of E with fire diameter is noticed. However, for the largest size, 35 m diameter, tested to date (Nedelka, et al., 1989) the trend shows E leveling off or even decreasing with increase in diameter of the fire. Unfortunately, the data scatter is too high and amount of data too low to make definitive predictions.

It should be noted that SEP is a derived quantity and its value depends upon the geometry of the fire used (especially the value used for the "height" of fire plume), assumptions made in calculating the atmospheric transmissivity and the general experimental errors. In fact, Nedelka, et al. (1989) have pointed out that in the Montoir tests (35 m diameter fire) when the results for E are calculated based on the actually observed flame radiating surfaces, one set of values is obtained (E=275 kW/m^2) whereas the same radiometer data gives a value of 175 kW/m^2 for SEP when the Thomas' flame height correlation is used for the radiating surface calculation! Therefore, when a fire thermal hazard calculation is made using equation 1, the value of E and the geometry used to obtain the value of E must be used in exactly the same way it was originally calculated from test data. Recently, for example, Hightower et al, (2004) have used a value of E=220 kW/m^2 to 350 kW/m^2 for large LNG pool fires (without indicating the source for these numbers), and using them with Moorhouse correlation for flame height. As already stated, Moorhouse correlation cannot be applied to large fires.

Smoke production in fires. Large diameter LNG fires seem to produce a significant amount of smoke (see Plate 1). This is similar to those observed in the burning of other higher hydrocarbon liquids (propane, butane, heavier oils, etc – see Plate 2). Two physical phenomena may contribute to the production of smoke, even in "clean burning" fuels such as LNG. The first is the lack of enough oxygen in the core of large diameter fires to burn the carbon produced by the pyrolysis of fuel vapor. This not only produces soot (carbon particles) but also lowers the overall heat release –and hence the temperature- resulting in the promotion of smoke production. The second phenomenon may be due to the lowering of the effective concentration of fuel and vapor in the core from the recirculation of burnt gases by the toroidal vortex that is prevalent in all large fires. The effect of smoke is to shield the emission of thermal radiation from the fire reducing, significantly, the thermal radiation hazard distance around large LNG or other fires. In addition, the formation (and recirculation) of smoke could result in less efficient combustion of the fuel and result in the lowering of the effective flame temperature. However, the reduction in the radiant emission out of the fire tends to increase the temperature of the gases; which particular effect dominates depends on the chemical properties of the fuel, chemistry of combustion, the physical dimensions and the hydrodynamics of gas flow within the fire. No model exists that considers all of these phenomena.

Soot is carbon particles (with diameters in the 3 nm to 30 nm) in a fire that are being oxidized and are "glowing"; in fact, the visibility of a fire is caused by the emission of radiation in the visible spectrum by the burning soot. When the carbon produces by pyrolysis is either partially oxidized or is not oxidized at all because of lower local temperature, carbon particles agglomerate to form long chain molecules of carbon or "smoke." Soot formation studies are extensively reported in the literature (see Narasimhan & Foster, 1965; Glassman, 1979; Hura & Glassman, 1988, Mrakstein, 1988; Fowler, 1988; McCaffrey & Harkleroad, 1988). However, there is very little work on the measurement of smoke production rates in large turbulent diffusion fires. Notarianni, et al (1993) measured the smoke production in crude oil fires of diameters from 0.085 m to 17.2 m and found that smoke yield (mass % of burnt fuel that is emitted as smoke) increases as the diameter of the fire increases. The data for the mass fraction smoke yield (Y in %) vs. fire diameter (D in meters) presented by these researchers can be correlated for crude oil fires as,

$$Y = 9.412 + 2.758 * \log_{10}(D) \tag{6}$$

It is anticipated that the constants in the equation will depend very critically on the fuel chemical composition and pyrolysis properties. Unfortunately, such data do not exist for LNG fires. Considine (1984) has discussed one approach to determining the effect of smoke in reducing the radiation by assuming an effective emissivity of the fire (in the regions of smoke production) to be about 0.3. This would typically make the LNG fire surface emissive power in the smoke regions to be about 65 kW/m^2. Delichatsios & Orloff (1988) are able to correlate measured radiation from optically thin flames with burner size and fuel flow rate by postulating the soot concentration to be proportional to a fuel residence time scale (which depends upon the degree of turbulence) and the reciprocal of the chemical formation time scale. This method is not applicable to large fires, which are optically thick and also radiate in the H_2O and CO_2 bands. McCaffrey & Harkleroad (1988) have presented soot data from small-scale experiments for a number of hydrocarbon fires in the form of specific extinction area (SEA) for soot; for propane SEA is found to be 124 m^2/kg and for crude oil fires it is 1000 m^2/kg. No direct data for the smoke yield, as a function of fire diameter exists for large fires of different fuels.

Modeling smoke effects on radiation The presence of smoke in a fire is to absorb thermal radiation emission and reduce the effective emissive power. The effective emissive power in the smoke obscured regions can be represented by,

$$E_{eff} = E_0 \, \tau_s \tag{7}$$

and,

$$\tau_s = \text{Transmissivity of smoke} = e^{-(k_m \rho_s L_b)} \tag{8}$$

where,

E_{eff}	=	Effective surface emissive power in the smoky regions of fire	
E_0	=	Maximum surface emissive power (generally at lower regions of fire unaffected by smoke)	
k_m	=	Specific soot extinction area	(m^2/kg)
ρ_s	=	density of smoke	(kg/m^3)
L_b	=	Beam length = 0.63 D, for cylindrical fires	(m)

A fire thermal radiation model for different diameter fires, which incorporates the results of correlations similar to those of Notarianni, et al (1993) and McCaffrey & Harkleroad (1988), the fire dynamics, smoke obscuration effects indicated above, and the intermittency of fire core exposure will be presented in a future publication (Raj, 2005). This model divides the fire radiation zone into two distinct zones (with height) and considers the pulsating nature of exposure in the second zone as a function of fire diameter and height. Equations to determine the relative heights of each zone and the variation of surface emissive power with height will be presented. It is calculated that for very large fires the smoke obscuration averaged SEP over the height of the fire (as given by the Thomas' correlation) is about 50% of the SEP at the base!

Fractional Energy Radiated by a Fire The hat flux to an object at a distance S from the center of a fire of diameter D is sometimes calculated using the point source model with the following equation

$$\dot{q}'' = \chi_R \frac{\left[\frac{\pi}{4} D^2 \dot{m}'' \Delta H_C\right]}{4\pi S^2} \qquad (9)$$

where,

$\dot{q}''$ = Radiative heat flux received by an object at distance S

χ_R = Fraction of combustion energy released that is radiated

$\dot{m}''$ = Mass burning rate per unit area

ΔH_C = Heat of combustion (lower) of the fuel burning

Based on radiation measurements in a limited number of field experiments with LNG fires of diameters less than 15 m, the value of χ_R is shown to vary between 12% and 32% (see Table 1). It is anticipated that for larger diameter fires the fraction radiated will decrease due to smoke obscuration effects, and simple geometry effects (ie, the heat generation rate is proportional to fire base area where as the emitting surface area increases as $d^{1.7}$). Data presented by McGrattan, et al (2000), <u>for heavier hydrocarbon fuels than LNG</u>, shows the following correlation for the fraction radiated with fire diameter,

$$\chi_R = 0.35 e^{-(D/20)} \quad ; \qquad \text{D = fire diameter in meters} \qquad (10)$$

This correlation, if applied to a 15 m diameter LNG fire, predicts $\chi_R = 0.17$, a value tantalizingly close to a measured value (in the China Lake experiments) of 0.125. However, this can only be considered as a coincidence; the application of the above correlation to a 35 m diameter fire (Montoir tests) yields $\chi_R = 0.06$ and application to a 100 m diameter fire results in $\chi_R = 2.36 \times 10^{-3}$. Of course, such predictions are unsubstantiated.

If it is assumed that the average surface emissive power (E) of LNG fire does not change with diameter (which as we have discussed earlier is not true), and assuming that the burning rate is also independent of diameter, it can be shown (with radiating height given by Thomas' equation) that

$$\chi_R = \frac{\pi D L E}{(\pi/4) D^2 \dot{m}'' \Delta H_C} = 220\left(\frac{E}{\dot{m}'' \Delta H_C}\right)\left[\frac{\dot{m}''}{\rho_a \sqrt{g D}}\right]^{\frac{2}{3}} \tag{11}$$

Fay (2003) has used the point source model to calculate LNG fire hazards. He assumed 15% for the fraction of energy radiated from a (340 m diameter) LNG fire on water caused by release of 14500 m^3 of LNG from a ship and calculated the hazard distance to 5 kW/m^2 heat flux to be 1900 m. The assumed value of 15% energy radiated from a 340 m size fire is incorrect and unsupportable on any measured data[2]. Using the China Lake test data of 12.5 % for the fraction of energy radiated by a 15 m fire together with the inverse cube root of diameter relationship from equation 11, it is calculated that for a 340 m diameter fire the fraction of combustion energy radiated is 4.4%. With this value for the fraction radiated, absent all other radiation reduction mechanisms for large fires discussed in this paper and neglecting atmospheric absorption, the hazard distance calculated by Fay would be reduced by a factor of 1.85 or hazard distance would become1030 m. Consideration of smokiness of large LNG fires, atmospheric absorption (by about a factor of 0.6 for this distance) results in a hazard distance of the order of 600 m, a reduction by a factor of 3 compared to Fay's predictions. Hence, caution should be exercised in using the point source model to estimate any fire hazard zones.

Thermal Hazard Heat Flux NFPA 59A Standard stipulates 5 kW/m^2 (1,600 Btu/hr ft^2) as a safe level of exposure at a property line that can be built upon next to a LNG storage facility. Also, this level has been set as (tolerable risk) threshold level for thermal exposure in the regulations of several other countries. Exposure of a bare human skin to this level of thermal radiation results in 2nd degree blisters in 30 seconds. A recent work (Raj, 2004) has compared this exposure level with exposures to thermal radiation in industrial and residential settings and finds that for an accident situation, this level may be acceptable. There is information in the literature that indicates that ordinary clothing worn by people affords an enhanced level of protection to 5 kW/m^2 radiation flux; in fact even the basic fabric extends reduces thermal intensity on the skin by a factor of about 5 and the duration time for 2nd degree burns is correspondingly increased by at least an order of magnitude. In addition, the buildings and indoor occupancies provide significant level of safety from exposure to an urban population; Considine (1988), McGrattan, et al (2000) have considered the effect of tallness of building and other deliberate obstruction panels in minimizing thermal radiation impact surrounding fuel storage plants. A realistic hazard criterion should be based on the likelihood of exposing larger than an *a priori* specified fraction of the population to the threshold heat flux level. Such a calculation should consider the effects of mitigating parameters such as clothing protection, shadows of buildings, ability of people to seek shelter within 30 s, etc., ie., a probit based model for hazard levels rather than a single number for heat flux level.

Conclusions

LNG fire thermal radiation hazard evaluation models currently used in regulations and those being used in several public presentations have been reviewed in the light of experimental data, both quantitative and qualitative. The following general conclusions are the result:

---

[2] On the basis of this number for the fraction of combustion energy emanating over the "nominal" surface of a cylindrical fire of diameter 340 m and flame height predicted by Thomas' equation, the SEP would be 750 kW/m^2.

1 In general, large LNG pool fires exhibit burning characteristics very similar to those seen in large higher hydrocarbon fires (kerosene, LPG, crude oil, JP-4, etc), which burn very sooty and reduced radiation output to the surroundings.

2 It is not possible at this time to predict the "critical diameter" of LNG fires beyond which they behave similar to higher hydrocarbon liquid fires; however, it can be argued that this value is in the 35 to 50 m range.

3 Newer generation models should take into account the significant soot formation and the intermittent burning characteristics of large LNG liquid pool diffusion fires.

4 Hazard calculations should also consider the effect of mitigating parameters such as peoples' clothing, shadow of buildings and ability of persons to seek shelter quickly. A probit based approach may be preferable.

Nomenclature

D	=	Diameter of the base of fire	(m)
E	=	Surface emissive power of a flame	(W/m^{2})
F	=	Geometric view factor between an object and the fire	
F_C	=	Combustion Froude number $= \dfrac{\dot{m}''}{\rho_a \sqrt{g\,D}}$	
g	=	Acceleration due to gravity = 9.8	(m/s^2)
ΔH_C	=	Heat of combustion (lower heat)	(J/kg)
k_m	=	Specific soot extinction area	(m^2/kg)
L	=	Length of visible fire plume	(m)
L_b	=	Mean beam length inside a fire plume = 0.63 D	(m)
$\dot{m}''$	=	Fuel burning rate per unit area	(kg/m^2 s)
$\dot{q}''$	=	Radiant heat flux received by an object	(W/m^2)
S	=	Distance from fire center to the radiation receiving object	(m)
U_{wind}	=	Wind speed	(m/s)
Y	=	Fraction by mass of burning fuel converted to smoke	
ρ_a	=	Density of air = 1.3	(kg/m^3)
ρ_s	=	density of smoke	(kg/m^3)
χ_R	=	Fraction of combustion energy that is radiated to surroundings	
τ	=	Transmissivity of the atmosphere for fire thermal radiation	
τ_S	=	Transmissivity of soot	

References

ABS Consulting, "Consequence Assessment Methods for Incidents Involving Releases from Liquefied natural Gas Carriers," Report# GEMS 1288209 to Federal Energy Regulatory Commission, Washington, DC, May 2004.

AGA, "LNG Safety Program, Interim Report on Phase II Work," IS-3-1, American Gas Association, Arlington, VA, 1974.

Burgess, D., and M.G. Zabetakis, "Fire and Explosion Hazards of LNG," US Bu. Of Mines Investigation Report # 6099, 1962.

Considine. M., "Thermal Radiation Hazard Ranges from Large Hydrocarbon Pool Fires," Report # SRD R297, Safety & Reliability Directorate, UK Atomic Energy Authority, Wigshaw Lane, Culcheth, Warrington, WA3 4NE, UK, Oct 1984.

Cox, G. and R. Chitty, "Some Source-Dependent Effects of Unbounded Fires," Combustion & Flame, v 60, p 219-232, 1985.

Delichatsios M.A., and L. Orloff, "Effects of Turbulence on Flame Radiation from Diffusion Flames," Twenty-Second Symp. (Int'l) on Combustion, The Combustion Institute, p 1271-1279, 1988.

Fay, J.A., "Model of Spills and Fires from LNG and Oil Tankers," J. Haz Mat, v B96, p171-188, Jan 2003

FERC, Federal Energy Regulatory Commission; http://www.ferc.gov/industries/gas/gen-info/horizon-lng.pdf, Jan, 2005.

Fowler, W.L., "An Investigation of Soot formation in Axisymmetric Turbulent Diffusion Flames at Elevated Pressure," Twenty-Second Symp. (Int'l) on Combustion, The Combustion Institute, p 425-435, 1988.

Haskestad. G., "Luminous Heights of Turbulent Diffusion Flames," Fire Safety Journal, v 5, p 103-108, 1983.

Hightower, M, L. Gritzo, A. Luketa-Hanlin, J. Covan, S.Tieszen, G. Wellman, M. Irwin, M. Kaneshige, B. Melof, C.Morrow, and D. Ragland, "Guidance on Risk Analysis and Safety Implications of a Large Liquefied Natural Gas (LNG) Spill Over Water," Sandia National Laboratory Rep.# SAND2004-6258, U.S. Department of Energy, Washington, DC, Dec 2004.

Hura, H.S. and I. Glassman, "Soot Formation in Diffusion Flames of Fuel/Oxygen Mixtures," Twenty-Second Symp. (Int'l) on Combustion, The Combustion Institute, p 371-388, 1988.

JGA, "A Study of Dispersion of Evaporated Gas and Ignition of LNG Pool Resulting from Continuous Spillage of LNG," Jour. Japan Gas Assoc., July 1976.

LNGFIRE3, "A Thermal Radiation Model for LNG Fires," GTI-04/0032, Gas Technology Institute, Chicago, Mar 2004.

Malalasekaera, W.M.G, H.K. Versteeg and K. Gilchrist, "A review of Research and Experimental Study on the Pulsation of Buoyant Diffusion Flames and Pool Fires," Fire and Materials, v 20, p 261-271, 1996.

Markstein, G.H., "Correlations for Smoke Points and Radiant Emission of Laminar Hydrocarbon Diffusion Flames," Twenty-Second Symp. (Int'l) on Combustion, The Combustion Institute, p 363-370, 1988.

May, H.G. and W. McQueen, "Radiation from Large Liquefied Natural Fires," Comb. Sci Tech, V 7, p 51-66, 1973.

McCaffrey, B.J., "Momentum Implications for Buoyant Diffusion Flames," Combustion and Flame, v 52, p 149-167, 1983.

McCaffrey, B.J. and M. Harkleroad, "Combustion Efficiency, radiation, CO and Soot Yield from a variety of Gaseous, Liquid and Solid Fueled Buoyant Diffusion Flames," Twenty-Second Symp. (Int'l) on Combustion, The Combustion Institute, p 1251-1261, 1988.

McGrattan, K.B, H.R. Baum and A. Hamins, "Thermal Radiation from Large Pool Fires," NISTIR 6546, National Institute of Standards & Technology, U.S. Department of Commerce, Washington, DC, Nov 2000.

Mizner, G.A., and J.A. Eyre, "Large Scale LNG and LPG Pool Fires," Inst. Chem. Eng Symp, Ser 71, Manchester, April 1982.

Moorhouse, D.J, "Scaling Criteria Derived from Large Scale Experiments – The Assessment of Major Hazards," Inst. Chem. Eng. Symp. Ser No. 71, p 165-179, Manchester, 1982a.

Moorhouse, J., and M.J. Pritchard, "Thermal Radiation Hazards from Large Pool Fires and Fireballs – A Literature Review," Inst. Chem. Eng. Symp Ser No. 71, p 397-428, 1982b.

Murgai, M.P., "Natural Convection from Combustion Sources," Oxford & IBH Publishing, New Delhi, 1976.

Nadelka, D., J. Moorehouse and R.F. Tucker, "The Montoir 35 m Diameter LNG Pool Fire Experiments," TRCP.3148R, 9th Intl. Conf & Expo on LNG, LNG IX, Nice, France, 1989.

Narasimhan, K.S. and P.J. Foster, "The rate of Growth of Soot in Turbulent Flow with Combustion Products and Methane," Tenth Symp. (Int'l) on Combustion, The Combustion Institute, p 253-257, 1965.

Notarianni, K.A., D.D. Evans, W.D. Walton, D. Madrzykowski and J.R. Lawson, "Smoke Production from Large Oil Pool Fires," Interflam '93, Fire Safety, Int'l Fire Conference, 6th, Oxford, England; Interscience Communications Ltd., London, England (Ed: Franks, C.A.), p 111-119, March 1993.

Thomas, P.H., "The Size of Flames from natural Fires," 9th Symp (int'l) Combustion, Academic Press, New York, p 844-859, 1963.

Raj, P.K. "Calculations of Thermal Radiation Hazards from LNG Fires - A Review of the State-of-the-Art," Paper presented at the AGA Transmission Conference, ST. Louis, Missouri, Paper number 2, Session 18, May 18, 1977.

Raj, P.K., A.L. Moussa and K. Aravamudan, "Experiments Involving Pool and Vapor Fires from Spills of LNG on Water," NTIS # AD-A077073, USCG Report, Washington, DC 20590, 1979

Raj, P.K. "Analysis of JP-4 Fire Test Data and Development of a Simple Fire Model,"; ASME Paper 81-HT-17, 20th Joint ASMEAICHE National Heat Transfer Conference, Milwaukee, August 1981

Raj, P.K. "LNG Fires – Combustion and Radiation, Vol III," Proceedings of MIT-GRI LNG Safety Workshop (March 22-24, 1982) at MIT, Cambridge, MA 02139, Published by the Gas Research Institute, Chicago, July 1982.

Raj, P.K., "Thermal Radiation Hazard Heat Flux for Establishing Safety Zones Around LNG & Other Hydrocarbon Fires," A white paper developed for the 59A (LNG) Committee, National Fire Protection Association, Quincy, MA, Nov 2004.

Raj, P.K. "A Model for Assessing Thermal Radiation Hazards from a Large Diffusion Hydrocarbon Fires," paper in preparation for publication, 2005.

Sandia; see Hightower, et al, 2004.

.

QRA of a Typical LNG Re-gasification Facility – What are the Historical Main Contributors to 3rd Party Risk?

Tor Egil Nielsen, President,
Scandpower Risk Management Inc.
Houston, Texas

AIChE Spring Meeting, April 2005
5th Topical Conference on Natural Gas Utilization
Atlanta April 10-14, 2005

ABSTRACT

Based on QRA of a typical LNG re-gasification facility, this paper will present the main contributors and their relative importance for 3rd party risk. This risk will be based on the frequency for different releases and their hazard zones from the main components in the facility. The hazard ranges can be assessed by different software tools/methods. Which consequence modeling tools to be applied is dependant on how good simpler tools are able to correctly simulate the cases in question and if the specific case is judged to qualify for advanced consequence modeling. A CFD dispersion and fire/smoke simulation tool will be presented as an advanced option for consequence assessments. The LNG industry has a good safety record and the number of LNG facilities is relatively low. From this, the historical accident statistic for LNG specific service is scarce. This paper will present the applied failure frequencies and discuss the quality of these frequencies.

As it is impossible to quantify the frequency of terrorist attacks, this will not be addressed in the paper. However, the maximum hazard ranges presented in the paper will be equal to those that can origin from such attacks (on single pieces of equipment).

Abbreviations used in the paper:

QRA	Quantitative Risk Analysis
3rd party	Public outside the property line of the LNG Plant
LNG	Liquefied Natural Gas
HC	Hydrocarbon
QA/QC	Quality Assurance/Quality Control
BOG	Boil off gas
LEL	Lower Explosion Limit; 50 000 ppm (5 by vol% in air) for methane
UEL	Upper Explosion Limit; 150 000 ppm (15 by vol% in air) for methane
PERC	Powered Emergency Release Couplings
CFD	Computational Fluid Dynamics

INTRODUCTION

Most LNG releases in an LNG re-gasification facility will not have the potential to harm 3rd party personnel outside the facility boundaries. This is because the facility siting is usually in compliance with the regulations given in NFPA59A. These regulations (or similar requirements in other parts of the world) say that design spills shall be collected in spill sumps and that gas clouds from these spills shall not have a concentration above ½ LEL on neighboring sites "which can be built upon". There are also requirements to maximum heat loads on neighboring areas from design LNG spills. These requirements will handle most of the LNG related scenarios for 3rd party exposure, but employees at the facility will still be exposed to the risk from these scenarios.

Since most LNG releases are assumed to be "design spills" that will be collected by the spill sumps, the following other scenarios will be presented and discussed in this paper:

-LNG releases from the LNG vessel

-LNG releases from the loading arms

-LNG releases from the pipelines between the loading arms and the LNG facility

-LNG releases from the LNG tanks

-Gas releases from or downstream of boil off compressors

-Gas releases from or downstream of the LNG vaporizers (includes export gas pipeline)

If an LNG facility is not designed according to NFPA59A (or with similar requirements to design spills), a more detailed analysis is necessary to determine 3rd party risk from different LNG spills in the facility.

WHAT IS THE MAIN LAYOUT OF A "TYPICAL" LNG TERMINAL?

The re-gasification facilities covered by this paper are import terminals and not equipment in connection with peak shaving facilities. A typical LNG re-gasification facility can look quite different in real life. However, if potential related facilities are disregarded, the size and the location compared to the jetty are the two principal parameters that decide 3rd party risk. This is because 3rd party exposure, wind data and topography have to be assumed equal for a generic analysis focusing on hazard ranges and frequencies for the different release scenarios.

In principle, there are two options for location of an LNG re-gasification facility. One option is where all parts of the facility are within the plant boundaries. The other option is where the berth and the main facility are separated by "public" land. In addition, the berth does not need to be at the shore line, but can be located at different distances from shore. Figure 1 below presents the two main options for the layout.

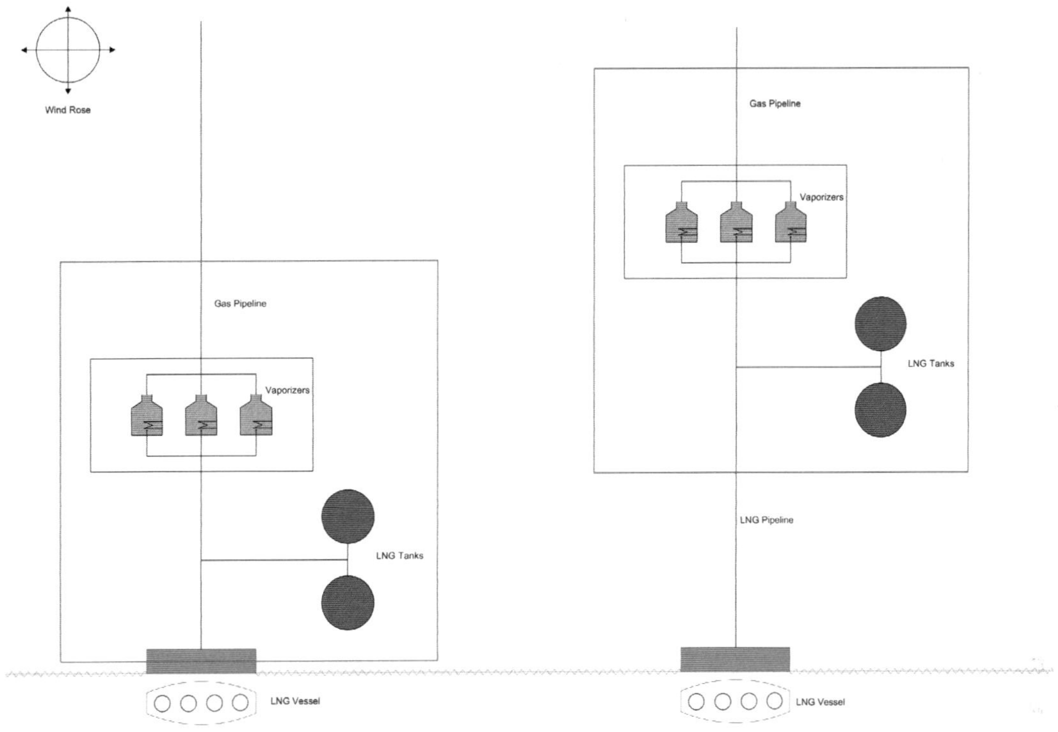

Figure 1. The two options for main layout of an LNG re-gasification facility

Which main layout option is best regarding 3rd party risk can not be concluded without knowing the 3rd party exposure in the area, but in general it can be stated that the area between the berth and the facility for option two will typically be exposed to releases from the LNG vessels, the loading arms, the pipelines, the LNG tanks and the gas releases from/ downstream of the boil off compressors and the vaporizers, i. e. all the main components included in this paper. One potential benefit from layout option no. 2 is that it can increase the distance from the loading arms to potentially heavily populated areas on the opposite side of the LNG facility.

In an actual analysis, the location of the facility in relation to wind direction data will significantly affect 3rd party exposure. Since no such data can be utilized for a generic risk comparison, it is assumed an even distribution on wind directions, i. e. it is equally probable that the wind comes from any direction.

TYPICAL EQUIPMENT IN A RE-GASIFICATION FACILITY

The HC containing equipment in a re-gasification facility is fairly simple as the process needed is basically to add heat to the LNG in heat exchangers (vaporizers) to transform the liquid to gas before the gas is exported through a pipeline. The gas can also be used at the site in connection with power generation facilities, but the equipment in connection with such is not included in this paper. Table 1 below presents typical equipment, amount and sizes for a re-gasification unit.

Table 1 Typical main equipment in a re-gasification facility, amount and dimensions

Type equipment	Typical amount	Typical dimension	Comment
LNG vessel	1-2 visits per week	125.000m3	4-5 tanks on LNG vessel
Loading arms w/valves	2-4		Loading, vapor return
LNG loading line	1	32 inch	
LNG circulation line	1	8 inch	
Gas return line	1	12 inch	
LNG tanks	2	10.000-140.000m3	
Valves, flanges	20-60		
Pumps	8-12		Most LNG pumps are submerged
Boil Off Compressors	2		
Vaporizers	3-9		
Gas export pipeline	1	30-40 inch	

LEAK FREQUENCIES

The leak frequencies are important in estimating the risk since the risk from a scenario is linearly scalable with the frequency for that scenario. A risk analysis for an LNG terminal will usually include all release scenarios. Since this paper aims at presenting the risk for 3rd party, many of the scenarios can be disregarded as the maximum extension of the gas cloud from the scenario will not reach outside the facility property line. Also, smaller releases in connection with the loading arms, e. g. PERC releases will have short hazard ranges.

To focus on the important risk contributors, releases with a hazard range of less than 250 meters are not included in this paper since they will typically not extend outside the property fence line. One exception to this are releases from the pipelines between the loading berth and the re-gasification facility (for main layout option 2) since the public can come closer to these lines. The exclusion of scenarios with a hazard range of less than 250 m can also be seen in connection with the requirements to maximum hazard ranges from design spills as given in NFPA59A. This standard says that ½ LEL gas clouds from a 10 minutes maximum pump rate spill in the spill sumps shall not extend outside the facility boundary. Hazard ranges from such spills are typically around 300 meters.

Based on publicly available equipment failure data sources, Scandpower Risk Management has compiled a set of failure frequencies to be used as the starting point for all frequency assessments. As there is not sufficient failure data available to establish failure rates based on LNG service, other data have to be utilized. Such data can then be adjusted for LNG service based on e. g. expert judgment or a review of failure causes. Scandpower has in some studies used data for gas pipelines and HC process equipment and reduced the failure rate for LNG service based on the assumed high specifications and QA/QC procedures for LNG equipment. In addition, reductions have been done due to the lower probability for e. g. the failure cause corrosion since the LNG contains no water. In this paper, no such reductions are done. Since it is impossible to establish the correct adjustment factors, adjusting or not is a balance between being conservative for the purpose of a QRA (measured towards acceptance criteria) or using data you believe are more correct.

For LNG tanks, the term "secondary containment" as used for single containment tanks in Table 2 needs to be explained. In a study presented in Ref. [1], the single containment tank design had cold resistant material in the lower section of the outer layer of the actual tank. This layer is assumed to contain a part of the smaller releases from the primary containment. The actual secondary containment for single containment tanks is the bunding wall around the tanks. Figure 2 and 3 present single and full containment LNG tanks.

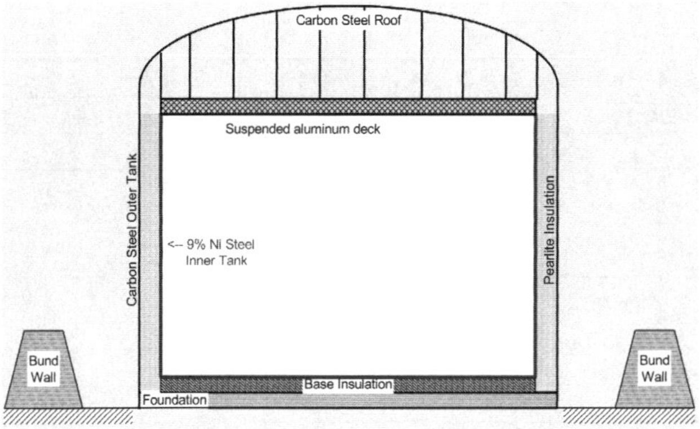

Figure 2: Single Containment Type LNG Tank

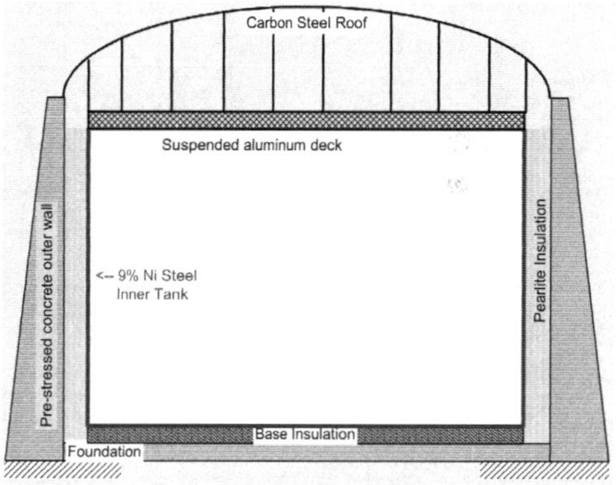

Figure 3: Full Containment Type LNG Tank

Table 2 below presents the annual frequency from one piece of equipment for scenarios that have a significant hazard range. As the loading arm leak frequency is derived per operation and not per year, it is assumed 80 loading operations per year to obtain the annual leak frequency. Leak contributions from valves and flanges in connection with the equipment are included in the leak frequencies. It should be noted that including the small and intermediate releases from the pipelines bias the frequency distribution significantly even if the actual risk from these scenarios is low. Also, the percentages are somewhat wrong since both the full and single containment tank options are included, while for an actual facility one of the designs is chosen. From this, only one tank design should have been included in the table. Nevertheless, the percentages will give a good overview of the differences in frequencies for the different scenarios.

Table 2 Frequencies of significant leak scenarios and the relative contribution compared to the sum of all listed frequencies

Equipment	Scenario	Frequency	% of total frequency
Ship at jetty	LNG tanker structural failure	1,5E-06	0,1
Loading arms	Rupture (outflow 1 minute)	7,6E-04	28,7
	Rupture (outflow 5 minutes)	4,0E-05	1,5
Unloading line 32" (1 km)	Small LNG (i.e.) gas release	3,1E-04	11,6
	Intermediate - LNG	5,7E-05	2,2
	Large – LNG	1,5E-05	0,6
Circulation line 8" (1 km)	Small LNG (i.e.) gas release	3,1E-04	11,6
	Intermediate - LNG	5,7E-05	2,2
	Large – LNG	1,5E-05	0,6
Vapor return 12" (1 km)	Small – gas	3,1E-04	11,6
	Intermediate - gas	5,7E-05	2,2
	Large – gas	1,5E-05	0,6
Single containment tank	LNG into 2nd containment	9,7E-05*	3,7
	LNG to surroundings	4,3E-05*	1,6
Full containment tank	LNG into 2nd containment	1,2E-04*	4,5
	LNG to surroundings	0*	0,0
Boil-off gas condenser	Large – LNG	1,9E-04	7,0
	Rupture - LNG	2,2E-06	0,1
BOG compressor suction pre-heater	Large – GAS	3,9E-05	1,5
BOG compressors and after-cooler	Large – GAS	3,9E-05	1,5
HP/MP vaporizer	Large – LNG	6,2E-05	2,3
	Large – GAS	1,1E-04	4,3
Gas pipeline (500 m)	Large – GAS	7,6E-06	0,3
Total		2,7E-03	100,0

* Figures taken from Ref. 1

CONSEQUENCE CALCULATIONS

For the consequence calculations, the computer code TRACE from SAFER systems[2] has been used.

The hazard ranges from the consequence modeling will be dependant on the release scenario, i.e. the (transient) release rate and release duration, which again is dependant on the hole size, pressure, volume in the process segment etc. The evaporation rate from the release scenario (if the release is LNG) is dependant on how big the pool gets (bunding, terrain) and the heat transfer to the LNG. The dispersion is based on meteorological parameters (and on topography and geometry in real life, but this can not be reflected properly by Trace). As there is an infinite number of potential scenarios, some taken from actual studies will be presented here.

Based on defined release scenarios, hazard ranges for defined gas concentrations can be calculated at different points in time. Also, the areas that have been exposed to a given gas concentration at any point in time, the so-called "footprint" can be presented. For the 3rd party risk comparison from the different scenarios, the footprint is less interesting than the maximum cloud size since the footprint can be relatively large even if the gas cloud establishing the footprint is relatively small. The release scenario, weather conditions, etc. will affect the relationship between maximum cloud size and the footprint area from the cloud. Figure 4 below presents printouts from a catastrophic tank failure at different points in time.

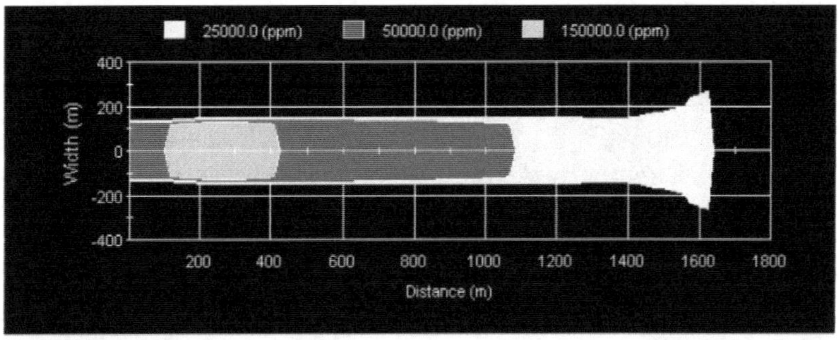

21 min 16 sec

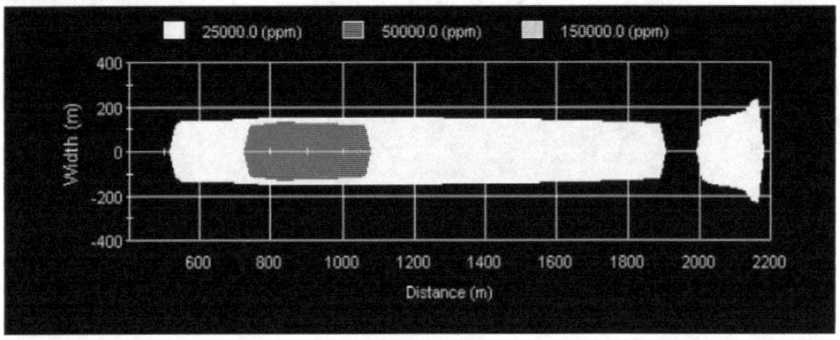

28 min 22 sec

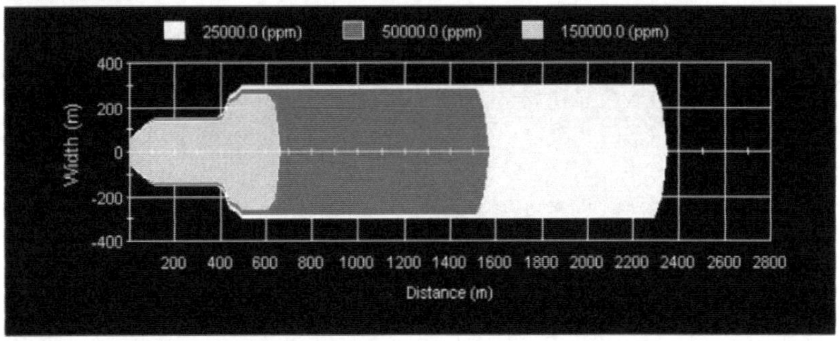

Footprint: 34 min 28 sec (last available)

Figure 4 Example of Trace printouts from a catastrophic failure of an LNG tank (140.000 m3 LNG into 10.000 m2 pool, normal soil).

For 3rd party risk, it is actually the maximum gas cloud outside the property fence line that is of interest since personnel inside the fence line is 1st party. It can also be argued that the area above UEL, i. e. the green area in the printouts should not be included in the risk calculation since it cannot explode. However, it is usually the heat loads that are most dangerous for personnel and the gas in the green areas will burn as sufficient oxygen becomes available during the fire.

Another element that affects 3rd party risk is the size of the gas cloud outside the property line at the time of ignition, i. e. the gas cloud might not have reached its maximum size when it ignites and hence a smaller exposure area results. This effect applies for all scenarios, but to a larger degree for the big gas clouds. Ref. [3,4,5] give more information on ignition modeling. In most situations the overall ignition probability is less than 0,5. This means that there is a higher probability that the gas cloud fully develops than not. Without discussing this further, the maximum gas cloud size will be used as the area exposure for all scenarios.

Table 3 below presents the approximate maximum cloud sizes for the different scenarios. The "Scenario" column gives an indication on how the scenario has been modeled in Trace. The "LNG leak to spill sump" indicates that the scenario is assumed to be collected by the spill drainage system and that it will end up in the spill sumps. These spills should not give gas clouds with a gas concentration above ½ LEL outside the property line.

Table 3 Scenario overview and approximate dimensions of resulting maximum gas clouds.

Equipment	Scenario	Length (m)	Width (m)	Area (m2)
Ship at jetty	LNG, 500.000 kg on water	900	1400	1260000
Loading arms	LNG, 1 minute on water, 1400 kg/s	200	1000	200000
	LNG, 5 minutes on water, 1400 kg/s	500	1000	500000
Unloading line 32"	LNG, 1,8 kg/s	110	15	1650
	LNG, 5300 kg	350	60	21000
	LNG, 22000 kg	330	80	26400
Circulation line 8"	LNG, 1,8 kg/s	110	15	1650
	LNG, 5300 kg	350	60	21000
	LNG, 22000 kg	330	80	26400
Vapor return 12"	Gas, 0,06 kg/s	8	1	8
	Gas, 2,4 kg/s	125	16	2000
	Gas, 1500 kg	75	55	4125
Single cont. storage tanks	LNG into 2nd containment	70	60	4200
	LNG to surroundings	2000	2000	4000000
Full cont. storage tanks	LNG into 2nd containment	70	60	4200
	LNG to surroundings	2500	2500	6250000
Boil-off gas condenser	LNG leak to spill sump			
	LNG leak to spill sump			
BOG compressor suction pre-heaters	Gas, 4600 kg	80	50	4000
BOG compressors and after-coolers	Gas, 4600 kg	80	50	4000
HP/MP vaporizers	LNG leak to spill sump			
	Gas, 4600 kg	80	50	4000
Gas pipeline	Gas, 4600 kg	80	50	4000

COMPARATIVE RISK ASSESSMENT METHODOLOGY

Risk is a combination of the probability of something happening and the consequences of the event. Since this is a generic study, it is not known what the consequences for 3rd party personnel will be since there are no exposure levels available. Nevertheless, in general it can be assumed that a large gas cloud will expose more people than a small gas cloud. To be able to compare the consequences for different release sizes, it is simply used that the number of 3rd party personnel exposed to the gas cloud is linear with the size of the gas cloud outside the re-gasification facility.

The ignition probability is an important factor when it comes to the risk level since an un-ignited cloud is virtually harmless. In general, it can be stated that a large gas cloud has a higher probability of finding an ignition source than a small gas cloud based on area/volume exposure. The duration of exposure will also be a factor due to the potential presence of intermittent ignition sources like e. g. traffic lights, road vehicles, trains, etc. Due to lack of information on intermittent ignition sources, these can not be specifically addressed in a generic risk comparison. However, with the relatively long duration of the gas clouds, it can be assumed that many of the potential intermittent ignition sources will be "active" at some point in time when the gas cloud passes by. From this, the error in disregarding the duration of the scenarios is assumed to be small. Due to the simplicity and the presumed small error involved, also the ignition probability will be scaled linearly with the size of the gas cloud. Ref. 3, 4, 5 give more information regarding ignition modeling.

Since a gas cloud (with a gas concentration within the ignitable range) is moving with the wind and gets diluted, it will have different sizes and expose different areas at different points in time. To deal with this, the maximum size of the gas cloud is used, independent on when this occurs. Use of the maximum sized gas clouds will handle the exposure levels for personnel and ignition sources for the different gas clouds, but will disregard the time effect especially with regard to potential intermittent ignition sources that are infrequently active.

On the frequency side of the risk, it is used that the risk is linearly scalable with the frequency, e. g. if you double the frequency of a scenario, you also double the risk from that scenario.

Having established a release frequency and an area exposure level for the different releases, a risk figure combining the two can be established by simply multiplying the frequency with the area squared. The resulting number will give a representation of the relative difference in risk for different scenarios, given that the exposed population and ignition sources are evenly spread out around the facility. This means that e. g. a scenario happening once every 1.000 years exposing an area of 1 square mile will have the same risk as a scenario happening once every 250 years exposing an area of 0,5 square mile.

The comparative generic risk assessment methodology does not work for moving targets like e. g. a car passing the facility. This is because the duration of the scenario will have to be reflected to address this risk, i. e. what is the probability that a gas cloud is present or "arrives" at the road while a car is passing the facility.

RISK RESULTS

The risk from a facility is dependant on the amount of equipment included in the facility and the number of LNG ships unloading per year. The risk results in the following is based on 80 loading operations per year, two LNG tanks, two boil-off compressors, six vaporizers and 500 meters of gas export pipeline. For the loading arms, the leak frequency is derived based on number of loading operations, hence the number of loading arms in operation can not be reflected. For main layout option 2 with pipelines between the loading arms and the facility, one km each of loading line, circulation line and vapor return line is included. The risk from the two main layout options is presented for both single containment and full containment tanks.

Table 4 presents the relative risk contributions for the different options. It should be noted that the presented relative risk figures are "dimensionless", i. e. they only represent the relative risk from the different scenarios given that the exposed areas has an evenly spread population and ignition sources. All the figures in Table 4 are comparable, i. e. they are calculated based on the annual frequencies and the gas cloud sizes as given in Table 3.

Table 4 Relative risk from the different release scenarios for the two main layout options and tank designs

Equipment	Scenario	Option 1, single	Option 1, full	Option 2, single	Option 2, full
Ship at jetty	LNG tanker structural failure	238	238	238	238
Loading arms	Rupture (outflow 1 minute)	3040	3040	3040	3040
	Rupture (outflow 5 minutes)	1000	1000	1000	1000
Unloading line 32"	Small LNG (i.e.) gas release			0,08	0,08
	Intermediate - LNG			2,51	2,51
	Large - LNG			1,06	1,06
Circulation line 8"	Small LNG (i.e.) gas release			0,08	0,08
	Intermediate - LNG			2,51	2,51
	Large - LNG			1,06	1,06
Vapor return 12"	Intermediate - gas			0,02	0,02
	Large - gas			0,03	0,03
Single cont. storage tanks	LNG into 2nd cont.	0,34		0,34	
	LNG to surroundings	137600		137600	
Full cont. storage tanks	LNG into 2nd cont.		0,42		0,42
BOG compressor suction pre-heaters	Large - GAS	0,12	0,12	0,12	0,12
BOG compressors and after-coolers	Large - GAS	0,12	0,12	0,12	0,12
HP/MP vaporizers	Large - GAS	1,08	1,08	1,08	1,08
Gas pipeline	500 m included	0,01	0,01	0,01	0,01

From Table 4 it is easy to see that there are only two significant risk contributors for 3rd party risk, namely the loading arms and single containment tanks.

DISCUSSION OF UNCERTAINTY / SENSITIVITIES

This section will focus on loading arm releases and tank releases as that is where most of the risk stems from.

For loading arms, due to lack of incidents specifically for LNG loading arms, the release frequencies are derived from incidents with all kinds of cargo transfers from tankers by loading arms or hoses. From Ref. [6],[7], it can be derived an overall cargo release frequency of 8E-5 per cargo transfer. This frequency is including what is termed significant releases. As LNG cargo transfer operations are assumed to be performed with stricter procedures and with generally better equipment than the basis for the statistic, the overall failure rate has been reduced to 1E-5 per cargo transfer. This reduction is also because many LNG loading operations are performed with PERC that are assumed to reduce the frequency of significant releases. Independent on how it is looked at, it is a fact that the release frequency for loading arms is uncertain. However, the frequencies used by Scandpower are in line with frequencies used by other risk analysis companies.

As for the loading arms, the accident statistics for LNG tanks is scarce. In Ref. 1 the overall failure frequency was assessed based on only 3 incidents specifically from LNG tanks. The three incidents are reported to have released 310 m3 of vapor, 466,000 kg of vapor and 10,000 kg of LNG. Since there are only 3 incidents available, no valid cause distribution can be established. In order to account for this, the cause distribution for all cryogenic tanks was applied for the LNG tanks in order to establish the relevant release scenarios from LNG tanks. At the end of the day, the frequency for LNG tank releases will be uncertain due to the low number of historic incidents. However, due to the large exposure areas from significant LNG releases into the bunding for single containment tanks, it can be concluded that this is a main contributor for 3rd party risk.

In this paper, a release rate of 1440 kg/s for 1 and 5 minutes have been used for the loading arms. In another study performed by Scandpower, the full flow release rate from the loading arms was 600 kg/s for 1 and 10 minutes (quick and delayed isolation, respectively). The resulting maximum gas cloud sizes were approximately equal in length for the 1 minute release cases, but the width was approximately the half for the 600 kg/s case. For the longer duration releases, it seems like the width of the cloud does not increase compared to the 1 minute scenario. However, the length of the cloud seems to be somewhat linear with the duration of the release. This can to some degree be explained by a slow dilution of the “front” of the cloud while moving downwind in stable atmosphere and the source of the cloud being fed by the LNG pool sustained by the leak. From this, it seems like both the release rate and the release duration significantly affect the cloud size.

In Ref. [8] results from different studies modeling LNG releases on water are presented. These results differ significantly, mostly due to differences in defining the scenario, but also due to differences in the results from the gas dispersion modeling.

Without discussing which tool gives the correct results, it can be concluded that there are uncertainties involved in using non-CFD tools. Even if a CFD tool will give more correct gas dispersion simulations, such results would probably not affect the conclusions in this paper.

CONCLUSIONS

Most of the third party risk from an LNG re-gasification facility is handled through the requirements in NFPA 59A or similar regulations. Of the scenarios not included in that standard, only releases from two different areas, i. e. loading arms and single containment tanks contributes significantly.

If an LNG re-gasification facility is in the vicinity of populated areas, the use of full containment tanks will lower the risk significantly, maybe as much as a factor 30 lower than for use of single containment tanks.

For a facility with full containment tanks, the biggest remaining contributor to the risk is loading arm failures. In this paper it is assumed that the 1 and 5 minutes full flow scenarios will not be collected, but will end up in the water. From this, risk reducing measures should be focused at reducing the probability of these scenarios and providing designs that will prevent the LNG from reaching the water. Reducing the probable duration of the scenarios will also reduce the risk, but a significant reduction is needed to give a significant risk reduction.

Even if the risk figures presented here are coarse and not site specific, the large difference in risk contributions should still give a good indication to where potential risk reducing measures can be most efficiently implemented. Also, the results can be used in connection with selection of single or full containment LNG tanks for the facility.

---

[1] Tor Egil Nielsen, Devendra Agrawal: Comparative QRA of Single and Full Containment LNG tank designs, paper presented at AIChE Spring National Meeting, New Orleans 2004

[2] TRACE, SAFER Systems, USA (www.safersystem.com).

[3] WS Atkins Safety & Reliability, for the HSE: Ignition probability of flammable gases, Contract Research Report 146/1997.

[4] WS Atkins Consultants Ltd, for the HSE: A model for the ignition probability of flammable gases, Phase 2, Contract Research Report 203/1998.

[5] Frank P. Lees, Loss prevention in the process industries, 2nd Ed. 1996, Chpt. 16.10, Ignition Models.

[6] E. Borill, J.H. Gould, K.W. Blyting, A.N. Lelland: "Incident Probabilities on Liquid Gas Ships", Report AEA/CS/HSE R1014, June 1994.

[7] R.G.J. Robinson, A.N. Elland: "Marine Incidents in Ports and Harbours in Great Britain, 1988 - 1992", Report AEA/CS/HSE R1051, March 1996.

[8] Sandia Report SAND2004-6258: “Guidance on Risk Analysis and Safety Implications of a Large Liquefied Natural Gas (LNG) Spill Over Water”, December 2004.

Quantitative Risk Evaluations of Apparatuses and Equipments of LNG Plants and its Specific Evaluation Methods

Takenori Makita
Takashi Sanjo, Kazu Sato, Masafumi Katagiri
Nihonkai LNG Co. Ltd.
Niigata, Japan

Shigemi Ochiai
Jonquil Consulting Inc.
Tokyo, Japan

AIChE Spring Meeting, April 2005
5th Topical Conference on Natural Gas Utilization
Atlanta April 10-14, 2005

1. Introduction

The Ministry of Economy, Trade and Industry (METI) has revised its policies regarding the safety laws for high pressure gas equipments from the conventional regulative administrative guidance to an execution of an autonomous control as their objective has been modified "to secure public safety by promoting autonomous safety activities." Therefore the maintenance of self-imposed standards that specifically control individual apparatuses and equipments have become an indispensable issue, because the basic articles of the safety law states that significant inspections, which are based on scientific and technical back grounds, should be executed whenever necessary under one's own responsibility for deteriorative and dilapidated apparatuses and equipments in high pressure gas plants.

With this in mind, this terminal that is an important energy base on the Japan Sea has, fortunately, never experienced a disaster in the past twenty years. However, as data that predict the deterioration and decrepitude of apparatuses and equipments can be found here and there, it is important to draw up an appropriate safety plan, considering these deteriorations and dilapidations, to secure a safer plant. Although there are no codes nor standards that quantitatively designate the deterioration and decrepitude of LNG apparatuses and equipment, it is possible to apply the Risk-based Maintenance (RBM) of the American Society of Mechanical Engineers (ASME)[*1] and the American Petroleum Institute (API)[*2] and focus on the performances of the material properties used for the components to induce methods to reduce qualitative risks. By applying this new method to the low pressure LNG pump[*3] on evaporative-type LNG terminals, it was found that by organizing the repair and check-up data, an appropriate safety plan for equipments could be established without tests or analysis of special material properties such as metallographic materials.

This paper is expected to deliver a safer terminal and to avoid public disasters by applying the quantitative risk evaluation method, which was designed by focusing on the material properties that compose LNG pumps, to other apparatuses and equipments as well as by applying and using these methods to similar terminals.

2. Standards of API and ASME and the Quantitative Risk Evaluation Method for Apparatuses and Equipments

Based on the risk index obtained by combining the probability of leaks occurring in the pressured equipment and the extent of its influence, the RBM is a rational way to reduce the risk and cost at the same time while carrying out inspections. To calculate the probability of leak occurrence, evaluation modules, such as the progress pace and receptivity of the various deteriorations occurring on the equipments, the deterioration evaluation ability by non-demolition tests, the condition of the equipments, and the monitoring data, are prepared. RBM was created by summarizing the 170 major accidents that occurred in the US petroleum and the petrochemical industries during the past thirty years from 1961 to 1991. Approximately 40% of the accidents were caused by mechanical strain on the equipments among which 80% were due to damages in the pressurized parts such as the pipes, drums, columns, receptors, tanks, pumps, and cooling units. RBM was introduced to effectively reduce the mechanical damages as the number of occurrences and the cost increased year by year.

RBM was originally used as a probabilistic risk assessment (PRA) in the US nuclear power generation industry to analyze malfunctions but from the end of the 1980s to the beginning of the 1990s it has been applied to execute economical safety control management as well as to secure the reliability and safety of petroleum and petrochemical plants.

With this in mind, surveys on domestic and international codes, standards and other similar criterions have been conducted to apply RBM to evaporative-type LNG terminals. However, the reduction of risks in the deterioration, cracks, defects and welding defects were evaluated quantitatively as it was impossible to find a direct method to assess quantitative risks of apparatuses and equipments.

3. New Quantitative Risk Evaluation of Apparatuses and Equipment based on Assessments of Material Properties in Components

The flow of the quantitative risk evaluation shown in Figure 3-1 consists of a workflow from STEP 1 to STEP 7 where the workflow is completed at each step.

STEP 1: Basic Table for quantitative risk evaluation of apparatuses and equipment

↓

STEP 2: Development of components

↓

STEP 3: Criterions for quantitative risk contents and evaluation

↓

STEP 4: Quantitative risk contents and degree of effects to maintenance

↓

STEP 5: Degree of effects to maintenance and development of costs

↓

STEP6: Relationship between the total cost of apparatuses and equipment and maintenance

↓

STEP7: Quantitative risk evaluation of apparatuses and equipment and recommendation to corrective action of maintenance

Figure 3-1 Quantitative Risk Evaluation Method

STEP 1 applies the RBM risk evaluation matrix as shown in Figure 3-2, which is the basic Table for quantitative risk evaluation for apparatuses and equipments. The vertical axis depicts the influence rate of deterioration, dilapidation and such that is caused by the change in environment and operational conditions. This is because the axis originally showed the impact of damages brought about by the apparatuses and equipments. The horizontal axis depicts the occurrence rate of deterioration and decrepitude. This is how the basic Table for quantitative risk evaluation for apparatuses and equipments were made.

By focusing on the parts that compose the apparatuses and equipment, STEP 2 designates the characteristics of the parts by developing the material properties of the parts. The main point is to check the test results and repair records from the beginning of the operation to cover every sign of deterioration, decrepitude and such.

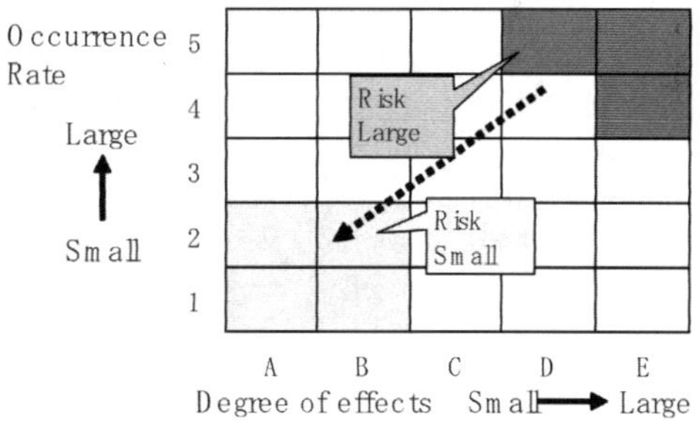

Figure 3-2 RBM Risk Evaluation Matrix

To clarify the influences and the rate of occurrences in the basic Table that was prepared in STEP 1, STEP 3 conducted surveys on; a) The mechanisms regarding the occurrence of damages, b) The technologies to diagnose deterioration, c) Investigations on examples of existing damages, d) Standards, documents, and references related to evaluation techniques of the cracked and fatigue parts.

From this survey as well as from the fact-finding survey by Dr. Davies[*4] and Dr. Dolan[*5] on the causes of damages, it was found that erosion and corrosion was the biggest factor followed by fatigue, brittle failure, loading, creep and friction. Tests conducted in similar environments showed matching results.

Therefore, it can be concluded that erosion and corrosion has the greatest impact, followed by fatigue, brittle failure, loading, creep and friction.

Table 3-1 Degree and Contents of Influences

A	Creep and friction
B	Loading
C	Brittle failure
D	Fatigue
E	Erosion and corrosion

Table 3-1 is a summary of these results. Focus was placed on the material properties of stainless steel and aluminum, which is often used in LNG plants, to grasp the status of the occurrence rate. Although stainless steel is often used as an anti-corrosion material, according to the studies by Dr. Y. Kitamura and Dr.T. Suzuki[*6]

corrosion in stainless steel can occur when; a) the welding process is inadequate b) there is no reflux of precipitation in the salt-level c) when water enters through the welding between the cold insulator and pipes. In addition, similar results were found in past inspections records and repair results conducted by our company. Aluminum alloy proved to show a prominent decrease in tenacity as intergranular fractures tend to generate in low temperatures. It is reported that the tenacity in the material itself decreased and fractures generated specifically after conducting a Charpy shock test in low temperatures. This is due to the transition of slide stress to intergranular fractures, which proves that brittle fracture progresses in the aluminum

Table 3-2 Damage Occurrence Rate of Stainless Steel and Aluminum Alloy

Size	Stainless Steel
5	Welding seams (cold insulator, indoor), erosion and corrosion, temperature change (normal – low), precipitation of highly dense salt, shut-down, low cycles
4	Welding seams, materials and cold insulators, erosion and corrosion, precipitation of highly dense salt, temperature change (normal – low)
3	Welding seams, materials and cold insulators, erosion and corrosion
2	Welding seams, materials and cold insulators
1	Welding seams, and no abnormalities in structural tests
Size	Aluminum Alloy
5	Number of start-ups and shut downs are over 4,000 times, erosion and corrosion, welding seams, temperature difference (low and normal), low cycle, shock waves (closed check valves), precipitation of dense salt
4	Number of start-ups and shut downs are over 4,000 times, erosion and corrosion, welding seams, temperature difference (low and normal), low cycle
3	Number of start-ups and shut downs are over 4,000 times, erosion and corrosion, welding seams, temperature difference (low and normal)
2	Number of start-ups and shut downs are over 4,000 times, erosion and corrosion, welding seams
1	Number of start-ups and shut downs are over 4,000 times, no abnormalities in structural tests

alloy used in extremely low temperatures[*7]. Although these characteristics of aluminum alloy have not been detected from our safety inspection records on our apparatuses and equipments, it is necessary to consider this possibility as suggested in the researches carried out by Dr. T. Kobayashi et al [*7]. Table 3-2 summarizes the causes of damages and its occurrence rate for stainless steel and aluminum alloy.

With this status Table on material property damages complied in STEP 3, STEP 4 inspects the effects this will have on maintenance. It is necessary to find the relationship between maintenance and the damage status of the components because although the effects in STEP 3 may be small, the actual labor input on maintenance may be great. Table 3-3 shows the effects on maintenance and is based on the actual applications carried out by our company as the workload.

Table 3-3 Effects on Maintenance

5	Fundamental check-ups for pressure container repairs (piping damages, cracks in pots, cracks in panel tubes and so forth)
4	Fundamental check-ups for barge of inner fluid
3	Fundamental check-ups for valve blocks (edge cutting, air locks and so forth)
2	Check-ups during shutdown (shutoff locks, cut off power, repairs on switches, no valve blocks)
1	Check-ups without shutdown

Components that were evaluated as having low occurrences of damages and have little effects on maintenance are not included in STEP 5.

STEP 5 inspects the maintenance costs for components that have high occurrences of damages and great effects on maintenance that were abstracted in STEP 4. By leaving the horizontal axis regarding the effects to maintenance found in STEP 4 as it is, the vertical axis on the occurrence rate of damages is substituted with the scale of costs. Table 3-4 shows the breakdown of the scale of costs based on the company's present standard at 100%.

Table 3-4 Scale of Costs

5	Over 125%
4	100 – 125%
3	Current Status 100% (scheduled check-ups)
2	Over 25% - under 99%
1	0 – 25%

Regarding the components that were found to have large scale of costs as well as great effects on maintenance in STEP 5, STEP 6 inspects how much the cost will affect the apparatuses and equipments as a whole. By keeping the horizontal axis on the effects on maintenance in STEP 5 as it is, the vertical axis on the scale of costs in Table 3-4 is substituted with the total costs of apparatuses and equipment. Whereas the scale of costs in STEP 5 indicated the construction fees to exchange components, the total costs of apparatuses and equipment defines the future costs necessary to maintain the apparatuses and equipment and at the same time this evaluates the occurrence rate of damages shown in Table 3-5 as well as its specific corrective actions. In addition, it confirms whether regulations, control, costs, and corrective data as written in Table 3-6 are available to inspect a specific executable method. If the occurrence rate can be reduced, costs can also be reduced as Table 3-5 simultaneously evaluates the total costs of apparatuses and equipment.

Table 3-5 Evaluation and Corrective Actions of Damage Occurrence Rates

Name	No.	Name	Specs	Units	Evaluation and corrective actions of damage occurrence rates				
					Law	Control	Cost	Corrective data	Specific method
	12	Impeller	AC4C-T6	3					
	13	Buffer plate	SUS3166	1					
	14	Impeller spacer	BC6	1					
	15	Lock hub	A6061-T6	1					
	16	Balance drum combined	SUS316/BC6	1					
	19	Wearing	BC6	3					
	34A	Orifice	SUS304	1					

STEP 7 completes the tasks between STEP 1 to STEP 6 as well as recommends corrective actions for maintenance.

Table 3-6 Assessment Criterions for Damage Occurrence Rates

Regulations	◎:apparatuses and equipment regulated by law ○yes: periodical check-ups ○: autonomous regulations -: not necessary
Control	◎: could not monitor or discover from the items on the right ○yes: trip ○:alarm -:patrol
Cost	◎: great damages ○yes: damaged ○:little damages -:none
Corrective data	◎: unavailable ○yes: available through research ○: available through experience and documents -: available

4. Development of Low Pressure LNG Pumps for Quantitative Risk Evaluation Inspections

A specific example of an inspection for a low pressure LNG pump was developed from the flow indicated in Figure 3-1. The materials were distinguished by using different colors on the LNG pump layout drawing in STEP 2 and records of the past inspections and repairs of each component (visual checks, PT inspections, measurement of size, state of condition, replacement of parts and so on), and whether abnormalities occurred during the inspections, were entered into the Table. Other details about the usage of the components such as fatigue, erosion, friction, and so on were also entered to create a Table that covers everything about the usage of the low pressure pump.

To calculate the occurrence rate of damages for STEP 3, 19 reports predicting deterioration of materials were found among our operational reports. These results were cross-checked with the component lists that were complied in STEP 2 to calculate the occurrence rate for damages. a) The inner components of the stainless steel pumps were given low assessments as many were not in a state to accumulate salt levels. b) The inner components of the aluminum pumps were given low assessments as ruptures and extensions were minimal because the number of start ups and shutdowns were not enough to incur repeated fatigue. c) The inner stainless steel bearings were given high assessments as deterioration from past frictions was detected. d) Pots and pipes were given high assessments as accumulation of salt concentration was possible in the welding seams. e) Nuts and bolts were given high assessments as they are placed in areas where the possibility of salt concentration accumulation is great. Furthermore they are in an environment where load is applied.

With reference Table 3-3, STEP 4 evaluated how each component affects the maintenance and the results were; a) The inner components of the low pressure LNG pump were given high assessments because the pump must be stopped during maintenance and the inner fluid must be purged. However, it was removed from STEP 5 because the damage occurrence rate was low. b) The maintenance effects on bearings are high but since they are exchanged during the periodical repairs, the damage occurrence rate was low. c) Since pots, pipes, nuts and bolts are pressurized parts, LNG will leak if these parts are damaged. This was given high assessments as repairs involve legal procedures and a long term shutdown is anticipated.

STEP 5 evaluated the costs to repair the components that were given high assessments in STEP 4. With reference to Table 3-4, calculations were carried out on the material and labor costs based on the present costs of periodical check-ups. As a result, pots, pipes, nuts and bolts, and bearing were given high assessments since they must be replaced when damaged.

To find the appropriate maintenance methods for pots, pipes, nuts and bolts, and bearings which were given high assessments in STEP 5, STEP 6 evaluated the level of priorities from the perspective of both the damage occurrence rates and the reduction of maintenance costs, with reference to Tables 3-5 and 3-6. As a result; a) By carrying out nondestructive tests during periodical check-ups to check the deterioration of material properties in pots and pipes, a corrective action plan was devised to simultaneously reduce both maintenance costs and damage occurrence rates. b) A corrective action plan was devised for bearings by carrying out a cooperative research with the bearing manufacturers to find whether it is possible to replace the materials to those that have a longer life. c) A corrective action plan to reduce damage occurrence rates in nuts and bolts were devised by replacing the parts, depending on the degree of decrepitude, after carrying out tests on the materials.

5. Conclusion

Ten years has passed since this high pressure gas equipment has been in operation. Costs for repairs will increase as this equipment deteriorates and dilapidates overtime. With the revision in the administrative regulations to an autonomous control, it has become indispensable for enterprises that possess high pressure gas equipment to secure a thorough safety control to continue to develop their business as well as to fulfill their social responsibilities.

With this in mind, safety will be secured on plants by applying appropriate maintenance methods which derived from the RBM methods based on the API and ASME standard which focus on the material properties of the apparatuses and equipments. To promote the advantages of this new method, it will be applied on apparatuses and equipments other than low pressure LNG pumps from fiscal year 2005.

References

(1) ASME Publishing Catalog 2002

(2) API Publications Programs Services 2002

(3) Makita and Ochiai etc. The Manual for Risk Management System for Use. During Plant Disasters and Specific Risk Management Methods AIChE Conference 2004

(4) G.J. Davis Performance in Services in E.J. Bradbury (ed.) Essential Metallurgy for Engineers, Van Nostrand Reinhold (UK), London, 1985

(5) T.J. Dolan Analyzing Failure of Metal Components Metal Eng. Quart., Vol. 12 (4), 1972

(6) T. Suzuki and Y. Kitamura Technologies on Corrosion Resistance September 10, 2002 Taninn Shokan

(7) T. Kobayashi Strengths and Toughness of Materials Springer 2004

Biography of Speaker

Shigemi Ochiai, Ph.D President, Jonquil Consulting Inc.

2-17-18 Honmachi Shibuya-Ku Tokyo

I was born in Tokyo in 1952. I obtained my Ph.D. in the field of Civil Engineering at the University of Wales. I had worked in the Chiyoda Chemical Engineering Company for 16 years. By utilizing its experience, I found the Jonquil Consulting Inc. in 1990. I became president also CEO. Now I have offered consulting business and a lecturer of risk management to the companies.

Paper #109e

LNG Codes & Process Safety

Dr. C Dale Zinn
Consulting Engineer
Austin, Texas
cdalezinn@gmail.com

Prepared for Presentation at
AIChE Spring National Meeting
Topical Conference on Natural Gas Utilization
Atlanta GA
13 April 2005

ABSTRACT

Due to the potential wide area impact of an unplanned LNG (Liquefied Natural Gas) release, the National Fire Protection Association (NFPA) and later the U.S. Department of Transportation issued standards and regulations which included specific methodologies for analyzing the consequences of an accidental LNG release.

The concept of a "design spill" was defined in order to evaluate the consequences of an accidental LNG release, with particular attention to the influence of safety protective measures. In addition to specifying the magnitude of the unplanned release, the design spill also specified some of the parameters for estimation of the downwind LNG vapor dispersion and the extent of the radiant heat from an LNG pool fire. The "design spill" was an important concept because the LNG rules were the first (and still quite unique) to define "exclusion zones" for purposes of establishing the allowable sites for establishing LNG facilities.

Since the NFPA-59A and 49-CFR-193 standards were the first in-depth regulations to address consequence analysis estimation for petroleum components, it is particularly important to consider their details in light of the OSHA 40CFR1910.119 Process Safety Management rules and the EPA 40-CFR-68 regulations; Risk Management for Chemical Accidental Release Prevention, which specifically address consequence analysis as a part of process safety management.

INTRODUCTION

The major United States regulations and standards governing LNG plant siting, design, construction, inspection, testing and process safety issues are 49-CFR-193, "Liquefied Natural Gas Facilities; Federal Safety Standards" [1] and NFPA-59A, "Production, Storage and Handling of Liquefied Natural Gas (LNG)" [2]. The 49-CFR-193 standard is the U.S. federal government regulation. NFPA-59A is an industry consensus standard issued by the National Fire Protection Association.

The federal regulations and industry standards for the design, construction, and operation of LNG facilities have been developed utilizing good engineering practices and incorporating many years of operating experience. These regulations and standards are intended to achieve process safety objectives by focusing on preventing accidental LNG containment failures and minimizing the impact of fires resulting from an accidental LNG containment release. A combination of siting separation distances, design practices, and hazard control systems are among the methods used to strive for the process safety goals.

As stated in the preamble to the 49-CFR-193 regulation announcement, foremost among the LNG safety issues are;

- Protection of persons and property near an LNG facility from thermal (heat) radiation caused by a fire resulting from a major spill of LNG.

- Protection of persons and property near an LNG facility from dispersion and delayed ignition of a natural gas cloud arising from a major spill of LNG.

- Reduction of the potential for a catastrophic spill of LNG.

The approach utilized to determine adequate siting separation distances and to specify hazard control systems is based upon the *design spill* concept. This concept hypothesizes accidental LNG releases which could occur at the facility based upon historical experience, regulatory requirements and the specific features of the facility. Potential hazards associated with the spills are then quantified using consequence analysis predictive techniques, which are widely accepted throughout the LNG industry. The design spill concept was broadly based upon the nuclear industry concept of the design basis accident.

LNG FIRE HAZARDS

For petroleum-based liquids and gases, the most recognized hazards are those associated with the flammability. However, LNG presents a few special hazards due to its low temperature characteristics.

In order for liquids to burn, they must first be changed to vapor. If LNG is released from containment, it vaporizes readily. Whenever LNG is released from its containment system, the liquid will be heated by the surroundings, thus causing the liquid gas to vaporize. The vapor generated by this boiling liquid will start to mix with the surrounding air and will be carried downwind with the air, thus creating a vapor cloud. As the vapor continues to be carried downwind, it will mix with additional air and be further diluted. Some portion of the vapor cloud will be within the flammable limits (about 5-15% by volume). If this flammable portion encounters a source of ignition, the vapor cloud may ignite. The flame might then propagate through the cloud, back to the source of the vapor, particularly if the flammable portion of the cloud is continuous. This simple burn-back of an unconfined vapor cloud can cause secondary fires by igniting materials in the path of the flame and can cause severe burns to persons caught within the cloud. Damage to equipment will generally be limited, since the time of exposure to the fire will be relatively short.

LNG presents about the same thermal characteristics as natural gas. In well ventilated areas, natural gas burns with a low laminar burning velocity and has a high ignition energy relative to other hydrocarbon fuels. Natural gas vapors in open areas have not produced unconfined vapor cloud (UVCE) explosions, which are more prevalent with higher chain hydrocarbons. Whether any flammable vapor cloud merely burns back to the vapor source, or undergoes an explosion, depends on many factors: the chemical structure of the vapor molecules, the size and concentration of the vapor cloud, the strength of the ignition source, and the degree of confinement of the vapor cloud. The conditions needed to produce an unconfined vapor cloud explosion of natural gas are generally not present in an LNG facility, so such explosions are not considered as potential hazards.

In closed areas, ignition of a flammable natural gas mixture may exhibit an explosion, with the resulting damaging overpressures. If LNG or its vapor leaks into a control room, compressor building, etc., then the possibility of an explosion is markedly increased. This is due to the pressure increase within the enclosure caused by the fire heating the air, and the increase in gas volume during combustion. Most buildings will withstand very little internal pressure and when the pressure limit is reached, the building literally explodes. The degree of confinement necessary to produce explosive behavior is currently a major research topic, particularly focused on the crowded conditions of offshore facilities. LNG facilities generally do not exhibit this crowded piping and equipment configuration.

An LNG leak or spill of sufficient size may result in an accumulation of liquid on the ground. If ignited, the resulting fire is known as a pool fire. Ignition can occur at the pool location (either immediately or after some delay), or the pool can be ignited by a vapor cloud fire. Objects directly contacted by the flame above the pool can be severely damaged or destroyed, and exposed personnel would receive extensive burn injuries. Objects and personnel outside the actual flame volume can also be damaged or injured by the radiant heat emitted by the flame. Compared to a vapor cloud fire, the effects are more localized, but of longer duration.

When a flammable liquid is accidentally released from pressurized containment, the leak may take the form of a spray of liquid droplets and vapor. If ignited, the resulting fire is termed a torch fire. Such a fire can also result from a pressurized vapor leak. Torch fires present the same types of hazards as pool fires, i.e., direct flame contact and radiant heating. However, the radiant heating power of a torch fire is often greater than that of a pool fire of similar size.

A BLEVE (Boiling Liquid Expanding Vapor Explosion) is the catastrophic failure of a pressurized container when its contents are above their boiling point temperature. The most common type of BLEVE occurs when an uninsulated pressurized vessel is exposed to an adjacent fire. The fire increases the internal pressure and weakens the vessel until it can no longer contain the pressure. The vessel then ruptures violently, and parts of it may be propelled great distances. The released liquid flashes and atomizes immediately, often resulting in a large fire-ball. The fireball can cause very widespread damage due to flame contact and thermal radiation. Although the fireball lasts only a few seconds, its effects can be devastating.

The probability of a BLEVE of an LNG storage tank is extremely small, since the main tank is protected by the outer tank and insulation that would retard the heat transfer from a fire to reach the main tank. Furthermore, most LNG storage tanks are designed for relatively low operating pressures. Therefore, if exposed to fire, these tanks will not BLEVE since they will fail at a fairly low internal pressure and, at the time of failure; the LNG would not be heated sufficiently to cause any significant quantity of liquid to flash to vapor.

A BLEVE is only considered a potential hazard for LNG facilities in special circumstances, such as unattended remote locations and process configurations with the potential for rapid over-pressurization of the storage tank (dead head discharge pressure of a pump above storage pressure limit). Smaller LNG storage vessels built to withstand pressure transfer conditions, other LNG vessels or containers (e.g., a liquid knockout drum or recondenser) within the plant, and LNG transport vessels might have design pressures large enough that they could potentially BLEVE. However, the vessel and piping insulation required by economic considerations may be sufficient to retard heat input from adjacent fires to preclude BLEVE conditions. LPG pressurized storage vessels are more susceptible to BLEVE, because they usually do not have the benefit of insulation to retard heat input from nearby flames.

Neither NFPA-59A nor 49-CFR-193 addresses the potential for BLEVE as a credible LNG accident.

Natural gas processing facilities which make, but do not store LNG, such as nitrogen or helium separation plants, are not covered by the federal regulations.

LNG USA REGULATORY BACKGROUND

With LNG industry growth in the early 1960s, the safety critical issues of LNG began to be generally recognized. The potential for wide area damage from a significant accidental LNG release [9] and review of the 1944 Cleveland LNG disaster [11] started an industry movement to develop special safety standards for LNG. The first edition of the NFPA-59A standard was adopted in 1967. Several subsequent editions broadened the scope, with the latest major revision to be issued in 2005.

In 1972, the LNG regulations were contained in 49-CFR-192, the natural gas transportation (pipeline) standards. An interim measure adopted in October of 1972 stated that then current edition of NFPA-59A was considered as the federal standard until specific federal regulations were developed.

One of the main reasons for promulgating federal LNG standards was a perceived inadequacy with the earlier editions (1971/1972) of the NFPA standard. Many of the concerns with the early NFPA-59A standard have been overcome with the release of subsequent editions by NFPA.

Most LNG facilities in the USA were built prior to the 1980 adoption of 49-CFR-193 and conform to earlier editions of the NFPA-59A standard.

The U.S. Coast Guard in 33 CFR Part 127 is responsible for facility site selection as it relates to nearby vessel traffic and matters pertaining to the loading/unloading docks. The federal co-ordination of regulatory overlap is covered by a formal interagency Memorandum of Understanding.

PROCESS SAFETY MANAGEMENT AND RMP RULES

Most other petroleum processing facilities, like a gas processing plant or a refinery that handle large quantities of flammable liquids or gases, are covered under the Occupational Safety and Health Administration (OSHA) process safety management (PSM) standard (29 CFR 1910.119) and the Environmental Protection Agency (EPA) risk management program (RMP) rule (40 CFR 68).

The PSM standard emphasizes the management of hazards through a comprehensive program that integrates technologies, procedures, and management practices. The standard has fourteen elements that are fundamental to running a safe chemical operation.

One of the key elements mandated is the Process Hazard Analysis (PHA), "an organized and systematic effort to identify and analyze the significance of potential hazards associated with the processing or handling of highly hazardous chemicals," according to OSHA's PSM compliance guidelines document. The PHA leads to the development of operating, maintenance, and training procedures, along with emergency response and incident investigation elements. A formal Management of Change program is required to maintain the safety level up to the original design intent.

The EPA Risk Management Plan builds on the OSHA PSM regulation, adding a "worse case consequence analysis" and also requires reporting of incidents.

However, both of those regulations specifically exempt LNG facilities that are regulated under the DOT regulations in 49 CFR Parts 193, being precluded by the concept of sufficient regulation by another (DOT) federal agency.

It is interesting to note that the offshore petroleum production industry is also exempt due to their regulations under the Minerals Management Service (MMS). However, MMS has adopted the language of PSM through voluntary rules which are codified in the API 14 series.

An important point is that neither the OSHA PSM rules or the EPA risk management rules have siting restrictions.

THE EUROPEAN PROCESS SAFETY REGULATORY INITIATIVES

After the occurrence of the Seveso incident, the European Union adopted a new regulation aiming to avoid similar events known as "Seveso I". This Directive 82/501/EEC on the major-accident hazards of process industrial activities was adopted in 1982 with the objective of preventing and responding to major industrial incidents, and to reduce the potential consequences on the environment and the public. Since 1999 the Directive 88/610/EEC "Seveso II" and called the Control of Major Accident Hazards (COMAH) requires industrial and storage activities to follow process safety and risk management measures in order to prevent the occurrence of more incidents.

The specific language for European LNG rules are based originally upon British B777 codes, but now the latest regulation is EN1473-1997. It contains many concepts which are similar to the USA PSM rules, and beyond.

MAJOR DIFFERENCES BETWEEN PSM AND THE USA LNG CODES

NFPA 59A and DOT 193 do not explicitly require a safety management system; however, require control of activities that are typically controlled by a safety management system. Specifically missing are Process Hazards Analysis and Management of Change.

Unique features of the LNG rules, however, are the siting "exclusion" zones. More analysis of this unique feature is contained herein.

Another unique feature of the USA LNG rules is the specification of distinct consequence software models. The use of alternate models which take into account the same physical factors and have been validated by experimental test data could be permitted, subject to the Administrator's approval. However, no one has been able to get Administrative approval, since no procedure to get approval exists. Similar Administrator approval language is used throughout the LNG rules, again with no apparent method to achieve this approval.

CONSEQUENCE ANALYSIS FOR EXCLUSION ZONE DETERMINATION

As noted in the previous discussion of LNG hazards, a plant that handles LNG presents two main types of flammability hazards that could potentially impact the facility and the adjacent area: vapor cloud fires and pool fires. In order to adequately address the potential impact of LNG spills within the facility, a method for quantifying the vapor cloud fire and pool fire hazards was required.

The first step in quantifying the accident scenario is to postulate parameters of the accidental breach of containment, including rate of release, time extent before the release is stopped, local geographical detail, and specific aspects of the process equipment. Then, the extent of a vapor cloud (delayed ignition) or pool fire (immediate ignition) scenario could be estimated.

It is not feasible to look at the hazards from all possible spills within the plant. Instead, a small number of rather large volume spills are investigated. These "*design spills*" are based on requirements of NFPA-59A/DOT 193, historical experience, and any unique features of the facility and surrounding area. *Design spills* are generally much larger than the accidental spills that might be expected to occur (e.g., valve or seal leaks). Therefore, they represent a reasonable upper bound on spills that could occur and whose hazards could be controlled.

The design of the secondary containment or impoundment area is a critical factor in quantifying the impact of the *design spill*. Hence, the standards and regulations focus on determining the thermal and vapor dispersion exclusion zone based upon the impoundment design.

DOT193 sets *design spill* requirements specifically for each major plant area, recognizing the differences in potential accident scenarios due to process equipment characteristics. The plant facilities are; LNG storage tanks, process areas (i.e., liquefaction, compressors, pumps, etc.), vaporization areas, and transfer areas (i.e., ship, railcar or tank truck loading areas, etc.).

Major failures of permanent plant piping .are not included in the NFPA59A *design spill* concept, but were included in DOT 193 regulations prior to the code/standard harmonizing project of the late 1990s.

<u>Design Spills for LNG Storage Tanks</u>

Figure 1 presents some of the variations in secondary containment design for LNG storage tanks. Another tank design feature which has a significant impact on the *design spill* is the location of the tank connection; either over-the-top or bottom connections.

For LNG tanks with side or bottom connections, the NFPA-59A section 2-2.3.3(a) design spill into the impounding areas serving the tanks is defined as the largest flow from any single line which would be severed. No minimum leak flow time for this design spill is specified, but the leak flow rate is specified by a formula which includes pipe diameter and liquid head. Therefore, the leak flow time for this design spill will be based upon the ability to drain the tank.

If an LNG tank with side or bottom connections is fitted with an internal shutoff valve, then the *design spill* may be assumed to continue for one (1) hour at a flow rate based upon a full liquid head. There is really no rational explanation for the one hour limitation.

For LNG tanks with no side or bottom connections (i.e., all pipe connections to the tank pass through the dome), the NFPA-59A *design spill* for the impounding areas serving the tanks is defined as "the largest flow from any single line which would be pumped into the impounding area with the container withdrawal pump(s) considered to be delivering full rated capacity." The minimum flow time for this *design spill* is 10 minutes "provided there exists demonstrable surveillance and shutdown provisions." Spills of this magnitude can be readily detected and the pumps can be stopped quickly. Therefore, the flow time for this design spill will be 10 minutes. In actuality, a spill of this size could be detected and stopped in much less than 10 minutes.

For an inground (buried) tank, the hole in the ground that is occupied by the tank is the impounding area that serves the tank. Therefore, the *design spill* has no real meaning when applied to inground tanks.

The 49-CFR-193 regulation is essentially the same as NFPA-59A in the design spill concept as it applies to storage tanks. It should be noted that both standards use the concept of continuous, not instantaneous, releases for the design spill evaluation. LNG studies [3, 7, 5, 11] have indicated that instantaneous spill assumption could only be a realistic for very small spill times, less than 2 minutes duration.

Controversy over Design Spills for Process Areas

NFPA-59A *design spill* at process areas, vaporizers, and transfer areas must be based on the discharge rate from "any single accidental leakage source." Leakage sources might include gasket failures at flanged connections, pump or compressor seal leaks, etc., and might or might not include full pipe ruptures, depending on the probability of such an occurrence. Total flow time for these design spills is 10 minutes "or a lesser time period based on demonstrable surveillance and shutdown provisions." The spill detection and emergency shutdown systems for the LNG facility may be sufficient to allow a reduction in the total spill time.

The time to detect the leak is somewhat dependent on the leak rate. Since larger leaks produce larger visible vapor clouds and larger accumulations of liquid, they are generally detected more quickly. The isolation time is set by the rate of closure of ESD valves. Generally, this can be done in 30-45 seconds or less without creating harmful hydraulic surges within the liquid transfer system.

The spill volume is equal to the spill rate times the spill time plus any drainage of LNG from the piping following an emergency shutdown. The drainage volume is controlled by the diameter and length of pipe between the ESD valves, excess flow valves, check valves, vertical pipe runs, or other items that would prevent liquid from flowing to the leak site.

Spill Volume = [(Spill Time) x (Flow Rate)] + Pipe Drainage

The DOT49-CFR-193 regulations mirror the NFPA-59A standard in the design spill concept for storage tank areas, but vary in the process and transfer piping areas. The NFPA-59A historically did not require *design spill* considerations for plant piping which conveys LNG between processing, storage and transfer areas. However, 49-CFR-193 expanded the definition of transfer to include transfer permanent piping, thereby requiring consideration of *design spill* concepts to plant piping. This means that piping impoundment or grading is essentially required to insure the vapor cloud exclusion zone defined in the code. Many LNG facilities use earthen or concrete trenching to isolate the LNG rundown piping from adjacent plant areas.

See the appendix for details on some recent interpretation controversies regarding the design spill concepts.

Vapor Dispersion Exclusion Zone Estimates

An un-ignited accidental release of LNG will develop a flammable vapor cloud, which can travel downwind until sufficient heating and mixing with the turbulent atmosphere dilutes the vapor concentration below the limits of flammability. From a safety standpoint, interest is focused on the area covered by the un-ignited flammable vapor cloud as it travels downwind. The rate at which vapor is generated by a liquid pool becomes the source for transient heavier-than-air dispersion model. The pool vaporization models are described in the LNG literature [5]. To estimate the downwind travel distance of the flammable cloud, several factors regarding type of accident (*design spill* conditions), surrounding area and local weather conditions must be considered. An overview of the currently available transient heavier-than-air vapor dispersion models is given the AIChE CCPS book on vapor dispersion [4].

The NFPA-59A standard suggested using the downwind distance of the flammable mixture to the lower flammable limit (LFL), while the 49-CFR-193 regulation always required the end point be the 1/2 LFL distance. The NFPA 59A adopted the more conservative DOT 193 definition in the latest update. Both NFPA-59A and 49-CFR-193 default to the Pasquill-Gifford atmospheric stability category F with a wind speed of 4.5 mph (2 m/sec), the 49-CFR-193 adds 50% relative humidity and 32^0 F (0^0 C) atmospheric temperature. Vapor control systems (channel flow, dikes, vapor fences, etc.) are no longer permitted in the defined hazard exclusion zones as estimated with the vapor dispersion calculation model, except as requested by a variance procedure.

The vapor dispersion estimates assume a flat topography. The distance to LFL represents the travel of a continuously flammable vapor cloud. Due to gusting of the wind, pockets of flammable vapor may break away from the continuous cloud and be carried beyond the continuous cloud. Therefore, a conservative estimate of the downwind flammable distance is made by assuming that these pockets will have dissipated when the cloud concentration is below 1/2 the LFL. Because of these assumptions used in vapor cloud prediction, the resulting cloud coverage length should be considered worst case possibilities. Under most atmospheric conditions the vapor clouds would be shorter than those presented herein. It also must be recognized that if the cloud is ignited before reaching its maximum downwind distance then the result will be a flash fire back to the source of LNG release.

The vapor cloud exclusion zones are estimated for each *design spill* in each identified process area or storage area. The hazard control concepts of spilled liquid channeling to a sump, thus reducing vapor travel (in comparison with the large area covered by a spill into a flat diked area) is a standard LNG storage area feature. Smaller impoundments with sloped or graded surfaces with curbing are suggested to reduce estimated downwind vapor travel from unplanned LNG *design spills* in process areas.

Pool Fire Thermal Exclusion Zone Estimates

Quantification of pool fire hazard exclusion zone involves calculating the distance from the fire at which thermal radiation levels are hazardous to people and equipment. If directly contacted by flames, personnel can be severely injured and plant equipment can be badly damaged.

Prior to the regulatory/standard harmonizing effort in 1999, the NFPA59A and DOT49CFR193 had different approaches to the estimation of thermal exclusion zones around an LNG impoundment area. In both cases, the design basis accident scenario assumes a fully-involved tank dike or impoundment area fire.

NFPA59A provided simple formulas, with only pool surface area of the impoundment as the variable, to define the minimum separation distance to:

- Nearest edge of property line from Tank Impoundment [1,600 BTUs/Hr-Sq Ft]
- Nearest point of assembly from any LNG Impoundment [1,600 BTUs/Hr-Sq Ft]
- Nearest building from any LNG Impoundment [3,000 BTUs/Hr-Sq Ft]
- Nearest edge of property line from any LNG Impoundment [10,000 BTUs/Hr-Sq Ft]

The DOT49CFR193 uses the same basic concept, but previously defined four levels of incident radiant flux (1,600; 4,000; 6,700; and 10,000 BTUs/Hr-Sq Ft) on offsite targets. While the exclusion zone distance formula used by both standards is the same, the safety or modification factors were different. The thermal radiation exclusion zones apply to all LNG spill impoundment areas, but the focus is typically on the large volume LNG storage area.

Instead of using the exclusion zone formulas, fire radiation profiles may be computed for the *design spills* using computational models which are based upon accepted correlations [3], local wind speed, ambient temperature, and atmospheric relative humidity. When computational models are used, NFPA-59A specified zero wind speed, 70°F ambient temperature and 50% relative humidity as the design basis conditions, whereas DOT193 required the use of the highest wind speed over 5% of the recorded time..

The accuracy of the pool fire models is a direct function of the accuracy of the model to predict the basic flame geometry and surface radiative flux. Only well established techniques for persisting flame length and bending angle have been used. The Thomas flame height correlation [12] and the Welker and Sliepcevich bending angle correlation [10], within the experimental accuracy of the data, adequately predicted the experimental results. Flame heights and bending angles from LNG fires were compared with results from the Thomas correlation and the Welker and Sliepcevich correlation in 1974 [11]. The conclusion of the authors was that, within experimental error, the correlation adequately predicted the experimental results.

Recent concern over the large fires that could result from LNG tanker incidents has focused on the accuracy of the pool spreading correlations and the surface flux of large fires.

CURRENT PROBLEMS WITH THE DESIGN SPILL CONCEPT

The identified problems associated the design spill concept are listed below;

- Recent updates have caused some confusion in the language of the design spill concept (see appendix).
- The vapor hold up concept issue
- The inability of code specified consequence models to address high pressure containment breach scenarios, jet mixing for vapor dispersion and torch fires.
- Difficulty in applying code specified consequence models to elongated trench designs
- Inability to easily apply updated modern consequence models due to the lack of an Approved Administrators variance methodology or procedure.
- Emergency personnel concern over the 5 kw/m^2 thermal hazard zone
- Impact of Terrorism consequence scenarios

SUMMARY

The NFPA-59A and 49-CFR-193 LNG standards developed consequence analysis requirements which provide designers with an procedure to analyze the process safety of their LNG facility designs, but also allowed for creative engineering solutions to mitigate the impact of catastrophic accidents.

Modifications and patching the regulations have caused more confusion than simplification, hence it is time for a complete review and revision of the "design spill" concept with it's important exclusion zone

REFERENCES

1. .NFPA-59A, Production, Storage, and Handling of Liquefied Natural Gas (LNG) 1990 Edition. National Fire Protection Association. Quincy, Massachusetts. Latest edition.

2. Title 49 Code of Federal Regulations Part 193, Liquefied Natural Gas Facilities: Federal Safety Standards. U.S. Government Printing Office. Washington, D. C. October 23, 1980.
3. West, H.H. & L.E. Brown, "Analyze Fire Protection Systems", Hydrocarbon Processing, August 1977.
4. "Guidelines for the Use of Vapor Cloud Dispersion Models", Center for Chemical Process Safety, American Institute of Chemical Engineers, New York, 1987.
5. "Evaluation of LNG Vapor Control Methods", American Gas Association, Washington D.C., October 1974.
6. Havens, J. A., "A Description and Assessment of the SIGMET LNG Vapor Dispersion Model." U.S. Coast Guard Report CG_M_3_79, February, 1979.
7. Raj, P. Phani, and A. Kalelkar (1973), "Fire Hazard Presented by a Spreading Burning Pool of LNG on Water." From Proceedings of Combustion Institute, Western Section Meeting, Los Angeles, California, October, 1973.
8. Colenbrander, G.W. and J.S. Puttock (1983), "Dense Gas Dispersion Behavior: Experimental Observations and Model Developments", International Symposium on Loss Prevention and Safety Promotion in the Process Industries, Harrogate, England, September, 1983.
9. Welker, J., H. Wesson, and C. Sliepcevich, "LNG: To Burn or Not To Burn", Proceeding of the American Gas Association Distribution Conference, May 1969.
10. Welker, J.R., and C.M. Sliepcevich (1970), "Susceptibility of Potential Target Components to Defeat by Thermal Action", Report AD-875-935L, Edgewood Arsenal, Maryland, 1970.
11. American Gas Association (1974), "Project IS 3-1: LNG Safety Program, Interim Report on Phase II Work", Battelle Columbus Laboratories, July 1, 1974.
12. Thomas, P.H. (19630), "The Size of Flames from Natural Fires", Proceedings of the Ninth International Symposium on Combustion, Academic Press, New York, New York, 1963, pp, 844-859.
13. Response to Weaver's Cove's Rebuttal of Jerry Havens' Two Part Report of August 25, 2004, to the City of Fall River, Massachusetts: Analysis of Draft Environmental Impact Statement for the Weaver's Cove LNG Project, filed in the FERC
14. Effects of secondary containment on source term modeling, WS Atkins for the Health & Safety Executive, contract research report 324/2001, HMSO, London 2001.
15. LNG Vapor Cloud Exclusion Zones for Spills into Impoundments, Jerry Havens and Tom Spicer, Chemical Hazards Research Center, University of Arkansas, AICHE Spring Meeting 2005
16. King, R., Safety in the process industries, Butterworth-Heinemann, 1980, Ch. 23.
17. Takala, J., Chemical Safety Information for developing countries, The Science of the Total Environment, 188 Supl., S112-S129, 1996.
18. Lees, F.P., Loss Prevention in the Process Industries, 3rd edition, Butterworth-Heinemann, 2005.
19. July 31, 1978, by the General Accounting Office titled "Liquefied Energy Gases

APPENDIX

Recent Controversy Over Vapor Dispersion Exclusion Zone Estimates

The estimation of LNG vapor cloud exclusion distances in the USA LNG DOT193 regulations (49CFR193) requires modeling the hazard scenarios of LNG flowing into spill containment sumps and storage tank diked areas. Exclusion distances are demanded to meet the public safety intent of Congress.

In several recent FERC published DEIS documents, the exclusion zone calculations resulting from a spill into a vaporization area and transfer area may have been substantially underestimated due to FERCs misunderstanding of DOT 193 rules.

The pre 2000 DOT193 rules, in no uncertain terms, specified that the exclusion zone estimates for all LNG spill secondary containment areas be based upon *Vaporization results from the spill caused by an assumed rupture of a single transfer pipe (or multiple pipes designed to deliver the same flow) which has the greatest overall flow capacity, discharging at the maximum potential capacity*

The exact same spill concept is used to define the spill containment volume.

The 1998 NPRM which proposed to harmonize the DOT193 and NFPA59A language **did not propose** reducing the exclusion zone calculation requirement for the vaporization and transfer area spill containment to the NFPA59A "accidental leakage source" There was no mention of this deletion in the NPRM. Unfortunately, there was an inadvertent language error in the final DOT193 Amendment announcement in the March 2000 Federal Register., which lead to ambiguity in this specific area.

But the DOT made it absolutely clear in their NPRM announcement and the DOT193 Amendment announcement that they did intend to maintain the **current level of safety**.

In fact, FERC does point out in their DEIS documents that some confusion on the design spill for vaporizer area exists, but due to their lack of co-ordination with DOT Office of Pipeline Safety technical staff, they appear to have misrepresented DOT intention. Furthermore, Some FERC DEIS documents do not provide transfer area exclusion zone estimates as required by DOT code.

For the vaporization area spill containment, FERC has used the older NFPA59A concept of an accidental spill from an "accidental leakage source". This has been defined by FERC to be an LNG spill from the rupture of a 1'' to a 3" connection. The DOT193 intent would demand the LNG spill rate from the largest pipe at full capacity. This spill rate would be the same as the 10 minute spill used to size the containment volume.

DOT has noted some other ambiguities in the March 2000 Amendment , as evidenced by the corrections to the 193.2005 section recently published on 10 March 2004 in the Federal Register.

Pertinent details of the pre 2000 DOT 193, the December 1998 NPRM and the March 2000 Amendment are contained herein.

The pre-2000 DOT 193 code specifically stated:

> *"§Sec. 193.2059 Flammable vapor-gas dispersion protection.*
> *(c) Computing dispersion distance. A minimum dispersion distance must be computed for the impounding system.*
> *(d) Vaporization design rate. In computing dispersion distance under paragraph (c) of this section, the following applies:*
> > *(1) Vaporization results from the spill caused by an assumed rupture of a single transfer pipe (or multiple pipes that lack provisions to prevent parallel flow) which has the greatest overall flow capacity, discharging at maximum potential capacity,...."*
> > > *i) The rate of vaporization is not less than the sum of flash vaporization and vaporization from boiling by heat transfer from contact surfaces during the time necessary for spill detection, instrument response, and automatic shutdown by the emergency shutdown system but, not less than 10 minutes, plus, in the case of impounding systems for LNG storage tanks with side or bottom penetrations, the time necessary for the liquid level in the tank to reach the level of the penetration*

or equilibrate with the liquid impounded assuming failure of the internal shutoff valve.

(ii) In determining variations in vaporization rate due to surface contact, the time necessary to wet 100 percent of the impounding floor area shall be determined by equation C-9 in the report ``Evaluation of LNG Vapor Control Methods," 1974, or an alternate model which meets the requirements of paragraphs (ii) through (iv) in Sec. 193.2057(c)(2).

(iii) After spill flow is terminated, the rate of vaporization is vaporization of the remaining spillage, if any, from boiling by heat transfer from contact surfaces that are reducing in area and temperature as a function of time.

(iv) Vapor detention space is all space provided for liquid impoundment and vapor detention outside the component served, less the volume occupied by the spilled liquid at the time the vapor escapes the vapor detention space.

(2) The boiling rate of LNG on which dispersion distance is based is determined using the weighted average value of the thermal properties of the contact surfaces in the impounding space determined from eight representative experimental tests on the materials involved. If surfaces are insulated, the insulation must be designed, installed, and maintained so that it will retain its performance characteristics under spill conditions.

(e) Planned vapor control. An LNG facility need not have a dispersion exclusion zone if the Administrator finds that compliance with paragraph (a) of this section would be impractical and the operator prepares and follows a plan for controlling LNG vapor that is found acceptable by the Administrator. The plan must include circumstances under which LNG vapor is controlled to preclude the dispersion of a flammable mixture from the LNG facility under all predictable environmental conditions that could adversely affect control. The reliability of the method of control must be demonstrated by testing or experience with LNG spills.

The 22 December 1998 NPRM contained the exact same paragraph §(d)(1) language, but proposed to eliminate the complex insulation of contact surface conditions.

"§Sec. 193.2059 Flammable vapor-gas dispersion protection.

* * * *

*(c) * * **

(2) Dispersion conditions are a combination of those which result in longer predicted downwind dispersion distances than other weather conditions to the site at least 90 percent of the time, based on U.S. Government weather data, or as an alternative where the model used gives longer distances at lower wind speeds, Atmospheric Stability (Pasquill Class) F, wind speed = 4.5 miles per hour (2.01 meters/sec) at reference height of 10 meters, relative humidity equals 50.0 percent, and atmospheric temperature = 80 deg. F(27 deg. C).

(3) The elevation for contour (receptor) output H = 0.5 meters.

(4) A surface roughness factor of 0.03 meters shall be used. higher values for the roughness factor may be used if it can be shown that the terrain both upwind and downwind of the vapor cloud has dense vegetation and that the vapor cloud height is more than ten times the height of the obstacles encountered by the vapor cloud.

*(d) * * **

(1) Vaporization results from the spill caused by an assumed rupture of a single transfer pipe (or multiple pipes designed to deliver the same flow) which has the greatest

overall flow capacity, discharging at the maximum potential capacity, in accordance with the following conditions:

(i) The rate of vaporization is not less than the sum of flash vaporization and vaporization from boiling by heat transfer from contact surfaces during the time necessary for spill detection, instrument response, and automatic shutdown by the emergency shutdown system but, not less than 10 minutes plus, in case of impounding systems for LNG storage tanks with side or bottom penetration, the time necessary for the liquid level in the tank to reach a level of penetration or equilibrate with the liquid impounded. In the case of storage tanks with an internal shutoff valve, the time necessary for spill detection and response of not less than one (1) hour must be used.

* * * * *

(2) If surfaces are insulated, the insulation must be designed, installed, and maintained so that it will retain its performance characteristics under spill conditions.

Unfortunately, after the code harmonizing project of 1999/2000, the same section of the 193 DOT code was changed to read;

§193.2059 Flammable vapor-gas dispersion protection.
...(c) The design spill shall be determined in accordance with section 2-2.3.3 of ANSI/NFPA 59A. [Amdt. 193-17, 65 FR 10959, Mar. 1, 2000]

Paragraph (d) was completely dropped. This is an obvious oversight error, since only a subparagraph was intended to be deleted per the NPRM.

The NFPA59A (2001) equivalent section 2-2.3.3 referenced above specifically stated :

§2.2.3.3 The spacing of an LNG tank impoundment to the property line that can be built upon shall be such that, in the event of an LNG spill specified in 2.2.3.5, an average concentration of methane in air of 50 percent of the lower flammability limit (LFL) does not extend beyond the property line that can be built upon, in accordance with calculations using one of the following:
..(a) ... DEGADIS..
..(b) ... FEM3A..
..(c) A model that incorporates the following:
(1) Takes into account physical factors influencing LNG vapor dispersion, including, but not limited to, gravity spreading, heat transfer, humidity, wind speed and direction, atmospheric stability, buoyancy, and surface roughness
(2) Has been validated by experimental test data appropriate for the size and conditions of the hazard to be evaluated
(3) Is acceptable to the authority having jurisdiction

The computed distances shall include calculations based on one of the following:
(1) The combination of wind speed and atmospheric stability that can occur simultaneously and result in the longest predictable downwind dispersion distance that is exceeded less than 10 percent of the time
(2) The Pasquill–Gifford atmospheric stability, Category F, with a 4.5-mph (2-m/sec) wind speed

The computed distances shall be based on the actual liquid characteristics and the maximum vapor outflow rate from the vapor containment volume (the vapor generation rate plus the displacement due to liquid inflow).
The effects of provisions for detaining vapor or otherwise mitigating flammable vapor hazards (e.g., impounding surface insulation, water curtains, or other methods) shall be permitted to be considered in the calculation where acceptable to the authority having jurisdiction.

2.2.3.5 The design spill shall be determined in accordance with Table 2.2.3.5 Design Spill
An excerpt from that table is:

For "impounding areas serving only vaporization, process and LNG transfer areas", the design spill is "the flow from any single accidental leakage source".

As an interesting sidelight, the NFPA language above, contains the word "ONLY". Hence sub impounding areas within storage tank dikes may not actually be covered by this language. Unfortunately, there is not language anywhere in either code that specifically covers the subdike issue.

Excerpts from the Amendments to DOT 193
published in the Federal Register Amdt. 193-17, 65 FR 10959, Mar. 1, 2000

... The incorporation by reference of this standard will allow the LNG industry to use the latest technology, materials, and practices while maintaining the current level of safety.

The current Federal safety standards for LNG facilities were developed as a requirement of the Pipeline Safety Act of 1979, now re-codified in 49 United States Code Section 60103. In 1979, Congress determined that the public would be better served if the US Department Of Transportation (DOT) developed its own standards for the LNG industry. Prior to July 1, 1976, no Federal standards for LNG facilities existed. The current standard, which addresses LNG facilities used in gas pipeline transportation, was issued as a Final Rule on February 11, 1980 [45 FR 9203] and now appears at 49 CFR Part 193. Between July 1, 1976, and February 11, 1980, LNG facilities were required to comply with ANSI/NFPA 59A (1972 edition) and Part 192.

A report issued on July 31, 1978, by the General Accounting Office titled "Liquefied Energy Gases" highlighted some of the safety concerns in the transportation and storage of LNG. Foremost among those were:

(1) protection of persons and property near an LNG facility from thermal radiation caused by ignition of a major spill of LNG,
(2) protection of persons and property near an LNG facility from dispersion and delayed ignition of a natural gas cloud arising from a major spill of LNG, and
(3) reduction of the potential for a catastrophic spill of LNG.

RSPA identified many deficiencies in the pre-1980 LNG standards which needed to be corrected to reduce the potential for a major spill of LNG and provide an acceptable level of safety. Because of the difference in format and the need for regulatory language to facilitate enforcement, a few sections of ANSI/NFPA 59A were rewritten for their adoption in Part 193.

There have been significant changes in the ANSI/NFPA 59A since 1980. The 1996 edition of the ANSI/NFPA 59A includes the latest developments in LNG facility design and safety. Many of these developments have not been incorporated into current Part 193. The format and language of the ANSI/NFPA 59A has also changed significantly, over the years, to facilitate enforcement. ANSI/NFPA 59A is revised on a regular basis, and the revision process includes input from a wide variety of experts and a broad representation of interests.

RSPA has been very active in incorporating by reference voluntary consensus standards in its pipeline safety regulations. RSPA has participated for many years on several voluntary committees that develop consensus standards, including the ANSI/NFPA 59A technical committee. The existing Part 193 references provisions of ANSI/NFPA 59A in eight different locations. Recent amendments to the LNG regulations (February 25, 1997; 62 FR 8402 and August 1, 1997; 62 FR 41311) have brought Part 193 closer to ANSI/NFPA 59A. Unlike older editions of the ANSI/NFPA 59A, text in the current standard is in a regulatory format that makes it more suitable for incorporation by reference. RSPA is adopting the 1996 version of the ANSI/NFPA 59A. When the standard is revised in the future, RSPA will incorporate by reference the revised versions, as appropriate.

RSPA published an NPRM [63 FR 70737; December 22, 1998], proposing to replace most LNG requirements for siting, design, construction, equipment, and fire protection in Part 193 by referencing the American National Standards Institute (ANSI), National Fire Protection Association (NFPA) Standard 59A (1996 edition), titled "Standards for the Production, Storage and Handling of Liquefied Natural Gas (LNG)"

Section 193.2059 - Flammable vapor-gas dispersion protection

In the NPRM we proposed to:

(1) retain minimum 10 minute spill duration for vaporization design rate;
(2) delete planned vapor control;
(3) retain 2.5% lower flammable concentration limit at the outer boundary of flammable vapor; and
(4) add one hour time duration necessary for spill detection and response for tanks with an internal shutoff valve.

AGA, NEGA, NFPA, two operators and the Iowa Utility Board each offered comments against one or more of those requirements. AGA, NEGA and one operator commented that NFPA standard 59A does not set a 10 minute spill duration limit so that operators can take advantage of technology by using controls that can provide response time in less than 10 minutes.

NEGA said that by deleting planned vapor control to mitigate the emerging vapor from a design LNG spill increases burden on the operator and denies the operator alternative credit.

The Iowa Utility Board supported the proposal to retain the 2.5% lower limit for gas concentration. NFPA said that the 5% lower flammability limit is sufficient because the model takes concentration variations into account, and our requirement is too conservative.

One operator said there is no rationale for a one hour response time for spill detection for a tank with an internal shutoff valve.

Response

(1) We agree with the commenters that with the current technology and control system operators can respond to spills in less than 10 minutes. We have revised this requirement to agree with the ANSI/NFPA 59A standard that 10 minute spill time can be reduced if the operator can demonstrate by instrument surveillance and emergency shutdown system that less than 10 minutes is needed to respond to spills.

(2) We have deleted, as we proposed in the NPRM, the planned vapor control requirement from the regulations. We do not believe, any facility would opt for this alternative. In this final rule planned vapor control requirement will still be allowed as an alternative through a waiver.

(3) We have retained the requirement for 2.5% lower flammable limit (LFL) concentration at the outer boundary of flammable vapor to provide a reasonable margin of safety. The DEGADIS model predicts only average concentration of LNG. Because vapor does not disperse uniformly, pockets of 5% LFL concentration could be adjacent to the average distance line predicted by the model. In other words, the model can under predict the actual concentration of LNG. Because many assumptions go in the formula, the distances predicted are not always accurate. Using a 2:1 safety margin was suggested by those who developed this model. On August 19, 1999, the NFPA 59A committee discussed this issue in great detail and voted to revise ANSI/NFPA 59A standard to require a 2.5% LFL in lieu of 5% LFL. Therefore, we see no need to revise the current concentration level in the regulations. In this final rule, we are allowing use of the FEM3A vapor dispersion model as an alternate to DEGADIS. The FEM3A model accounts for additional cloud dilution which may be caused by the complex flow patterns induced by tank and dike structures. Dispersion distances are calculated in accordance with this model described in Gas Research Institute report GRI-96/0396.5, "Evaluations of Mitigation Methods for Accidental LNG Releases. Volume 5: Using FEM3A for LNG Accident Consequence Analyses."

(4) ANSI/NFPA 59A standard also requires a one hour duration for spills from tanks fitted with internal shutoff valves. We have referenced ANSI/NFPA 59A for determining design spills.

Proposed requirement on determining Vaporization design rate under 193.2059(d) has been deleted in this rule to allow operators more flexibility in computing.

Hence,
9. Section 193.2059 is amended to read as follows:
....
§ 193.2059 Flammable vapor-gas dispersion protection.
Each LNG container and LNG transfer system must have a dispersion exclusion zone in accordance with section 2-2.3.2 of ANSI/NFPA 59A with the following exceptions:
(a) Flammable vapor-gas dispersion distances must be determined in accordance with the model described in the Gas Research Institute report GRI-89/0242, "LNG Vapor Dispersion Prediction with the DEGADIS Dense Gas Dispersion Model." Alternatively, in order to account for additional cloud dilution which may be caused by the complex flow patterns induced by tank and dike structure, dispersion distances may be calculated in accordance with the model described in the Gas Research Institute report GRI 96/0396.5, "Evaluation of Mitigation Methods for Accidental LNG Releases. Volume 5: Using FEM3A for

LNG Accident Consequence Analyses". The use of alternate models which take into account the same physical factors and have been validated by experimental test data shall be permitted, subject to the Administrator's approval.
(b) The following dispersion parameters must be used in computing dispersion distances:
(1) Average gas concentration in air = 2.5 percent.
(2) Dispersion conditions are a combination of those which result in longer predicted downwind dispersion distances than other weather conditions at the site at least 90 percent of the time, based on figures maintained by National Weather Service of the U.S. Department of Commerce, or as an alternative where the model used gives longer distances at lower wind speeds, Atmospheric Stability (Pasquill Class) F, wind speed = 4.5 miles per hour (2.01 meters/sec) at reference height of 10 meters, relative humidity = 50.0 percent, and atmospheric temperature = average in the region.
(3) The elevation for contour (receptor) output H = 0.5 meters.
(4) A surface roughness factor of 0.03 meters shall be used. Higher values for the roughness factor may be used if it can be shown that the terrain both upwind and downwind of the vapor cloud has dense vegetation and that the vapor cloud height is more than ten times the height of the obstacles encountered by the vapor cloud.
(C) The design spill shall be determined in accordance with section 2-2.3.3 of ANSI/NFPA 59A.

To illustrate the DOT attempt to overcome recognized ambiguities resulting from the Harmonizing code effort.

Excerpts from the Federal Register: March 10, 2004 (Volume 69, Number 47) [Page 11330-11337]

Background

On March 1, 2000, we published a final rule document amending the safety regulations in 49 CFR part 193, which apply to LNG facilities used in gas pipeline transportation (65 FR 10950). That document replaced many part 193 siting, design, construction, equipment, and fire protection requirements with references to a consensus standard, NFPA 59A, ``Standard for the Production, Storage, and Handling of Liquefied Natural Gas (LNG)'' (1996 edition). Until then, part 193 referenced NFPA 59A (1996 edition) in only a few instances concerning siting, design, and fire protection.

An amendment to Sec. 193.2005, ``Applicability,'' inadvertently implied that LNG facilities existing on March 31, 2000 (hereafter, ``existing LNG facilities''), were exempt from part 193 operation, maintenance, and fire protection standards. After recognizing this ambiguity, we published a notice of proposed rulemaking (NPRM) to revise Sec. 193.2005 (68 FR 23272; May 1, 2003). In the NPRM, we also proposed to revise incorrect cross-references that resulted from the March 1, 2000, final rule to establish minimum standards for fire drills used in fire protection training, and to require that operators review their part 193 plans and procedures at least once a year. We further proposed to update all part 193 references to NFPA 59A to the 2001 edition of that standard.

Presenters Resume

Dr. Zinn (born and raised in Hope Arkansas like another famous American!) graduated with a PhD in Industrial Engineering from the University of Oklahoma. Thereafter, Dr. Zinn was a faculty member at the University of Texas. He is a registered Professional Engineer in Texas.

Dr. Zinn began his engineering career with providing process hazards analysis, risk assessments, safety systems design reviews, and site safety audits for liquefied natural gas (LNG) and liquefied petroleum gas (LPG) facilities with Energy Analysts (later renamed Quest) and then with Jones & Neuse of Austin Texas (later renamed RMT)

During the past thirty years, he has provided process safety engineering services for peaks having & baseload production facilities, receiving terminals, storage, and transportation systems, both domestically and internationally. Dr Zinn has taught LNG / LPG process safety courses worldwide.

LNG Fire Fighting at Texas A&M: Training and Research

M. Sam Mannan and Harry H. West
Mary Kay O'Connor Process Safety Center
Chemical Engineering Department
Texas Engineering Extension Service
Texas A&M University

Kirk Richardson and Mike Wisby
Brayton FireTraining School
Emergency ServicesTraining Dept

AIChE Spring Meeting, April 2005
5th Topical Conference on Natural Gas Utilization
Atlanta April 10-14, 2005

Abstract

The Brayton Fire Training School at Texas A&M has been training LNG fire fighters for over 25 years. The older LNG area has recently been replaced with new facilities for LNG fire fighting and vapor control training, with the first LNG training sessions conducted in October. The LNG area features an ability to demonstrate the proper use of various fire mitigation apparatus, including dry chemical, foam and vapor control systems.

Various LNG Fire research and development projects that benefit from the capability of actual LNG fire conditions, include

- Testing and demonstration of gas detection equipment and fire thermal radiation sensors.
- Updating foam application rate information for system designers, particularly with all the new formulation foams now available.
- Development and validation of long trench consequence models for vapor dispersion and fire scenarios.
- Preliminary tests of experimental instrumentation systems for the proposed industry sponsored large scale LNG fires tests. Experimental design ideas are underway.

Introduction

BP and Texas A&M developed the new LNG Fire training facility in College Station, Texas. BP is funding the construction of the main LNG fire area, replacing the old LNG fire training area that was used for the past 25 years. In October the first class of industrial emergency responders attended LNG courses. Over 10 LNG fire training sessions are currently scheduled for 2005, with some dedicated to specific company staff and others on an open enrollment basis.

To increase the versatility of the facilities, donations of equipment were also made by Angus Fire, Hale Products, International Paint, Knowsley SK, Micropack Detection and Zellweger Analytics.

Facilities

This LNG fire area is the world's first full integrated detection and suppression LNG testing ground where workshops, training, equipment tests and experiments on LNG vapor clouds, fire and fire control methods can be conducted.

At present, the Texas A&M Fire school has the following concrete lined pits dedicated to LNG training sessions;
2 Square pits
    Small: 10 ft. x 10 ft. x 4 ft.
    Large: 33 ft. x 21 ft. x 4 ft.
An L Shaped Trench pit with dimensions: 26 ft. x 4 ft. x 4 ft.
Several larger and deeper pits are under design and construction.

Each training session typically consumed 20,000 gallons of LNG transported to the site by tanker truck.
Classes and demonstrations. illustrate un-ignited LNG vapors, vapor cloud control, ignition and LNG pool fires.

Sitting at the edge of the large pit are two Angus Turbex fixed high expansion foam generators, specifically designed to control LNG spills. Medium and low expansion foams are also available using mobile appliances. The newer compressed air foam is also available.

A variety of instrumentation is available, including Micropack visual flame detection cameras and several types of thermal & gas detection sensors. A new style of fire hydrant with oscillating water monitors can be used to give protection to nearby exposures. Several fireproofing materials are also available for testing.

The Experimental Program

Although there are some available engineering data on dry chemical water curtains and foam application rates that can be used to design fire response systems, very little quantitative data has been published in the last 20 years. Chemical engineering graduate students from the Mary Kay O'Connor Process Safety Center will design scientifically based experiments, gather the data and provide analysis, thereby validating older information as well as extending the fire control data base to new formulations and equipment.

Foam has a top priority in the experimental design program, due to the rapid advancement in foam formulations. Foam has a big advantage in two ways; the dispersion of vapors. Foam is a gentle way of warming LNG vapors, causing the escaping vapors to rise upward away from people and nearby ignition sources. If the vapors should ignite, foam can be an effective fire control mechanism. The foam bubbles insulate the LNG from the hot flame above, controlling the release of vapor. As the heat breaks down the foam blanket, more foam must be applied.

Selection of foam concentrate, expansion ratio, start-up delay time and drainage properties from the wide range of high expansion foams available is very important for LNG applications because the foams that drain much more quickly or are too slow in initial response do not have the capability to provide fire control. Similarly, weak foams are blown away in the wind.

Preliminary qualitative results indicate that the compressed air foam significantly reduce the thermal hazard zone as compared with certain a high expansion foam. More quantitative experiments are planned.

Measurements of dry chemical application rates, combined dry chemical & foam applications, various foam application rates and water curtain effectiveness are planned.

Example of a Specific Experiment: Vapor hold up

The estimation of LNG vapor cloud exclusion distances in the USA LNG DOT/OPS regulations (49CFR193, and NFPA section 2) requires modeling the hazard scenarios of LNG flowing into spill containment sumps and storage tank diked areas.

LNG analysts use a dense gas dispersion model to predict the downwind travel of LNG vapors. The DOT/OPS rules specify DEGADIS and FEM3A consequence models. DEGADIS or any other proprietary model is typically based upon vapor dispersion over flat terrain. Hence, the spill source term model must provide an LNG vapor or LNG vapor mixture concentration to the DEGADIS input file.

Figure 1 provides a sketch of the two basic vapor hold up models currently in use; namely the plug flow model and the partial mixing model. The controversy on vapor hold-up models is clearly stated by Dr. Havens (1,2).

Project Outline

A series of experiments are proposed in conjunction with the LNG training facilities at the Texas A&M Brayton fire school. Using LNG being spilled into the available test pits during LNG vapor training exercises, graduate students will obtain experimental data on LNG vapor hold under a range of conditions. Figure 2 presents a sketch of the preliminary experimental design.

Analysis of the experimental data using FEM3A, FLACS and FLUENT Computational Fluid Dynamic (CFD) consequence computer modeling tools are envisaged. The raw data from these experiments will also be forwarded to Professors Spicer and Havens at the University of Arkansas to aid in their wind tunnel dispersion work.

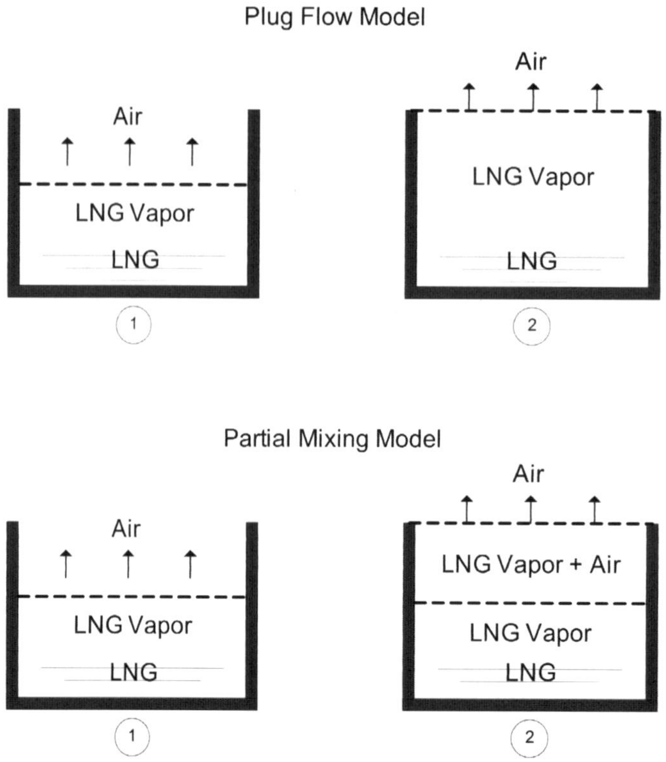

Figure 1 LNG vapor hold up models

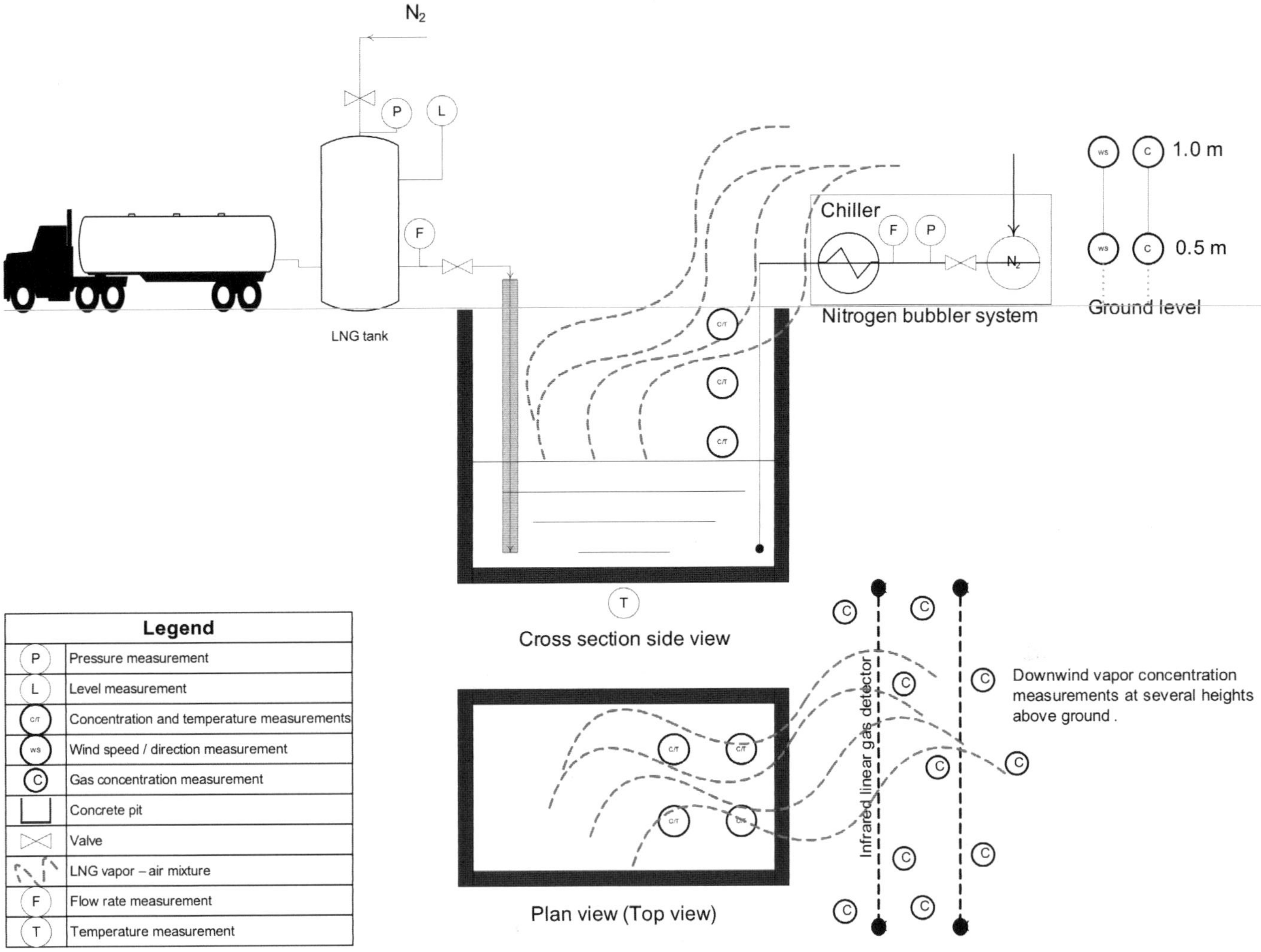

Figure 2 Sketch of the experimental design

Another Example Project: Preparation for Large Pool Fire test

The estimation of LNG thermal radiation exclusion distances in the USA LNG DOT/OPS regulations (49CFR193, and NFPA section 2) requires modeling several hazard scenarios, including LNG flowing into spill containment sumps and storage tank diked areas. Hazard analysis of LNG Tanker accidents requires a prediction of the potential vulnerability zones from a fire which may result from the maximum pool to which a significant LNG spill might attain whilst burning, following a tank puncture.

LNG analysts use a tilted solid cylinder fire model to predict the downwind distance to several thermal radiation hazard levels. The DOT/OPS rules and NFPA rules specify FIRE3 consequence model. A separate pool spreading model is used to define the fire pool size. Hence, the spill source term coupled with the pool spreading model must provide an LNG liquid pool size to the LNGFIRE3 input file.

Figure 3 provides a sketch of the basic tilted cylinder fire model currently is use.

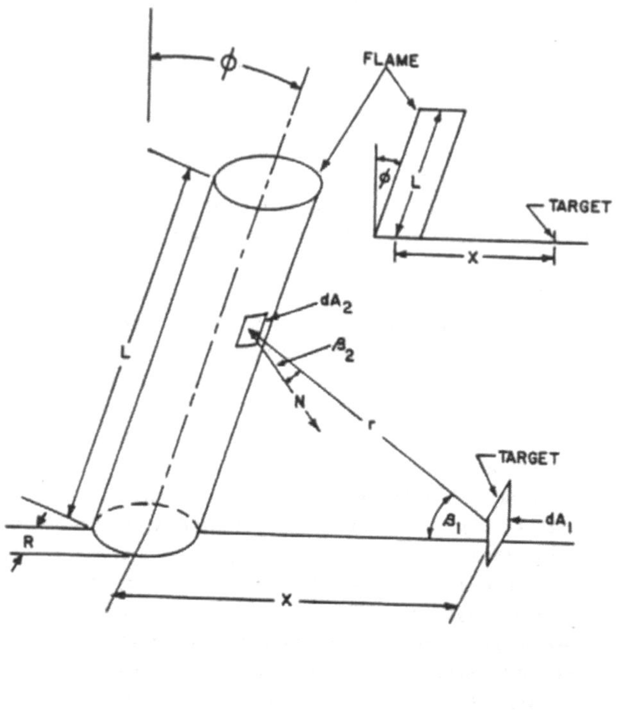

Figure 3 The solid tilted cylinder representation for Pool Fires

Computational Fluid Dynamics (CFD) models, such as FLUENT or VULCAN have not been used in LNG hazard analysis, but it is anticipated that future safety analysts may begin to use these more complex models

Large Fires (up to 1000 meter diameter or even larger) have not been witnessed in the LNG industry. There is some controversy about the surface flux that is emitted from very large LNG fires. Gaz de France reported higher flux values at their large Montour fire tests (35 m^3), but then noted that it may have been associated with certain instrument problems.

Project Outline

Phase I

LNGFIRE3 parametric studies

Develop tables of results for FERC cargo tank spill 1 m diameter hole only at first, then add other hole size info later. Varying all available parameters, including evaporation rate, pool size, etc. For purpose of analysis, identify target distance as one meter above grade level, 5 kW/m2. Then develop some curves for other target height and distances later.

Phase II

Compare with published results, Fay, Quest Canary, Havens, Koopman, FERC modified ABS report, ABS original study, others as appropriate, including various recently submitted DEIS from FERC and USCG. Many of these reports are available to the Texas A&M students on the MKOPSC server.

Phase III

Simultaneously, a literature survey on all fire thermal tests and models, including a survey of LNG safety analysts and commercial models will be conducted to identify the range of current opinion on modeling this scenario. Identify other Pool Fire models, used for other hydrocarbon fires, including the HSE models.

Phase IV

From the literature survey, develop available raw fire experimental data as a database that may be used for model validation purpose.

Phase V

Analysis of the experimental data using FLUENT Computational Fluid Dynamic (CFD) consequence computer modeling tools. Other CFD models may be used as available, such as VULCAN, used by Sandia.

Phase VI

A series of experiments are proposed in conjunction with the LNG training facilities at the Texas A&M Brayton fire school. The key factor would be the evaluation of various types and manufactures of thermal radiation sensors, thereby understanding the range of accuracy one may anticipate for large gas fires.

Phase VII

Detailed experimental design including cost estimates will be developed for a cold gas, more controlled fire experiment prior to any large scale LNG spills. The advantage of this step is to avoid the problems of the Montour test. The experiment will be modeled after the pool fire test design of experiment using Welker type multiple small orifice to simulate the large pool fire.

References

Response to Weaver's Cove's Rebuttal of Jerry Havens' Two Part Report of August 25, 2004, to the City of Fall River, Massachusetts: Analysis of Draft Environmental Impact Statement for the Weaver's Cove LNG Project, filed in the FERC

Effects of secondary containment on source term modeling, WS Atkins for the Health & Safety Executive, contract research report 324/2001, HMSO, London 2001.

LNG Vapor Cloud Exclusion Zones for Spills into Impoundments, Jerry Havens and Tom Spicer, Chemical Hazards Research Center, University of Arkansas, AICHE Spring Meeting 2005.

Biography of Speaker

Dr. Harry West is an independent consulting chemical engineer, specializing in process safety, process automation technology and LNG issues. He began his process safety and automation safety interest over 30 years ago in the developing LNG industry. Dr. West was President of EDG Engineering, a worldwide SCADA/DCS system integrator, ranked 11th largest in the early 1980s. Dr. West was also a member of the founding group of ChemShare (a flow-sheet simulation software company now renamed WinSim), a co-founder of Energy Analysts (renamed Quest Consultants, an international process safety consultancy) and a co-founder of Environmental Abatement Associates.

Dr. West's academic career began as a Chemical Engineering faculty member at the University of Oklahoma and the University of Pittsburgh. Currently Dr. West is an Adjunct Professor of Chemical Engineering in the Process Safety center of the Chemical Engineering Department at Texas A&M University. He also regularly teaches engineering seminars at the Institute Technology Bandung in Indonesia.

Safety Implications of a Large LNG Spill over Water

Michael Hightower
Distinguished Member of the Technical Staff
Sandia National Laboratories
P.O. Box 5800, MS0708
Albuquerque, New Mexico 87185

AIChE Spring Meeting, April 2005
5th Topical Conference on Natural Gas Utilization
Atlanta April 10-14, 2005

Background

The increasing demand for natural gas in the U.S. could significantly increase the number and frequency of marine LNG imports. While many studies have been conducted to assess the consequences and risks of potential LNG spills, the increasing importance of LNG imports suggests that consistent methods and approaches be identified and implemented to help ensure protection of public safety and property from a potential LNG spill.

For that reason, the U.S. Department of Energy (DOE), Office of Fossil Energy, requested that Sandia National Laboratories (Sandia) develop guidance on a risk-based analysis approach to assess and quantify potential threats to an LNG ship, the potential hazards and consequences of a large spill from an LNG ship, and review prevention and mitigation strategies that could be implemented to reduce both the potential for and the risks of an LNG spill over water. Specifically, DOE requested:

- An in-depth literature search of the experimental and technical studies associated with evaluating the safety and hazards of an LNG spill from an LNG cargo tank ship;
- A detailed review of four recent spill modeling studies related to the safety implications of a large-scale LNG spill over water;
- Evaluation of the potential for breaching an LNG ship cargo tank, both accidentally and intentionally, identification of the potential for such breaches and the potential size of an LNG spill for each breach scenario, and an assessment of the potential range of hazards involved in an LNG spill; and
- Development of guidance on the use of modern, performance-based, risk management approaches to analyze and manage the threats, hazards, and consequences of an LNG spill over water to reduce the overall risks of an LNG spill to levels that are protective of public safety and property.

To support this effort, Sandia worked with the U.S. DOE, the U.S. Coast Guard, LNG industry and ship management agencies, LNG shipping consultants, and government intelligence agencies to collect background information on ship and LNG cargo tank designs, accident and threat scenarios, and standard LNG ship safety and risk management operations. The information gathered was used to develop accidental and intentional LNG cargo tank breach scenarios, for modeling of potential spill hazards, and as the basis for analysis to determine the extent and severity of LNG spill consequences. Based on analysis of the modeling results, three consequence-based hazard zones were identified. In addition, risk reduction and mitigation techniques were identified to reduce impacts on public safety and property.

The results from the Sandia study were reviewed by an external peer review panel and made available to the public in a recent Sandia report "Guidance on Risk Analysis and Safety Implications of a Large Liquefied Natural Gas (LNG) Spill Over Water" [Hightower 2004]. The overall role of the report was to provide communities and agencies with suggestions and guidance on how to consider various issues, including terminal location and site conditions, operational conditions, environmental effects, and safety and security issues and measures to achieve appropriate levels of protection of public safety and property for LNG imports. This paper provides an overview of the conclusions and recommendations from that study.

Safety Analysis and Risk Management of Large LNG Spills over Water

Several analysis techniques are available to help identify the possible hazards to the public from activities that occur in modern society such as hazardous material manufacturing, transportation, or use. These include vulnerability analyses, safety analyses, and consequence analyses. These techniques have become elements of modern, performance-based risk analysis and risk management approaches. In modern risk analysis, the risks associated with an event are commonly defined as a function of the following four elements:

- The probability of an event — such as an LNG cargo tank breach and spill;
- The hazards associated with the event — such as thermal radiation from a fire due to an LNG spill;
- The consequences of the event — such as the thermal damage from a fire, and
- The effectiveness of systems for preventing the event or for mitigating the hazards and consequences — such as any safety/security systems.

Therefore, discussions of the safety issues of hazardous operations, such as LNG imports and transportation, should not only include discussions of vulnerability and safety analyses, but also include risk management considerations that can help reduce the possibility or impacts of an event and improve the overall safety and protection of the public and property from the event. In the sections below we discuss the results of the vulnerability and safety analyses conducted, as well as discuss risk and safety management options that should be considered to improve overall public safety from possible spills during LNG import operations.

<u>LNG Breach, Spill, and Hazard Analyses</u>

Currently, the potential for an LNG cargo tank breach, whether accidental or intentional, the dynamics and dispersion of a large spill, and the hazards of such a spill, are not fully understood, for two primary reasons. First, the combination of current LNG ship designs and safety management practices for LNG transportation have reduced LNG accidents to the extent that there is little historical or empirical information on the consequences of breaches or large spills. Second, current experimental data on LNG spill dynamics and its dispersion over water address spill sizes that are more than a factor of one hundred smaller than spill sizes currently being postulated for some intentional events. Variations in site conditions, LNG ship designs, and environmental conditions further complicate hazard predictions.

The lack of large-scale experimental data forces analysts to make many assumptions and simplifications in calculating the breach of an LNG cargo tank, the resulting spill dispersion, and associated thermal hazards. For example, an evaluation of four recent LNG spill studies showed significant differences in thermal hazard estimates due to the differences in assumptions and modeling approaches used in each analysis [Lehr and Simicek-Beatty 2003][Fay 2003][Quest 2003][Vallejo 2003].

Although current spill assessment and modeling techniques and validation of models against large-scale LNG spill data have limitations, the guidance provided is applicable to performance-based hazard and risk management approaches. Such approaches can be used in conjunction with existing spill and hazard analysis techniques, and safety and security methods, to assess and reduce the risks to both public safety and property caused by an LNG spill over water. Guidance is provided on the use of existing analysis techniques applied to site-specific conditions for increasing confidence in the management of hazards and risks. As additional LNG spill data are obtained and hazard analysis models are improved, they can be incorporated into future risk analysis guidance.

LNG Cargo Tank Breach Analysis

Based on available information, a range of historically credible and potential accidental and intentional events was identified that could cause an LNG cargo tank breach and spill. Modern finite element modeling and explosive shock physics modeling were used to estimate a range of breach sizes for credible accidental and intentional LNG spill events, respectively.

From these analyses, the sizes of LNG cargo tank breaches for accidents were estimated to be less than 2 m^2 (1.6 m diameter). For intentional events, the size of the hole depends on its location on the ship and the source of the threat. Intentional breaches were estimated at 2 m^2 to approximately 12 m^2 (1.6 m – 4 m diameter) with nominal sizes of about 5 – 7 m^2 (2.5 m – 3.0 m diameter). These sizes are smaller than those used in many recent studies. Although smaller, the breach sizes estimated can still lead to large LNG spills.

Using structural fracture mechanics analyses, the potential for cryogenic damage to the LNG ship and other LNG cargo tanks was also evaluated. Based on these analyses, the potential for cryogenic damage to the ship cannot be ruled out, especially for large spills. The degree and severity of damage depends on the size and location of the breach. Sandia considered cryogenic damage to the ship's structure and concluded that releases from no more than two or three tanks would be involved in a spill that occurs due to any single incident. This cascading release of LNG was analyzed and is not expected to increase significantly the overall fire size or hazard ranges, but the expected fire duration will increase. Hazard analysis and risk prevention and mitigation strategies should consider this in assessing public safety and damage to property.

Spill and Dispersion Analysis

The variability in existing LNG spill and dispersion/thermal hazard modeling approaches is due to physical limitations in the models and the lack of validation with large-scale spill data.

Obtaining experimental data for large LNG spills over water would provide needed validation and help reduce modeling uncertainty. Because extrapolation of existing models will be necessary for analysis of potentially large spills, models should be used that invoke as much fundamental physics as possible. Based on the evaluations presented in the report, several types of models currently exist to assess hazards. Models should be used only where they are appropriate and understood to ensure that the results increase confidence in the analysis of the hazards and risks to public safety and property.

In higher hazard zones, where analysis reveals that potential impacts on public safety and property could be high and where interactions with terrain or structures can occur, modern, CFD models, as listed in Table 1, can be used to improve analysis of site-specific hazards, consequences, and risks. Use of these models is suggested because many of the simpler models have limitations that can cause greater uncertainties in calculating liquid spread, vapor dispersion, and fire hazards. CFD models have their own limitations and should be validated prior to use. Further refinement of CFD models will continue to improve the degree of accuracy and reliability for consequence modeling.

Table 1: Models for Improved Analysis of an LNG Spill in High Hazard Areas

APPLICATION	IMPROVED MODELING APPROACHES
Breach Analysis	Finite element codes for modeling accidental ship collisions & shock physics codes for modeling intentional breaches.
Tank Emptying	Modified orifice model that includes the potential for LNG leakage between hulls.
Structural Damage Modeling	Coupled spill leakage, fluid flow, and fracture mechanics codes for modeling ship structural damage & damage to LNG cargo tanks.
Spreading	CFD codes for modeling spread of cryogenic liquids on water.
Dispersion	CFD codes for modeling dispersion of dense gases.
Fire	CFD codes for modeling fire phenomena, including combustion, soot formation, and radiative heat transfer.

While these studies provide insight into appropriate models to use, additional factors should be considered in applying models to a specific problem. These include model documentation and support, assumptions and limitations, comparison and validation with data, change control and upgrade information, user support, appropriate modeling of the physics of a spill, modeling of the influence of environmental conditions, spill and fire dynamics, and model peer review.

Hazards Analysis and Public Safety Impacts

Current LNG spill and dispersion modeling and analysis techniques have limitations. In addition, variations exist in location-specific conditions that influence dispersion, such as terrain, weather conditions, waves, currents, and the presence of obstacles. Therefore, it is sensible to provide guidance on the general range of hazards for potential spills rather than suggest a specific, maximum hazard guideline.

To assess the general magnitude of expected hazard levels, a limited sensitivity analysis was performed using simplified models for a range of spill volumes. The spill volumes were based on potential breaches from credible accidental and intentional threats and approximated about 12,500 m^3 for each tank breached. While not conducted for a specific site, the analyses provide examples of general considerations for hazards and risks. The general results are presented in Table 2 for a set of possible breaches and an ensuing fire, and Table 3 for a spill without ignition and dispersion under nominal wind conditions of 2.33 m/s and nominal atmospheric stability. Most of the events considered should provide an ignition source and the likelihood of a large unignited release of LNG is unlikely. A vapor cloud dispersion of LNG will also have a tendency to create a more linear hazard zone rather than a generally circular hazard zone from an LNG spill fire. The 37.5 kW/m^2 and 5 kW/m^2 values shown in Table 2 are thermal flux values commonly recognized for defining hazard distances. The 37.5 kW/m^2 is a level suggesting severe structural damage and major injuries if expected for over 10 minutes. The 5 kW/m^2 is a level suggesting second-degree skin burns on exposed skin if expected for periods of over about 20 seconds, and is the value suggested as the protection standard for people in open spaces. These values were considered as defining the bounds between high, medium, and low hazard zones from a fire.

Table 2: Potential Thermal Hazard Distances for Several Possible Breaching Events

HOLE SIZE (m^2)	TANKS BREACHED	DISCHARGE COEFFICIENT	BURN RATE (m/s)	SURFACE EMISSIVE POWER (kW/m^2)	POOL DIAMETER (m)	BURN TIME (min)	DISTANCE TO 37.5 kW/m^2 (m)	DISTANCE TO 5 kW/m^2 (m
ACCIDENTAL EVENTS								
1	1	.6	3X10^{-4}	220	148	40	177	554
2	1	.6	3X10^{-4}	220	209	20	250	784
INTENTIONAL EVENTS								
5	3	.6	3 x 10^{-4}	220	572	8.1	630	2118
5*	1	.6	3 x 10^{-4}	220	330	8.1	391	1305
5	1	.9	3 x 10^{-4}	220	405	5.4	478	1579
5	1	.6	2 x 10^{-4}	220	395	8.1	454	1538
5	1	.6	3 x 10^{-4}	350	330	8.1	529	1652
12	1	.6	3 x 10^{-4}	220	512	3.4	602	1920

* nominal case

Table 3: Potential Lower Flammability Limit (LFL) Distances for Possible Vapor Dispersions

HOLE SIZE (m^2)	TANKS BREACHED	POOL DIAMETER (m)	SPILL DURATION (min)	DISTANCE TO LFL (m)
Accidental Events				
1	1	148	40	1536
2	1	209	20	1710
Intentional Events				
5	1	330	8.1	2450
5	3	572	8.1	3614

From the assessment conducted, thermal hazards will occur predominantly within 1600 m of an LNG ship spill, with the highest hazards generally in the near field (approximately 250 - 500 m of a spill). While thermal hazards can exist beyond 1600 m, they are generally lower in most cases.

The general hazard zones and safety guidance identified from the study are as follows:

- The pool sizes for the credible spills estimated could range from generally 150 m in diameter for a small, accidental spill to several hundred meters for a large, intentional spill. Therefore, high thermal hazards from a fire are expected to occur within approximately 250 – 500 m from the origin of the spill, depending on the size of the spill. Major injuries and significant structural damage are possible in this zone. The extent of the hazards will depend on the spill size and dispersion from wind, waves, and currents. People, major commercial/industrial areas or other critical infrastructure elements, such as chemical plants, refineries, bridges or tunnels, or national icons located within portions of this zone could be seriously affected.
- Hazards and thermal impacts transition to lower levels with increasing distance from the origin of the spill. Some potential for injuries and property damage can still occur in portions of this zone; but this will vary based on spill size, distance from the spill, and site-specific conditions. For small spills, the hazards transition quickly to lower hazard levels.
- Beyond approximately 750 m for small accidental spills and 1600 m for large spills, the impacts on public safety should generally be low for most potential spills. Hazards will vary; but minor injuries and minor property damage are most likely at these distances. Increased injuries and property damage would be possible if vapor dispersion occurred and a vapor cloud was not ignited until after reaching this distance.

Table 4 summarizes the results on expected hazard levels for several types of accidental and intentional spills. While the analyses included evaluations of the size and number of breaches, spill rate and discharge coefficient, burn rate, surface emissive power, and transmissivity, site-specific environmental conditions such as wind speed, direction, waves, and currents, were not specifically considered. Therefore, the distances to each of the different hazard zones are provided as guidance and will vary depending on site-specific conditions and location.

The upper part of Table 4 identifies the estimated hazard zones for public safety from potential accidents, where spills are generally small, while the lower part of Table 4 identifies the estimated hazard zones for public safety from intentional LNG spills, which can be larger absent application of aggressive prevention and mitigation approaches.

Many of the hazard zones identified in Table 4 are based on thermal hazards from a pool fire, because many of the events will provide ignition sources such that a fire is likely to occur immediately. In some cases, the potential exists for a vapor cloud to be created without being ignited. A vapor cloud from an LNG spill could extend to 2,500 m for nominal conditions, if an ignition source is not available. The potential thermal hazards within a vapor cloud could be high. Because vapor cloud dispersion is highly influenced by atmospheric conditions, hazards from this type of event will be very site-specific. In addition, latent or indirect effects, such as additional damage that could be caused by a damaged infrastructure (e.g. a refinery or power plant), were not directly assessed. These types of issues and concerns are site-specific and should be included as part of the overall risk management process.

Table 4: Guidance for Impacts on Public Safety from LNG Breaches and Spills

EVENT	POTENTIAL SHIP DAMAGE AND SPILL	POTENTIAL HAZARD	POTENTIAL IMPACT ON PUBLIC SAFETY*		
			High	Medium	Low
Collisions: Low speed	Minor ship damage, no spill	Minor ship damage	None	None	None
Collisions: High Speed	LNG cargo tank breach and small - medium spill	Damage to ship and small fire	~ 250 m	~ 250 – 750 m	> 750 m
Grounding: <3 m high object	Minor ship damage, no breach	Minor ship damage	None	None	None
Intentional Breach	Intentional breach and medium to large spill	Damage to ship and large fire	~ 500 m	~ 500 m – 1600 m	> 1600 m
	Intentional, large release of LNG	▪ Damage to ship and large fire	~ 500m	~ 500 m – 1600 m	> 1600 m
		▪ Vapor cloud dispersion with late ignition	~ 500 m	> 1600 m	> 2000 m

[a] Distance to spill origin, varies according to site
Low – minor injuries and minor property damage
Medium – potential for injuries and property damage
High – major injuries and significant damage to property

Safety Analysis Conclusions

The potential for damage that could result from accidents or intentional events was evaluated. While hazard distances and levels will vary based on site-specific conditions, a summary of the safety analysis conclusions is presented below.

General Conclusions

1. The most significant impacts to public safety and property exist within approximately 500 m of a spill, with much lower impacts at distances beyond 1600 m, even for very large spills.
2. Under certain conditions, it is possible that multiple LNG cargo tanks could be breached, either as a result of the breaching event itself, as a consequence of LNG-

induced cryogenic damage to nearby tanks, or from fire-induced structural damage to the vessel.

3. Multiple breach and cascading LNG cargo tank damage scenarios were analyzed. While possible under certain conditions, they are likely to involve no more than two to three cargo tanks at any one time. These conditions will not greatly change the hazard ranges noted in General Conclusion Number 1, but will increase expected fire duration.

Accidental Breach Scenario Conclusions

1. Accidental LNG cargo tank damage scenarios exist that could potentially cause an effective breach area of 0.5 to 1.5 m^2.
2. Due to existing design and equipment requirements for LNG carriers, and the implementation of navigational safety measures such as traffic management schemes and safety zones, the risk from accidents is generally low.
3. The most significant impacts to public safety and property from an accidental spill exist within approximately 250 m of a spill, with lower impacts at distances beyond approximately 750 m from a spill.

Intentional Breach Scenario Conclusions

1. Several credible, intentional LNG cargo tank damage scenarios were identified that could initiate a breach of between 2 m^2 (1.6 m diameter) to approximately 12 m^2 (4 m diameter), with a probable nominal size of 5 – 7 m^2 (2.5 m – 3.0 m diameter).
2. Most of the intentional damage scenarios identified produce an ignition source and an LNG fire is very likely to occur.
3. Some intentional damage scenarios could result in vapor cloud dispersion, with delayed ignition and a fire.
4. Several intentional damage scenarios could affect the structural integrity of the vessel or other LNG cargo tanks due to ignition of LNG vapor trapped within the vessel. While possible under certain conditions, these scenarios are likely to involve no more than two to three cargo tanks at any one time.
5. Rapid phase transitions (RPT) are possible for large spills. Effects will be localized near the spill source and should not cause extensive structural damage.
6. The potential damage from spills to critical infrastructure elements such as bridges, tunnels, industrial/commercial centers, LNG unloading terminals and platforms, harbors, or populated areas can be significant in high hazard zones.
7. In general, the most significant impacts on public safety and property from an intentional spill exist within approximately 500 m of a spill, with lower impacts at distances beyond approximately 1600 m from a spill, even for very large spills.

Guidance on Risk Management for LNG Operations over Water

Risks from a potential LNG spill over water could be reduced through a combination of approaches, including 1) reducing the potential for a spill, 2) reducing the consequences of a spill, or 3) improving LNG transportation safety equipment, security, or operations to prevent or mitigate a spill.

For example, a number of international and U.S. safety and design standards have been developed for LNG ships to prevent or mitigate an accidental LNG spill over water. These standards are designed to prevent groundings, collisions, and steering or propulsion failures. They include traffic control, safety zones around the vessel while in transit within a port, escort by Coast Guard vessels, and coordination with local law enforcement and public safety agencies. In addition, since September 11, 2001, further security measures have been implemented to reduce the potential for intentional LNG spills over water. They include earlier notice of a ship's arrival (from 24 hours to 96 hours), investigation of crew backgrounds, at-sea boardings of LNG ships and special security sweeps, and positive control of an LNG ship during port transit.

Proactive risk management approaches can reduce both the potential for and hazards of such events. These include:

- Improvements in ship and terminal safety/security systems,
- Modifications and improvements in LNG tanker escorts, vessel movement control zones, and safety operations near ports and terminals,
- Improved surveillance and searches,
- Redundant or offshore mooring and offloading systems, and
- Improved emergency response coordination and communications.

Risk prevention and mitigation techniques can be important tools in reducing both the potential for and the hazards of a spill, especially in zones where the potential impact on public safety and property can be high. However, what might be applicable for effective risk reduction in one location might not be appropriate at another. Table 5 provides examples of how implementation of different strategies, alone or in combination, can be used to reduce certain threats, mitigate consequences of a spill, or reduce hazard analysis uncertainties.

Table 5: Representative Options for LNG Spill Risk Reduction

IMPACT ON PUBLIC SAFETY	REDUCTION IN EVENT POTENTIAL (Prevention)	IMPROVE SYSTEM SECURITY AND SAFETY (Mitigation)	IMPROVED HAZARD ANALYSIS (Reduce Analytical Uncertainties)	RESULTANT RISK REDUCTION
High and Medium	▪ Early off-shore interdiction ▪ Ship inspection ▪ Control of ship, tug and other vessel escorts ▪ Vessel movement control zones (safety/security zones) ▪ One-way traffic	▪ Harbor pilots ▪ Ship and terminal safety and security upgrades ▪ Expanded emergency response and fire fighting to	▪ Use of validated CFD models for LNG spill and thermal consequence analysis for site specific conditions ▪ Use of CFD and structural dynamic models for spill/structure interactions	Combination of approaches to reduce risks to acceptable levels

	▪ LNG offloading system security interlocks	address fires, vapor clouds, and damaged vessels		
Low	Use of existing best risk management practices on traffic control, monitoring & safety zones	Use of existing best risk mitigation practices to ensure risks remain low	Use of appropriate models to ensure hazards are low for site-specific conditions	Combination of approaches to ensure risks are maintained at acceptable levels

Risk identification and risk management processes should be conducted in cooperation with appropriate stakeholders, including public safety officials and elected public officials. Considerations should include site-specific conditions, available intelligence, threat assessments, safety and security operations, and available resources. This approach should be performance-based and include identification of site-specific hazards and risks and protection needs.

Based on the safety and risk mitigation and prevention evaluations, the following guidance is provided to assist risk management professionals, emergency management and public safety officials, port security officials and other appropriate stakeholders in developing and implementing appropriate safety and risk management strategies and processes for LNG import operations. For both accidental and intentional spills:

- Use effective security and protection operations that include enhanced interdiction, detection, delay procedures, risk management procedures, and coordinated emergency response measures, to reduce the risks from a possible breaching event;
- Implement risk management strategies based on site-specific conditions and the expected impact of a spill on public safety and property. Less intensive strategies would often be sufficient in areas where the impacts of a spill are low.
- Where analysis reveals that potential impacts on public safety and property could be high and where interactions with terrain or structures can occur, modern, validated computational fluid dynamics (CFD) models can be used to improve analysis of site-specific hazards.

Guidance on Risk Management for Accidental LNG Spills

Zone 1: These are areas in which LNG shipments transit narrow harbors or channels, pass under major bridges or over tunnels, or come within approximately 250 meters of people and major infrastructure elements, such as military facilities, population and commercial centers, or national icons. Within this zone, the risk and consequences of an accidental LNG spill could be significant and have severe negative impacts. Thermal radiation poses a severe public safety and property hazard, and can damage or significantly disrupt critical infrastructure located in this area.

Risk management strategies for LNG operations should address both vapor dispersion and fire hazards. Therefore, the most rigorous deterrent measures, such as vessel security zones, waterway traffic management, and establishment of positive control over vessels are options to be considered as elements of the risk management process. Coordination among all port security stakeholders is essential. Incident management and emergency response measures should be carefully evaluated to ensure adequate resources (i.e., firefighting, salvage, etc.) are available for consequence and risk mitigation.

Zone 2: These are areas in which LNG shipments and deliveries occur in broader channels or large outer harbors, or within approximately 250 m – 750 m of major critical infrastructure elements like population or commercial centers. Thermal radiation transitions to less severe hazard levels to public safety and property. Within Zone 2, the consequences of an accidental LNG spill are reduced and risk reduction and mitigation approaches and strategies can be less extensive.

Within Zone 2, the consequences of an accidental LNG spill are reduced and risk reduction and mitigation approaches and strategies can be less extensive. In this zone, risk management strategies for LNG operations should focus on approaches dealing with both vapor dispersion and fire hazards. The strategies should include incident management and emergency response measures such as ensuring areas of refuge (e.g. enclosed areas, buildings) are available, development of community warning signals, and community education programs to ensure persons know what precautions to take.

Zone 3: This zone covers LNG shipments and deliveries that occur more than approximately 750 m from major infrastructures, population/commercial centers, or in large bays or open water, where the risks and consequences to people and property of an accidental LNG spill over water are minimal. Thermal radiation poses minimal risks to public safety and property.

Within Zone 3, risk reduction and mitigation strategies can be significantly less complicated or extensive. Risk management strategies should concentrate on incident management and emergency response measures that are focused on dealing with vapor cloud dispersion. Measures should ensure areas of refuge are available, and community education programs should be implemented to ensure that persons know what to do in the unlikely event of a vapor cloud.

Guidance on Risk Management for Intentional LNG Spills

Zone 1: These are areas in which LNG shipments occur in narrow harbors or channels, pass under major bridges or over tunnels, or come within approximately 500 meters of major infrastructure elements, such as military facilities, population and commercial centers, or national icons. Within this zone, the risk and consequences of a large LNG spill could be significant and have severe negative impacts. Thermal radiation poses a severe public safety and property hazard, and can damage or significantly disrupt critical infrastructure located in this area.

Risk management strategies for LNG operations should address vapor dispersion and fire hazards. The most rigorous deterrent measures, such as vessel security zones, waterway traffic management, and establishment of positive control over vessels are elements of the risk management process. Coordination among all port security stakeholders is essential. Incident management and emergency response measures should be carefully evaluated to ensure adequate resources (i.e., firefighting, salvage) are available for consequence and risk mitigation.

Zone 2: These are areas in which LNG shipments and deliveries occur in broader channels or large outer harbors, within approximately 500 m – 1.6 km of major critical infrastructure elements, such as population or commercial centers. Within Zone 2, the consequences of even a large LNG spill are reduced. Thermal radiation transitions to less severe hazard levels to public safety and property.

Risk management strategies for LNG operations that occur in this zone should focus on vapor dispersion and fire hazards. The strategies should include incident management and emergency response measures that ensure areas of refuge (enclosed areas, buildings) are available, the development of community warning procedures, and education programs to ensure that communities are aware of precautionary measures.

Zone 3: This zone covers LNG shipments and deliveries that occur more than approximately 1.6 km from major infrastructures, population/commercial centers, or in large bays or open water, where the risks and consequences to people and property of a large LNG spill over water are minimal. Thermal radiation poses minimal risks to public safety and property.

Risk reduction and mitigation strategies can be significantly less complicated or extensive than Zones 1 and 2. Risk management strategies should concentrate on incident management and emergency response measures for dealing with vapor cloud dispersion. Measures should ensure that areas of refuge are available, and community education programs should be implemented to ensure that persons know what to do in the unlikely event of a vapor cloud.

References

[Fay 2003] Fay, J.A., "Model of spills and fires from LNG and oil tankers." Journal of Hazardous Materials, B96-2003, 171-188, 2003.

[Hightower 2004] Hightower, M. et al., "Guidance on Risk Analysis and Safety Implications of a Large Liquefied Natural Gas (LNG) Spill Over Water," SAND2004-6258, Sandia National Laboratories, Albuquerque, NM, December 2004.

[Lehr and Simecek-Beatty 2003] Lehr, w. and Simecek-Beatty, D., "Comparison of Hypothetical LNG and Fuel Oil Fires on Water." DRAFT report by the National Oceanic and Atmospheric Administration [NOAA] , Office of Response and Restoration, Seattle, WA, 2003.

[Quest 2003], "Modeling LNG Spills in Boston Harbor." Copyright© 2003 Quest Consultants, Inc., 908 26th Ave N.W., Norman, OK 73609; Letter from Quest Consultants to DOE (October 2, 2001); Letter from Quest Consultants to DOE (October 3, 2001); and Letter from Quest Consultants to DOE (November 17, 2003).

[Vallejo 2003] "Liquefied Natural Gas in Vallejo: Health and Safety Issues." LNG Health and Safety Committee of the Disaster Council of the City of Vallejo, CA, January 2003.